Devendra K. Chaturvedi

Soft Computing

Studies in Computational Intelligence, Volume 103

Editor-in-chief
Prof. Janusz Kacprzyk
Systems Research Institute
Polish Academy of Sciences
ul. Newelska 6
01-447 Warsaw
Poland
E-mail: kacprzyk@ibspan.waw.pl

Further volumes of this series can be found on our
homepage: springer.com

Vol. 79. Xing Cai and T.-C. Jim Yeh (Eds.)
*Quantitative Information Fusion for Hydrological
Sciences,* 2008
ISBN 978-3-540-75383-4

Vol. 80. Joachim Diederich
Rule Extraction from Support Vector Machines, 2008
ISBN 978-3-540-75389-6

Vol. 81. K. Sridharan
Robotic Exploration and Landmark Determination, 2008
ISBN 978-3-540-75393-3

Vol. 82. Ajith Abraham, Crina Grosan and Witold
Pedrycz (Eds.)
Engineering Evolutionary Intelligent Systems, 2008
ISBN 978-3-540-75395-7

Vol. 83. Bhanu Prasad and S.R.M. Prasanna (Eds.)
*Speech, Audio, Image and Biomedical Signal Processing
using Neural Networks,* 2008
ISBN 978-3-540-75397-1

Vol. 84. Marek R. Ogiela and Ryszard Tadeusiewicz
*Modern Computational Intelligence Methods
for the Interpretation of Medical Images,* 2008
ISBN 978-3-540-75399-5

Vol. 85. Arpad Kelemen, Ajith Abraham and Yulan Liang
(Eds.)
Computational Intelligence in Medical Informatics, 2008
ISBN 978-3-540-75766-5

Vol. 86. Zbigniew Les and Mogdalena Les
Shape Understanding Systems, 2008
ISBN 978-3-540-75768-9

Vol. 87. Yuri Avramenko and Andrzej Kraslawski
Case Based Design, 2008
ISBN 978-3-540-75705-4

Vol. 88. Tina Yu, David Davis, Cem Baydar and Rajkumar
Roy (Eds.)
Evolutionary Computation in Practice, 2008
ISBN 978-3-540-75770-2

Vol. 89. Ito Takayuki, Hattori Hiromitsu, Zhang Minjie
and Matsuo Tokuro (Eds.)
Rational, Robust, Secure, 2008
ISBN 978-3-540-76281-2

Vol. 90. Simone Marinai and Hiromichi Fujisawa (Eds.)
*Machine Learning in Document Analysis
and Recognition,* 2008
ISBN 978-3-540-76279-9

Vol. 91. Horst Bunke, Kandel Abraham and Last Mark (Eds.)
Applied Pattern Recognition, 2008
ISBN 978-3-540-76830-2

Vol. 92. Ang Yang, Yin Shan and Lam Thu Bui (Eds.)
Success in Evolutionary Computation, 2008
ISBN 978-3-540-76285-0

Vol. 93. Manolis Wallace, Marios Angelides and Phivos
Mylonas (Eds.)
*Advances in Semantic Media Adaptation and
Personalization,* 2008
ISBN 978-3-540-76359-8

Vol. 94. Arpad Kelemen, Ajith Abraham and Yuehui Chen
(Eds.)
Computational Intelligence in Bioinformatics, 2008
ISBN 978-3-540-76802-9

Vol. 95. Radu Dogaru
*Systematic Design for Emergence in Cellular Nonlinear
Networks,* 2008
ISBN 978-3-540-76800-5

Vol. 96. Aboul-Ella Hassanien, Ajith Abraham and Janusz
Kacprzyk (Eds.)
*Computational Intelligence in Multimedia Processing:
Recent Advances,* 2008
ISBN 978-3-540-76826-5

Vol. 97. Gloria Phillips-Wren, Nikhil Ichalkaranje and
Lakhmi C. Jain (Eds.)
Intelligent Decision Making: An AI-Based Approach, 2008
ISBN 978-3-540-76829-9

Vol. 98. Ashish Ghosh, Satchidananda Dehuri and Susmita
Ghosh (Eds.)
*Multi-Objective Evolutionary Algorithms for Knowledge
Discovery from Databases,* 2008
ISBN 978-3-540-77466-2

Vol. 99. George Meghabghab and Abraham Kandel
Search Engines, Link Analysis, and User's Web Behavior,
2008
ISBN 978-3-540-77468-6

Vol. 100. Anthony Brabazon and Michael O'Neill (Eds.)
Natural Computing in Computational Finance, 2008
ISBN 978-3-540-77476-1

Vol. 101. Michael Granitzer, Mathias Lux and Marc Spaniol
(Eds.)
Multimedia Semantics - The Role of Metadata, 2008
ISBN 978-3-540-77472-3

Vol. 102. Carlos Cotta, Simeon Reich, Robert Schaefer and
Antoni Ligeza (Eds.)
Knowledge-Driven Computing, 2008
ISBN 978-3-540-77474-7

Vol. 103. Devendra K. Chaturvedi
Soft Computing, 2008
ISBN 978-3-540-77480-8

Devendra K. Chaturvedi

Soft Computing

Techniques and its Applications in Electrical Engineering

With 320 Figures and 132 Tables

 Springer

Dr. D.K. Chaturvedi
Convener, Faculty Training & Placement Cell
Reader, Department of Electrical Engineering
 & Faculty of Engineering
Dayalbagh Educational Institute
(Deemed University)
Dayalbagh
Agra, India
dkc.foe@gmail.com

ISBN 978-3-540-77480-8   e-ISBN 978-3-540-77481-5

Studies in Computational Intelligence ISSN 1860-949X

Library of Congress Control Number: 2008920252

Cover design: Deblik, Berlin, Germany

Printed on acid-free paper

9 8 7 6 5 4 3 2 1

springer.com

Foreword

Soft computing, as explained by Prof. Lotfi Zadeh, is a consortium of methodologies (working synergistically, not competitively) that, in one form or another, exploits the tolerance for imprecision, uncertainty, approximate reasoning and partial truth, to achieve tractability, robustness, low-cost solution, and close resemblance with human-like decision making. It has flexible information processing capability for representation and evaluation of various real life ambiguous and uncertain situations. Therefore, soft computing provides the foundation for the conception and design of high MIQ (Machine IQ) systems. Fuzzy logic having capability of handling uncertainty arising from, say, vagueness, incompleteness, overlapping concepts; neural networks providing machinery for adaptation and learning, genetic algorithms for optimization and learning, and probabilistic reasoning for inference were considered to be the basic ingredients of soft computing. Recently, the theory of rough sets of Prof. Z. Pawlak, having capability of handling uncertainty arising from granularity in the domain of discourse, is found to be another significant component.

Since the aforesaid characteristic features are important in designing intelligent systems and making any computer application successful, several research centres related to soft computing have been established over the world, dedicated to conduct research both for the development of its theory and demonstrating real life applications. The Department of Science & Technology (DST), Govt. of India, has established, in this line, the nation's first Soft Computing Research Center at Indian Statistical Institute, Kolkata, in 2004.

One of the challenges of soft computing research lies in judicious integration of the merits of its component technologies so that the resulting one has application-specific advantages, which can not be achieved using the individual techniques alone, in decision making. Among all such integrations, neuro-fuzzy hybridization is the most visible one realized so far. Recently, rough-fuzzy computing has drawn the attention of researchers as a strong paradigm for handling uncertainty in many real life problems, e.g., bioinformatics, web intelligence, and data mining & knowledge discovery. Since it has led to the concept of fuzzy-granulation, or f-granulation, it has a strong relevance to the Computational Theory of Perceptions (CTP), which provides the capability to compute and reason not using any measurements, but with perception-based information. The perception-based information is inherently imprecise and fuzzy-granular, reflecting the finite ability of the sensory organs (and finally the brain) to resolve details and store information. Typical daily life examples based on CTP include car parking, driving in city, cooking meal, and summarizing story. The next decade is expected to bear the testimony of many stimulating issues and solutions in this domain.

The present volume titled "Soft Computing and its Applications" by Dr. D.K. Chaturvedi, an experienced researcher from Faculty of Engineering, Dayalbagh Educational Institute, Agra, India, deals with the introduction and basic concepts of fuzzy logic, artificial neural networks and genetic algorithms, and their role in certain combinations. It also describes some applications like load forecasting problem and power system identification. I believe the chapters would help in understanding not only the basic issues and characteristic features of soft computing, but also the aforesaid problems of CTP and in formulating possible solutions. Dr. Chaturvedi deserves congratulations for bringing out the nice piece of work.

Kolkata, India
November 3, 2007

Sankar K. Pal
Director
Indian Statistical Institute

Preface

The modern science is still striving to develop consciousness-based machine. In the last century, enormous industrial and technological developments had taken place. Technology had developed laterally well up to the biggest giant-sized complexes and also to the smallest molecular nano mechanisms. Thus, having explored to the maxima of the two extreme fields, technology is exploring now vertically to reach the dizzy heights of soft computing, subtle soft computing, and the millennium wonder of reaching the almost unchartered height of evolving consciousness in computers (machines). This book makes its small and humble contribution to this new astounding scenario and possibly the greatest of all mechanical wonders, to transfer consciousness of man to machine. Prior to World War II, numerical calculations were done with mechanical calculators. Simulated by military requirements during World War II, the first version modern digital computers began to make their appearance in late 1940s and early 1950s. During that pioneering period, a number of different approaches to digital computer organization and digital computing techniques were investigated. Primarily, as a result of the constraints imposed by the available electronics technology, the designers of digital computers soon focused their attention on the concept of computer system architecture, which was championed by Dr. John Von Neumann, who first implemented it in the computer constructed for the Institute of Advanced Studies at Princeton. Because of the pervasiveness of the Von Neumann architecture in digital computers, during the 1950s and 1960s, most numerical analysts and other computer users concentrated their efforts on developing algorithms and software packages suitable to these types of computers. In 1960s and 1970s, there were numerous modifications and improvements to computers of the earlier generation. The "bottle neck" of Neumann computers was the memory buffer sizes and speeds on it. In the 1990s, there was a quantum leap in the size of computer memory and speeds. As a result of this, Supercomputers have been developed, which could do lacs of calculations within a fraction of a second. Supercomputers can also do all routine task and it could handle it better with multi-coordination than a human being, and thus reducing a series of simple logical operations. It could store vast information and process the same in a flash. It does not also suffer from the human moods and many vagaries of mind.

But, the super computers cannot infer or acquire any knowledge from its information contents. It cannot think sensibly and talk intelligently. It could not recognize a person or could not relate his family background.

On the Other hand, as human beings, we continuously evolve our value judgment about the information we receive and instinctively process them. Our judgment is based on our feelings, tastes, knowledge and experience. But computers are incapable of such judgments. A computer can be programmed (instructed), i.e. to generate poetry or music, but it cannot appraise or judge its quality.

Hence, there is a genuine and compulsory need for some other logic, which can handle such real life scenario. In 1965, Prof. Lofti A. Zadeh at the University of California, introduced an identification tool by which this degree of truth can be handled by fuzzy set theoretic approach. With the invention of fuzzy chips in 1980s fuzzy logic received a great boost in the industry.

Now in this twentyfirst century, along with fuzzy logic, Artificial Neural Network (ANN), and Evolutionary Algorithms (EA) are receiving intensive attention, in both academics and industry. All these techniques are kept under one umbrella called "soft computing." Enormous research had already been done on soft computing techniques to identify a model and control of its different systems.

This book is an introduction to some new fields in soft computing with its principal components of fuzzy logic, ANN, and EA, and it is hoped that it would be quite useful to study the fundamental concepts on these topics for the pursuit of allied research.

Intuitive consciousness/wisdom is also one of the frontline areas in soft computing, which has to be always cultivated by meditation. This is, indeed, an extraordinary challenge and virtually a new wondrous phenomenon to include such phenomena into the computers.

The approach in this book is

- To provide an understanding of the soft computing field
- To work through soft computing (ANN, fuzzy systems, and genetic algorithms) using examples
- To integrate pseudo-code operational summaries and Matlab codes
- To present computer simulation
- To include real world applications
- To highlight the distinctive work of human consciousness in machine.

Organization of the Book

This book begins with the introduction of soft computing and is divided into four parts.

The first part deals with the historical developments in the exciting field of neural science to understand the brain and its functioning (Chap. 2). This is followed by the working of ANN and their architectures (Chap. 3). The feed forward back-propagation ANNs are widely used in operations and control of the various industrial processes and plants, for modeling and simulation of systems, and for forecasting purposes. The ANN needs many pairs of input–output $(X-Y)$ as training and testing data. The relation between input and output, the size of neural network, type of neuron and connectivity of neurons among various layers generally contribute to training time of the neural network. The study has been conducted to observe the effect of range of normalization like 0–1, −1–1, 0–0.9, etc., the type of mapping of input–output pairs like $X-Y$, $X-\Delta Y$, $\Delta X-Y$, and $\Delta X-\Delta Y$ and their sequence of presentation, threshold, and aggregation functions used for different neurons (i.e. neuron structure) on training time. In addition, influence of noise in the input–output data on accuracy of learning and training time has been studied. The noise in input–output data has major contribution in generalization of ANN. The neural network model has been developed to study the above-mentioned issues for DC machines modeling to predict armature current and speed, and for short-term load forecasting problems (Chap. 4). Efforts have been taken in the past to reduce the training time of ANN by selection of an optimal network and modification in learning algorithms. A new (generalized) neuron model using neuro-fuzzy approach to overcome the problems of ANN incorporating the features of fuzzy systems at a neuron level had been developed and tested on various bench mark problems (Chap. 5). Taking benefit of the characteristics of the GN, it is used for various applications such as machines modeling, electrical load forecasting system, aircraft landing control system, load frequency controller, and power system stabilization problem (Chap. 6).

In the second part, the book concentrates on the introduction of fuzzy logic concepts and basics of fuzzy systems (Chap. 7). Fuzzy logic is applied to a great extent in controlling the process, plants, and various complex systems because of its inherent advantages like simplicity, ease in design, robustness, and adaptivity. It is established that this approach works very well especially when the systems are not transparent. Also, the effect of different connectives (like intersection, union, and compensatory operators as well as averaging operators), different implication methods, different compositional rules, different membership functions of fuzzy sets and their degrees of overlapping, and different defuzzification methods have been studied in the context of fuzzy system based modeling of electrical machines and load forecasting problems.

The third part lays the foundation for genetic algorithms (GA) and its variant (Chap. 9). In Chap. 10, the application of GA for load forecasting problems is discussed. The most difficult and crucial part of fuzzy system development is the knowledge acquisition. System dynamics technique (causal relationships) helps in the knowledge acquisition and representation of it. The integrated approach of systems dynamics technique and fuzzy systems has been used for socio-economic systems like HIV/AIDS population forecasting problem.

The last part of this book covers the synergism between different components of soft computing technology such as GA, fuzzy systems, and ANN. The GA-fuzzy system based approach is used for power system applications such as optimal electrical power flow problem, transmission pricing in deregulated environment and congestion management problems (Chap. 11). The GA-fuzzy (GAF) approach has also been used for load forecasting problems on long-term basis.

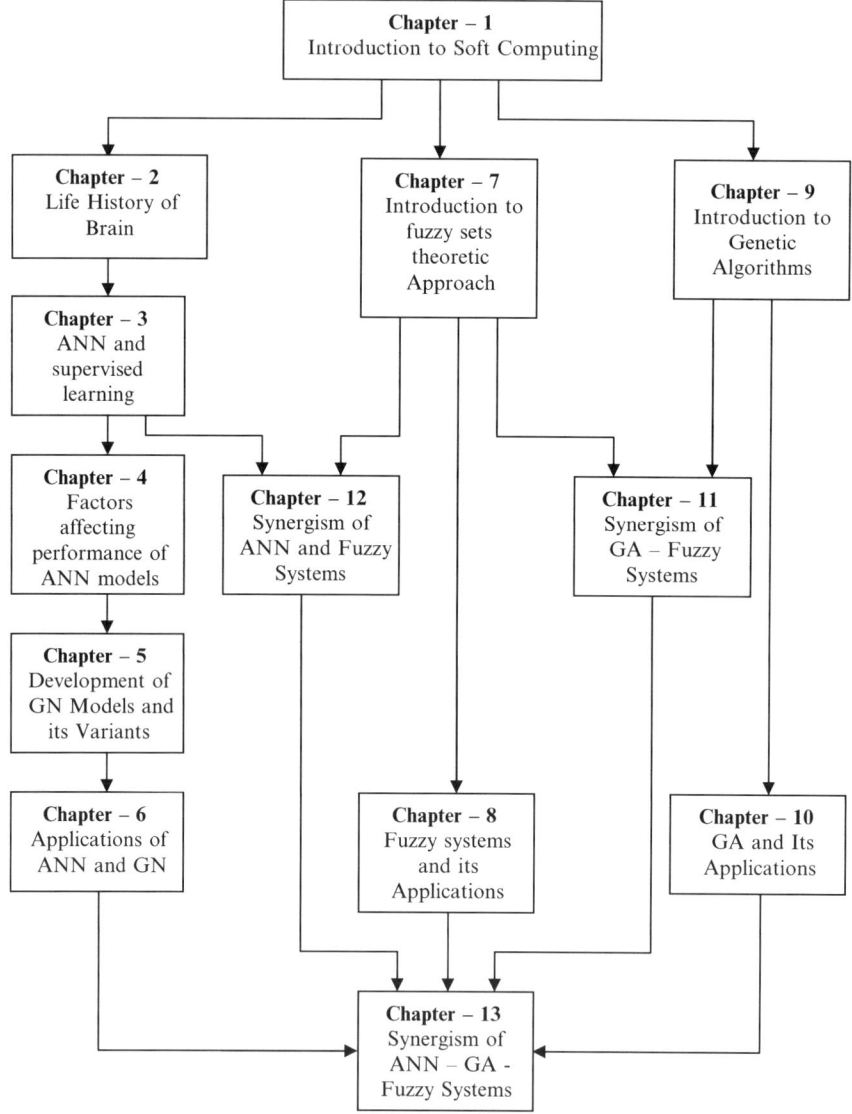

Fig. P1. Schematic outline of the book

Chapter 12 deals with the Adaptive Neuro-Fuzzy Inference System (AN-FIS). The back-propagation learning algorithm is generally used to train ANN and GN. The back-propagation learning has various drawbacks such as slowness in learning, stuck in local minima, and requires functional derivative of aggregation and threshold functions to minimize error function. Various researchers have suggested a number of improvements in simple back propagation learning algorithm developed by Widrow and Holf (1956). Chapter 13 deals with the synergism of feed forward ANN with GA as the learning mechanism to overcome some of the disadvantages of back-propagation learning mechanism to minimize the error function of ANN. GA optimization is slow and depends on the number of variables. To improve the convergence of GA, a modified GA is developed in which the GA parameters like cross-over probability (P_c), mutation probability (P_m), and population size (popsize) are modified using fuzzy system with concentration of genes. The ANN-GA-fuzzy system integrated approach is applied to different benchmark problems to test this approach. The schematic outline of the book is shown in Fig. P1.

Dayalbagh, Agra, India

D.K. Chaturvedi
Faculty of Engineering
Dayalbagh Educational Institute

Acknowledgements

The author would like to seize this opportunity with deep humility to thank and to express his profound and grateful veneration to Prof. P.S. Satsangi, Chairman, Advisory Committee on Education in Dayalbagh, Agra, India, who is the fountain source of my inspiration and intuition in soft computing techniques and graciously leading him verily as the only Kindly Light in the utter darkness of this new domain of profound research in the whole world. The author is also thankful to Prof. P.K. Kalra, Indian Institute of Technology, Kanpur, India, for the encouragement, support, and visionary ideas provided during my research. Author is also very deeply indebted to Prof. O.P. Malik, University of Calgary, Canada, for providing all the facilities in his laboratory to pursue my experimentation work during his BOYSCAST fellowship of Department of Science and Technology, Govt. of India during 2001–2002 and later in summer 2005 and now in 2007.

Also thanks to Mr. S. Krishnananda from Dayalbagh and Mr. Garwin Hancock, University of Calgary, Canada, for their good wishes and help always extended to me in preparing my scripts properly. Thanks go to my colleague Dr. Manmohan, Ashish Saini, and students Mr. Praveen Kumar, Himanshu Vijay, and V. Prem Prakash for their help.

Last but not least, I am grateful to my wife Dr. Lajwanti and daughters Miss Jyoti and Swati who had bare all the inconvenience in the period of this work.

During the course of writing this book many people have supported me in countless different ways. I express my deepest sense of gratitude to all of them, whether mentioned here or not.

Contents

1

Introduction to Soft Computing

1.1 Introduction

Soft computing (SC) is a branch, in which, it is tried to build intelligent and wiser machines. Intelligence provides the power to derive the answer and not simply arrive to the answer. Purity of thinking, machine intelligence, freedom to work, dimensions, complexity and fuzziness handling capability increase, as we go higher and higher in the hierarchy as shown in Fig. 1.1. The final aim is to develop a computer or a machine which will work in a similar way as human beings can do, i.e. the wisdom of human beings can be replicated in computers in some artificial manner.

Intuitive consciousness/wisdom is also one of the important area in the soft computing, which is always cultivated by meditation. This is indeed, an extraordinary challenge and virtually a new phenomenon, to include consciousness into the computers.

Soft computing is an emerging collection of methodologies, which aim to exploit tolerance for imprecision, uncertainty, and partial truth to achieve robustness, tractability and total low cost. Soft computing methodologies have been advantageous in many applications. In contrast to analytical methods, soft computing methodologies mimic consciousness and cognition in several important respects: they can learn from experience; they can universalize into domains where direct experience is absent; and, through parallel computer architectures that simulate biological processes, they can perform mapping from inputs to the outputs faster than inherently serial analytical representations. The trade off, however, is a decrease in accuracy. If a tendency towards imprecision could be tolerated, then it should be possible to extend the scope of the applications even to those problems where the analytical and mathematical representations are readily available. The motivation for such an extension is the expected decrease in computational load and consequent increase of computation speeds that permit more robust system (Jang et al. 1997).

The successful applications of soft computing and the rapid growth of the same suggest that the impact of soft computing will be felt increasingly in the

D.K. Chaturvedi: *Soft Computing Techniques and its Applications in Electrical Engineering*, Studies in Computational Intelligence (SCI) **103**, 1–10 (2008)
www.springerlink.com   © Springer-Verlag Berlin Heidelberg 2008

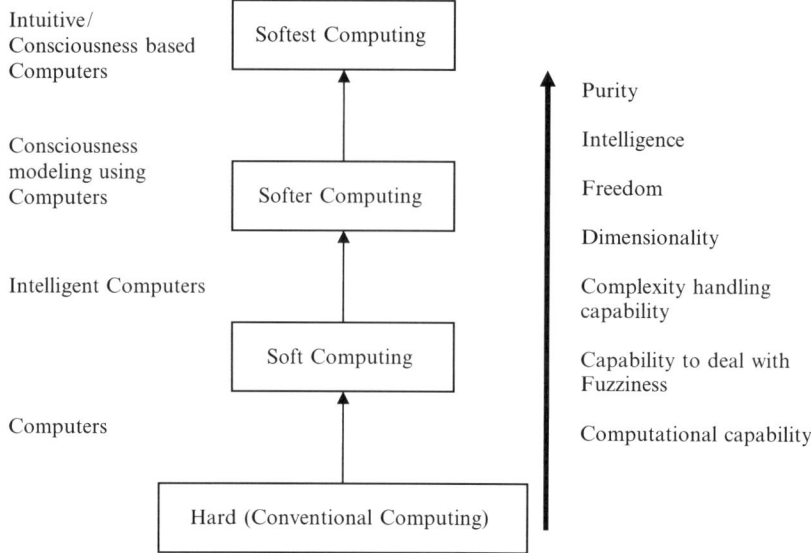

Fig. 1.1. Development soft computing

coming years. Soft computing is likely to play an especially important role in science and engineering, but eventually its influence may extend much farther. In many ways, soft computing represents a significant paradigm shift in the aims of computing – a shift which reflects the fact that the human mind, unlike present day computers, possesses a remarkable ability to store and process information which is pervasively imprecise, uncertain, lacking in categoricity and approximations in containing high quality engineering solutions.

In soft computing the problem or task at hand is represented in such a way that the "state" of the system can somehow be calculated and compared to some desired state. The quality of the system's state is the basis for adapting the system's parameters, which slowly converge towards the solution. This is the basic approach employed by evolutionary computing and neural computing.

Soft computing differs from conventional (hard) computing in many ways. For example, soft computing exploits tolerant of imprecision, uncertainty, partial truth, and approximation. In effect, the role model for soft computing is *the human mind*.

Soft-computing is defined as a collection of techniques spanning many fields that fall under various categories in computational intelligence. Soft-computing has three main branches: fuzzy Systems, evolutionary computation, artificial neural computing, with the latter subsuming machine learning (ML) and probabilistic reasoning (PR), belief networks, chaos theory, parts of learning theory and wisdom based expert system (WES), etc. as shown in Fig. 1.2.

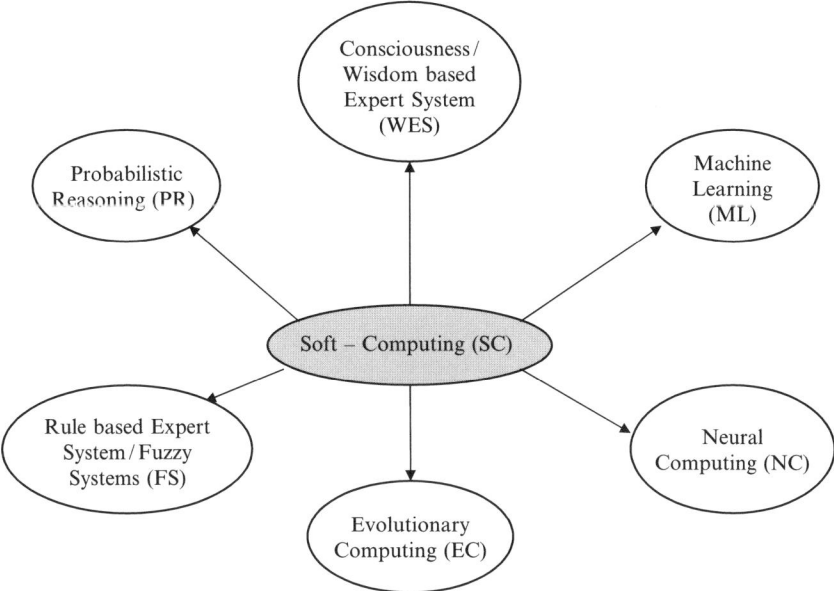

Fig. 1.2. Soft-computing techniques

Soft computing is a partnership in which each of the partners contributes a distinct methodology for addressing problems in its domain. In this perspective, the principal constituent methodologies in SC are complementary rather than competitive. Furthermore, soft computing may be viewed as a foundation component for the emerging field of conceptual intelligence.

1.2 Importance of Soft Computing

In many cases a problem can be solved most effectively by using WES, FS, NC, GC and PR in combination rather than exclusively. A striking example of a particularly effective combination is what has come to be known as "neuro-fuzzy systems". Such systems are becoming increasingly visible as consumer products ranging from air conditioners and washing machines to photocopiers and camcorders. Less visible but perhaps even more important are neuro-fuzzy systems in industrial applications. What is particularly significant is that in both consumer products and industrial systems, the employment of soft computing techniques leads to systems which have high machine intelligence quotient (MIQ). In a large measure, it is the high MIQ of SC-based systems that accounts for the rapid growth in the number and variety of applications of soft computing. One of the important features of SC is acquisition of knowledge/information from inaccurate and uncertain data. It is expected that combination or fusion of the elemental technologies will help to overcome the limitations of individual elements (Furuhashi 2001).

Soft-computing is often robust under noisy input environments and has high tolerance for imprecision in the data on which it operates.

1.3 Main Components of Soft Computing

Soft-computing, to some extent, draws inspiration from natural phenomena. Neural computing tries to mimic the animal brain, genetic algorithms is found on the dynamics of Darwinian evolution, while fuzzy logic is heavily motivated by the highly imprecise nature of human speech.

1.3.1 Fuzzy Logic

The human beings deal with imprecise and uncertain information as we go about our day to day routines. This can be gleaned from the language we use which contains many qualitative and subjective words and phrases such as "quite expensive", "very young", or "a little far", "expensive", etc. In human information processing, approximate reasoning is used and tried to accommodate varying degrees of imprecision and uncertainty in the concepts and tokens of information that we deal with.

Fuzzy systems are a generalization of stiff Boolean logic. It uses fuzzy sets which are a generalization of *crisp sets* in classical set theory. In classical set theory, an object could just be either a member of set or not at all, in fuzzy set theory, a given object is said to be of a certain "degree of membership" to the set. Hence, in fuzzy sets membership value of an object could be in the range 0–1, but in crisp set the membership value is always 0 or 1.

1.3.1.1 Historical Perspective of Fuzzy Logic

 Fuzzy logic was conceived in the USA by Prof. Lotfi A. Zadeh, Department of Electrical Engineering and Computer Sciences, University of California, Berkeley, in the early 1960s. But his early work with the concept was severely criticized by many of his colleagues in the field and did not gain much acceptance by the scientific community. By the early seventies, some European researchers had started applying fuzzy logic and made successful implementations of it in industrial process and control.

In the 1980s, Japanese researchers became interested in the successful applications of fuzzy logic in Europe. Also, some very prominent researchers in Japan further developed the theory of it; the most notable among them is Professor Michio Sugeno. The Japanese government and academic institutions, as well as the big Japanese firms, were involved not only in fuzzy logic R&D, but also in the mass marketing of fuzzy logic based products. This resulted in widespread use of simple fuzzy logic components to control various home appliances such as washing machines, handy cam, micro-wave oven and rice cookers. Even the bullet trains of Japan made use of such technology.

With the success of fuzzy logic applications in Europe and Asia, the United States have recently given fuzzy logic a second look, much more receptive to the "fuzzy" idea this time. Many applications of fuzzy systems have been flourished. These applications include areas in industrial systems, intelligent control, decision support systems, and consumer products. Fuzzy logic-based products now account for billions of US dollar business every year.

1.3.1.2 Fuzzy System

A fuzzy expert system consists of a *fuzzy rule base*, a *fuzzification module*, an *inference engine*, and a *defuzzification module* as shown in Fig. 1.3. The fuzzification module pre-processes the input values submitted to the fuzzy expert system. The inference engine uses the results of the fuzzification module and accesses the fuzzy rules in the fuzzy rule base to infer what intermediate and output values to produce. The final output of the fuzzy expert system is provided by the defuzzification module.

1.3.2 Artificial Neural Networks

Artificial neural networks (ANN), or simply *neural networks*, can be loosely defined as large sets of interconnected simple units which execute in parallel

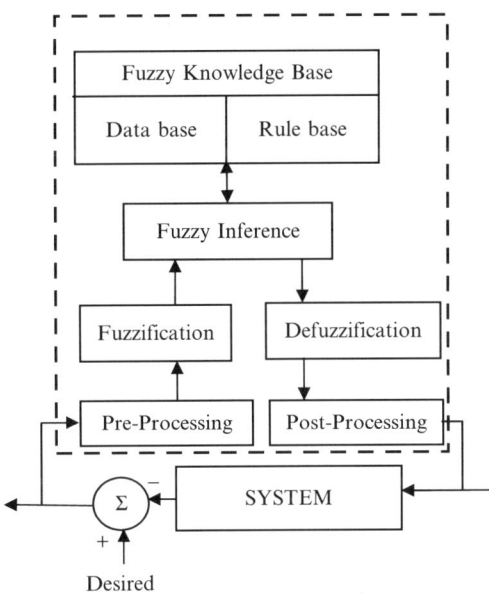

Fig. 1.3. Functional module of a fuzzy system

to perform a common global task. These units usually undergo a learning process which automatically updates network parameters in response to a possibly evolving input environment. The units are often highly simplified models of the biological neurons found in the animal brain.

Some basic characteristics of most neural network models:

- *Inherent parallelism* – practically all neural network models have some element of parallelism in the execution of their numerous components;
- *Similarity of components* – to a very a large extent, the basic components of a neural network look all alike and behave similarly;
- *Access to local information* – any given node's level of activation and eventual output will depend exclusively on its current states and the outputs of the other nodes to which it is connected; and
- *Incremental learning* – neural networks parameters undergo several small changes, which, over time, would come to settle on their final values.

Neural network models can be classified in a number of ways as mentioned below:

1. According to the network architecture, there are three major types of neural networks namely,
   a. Recurrent network,
   b. Feed forward network and
   c. Competitive networks.
2. According to the network structure:
   a. *Static (fixed) structure and*
   b. *Dynamic structure.*
3. According to the mode of learning
   a. *Supervised learning and*
   b. *Unsupervised learning.*
4. Neural network models can also be classified on the basis of their over-all task such as
   a. *Pattern association,*
   b. *Classification and*
   c. *Function approximation.*
5. According to the neuron structure

1. *Based on aggregation function*
   a. *Summation type neuron*
   b. *Product type neuron*
   c. *Combination of summation and product type*
   d. *Averaging or compensatory type.*
2. *Based on activation function used*
   a. *Sigmoid function*
   b. *Gaussian function*
   c. *Linear function, etc.*

1.3.2.1 Historical Perspective of ANN

In the early 1940s, McCulloch and Pitts' monumental work on *linear threshold units* was the first study in finite state machines that are designed to model the way neurons in the brain are interconnected. The first wave of neural network research towards producing intelligent machines thus began, under what was then known as "cybernetics". Just before the end of the 1940s, Hebb published his thoughts on "cell assemblies" and how nearby neurons reinforce each other in what is now known as "Hebbian learning".

The early 1950s saw among the first re-inforcement learning systems built by Marvin Minsky. By the early 1960s, Rosenblatt's celebrated work on Perceptrons opened up a second-wave of interest in neural networks. Much of the euphorbia based on the Perceptron's initial successes was, however soon dampened by a rigorous analysis of the limitations of Perceptrons conducted by Minsky and Papert. And so the 1970s was a calm period for neural network research, although many researchers continued their work on the sidelines, such as Stephen Grossberg, Teuvo Kohonen, and Shun-Ichi Amari.

The 1980s ushered in the third wave of neural network research, reinvigorated by Hopfield's "Hopfield networks" and the independent discoveries of "back-propagation" by Le Cun, Parker and the group of Rumelhart, Hinton and Williams.

The 1990s has been marked by numerous applications of neural network technology in all domains of computer work and intelligent hybrid systems. There is now further consolidation and handshaking among researchers with the flourishing of loosely grouped fields under *soft-computing*.

In recent years, the NN has been applied successfully to many fields of engineering such as aerospace, digital signal processing, electronics, robotics (Liu et al. 1989), machine vision, speech, manufacturing, transportation, controls (Miller et al. 1990; Miller 1994; Zhang et al. 1995; Gupta and Sinha 1996) and medical engineering (a number of papers appeared in IEEE Trans. On Bio-Medical Engineering). A partial list of NN industrial applications includes control applications (Potter Don et al. 1997); inverted pendulum controller (Jung Sooyong and Wen John 2003; Yongcai 2001; Yamakita et al. 1994; Wei et al. 1995]; robotics manipulators (Adam 2004), servo motor control (Ozcalik 2002); Automotive control (Miller et al. 1990); aircraft controls (Ming et al. 2004); Image processing and recognition and process identification (Narendra et al. 1995).

1.3.3 Introduction to Evolutionary Algorithms

The evolutionary algorithms (EA) which are inspired by the biological genetics found in living nature which is simple, powerful, domain free, and probabilistic approach to general problem solving technique. The phenomena incorporated so far in EA models include phenomena of natural selection as there are selection and the production of variation by means of recombination and mutation, and rarely inversion, diploid and others. Most genetic algorithms (GA) work with one large panmictic population, i.e. in the recombination step, each individual may potentially choose any other individual from the population as a mate. Then GA operators are performed to obtain the new child offspring. There are three important GA operators which are commonly used are as follows:

(1) Crossover
(2) Mutation
(3) Selection and survival of the fittest.

1.3.4 Hybrid Intelligent Systems

In many cases, hybrid applications methods have proven to be effective in designing intelligent systems. As it was shown in recent years, fuzzy logic, neural networks and evolutionary computations are complementary methodologies in the design and implementation of intelligent systems. Each approach has its merits and drawbacks. To take advantage of the merits and eliminate their drawbacks, many ways of integrating these methodologies have been proposed by researchers during the past few years. These techniques include the integration of neural network and fuzzy logic techniques as well as the combination of these two technologies with evolutionary methods. The merging of the ANN and FS can be realized in three different directions, resulting in systems with different characteristics given in the Chap. 12.

1. Neuro-fuzzy systems: provide the fuzzy systems with automatic tuning systems using ANN as a tool. The adaptive neuro fuzzy inference systems (ANFIS) are included in this classification.
2. Fuzzy neural networks: retain the functions of ANN with fuzzification of some of their elements. For instance, fuzzy logic can be used to determine the learning parameters of ANN like learning rate and momentum factor.
3. Fuzzy-neural hybrid systems: utilize both fuzzy logic and neural networks in a system to perform separate tasks for decouple subsystems. The architecture of the systems depends on a particular application. For instance, the NN can be utilized for the prediction where the fuzzy logic addresses the control of the system.

On the other hand, the ANN, FS and evolutionary computations can be integrated in various ways. For example, the structure and parameter learning problems of neural network can be coded as genes in order to search for optimal structures and parameters of neural network. In addition, the inherent flexibility of the evolutionary computation and fuzzy systems has created a large diversity and variety in how these two complementary approaches can be combined to solve many engineering problems. Some of their applications include control of temperature (Khalid and Omatu 1992), robot trajectory (Rabelo and 1992), automatic generation control (Zeynelgil et al. 2002), decision support system (Zeng and Trauth 2005), system identification (Narendra et al. 1995), load forecasting (Park 1991; Chaturvedi 2000); system failure (Naida et al. 1990), automotive control (Miller et al. 1990), and various other applications (Simpson 1990).

We have lot of successful stories of soft computing applications in various techno-socio-economical systems with the present conventional computers as mentioned above. Hence, there exist a fusion between soft computing and conventional (Hard) computing at algorithmic levels. It is the time to think about the fusion of these techniques (Ovaska et al. 2002) not only on the algorithmic level but also on the higher system level. In real world applications, such a fusion is always a concrete response to some needs to improve performance, reduce computational burden, or lower the total product/process cost. However, hard computing community seems to be more reluctant in applying the complementing SC techniques. This situation will change only after positive experience and the changing emphasis in engineering education.

1.4 Summary

Soft computing is an emerging approach to computing to construct intelligent systems. It provides the ability of parallel computing as in human brain, and also reasoning and learning in an uncertain and imprecise environmental conditions. It consists of several computing paradigms, mainly neural networks, fuzzy systems and evolutionary algorithms. Strengths and weaknesses of these paradigms are provided in Table 1.1.

Table 1.1. Strengths and weaknesses of main constituents of soft computing

	ANN	Fuzzy	Evolutionary Algorithms
Strengths	1. Learning capability and adaptability 2. Fault tolerant capability 3. Model free approach 4. Historical (numerical) data needed for training	1. Knowledge representation in the form of rules 2. Fault tolerant capability 3. Expert knowledge is required 4. Reasoning capability	1. Systematic random search 2. Provides multiple solutions
Weaknesses	1. Black box approach 2. It can only handle quantitative information 3. No reasoning capability	1. No learning capability 2. It can only handle qualitative information	1. Convergence is slow near optimal solution

The field of soft computing is evolving rapidly and synergism between various constituents of it is coming up with faster rate. For better future of mankind a highly automated and intelligent/wiser machine could be build.

1.5 Bibliography and Historical Notes

The comprehensive overview of state-of-the-art-theory and successful industrial applications of soft computing around the world is given in the edited book by Suzuki and his colleagues (2000). Computational intelligence, natural computing and evolutionary computation gaining popularity which is described by Shengxiang (2007). The concept of Pareto-optimality to machine learning, particularly inspired by the successful developments in evolutionary multi-objective optimization provided by Yaochu (2006). Complete overview on the main constituents of soft computing is nicely written by Yaochu (2003).

1.6 Exercises

1. Define the soft computing
2. Mention its various components
3. What are the merits and demerits of main constituents of soft computing?
4. What are the differences between ANN, fuzzy systems and GA?
5. Who has developed artificial neuron? Who did point out the drawbacks of that neuron?
6. Mention the historical background of fuzzy logic.

2

Life History of Brain

*In the structure of the human frame, the brain is the most extraor-
dinary organ. The functions of all its parts are, however, not quite
understood but the knowledge we possess of the functions of the brain
is of most superficial character and is quite incommensurate with the
economy of this wonderful apparatus.*

*Maharaj Sahab Pandit Brahm Sankar Misra
(Discourses on Radhasoami Faith, 1960)*

2.1 Introduction

The history of our quest to understand the brain is certainly as long as human
history itself. Use this extensive timeline to meander through some of the
high-lights (and low-lights) of this great journey of understanding. There are
many evidences of ancient civilization which show that people were conducted
surgery on head (Brain). In Hindu religion, Lord *Ganesh* had the head of
elephant, Incarnation of Lord *Narsingha*, etc. Today due to fast progress of
neuro-science, we are at the verge of understanding that how brain functions
and what is the relationship between mind and brain, which may provide a
basis for understanding consciousness.

At this juncture, let us discuss an episode from the Hindu Epic:
"Ramayana". It is said that demon king Ravana attained the status of
the Gods in heavens. It is said that he used his enormous power in his spir-
itual battery to fly his airplane called Puspak, by which he abducted Sitaji
to his palace in Lanka. In this regard, let us only confine ourselves to the
fact of harnessing consciousness by the demon king Ravana to propel his
plane, myriad millennium ago. This would indeed be a point to meditation
to top one's consciousness and explain methodology to harness consciousness
in machine.

Plato hypothesized that the brain was the seat of the soul and also the
center of all control. It is somewhat surprising that he came to this correct

D.K. Chaturvedi: *Soft Computing Techniques and its Applications in Electrical Engineering,*
Studies in Computational Intelligence (SCI) **103**, 11–22 (2008)
www.springerlink.com   © Springer-Verlag Berlin Heidelberg 2008

conclusion in spite of the fact he rejected experiment and observation, and believed that true knowledge came only from pure reasoning and thought such as that involved in mathematics. It is also mentioned that our pleasure, joys, laughter and jests as well as our sorrows, pains, grief and tears every thing is closely controlled by the brain condition.

2.2 Development of Brain with Age

A *baby's brain*

A baby's brain is a mystery whose secrets scientists are just beginning to unravel. The mystery begins in the womb – only 4 weeks into gestation the first brain cells, the neurons, are already forming at an astonishing rate: 250,000 every minute. Billions of neurons will forge links with billions of other neurons and eventually there will be trillions and trillions of connections between cells. Every cell is precisely in its place, every link between neurons carefully organized. Nothing is random, nothing arbitrary.

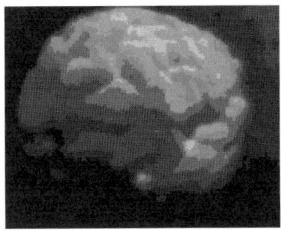

One way a newborn is introduced to the world is through vision. The eyes and the visual cortex of an infant continue to develop after birth according to how much stimulation she can handle. What happens to the brain when a baby is born with a visual abnormality? Infant cataracts pose an interesting challenge to scientists: How to remove the visual obstruction without compromising brain development.

Baby's brains are more open to the shaping hand of experience than at any time in our lives. In response to the demands of the world, the baby's brain sculpts itself. Scientists have begun to understand how that happens, but as Neurologist Carla Shatz says, "There's a great mystery left. Our memories and our hopes and our aspirations all of that is in there. But we only have the barest beginnings of an understanding about how the brain really works."

Child's brain

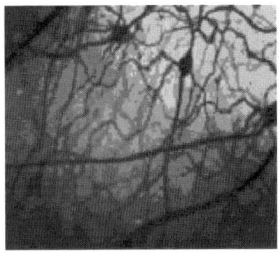

A child's brain is a magnificent engine for learning. A child learns to crawl, then walk, run and explore. A child learns to reason, to pay attention, to remember, but nowhere is learning more dramatic than in the way a child learns language. As children, we acquire language – the hallmark of being human.

In nearly all adults, the language center of the brain resides in the left hemisphere, but in children the brain is less specialized. Scientists have demonstrated that until babies become about a year old, they respond to language with their entire brains, but then, gradually, language shifts to the left hemisphere, driven by the acquisition of language itself.

Teenage brain

When examining the adolescent brain it is mystery, complexity, frustration, and inspiration. As the brain begins teeming with hormones, the prefrontal cortex, the center of reasoning and impulse control, is still a work in progress. For the first time, scientists can offer an explanation for what parents already know – adolescence is a time of rolling emotions, and poor judgment. Why do teenagers have distinct needs and behaviors? Why, for example, do high school students have such a hard time waking up in the morning? Scientists have just begun to answer questions about the purpose of sleep as it relates to the sleep patterns of teenagers.

A major challenge to the adolescent brain is schizophrenia. Throughout the world and across cultural borders, teenagers from as early as age 12 suffer from this brain disorder.

While adults spend about one third of their time sleeping, babies and toddlers sleep away half of their early childhood. It cannot be the terrible waste of time that it seems. Or can it be? Embarrassingly, scientists still cannot persuasively point out the biological function of sleep. Sex, eating, and sleeping constitute the triad of basic impulses of human beings. Yet, while the functions of the first two have been obvious for millennia, it is not clear why we crave to spend a third of our life in bed.

The first few hints for the function of sleep came from observations on animals. All mammals sleep, as do birds and even bees. One theory suggests that sleep is a simple protection mechanism, a way to keep animals quiet and still, so that they attract less attention, and thus are less noticeable to predators. This stillness is particularly important when the animal is most vulnerable, which for many animals, is during the dark of night. But comparing sleep patterns of different species suggests that this may be too simplistic explanation. Opossum, for example, sleep up to 20 h a day. Giraffes, Dolphins and whales also spend a very short time in sleeping. Some scientists even claim that dolphins let only half of their brain sleep at a time. Clearly, sleep is an opportunity to rest. Hence, many theorists have hypothesized that the main purpose of sleep is to enable the muscles and the brain to recuperate after a busy day. But measuring the electric activity of the brain unveils the shortcomings of this theory: A sleeping brain is far from dormant.

Adult brain

The adult brain is the apotheosis of the human intellect, but what of emotion? The science has changed the study of emotion: Emotion is now considered integral to our over-all mental health. In mapping our emotions, scientists have found that our emotional brain overlays our thinking brain.

There is a critical interplay between reason and emotion. We are well aware of how brain malfunctions can cause pain, depression, and emotional paralysis. We must also understand that the brain affects positive emotional responses such as laughter, excitement, happiness, and love. "Brain systems work together to give us emotions just as they do with sight and smell. If you lose the ability to feel, your life, and the lives of people around you, can be devastated." – Antonio R. Damasio

Aging brain

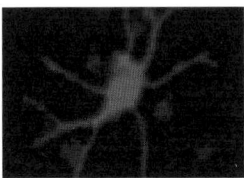 The latest discoveries in neuroscience present a new view of how the brain ages. Overturning decades of dogma, scientists recently discovered that even into our seventies, our brains continue producing new neurons. Scientists no longer hold the longstanding belief that we lose vast numbers of brain cells as we grow older. The normal aging process leaves most mental functions intact, and may even provide the brain with unique advantages that form the basis for wisdom. The aging brain is also far more resilient than was previously believed. Despite this, many people still suffer from the disease most associated with aging.

Mind illusion

 Vision has only partly to do with the retina, lens, and cornea. Mostly it depends on the state of brain. It is a reason for optical illusions. The process of seeing begins with the presence of light, an image being formed on the retina, and an impulse transmitted to the brain, but there are many other factors that play a part in how we perceive visually. Our perceptions are influenced by our past experiences, imagination, and associations.

2.3 Technologies for Study the Details of Brain

It's an exciting time in the study of the human brain and mind. Much of this advance in knowledge is the result of technological advances in brain imaging. It seems that almost every time one hears about some neurological experiment or advance in human neuroscience, the term "brain scan" appears. There are five most important technologies that allowed scientists to peer into the workings and structure of the living human brain.

2.3.1 Electro Encephalo Graph (EEG)

The electro encephalo graph (EEG) deserves as one of the first and still very useful ways of non-invasively observing human brain activity. An EEG is a recording of electrical signals from the brain made by hooking up electrodes to the subject's scalp. These electrodes pick up electric signals naturally produced by the brain and send them to galvanometers which detect and measure small electric currents that are in turn hooked up to pens, under which graph paper moves continuously. The pens trace the signals onto the graph paper.

Although it was known in the nineteenth century that living brains have electrical activity. An Austrian psychiatrist named Hans Berger was the first to record this activity in humans, in the late 1920s. EEGs allow researchers to follow electrical impulses across the surface of the brain and observe changes over split seconds of time. An EEG can show what state a person is in – asleep, awake, and anaesthetized because the characteristic patterns of current differ for each of these states. One important use of EEGs has been to show how long it takes the brain to process various stimuli. A major drawback of EEGs, however, is that they cannot show us the structures and anatomy of the brain or really tell us which specific regions of the brain do what.

2.3.2 Computerized Axial Tomography (CAT)

Developed in the 1970s, CAT (or CT) scanning is a process that combines many two-dimensional X-ray images to generate cross-sections or three-dimensional images of internal organs and body structures (including the brain). Doing a CAT scan involves putting the subject in a special, donut-shaped X-ray machine that moves around the person and takes many X-rays. Then, a computer combines the two-dimensional X-ray images to make the cross-sections or three-dimensional images. CAT scans of the brain can detect brain damage and also highlight local changes in cerebral blood flow (a measure of brain activity) as the subjects perform a task.

2.3.3 Positron Emission Tomography (PET)

It is developed in the 1970s to scan or observe blood flow or metabolism in any part of the brain. In a PET scan, the subject is injected with a very small quantity of radioactive glucose. The PET then scans the absorption of the radioactivity from outside the scalp. Brain cells use glucose as fuel, and PET works on the theory that if brain cells are more active, they will consume more of the radioactive glucose, and if less active, they will consume less of it.

A computer uses the absorption data to show the levels of activity as a color-coded brain map, with one color (usually red) indicating more active brain areas, and another color (usually blue) indicating the less active areas.

PET imaging software allows researchers to look at cross-sectional "slices" of the brain, and therefore observe deep brain structures, which earlier techniques like EEGs could not. PET is one of the most popular scanning techniques in current neuroscience research.

2.3.4 Magnetic Resonance Imaging (MRI)

MRI technique was a major breakthrough in 1977 in imaging technology. In an MRI, the subject is placed on a moveable bed that is inserted into a giant circular magnet. It is a non-invasive technique that does not involve exposure to radiation. It is usually painless medical test that helps physicians diagnose and treat medical conditions. MRI uses a powerful magnetic field, radio waves and a computer to produce detailed pictures of organs, soft tissues, bone and virtually all other internal body structures. The images can then be examined on a computer monitor.

2.3.5 Magneto Encephalo Graphy (MEG)

It is a new technology that measures the very faint magnetic fields that emanate from the head as a result of brain activity. In MEG, magnetic detection coils bathed in liquid helium are poised over the subject's head. The brain's magnetic field induces a current in the coils, which in turn induces a magnetic field in a special, incredibly sensitive instrument called a superconducting quantum interference device (SQUID). Of all the brain scanning methods, MEG provides the most accurate resolution of the timing of nerve cell activity–down to the millisecond.

2.4 Brain Functioning

Animals, lizards, frogs, fish, even birds have brains. But none of these creatures demonstrate the same capacity for learning, language, emotion and abstract thought that distinguishes the human species. Neuroscientists learned plenty about the functioning of the brain. But they admit there are aspects of brainpower that remain among humanity's most enduring mysteries. The brain performs a number of functions, many of which are related to the physical needs and actions of the body. For these functions, the brain can be thought of as the command centre of the human nervous system, much like the headquarters of a military unit. It receives information from its vast network of neurons throughout the body. Based on this information, it makes decisions and issues commands that stimulate muscles and give the body movement. Other brain functions are more like those of a university than a military headquarters. These functions give us the ability to read, write, talk and think about issues more broad than where the next meal is coming from.

2.5 Brain Structure

The brain is made of three main parts: the forebrain, midbrain, and hind-brain. The forebrain consists of the cerebrum, thalamus, and hypothalamus (part of the limbic system). The midbrain consists of the tectum and tegmentum. The hindbrain is made of the cerebellum, pons and medulla. Often the midbrain, pons, and medulla are referred to together as the brainstem. The "wrist" is the *brainstem*, connecting the brain to the spinal column, and the "fists" constitute the left and right hemispheres of the largest part of the brain, the *cerebrum*. It is associated with higher brain function such as thought and action. The cerebral cortex is divided into four sections, called "lobes": the frontal lobe, parietal lobe, occipital lobe, and temporal lobe. Figure 2.1 shows the cortex. Different areas or lobes of the cerebral cortex shown in figure play specific roles in human thought and activity. For example:

The *frontal lobes* control behavior, intellect and emotion, talking, self-monitoring, speaking (word finding), smell, abstract thinking and reasoning.

Temporal lobe is responsible for long-term memory storage, hearing, speech and understanding of language.

Parietal lobe is associated with movement, sense of touch, differentiation between size, shape colour, spatial perception, and visual perception.

Occipital lobe lies at the back of the brain and does vision function in the brain.

The cerebrum has an outer layer of *grey matter* arranged in folds (wrinkled texture). This outer layer, the *cerebral cortex*, is just a few millimeters thick but because of its numerous folds constitutes 40% of the entire brain mass. Essentially this makes the brain more efficient, because it can increase the

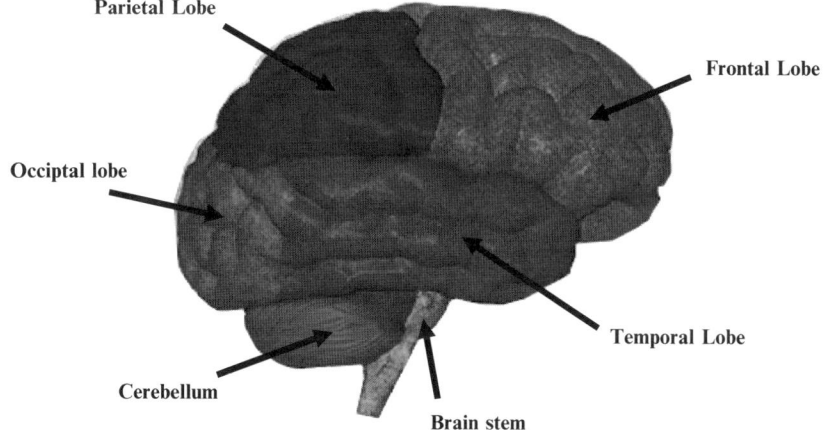

Fig. 2.1. Exposed view of brain

Table 2.1. Comparison between brain size and body size of human and different animals

Species	Brain length (cm)	Brain weight (g)	Body length (cm)	Body weight (g)
Camel	15	680	200	529,000
Dolphin		1,700	305	160,000
Human	15	1,400	100	62,000
Kangaroo	5	56	150	35,000
Baboon	8	140	75	30,000
Monkey	5	100	30	7,000
Raccoon	5.5	39	80	4,290
Cat	5	30	60	3,300
Rabbit	5	12	30	2,500
Squirrel	3	6	20	900
Frog	2	0.1	10	18

surface area of the brain and the amount of neurons within it. The outer cortex is divided into *gyri* (ridges) and *sulci* (valleys).

An obvious anatomical observation that one makes is that the brain is divided into two cerebral hemispheres. The size of these hemispheres is quite larger in humans than any other animal. The two hemispheres look mostly symmetrical yet it has been shown that each side functions slightly different than the other. In general, the right side of the brain controls movement in the left side of the body and the left side controls the right. However, there is some specialization. For example, language is more a function of the left hemisphere and recognition of shapes is more a function of the right.

The cerebellum, or "little brain", is similar to the cerebrum and below to it, has two hemispheres and has a highly folded surface or cortex. This structure is associated with regulation and coordination of movement, posture, and balance.

The limbic system, often referred to as the "emotional brain", is found buried within the cerebrum. Underneath the limbic system is the brain stem. This structure is responsible for basic vital life functions such as breathing, heartbeat, and blood pressure. Scientists say that this is the "simplest" part of human brains because animals' entire brains, such as reptiles (who appear early on the evolutionary scale) resemble our brain stem as given in Table 2.1.

2.6 Brainwaves to Study the State of Brain

It is well known that the brain is an electrochemical organ. Researchers have speculated that a fully functioning brain can generate as much as 10 W of electrical power. Electrical activity emanating from the brain is displayed in the form of brainwaves. There are four categories of these brainwaves, ranging from the least activity to the most activity.

1. *Delta waves*

   It is in the frequency range from 1 to 8 Hz. It is seen normally in babies and young children. It may be seen in drowsiness or arousal in older children and adults; it can also be seen in meditation.

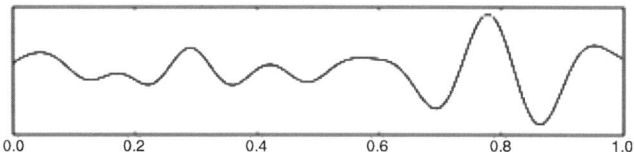

2. *Alpha waves*

   Alpha waves are shown when brain in non-arousal state. These waves are slower and higher in amplitude. Their frequency ranges from 9 to 14 Hz. A person who has completed a task and sits down to rest or meditates is usually in an alpha state.

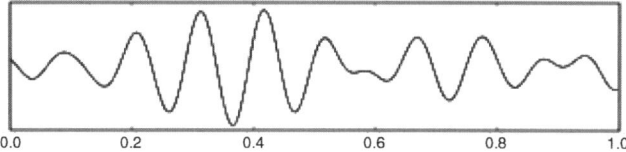

3. *Beta waves*

   When the brain is aroused and actively engaged in mental activities, it generates beta waves. These beta waves are of relatively low amplitude, and are the fastest of the four different brainwaves. The frequency of beta waves ranges from 15 to 40 Hz. Beta waves are characteristics of a strongly engaged mind. These brain waves are often associated with active, busy or anxious thinking and active concentration.

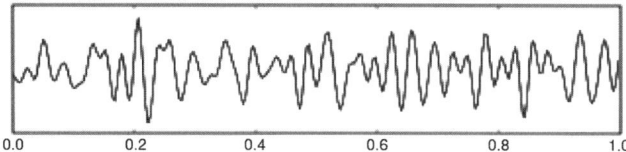

4. *Gamma*

   It is the frequency range approximately 26–100 Hz. It represents binding of different populations of neurons together into a network for the purpose of carrying out a certain cognitive or motor function.

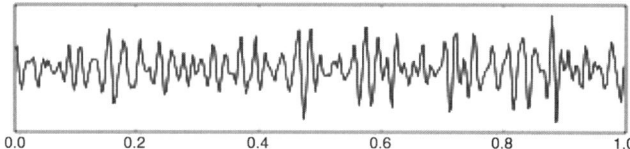

5. *Higher gamma waves*
   It is the frequency range approximately 100–130 Hz.

Do you know?

1. *The brain* is the main switching unit of the central nervous system; it is the place to which impulses flow and from which impulses originate.
2. The spinal cord provides the link between the brain and the rest of the body.
3. The brain has three main parts:
   a. The cerebrum
   b. The cerebellum
   c. The brain stem
4. The brain is a highly organized organ that contains approximately 100 billion neurons and has a mass of 1.4 kg and it is protected by a bony covering called the skull.
5. In order to perform the brain functions, it needs a constant supply of food and oxygen. If the oxygen supply to the brain is cut off even for a few minutes, the brain will usually suffer enormous damage. Such damage may result in death.

The cerebrum

1. The cerebrum is the control center of the brain. It is the largest and most prominent part of the human brain and 85% of the weight of a human brain is of cerebrum. It is responsible for all the voluntary (conscious) activities of the body.
2. It is the site of intelligence, learning and judgment. It functions in language, conscious thought, memory, personality development, vision, and other sensations.
3. The cerebrum takes up most of the space in the cavity that houses the brain (skull). It is divided into two hemispheres, the left and right cerebral hemispheres and a deep grove that separates the two hemispheres. The hemispheres are connected in a region known as the corpus callosum. The right and left cerebral hemispheres are linked by a bundle of neurons called a tract. The tract tells each half of the brain what the other half is doing.
4. The most obvious feature on the surface of each hemisphere are numerous folds. These folds and the groves increase the surface area of the cerebrum. The ridges are called gyri, and the grooves are called sulcus.

5. The cerebrum, which looks like a wrinkled mushroom, is positioned over the rest of the brain. It contains thick layers of gray matter.
6. Each hemisphere of the cerebrum is divided into four regions called lobes. These lobes are named as, frontal, parietal, temporal, and occipital lobes.
7. Scientist have discovered that the left side of the body sends its sensations to the right hemisphere of cerebrum, and the right side of the body sends its sensations to the left hemisphere. The right hemisphere is associated with creativity and artistic ability and left hemisphere is associated with analytical and mathematical ability.
8. The cerebrum consists of two surfaces, one is folded outer surface called the cerebral cortex and consists of gray matter and second is the inner surface called cerebral medulla, which is made up of white matter.

The cerebellum

1. The cerebellum is the second largest part of the brain, and is located at the back of the skull. It coordinates muscle movements and balances the body.
2. This is a small cauliflower shaped structure, and well developed in mammals and birds. Bird performs more complicated feats of balance than most mammals.

The brain stem

1. The *brain stem* connects the brain to the spinal cord and maintains life support systems. It controls vital body processes.

2.7 Summary

The human brain is responsible for overseeing the daily operations of the human body and for interpreting the vast amount of information it receives. The adult human brain weighs an average of 1.4 kg, or about 2% of the total body weight. Despite this relatively small mass, the brain contains approximately 100 billion neurons. Functioning as a unit, these neurons make up the most complex and highly organized structure on Earth. The brain is responsible for many of the qualities that make each individual unique-thoughts, feelings, emotions, talents, memories, and the ability to process information. Much of the brain is dedicated to run the body, the brain is responsible for maintaining Homeostasis by controlling and integrating the various systems that make up the body.

- Our brain is more complicated than any computer we can imagine. There are 100 billion nerve cells in your brain, and every nerve cell has many connections to other nerve cells. In fact, your brain has more connections in it than there are stars in the universe!

- Sleep deprivation also decreases brain activity and limits access to learning, memory, and concentration. People who consistently slept less than 7 h had overall less brain activity.
- Stress negatively affects brain function. Brain cells can die with prolonged stress.
- Every time you learn something new your brain makes a new connection. Learning enhances blood flow and activity in the brain.

2.8 Bibliography and Historical Notes

An excellent up to date introduction to neuroscience can be found in Bear et al. (1996). Other interesting historical perspectives can be gleaned from (Arbib 1987, 1995, 2003; Kandel 2000; Kandel and Schwartz 1982). Bridgeman (1988) introduced the contemporary theories of behaviour and mind in lucid manner. The historical development of brain and mind are provided in accessible manner by Blackmore. For the detailed and advanced treatment on neurons and their working look (Koch 1999). Neurobiological aspects of memory are given in Dudai (1989).

2.9 Exercises

1. Explain the salient points of human brain development.
2. What do you mean by mind illusion?
3. Explain the different parts of human brain and also mention various functions performed by these part.
4. Summarize the functions of the major parts of the brain.
5. Summarize the functions of the cerebrum, brain stem, and cerebellum.
6. Describe how the brain is protected from injury.
7. Compare the brain size and body size of human beings and different animals.
8. Write in brief about the emotional brain of human being. Where does it locate?
9. What do you mean by brain waves?
10. What are the different categories of these brainwaves?
11. Which type of brain waves often associated with active, busy or anxious thinking and active concentration?

3

Artificial Neural Network and Supervised Learning

3.1 Introduction

Artificial neural networks are biologically inspired but *not necessarily biologically plausible*. Researchers are usually thinking about the organization of the brain when considering network configurations and algorithms. But the knowledge about the brain's overall operation is so limited that there is little to guide those who would emulate it. Hence, at present time biologists, psychologists, computer scientists, physicists and mathematicians are working all over the world to learn more and more about the brain. Interests in neural network differ according to profession like neurobiologists and psychologists try to understanding brain. Engineers and physicists use it as tool to recognize patterns in noisy data, business analysts and engineers use to model data, computer scientists and mathematicians viewed as a computing machines that may be *taught* rather than programmed and artificial intelligentsia, cognitive scientists and philosophers use as sub-symbolic processing (reasoning with patterns, not symbols), etc.

A conventional computer will never operate as brain does, but it can be used to simulate or model human thought. In 1955, Herbert Simon and Allen Newell announced that they had invented a thinking machine. Their program, the logic theorist, dealt with problems of proving theories based on assumptions it was given. Simon and Newell later developed the general problem solver, which served as the basis of artificial intelligence (AI) systems. Simon and Newell believed that the main task of AI was figuring out the nature of the symbols and rules that the mind uses. For many years AI engineers have used the "top-down" approach to create intelligent machinery. The top-down approach starts with the highest level of complexity, in this case thought, and breaks it down into smaller pieces to work with. A procedure is followed step by step. AI engineers write very complex computer programs to solve problems. Another approach to the modelling of brain functioning starts with the lowest level, the single neuron. This could be referred to as a bottom-up approach to modelling intelligence.

D.K. Chaturvedi: *Soft Computing Techniques and its Applications in Electrical Engineering*, Studies in Computational Intelligence (SCI) **103**, 23–50 (2008)
www.springerlink.com © Springer-Verlag Berlin Heidelberg 2008

3.2 Comparison of Neural Techniques and Artificial Intelligence

Artificial intelligence (AI) is a branch of computer science that has evolved to study the techniques of construction of computer programs capable of displaying intelligent behavior. It is the study of computations that make it possible to perceive, reason, and act. Growth of artificial intelligence based on the hypothesis that thought processes could be modeled using a set of symbols and applying a set of logical transformation rules. It is important to note that any artificially intelligent system must possess three essential components:

1. A representation mechanism to handle knowledge which could be general or domain specific, implicit or explicit and of different level of abstraction.
2. An inference mechanism to get the appropriate conclusion for the given information or fact.
3. A mechanism for learning from new information or data without disturbing much to the existing set of rule.

The languages commonly used for AI model development are list processing language (LISP) and programming in logic (PROLOG).

The symbolic approach has a number of limitations:

- It is essentially sequential and difficult to parallelize.
- When the quantity of data increases, the methods may suffer a combinatorial explosion.
- An item of knowledge is represented by a rule. This localized representation of knowledge does not lend itself to a robust system.
- The learning process seems difficult to simulate in a symbolic system.

The ANN approach offers the following advantages over the symbolic approach:

- Parallel and real-time operation of many different components
- The distributed representation of knowledge
- Learning by modifying connection weights.

Both approaches are combined to utilize the advantages of both the techniques. A brief comparison of these techniques is given in Table 3.1.

3.3 Artificial Neuron Structure

The human nervous system, built of cells called neurons is of staggering complexity. An estimated 10^{11} interconnections over transmission paths are there that may range for a meter or more. Each neuron shares many characteristics with the other cells in the body, but has unique capabilities to receive, process, and transmit electrochemical signals over neural pathways that comprise the

Table 3.1. Comparison between ANN and AI

	ANN	AI
Type of information	Quantitative	Qualitative
Input	Measurements	Facts
Output	Predictions	Decision
Type of model	Mathematical	Logical
Requirement for model development	Historical data	Human experts
Adaptability	Learning capability	No learning capability
Flexibility	Re-trained for other problems	Completely change the knowledge base if problem changes
Model accuracy	Depends on learning	Depends on the knowledge acquired
Explanation	No explanation	Explanation depends on the depth of knowledge
Processing	Parallel and distributed	Sequential and logical
Representational structure of knowledge	Store global patterns or function information	Declarative (a collection of facts) or procedural (specifying an algorithm code to process information)

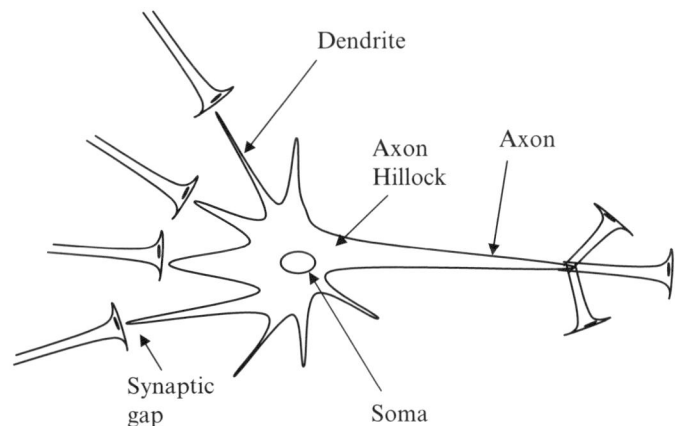

Fig. 3.1. Structure of biological neuron

brain's communication system. Figure 3.1 shows the structure of typical biological neurons. Biological neuron basically consists of three main components cell body, dendrite and axon. Dendrites extend from the cell body to other neurons where they receive signals at a connection point called a synapse. On the receiving side of the synapse, these inputs are conducted to the cell body, where they are summed up. Some inputs tend to excite the cell causing a reduction in the potential across the cell membrane; others tend to inhibit its

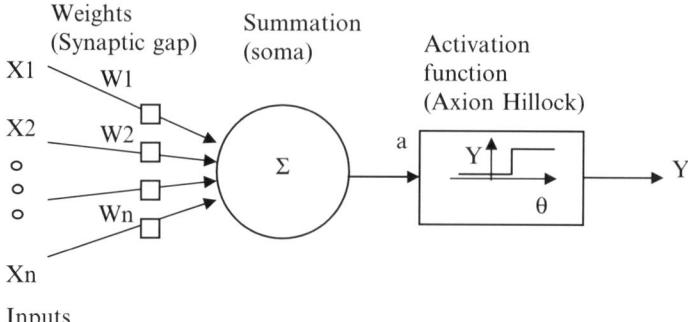

Fig. 3.2. Artificial neuron structure (perceptron model)

firing causing an increase in the polarization of the receiving nerve cell. When the cumulative excitation in the cell body exceeds a threshold, the cell fires and *action potential* is generated and propagates down the axon towards the synaptic junctions with other nerve cells.

The artificial neuron was designed to mimic the first order characteristics of the biological neuron. McCulloch and Pitts suggested the first synthetic neuron in the early 1940s. In essence, a set of inputs are applied, each representing the output of another neuron. Each input is multiplied by a corresponding weight, analogous to a synaptic strength, and all of the weighted inputs are then summed to determine the activation level of the neuron. If this activation exceeds a certain threshold the unit produces an output response. This functionality is captured in the artificial neuron known as the threshold logic unit (TLU) originally proposed by McCulloch and Pitts. Figure 3.2 shows a model that implement this idea. Despite of the diversity of network paradigms, nearly all are based upon this neuron configuration. Here a set of input labeled $X_1, X_2, \ldots, X_n$ is applied from the input space to artificial neuron. These inputs, collectively referred as the input vector "X" corresponds to the signal into the synapses of biological neuron. Each signal is multiplied by an associated weight $W_1, W_2, \ldots W_n$, before it is applied to the summation block.

The activation a, is given by

$$a = w_1x_1 + w_2x_2 + \ldots w_nx_n + \theta. \tag{3.1}$$

This may be represented more compactly as

$$a = \sum_{i=1}^{n} X_i W_i + \theta, \tag{3.2}$$

the output y is then given by $y = f(a)$, where f is a activation function.

In McCulloh–Pitts Perceptron model hard limiter as activation function was used and defined as:

$$y = \begin{cases} 1 & \text{if } a >= ß \\ 0 & \text{if } a < ß \end{cases}$$

The threshold ß will often be zero. The activation function is sometimes called a *step-function*. Some more non-linear activation functions also tried by the researchers like sigmoid, Gaussian, etc. and the neuron responses for different activation functions shown in Fig. 3.3 with the Matlab program.

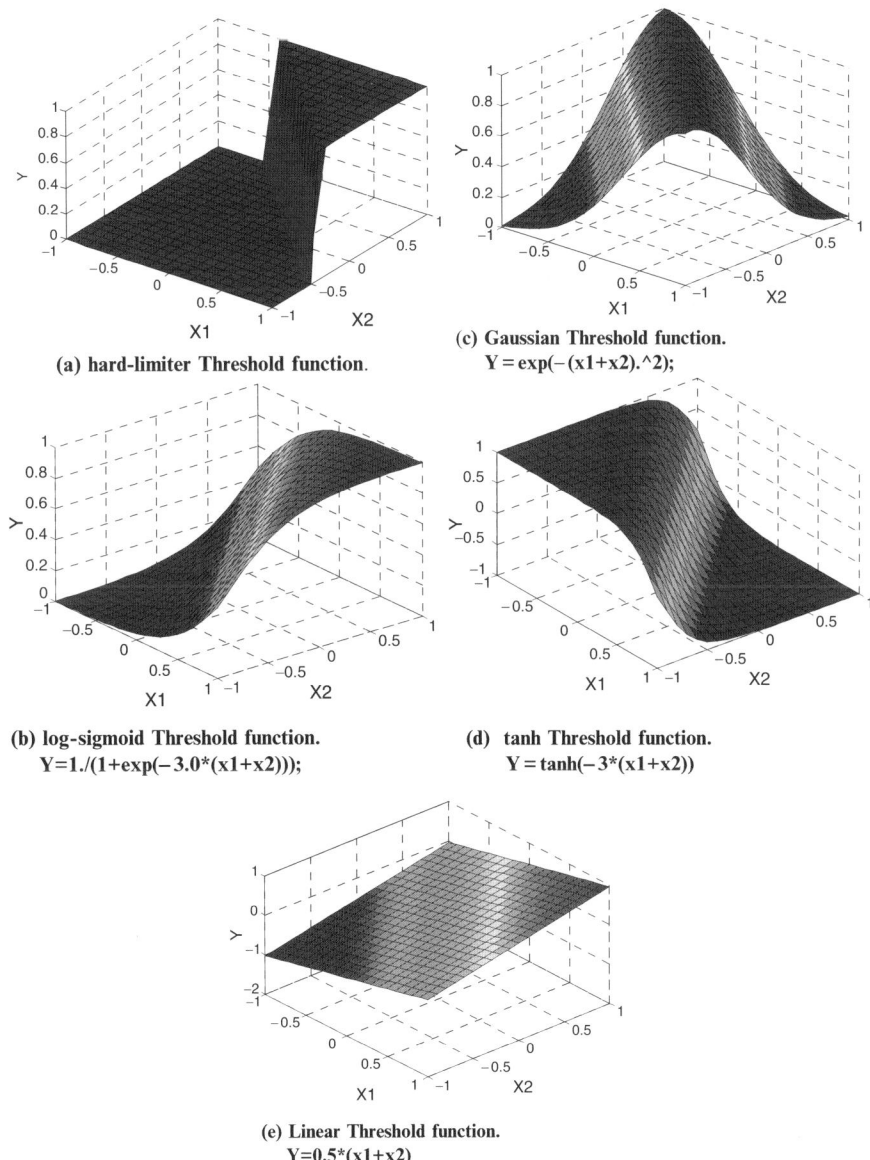

(a) hard-limiter Threshold function.

(c) Gaussian Threshold function.
Y = exp(− (x1+x2).^2);

(b) log-sigmoid Threshold function.
Y=1./(1+exp(− 3.0*(x1+x2)));

(d) tanh Threshold function.
Y = tanh(− 3*(x1+x2))

(e) Linear Threshold function.
Y=0.5*(x1+x2)

Fig. 3.3. Effect of different activation function on summation type simple neuron model

```
% 3-D surface generation program for simple neuron with
  different threshold functions
% Inputs
x1=-1:.1:1;
x2=-1:.1:1;
[m n]=size(x1);
Th=0;
for i=1:n
    for j=1:n
        sum=x1(i)+x2(j)-0.05;
        if sum>=Th      Y(i,j)=1 else Y(i,j)=0 end
    end
end
Y1=exp(-Y.^2);
surf(x1(1:n),x2(1:n),Y1)
```

3.4 Adaline

The next major development after the M & P neural model was proposed,
occurred in 1949 when Hebb (1949) proposed a learning mechanism for the
brain that became the starting point for artificial neural network learning
(training) algorithms. He postulated that as brain learns, it changes its connectivity patterns. More specifically, his learning hypothesis is as follows: "When
the axon of cell A is near enough to excite cell B and repeatedly or persistently
takes part in firing it, some growth process or metabolic change takes place in
one or both cells such that the A's efficiency, as one of the cells firing cell B, is
increased." Hebb further proposed that if one cell repeatedly assists in firing
another, the knobs of the synapse, are the junction, between the cells would
grow so as to increase the area of contact. The Hebb's learning hypothesis is
schematically shown in Fig. 3.4 (Levine 1983). This idea of learning mechanism was first incorporated in artificial neural network by (Rosenblatt 1958).
He combined the simple M & P model with the adjustable synaptic weights
based on Hebbian learning hypothesis to form the first artificial neural network with the capability to learn. The delta rule or the least mean squares
(LMS) learning algorithm, was developed by Widrow and Hoff (1960). This
model was called ADALINE for ADAptive LInear NEuron which is shown in
Fig. 3.5. This learning algorithm first introduced the concept of supervised
learning using a teacher which guides the learning process. It is the recent
generalization of this learning rule into the backpropagation algorithm that
has led to the resurgence in biologically based neural network research today.
This states that if there is a difference between the actual output pattern and
the desired output pattern during training, then the weights are changed to
reduce the difference. The amount of change of weights is equal to the error on

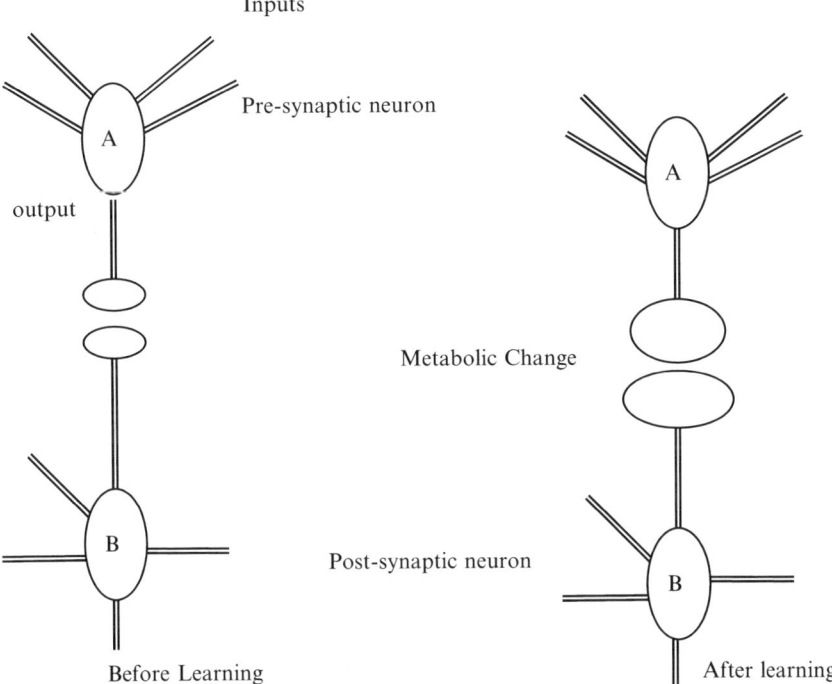

Fig. 3.4. Hebbian learning

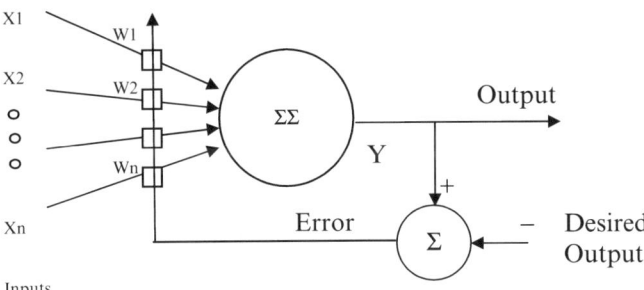

Fig. 3.5. ADLINE model

the outputs times the values of the inputs, times the learning rate. Many networks use some variation of this formula for training. In 1969 research in the field of artificial neural networks suffered a serious setback. Minsky and Papert published a book called Perceptrons (Minsky and Papert 1969) in which they proceed that single layer neural networks have limitations in their abilities to process data, and are capable of any mapping that is linearly separable. They pointed out, carefully applying mathematical techniques that the logical exclusive OR (XOR) function could not be realized by perceptrons. Further,

Minsky and Papert argued that research into multi-layer neural network could be unproductive. Due to this pessimistic view of Minsky and Papert, the field of artificial neural networks entered into an almost total eclipse for nearly two decades. Fortunately, Minsky and Papert's judgement has been disproved; all non-linear separable problems can be solved by multi-layer perceptron networks. Nevertheless, a few dedicated researchers such as Kohonen, Grossberg, Anderson, Hopfield continued their efforts. A renaissance in the field of neural networks started in 1982 with the publication of the dynamic neural architecture by Hopfield (1982). This was followed by the landmark publication "Parallel Distributed Processing" by McClelland and Rumelhart (1986) who introduced into the back-propagation learning technique for multi-layer neural networks. Back-propagation, developed independently by Werbos (1974), provides a systematic means for training multi-layer neural networks. This development resulted in renewed interest in the field of neural networks and since mid-nineties a tremendous explosion of research has been occurring. Most of the neural network structures used presently for engineering applications is feed-forward neural networks (static). These neural networks comprising of a number of neurons respond instantaneously to the inputs. In other words, the response of static neural networks depends on the current inputs and the weights. The absence of feedback in static neural networks ensures that networks are conditionally stable. However, these networks suffer from the following limitations:

(1) In feed forward neural networks, where the information flows from A to B, to C, to D and never comes back to A. On the other hand, biological neural systems almost always have feedback signals about their functioning.
(2) The structure of the computational (artificial) neuron is not dynamic in nature and performs a simple summation operation. On the other hand, a biological neuron is highly complex in structure and provides much more computational functions than just summation.
(3) The static neuron model does not take into account the time delays that affect the dynamics of the system; inputs produce an instantaneous output with no memory involved. Time delays are inherent characteristics of biological neurons during information transmission.
(4) Static networks do not include the effects of synchronism or the frequency modulation function of biological neurons. In recent years, many researchers are involved in developing artificial neural networks to overcome the limitations of static neural networks mentioned above. Instead of summation as an aggregation function, the product (II) is used as an aggregation function as shown in Fig. 3.6. The effect of different aggregation functions are also studied and shown in Fig. 3.7. The aggregation function could also be the combination of summation and product.

In product neuron the activation a, is given by

$$a = w_1x_1{}^*w_2x_2{}^* \ldots w_nx_n + \theta.$$

Fig. 3.6. Product type neuron

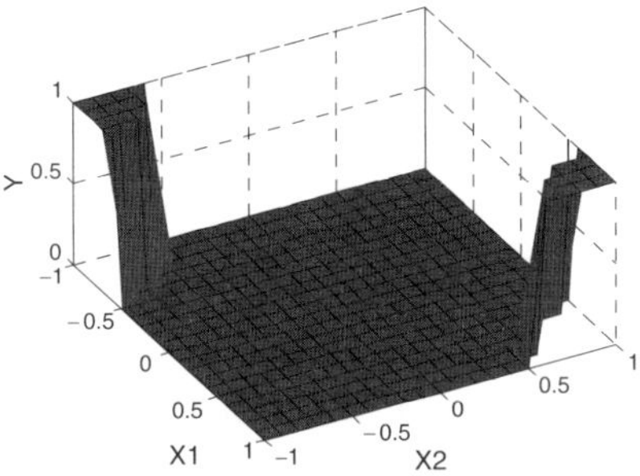

(a) Hard-limiter Threshold function.

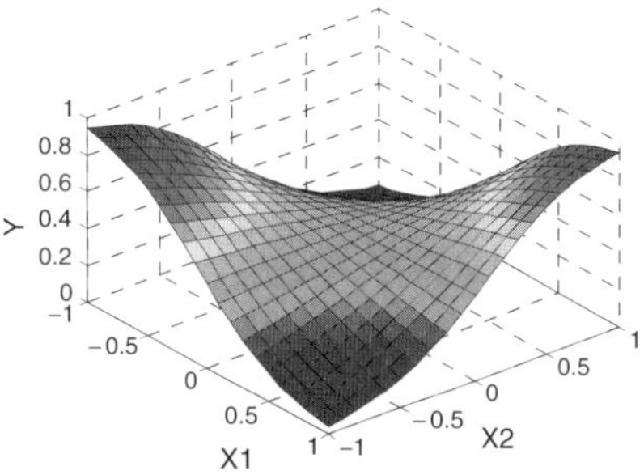

(b) Log-sigmoid Threshold function.

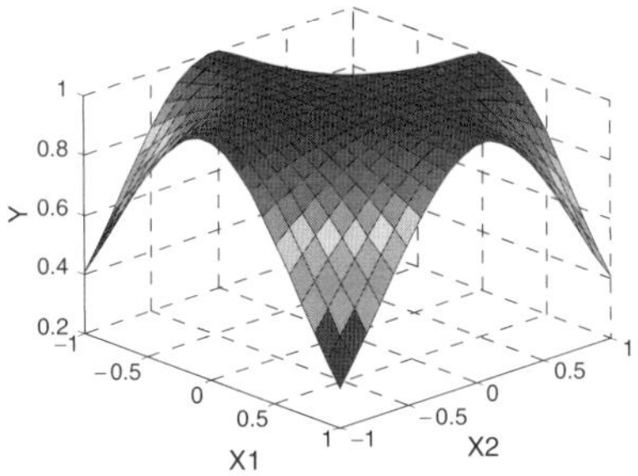

(c) Gaussian Threshold function.

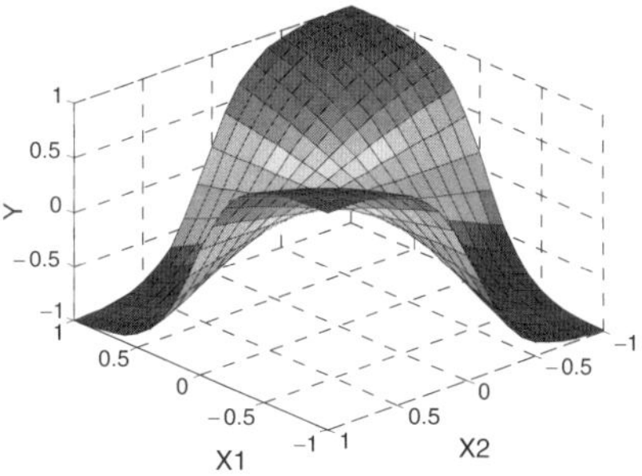

(d) Tanh Threshold function.

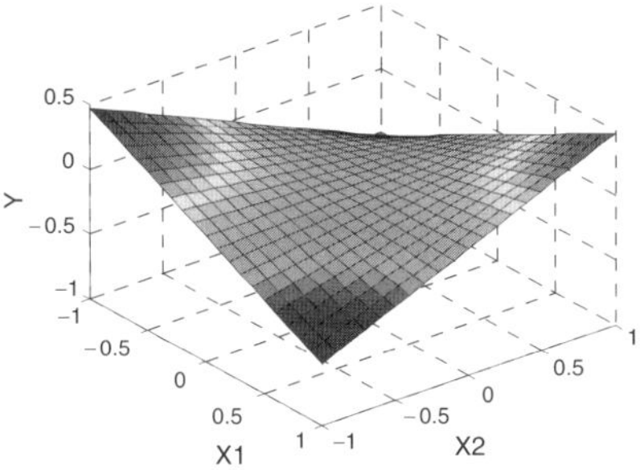

(e) Linear Threshold function.

Fig. 3.7. Effect of different activation functions on the product type neuron

This may be represented more compactly as

$$a = \prod_{i=1}^{n} X_i W_i + \theta,$$

the output y is then given by $y = f(a)$.

A major characteristic of perceptron (summation type neuron) is the linear separation due to its threshold function. It can identify linearly separable regions easily as shown in Fig. 3.8.

To be able to divide the area into two regions for the problem in Fig. 3.8a, only one perceptron is required, since the whole area is divided into two separate regions by a single line. However, the threshold area in Fig. 3.8b is formed by many lines, thus we need more than one perceptron in different layers to generate a solution to this problem as shown below in Fig. 3.9.

Here, θ_1, θ_2 and θ_3 are the threshold values and w_{1j}, w_{2j}, and w_{3j} are the interconnection weights for the processing elements in the hidden layer, 1, 2 and 3, respectively. Each of the processing elements is connected to the third layer, which is the output perceptron through equal weight of 1, and a threshold value of 2.5 to be able to perform the operation. X and Y represent the x-y co-ordinates of the point selected from the region specified in Fig. 3.8b. Each perceptron separates the area into two regions by a line, but the solution is the intersection of these areas. Therefore, one more perceptron is needed to combine the outputs of these perceptrons to identify the marked area. This special perceptron combines the outputs from other perceptrons with unity weighing and the threshold value of "n–0.5", where n is the number of separation lines created.

The perceptron architecture is also able to perform an Exclusive OR operation, for example, in identifying the truth table of the region given in Fig. 3.10.

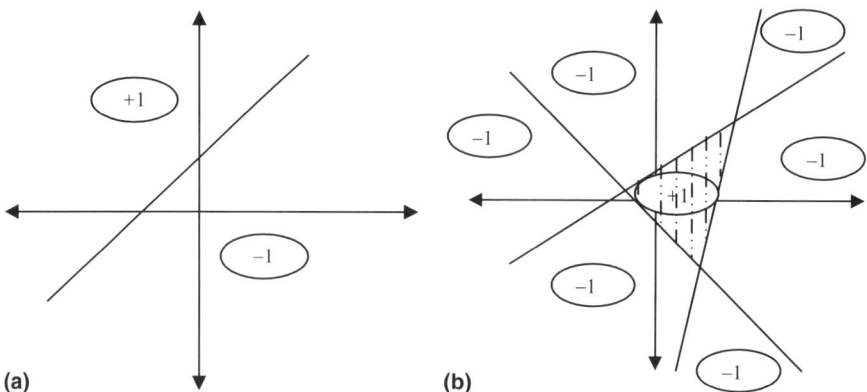

(a)   **(b)**

Fig. 3.8. (a) Two separate regions defined. (b) A selected region defined by a single perceptron intersection of areas created by perceptrons

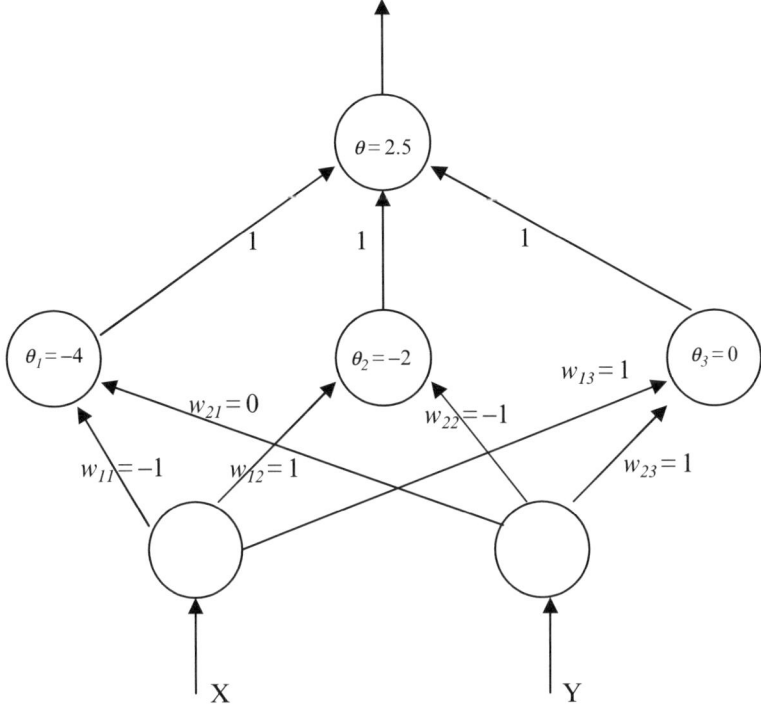

Fig. 3.9. A solution perceptron network

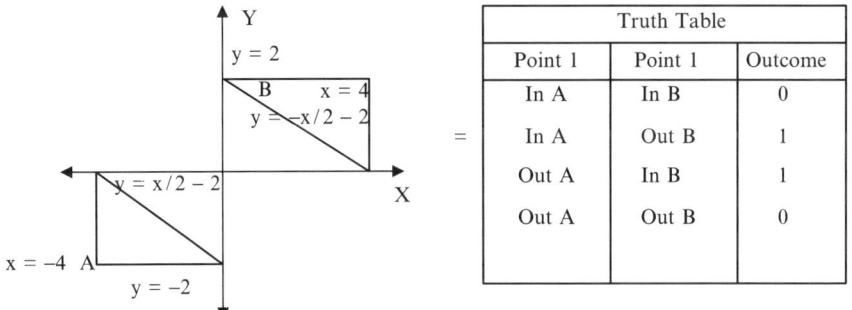

Truth Table		
Point 1	Point 1	Outcome
In A	In B	0
In A	Out B	1
Out A	In B	1
Out A	Out B	0

Fig. 3.10. The exclusive-OR problem

Lets identify the region A using a three-layer perceptron network. Hence we need to compute the weights, w_{ij}s associated with the interconnections between the input units, which represent the x-y co-ordinates of the points given in Fig. 3.8b and a corresponding threshold value, θ_i to be able to make the distinction between the two regions identified as $+1$ and -1 in the figure. These values also correspond to the output values of perceptron model as shown in Figs. 3.11–3.15.

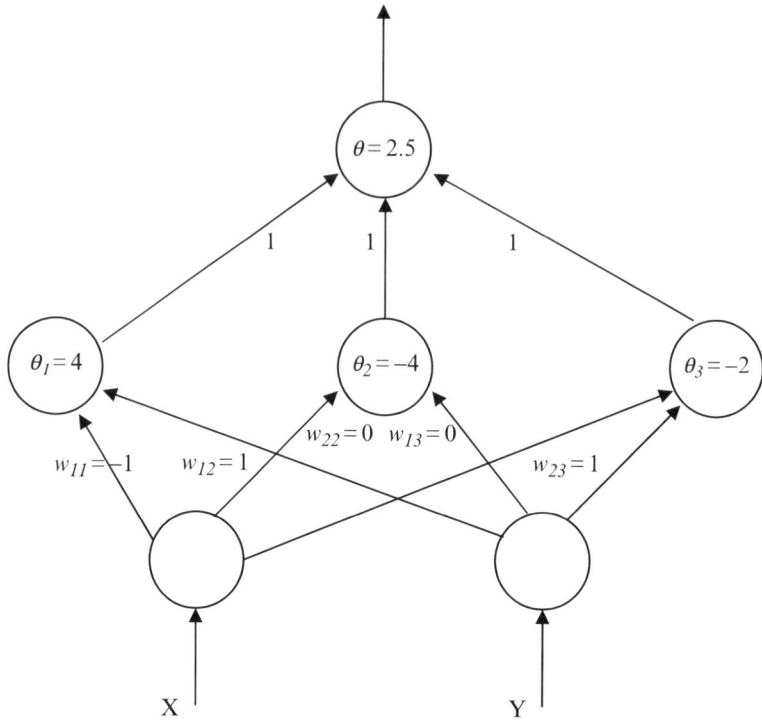

Fig. 3.11. Perceptron network for region A

$$y < -x/2 - 2; \quad x > -4; \quad y > -2$$
$$x/2 + y + 2 < 0; \quad x + 4 > 0; \quad y + 2 > 0$$
$$-x - 2y - 4 > 0$$

Similarly for region B;

$$y > -x/2 + 2; \quad x < 4; \quad y < 2$$
$$x + 2y - 4 > 0; \quad -x + 4 > 0; \quad -y + 2 > 0$$

To be able to produce result as in the truth table shown in Fig. 3.6, which is an Exclusive OR operation, we need another three-layer perceptron as shown in Fig. 3.9.

Thus the entire solution architecture is as shown above.

```
% Matlab Program for solving OR problem
clc; clear all;
X=[0 0 1 1;
   0 1 0 1];                    % Row wise inputs
D=[0 1 1 1];                    % Row wise output
```

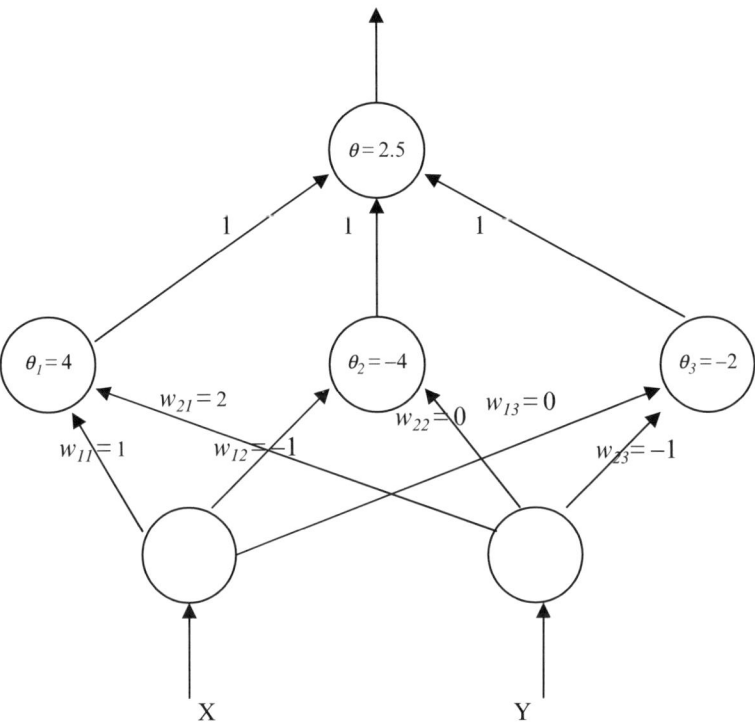

Fig. 3.12. Perceptron network for region B

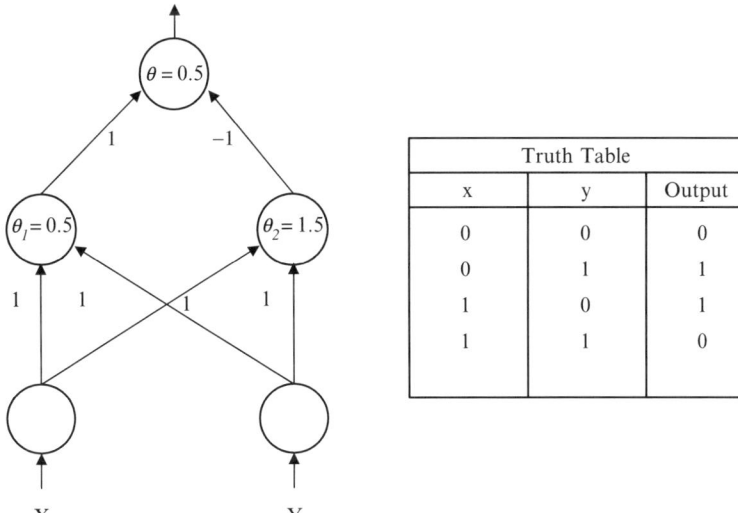

Truth Table		
x	y	Output
0	0	0
0	1	1
1	0	1
1	1	0

Fig. 3.13. Perceptron network for XOR

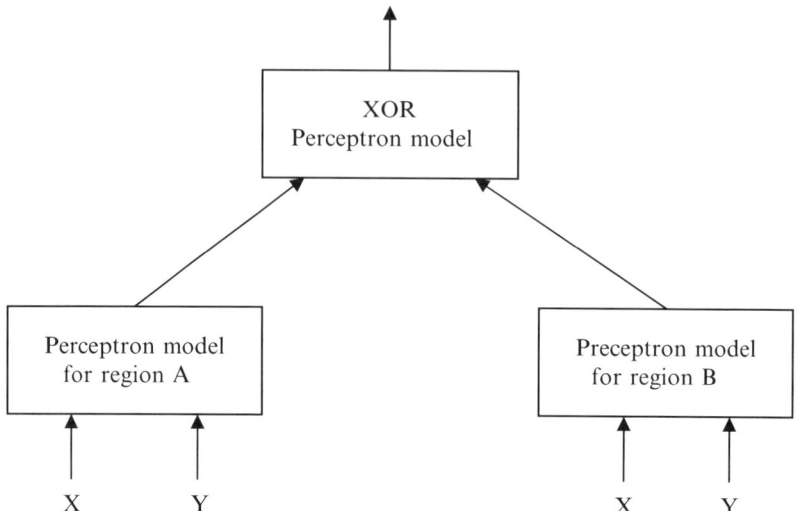

Fig. 3.14. Combined architecture for the subject problem

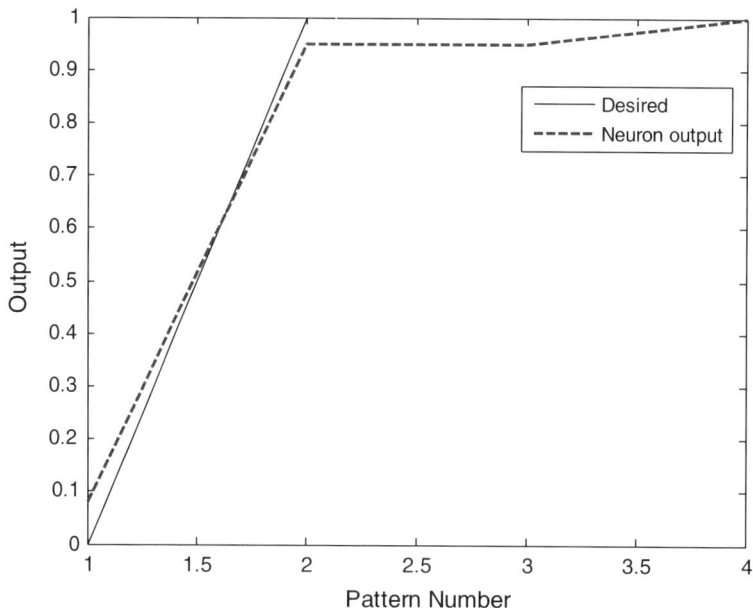

Fig. 3.15. Comparison of neuron output with actual results

```
% Initialization
W=randn(2,1);                    % Weight
B=randn(1,1);                    % Bias
LR=0.9;                          % learning rate
MF=0.1;                          % momentum factor
dW=zeros(2,1);dB=0;              % change in weight
for i=1:500
  wSum=W'*X+B;                   % Calculating activation
  O=1./(1+exp(-wSum));           % Output of neuron
  e=(D-O);                       % Error calculation
  sse=sum(e.^2)/2;               % Sum squared error
  de=(e.*O.*(1-O));              % Derivative of error
  dW=LR*X*de'+MF* dW;
  dB=LR*sum(de')+MF*dB;
  B=B+dB;                        % New value of bias
  W=W+dW                         % New value of weight
end
wSum=W'*X+B;                     % Testing
O=1./(1+exp(-wSum))
plot(D);
hold
plot(O, '--')
xlabel('Pattern Number')
ylabel ('Output')
```

The neurons or processing elements of an ANN are connected together and the overall system behaviour is determined by the structure and strength of these connections. This network structure consists of the processing elements arranged in groups or layers. A single structure of interconnected neurons produces an auto-associative system and is often used as a content-addressable memory. Connections between same neurons are referred as *lateral connections* and those that loop back and connected to the same neuron are called *recurrent connections*. Multi-layered systems contain input and output neuron layers that receive or emit signals to the environment and neurons, which are neither, called *hidden neuron layers*. This hidden layer provides networks with the ability to perform non-linear mappings as well as contributing to the complexity of reliable training of the system. Inter-field connections or connections between neurons in different layers can propagate signals in one of two ways:

- Feed-forward signals only allow information to flow along connections in one directions, while
- Feedback signals allow information to flow in either direction and/or recursively.

Artificial neural networks can provide content-addressable memory (CAM), which stores data at stable states in the weight matrix, and associative memory (AM), which provides output responses from, input stimuli. In this hetero-associative memory system, the ANN recall mechanism is a function g(.) that takes the weight matrix $\mathbf{W}$ and an input stimulus Input$_k$, and produces the output response Output$_k$. The two primary recall mechanisms are

1. Nearest-neighbour recall and
2. Interpolative recall.

Nearest-neighbour recall finds the stored input that closely matches the stimulus and responds with the corresponding output using a distance measure such as Hamming or Euclidean distance.

Interpolative recall takes the stimulus and interpolates the entire set of stored inputs to produce the corresponding output.

3.5 ANN Learning

Learning in an ANN is defined as any progressive systematic change in the memory (weight matrix) and can be supervised or unsupervised.

Unsupervised learning or self-organisation is a process that does not incorporate any external teacher and relies only upon local information and internal control strategies. Examples include:

- Adaptive Resonance Theory (ART1 and ART2) (Carpenter and Grossberg 1988)
- Hopfield Networks (Hopfield 1982)
- Bi-directional Associative Memory (BAM) (Kosko 1987, 1988)
- Learning Vector Quantisation (LVQ) (Kohonen 1988, 1997)
- Counter-Propagation networks (Hecht-Nielsen 1987, 1988, 1990).

Supervise learning, which includes:

- Back-propagation
- The Boltzmann Machine (Ackley et al. 1985)

It incorporates an external teacher and/or global information and includes such techniques as error-correction learning, reinforcement learning and stochastic learning. *Error-correction learning* adjusts the correction weight matrix in proportion to the difference between the desired and the computed values of each neuron in the outer layer. *Reinforcement learning* is a technique by which the weights are reinforced for properly performed actions and punished for inappropriate ones where the performance of the outer layer is captured in a single scalar error value. *Stochastic learning* works by making a random change in the weight matrix and then determining a property of the network called the *resultant energy*. If the change has made this energy value lower than it was previously, then the change is accepted, otherwise the change is accepted according to a pre-chosen probability distribution. This random acceptance of change that temporarily degrades the performance of the system allows it to escape from local energy minima in its search for the optimal system state.

In a multi-layered net using supervised learning, the input stimuli can be recorded into an internal representation and the outputs generated are then representative of this internal representation instead of just the original pattern pair. The network is provided with a set of example input–output pairs (a training set) and the weight matrix modified so as to approximate the function from which the training set has been derived. In the ideal case, the net would be able, after training, to generalise or produce reasonable results for input simulation that it has never been exposed to. Currently, the most popular technique for accomplishing this type of learning in an ANN is the multi-layered perceptron employing back-propagation. It evolved from Rosenblatt's perceptron; a two-layered supervised ANN, which provides nearest neighbour pattern matching via the perceptron error-correction procedure. This procedure effectively works to place a hyper plane between two classes of data in an n-dimensional pattern space. It has been shown that this algorithm will find a solution for any linearly separable problem in a finite amount of time.

3.6 Back-Propagation Learning

Error back-propagation through non-linear systems has existed in variational calculus for many years but the first application of gradient descent to the training of multi-layered nets was proposed by Amari (1967) who used a single hidden layer to perform a non-linear classification. Werbos (1974) discovered *dynamic feedback* and Parker and Chau (1987) talked about *learning logic*, but the greatest impact on ANN field came when Rumelhart, Hinton and Williams published their version of the *Back-propagation algorithm*.

One of the major reasons for the development of the back-propagation algorithm was the need to escape one of the constraints on two layer ANNs, which is that similar inputs lead to similar output(s). But, while ANNs like the perceptron may have trouble with non-linear mappings, there is a guaranteed learning rule for all problems that can be solved without hidden units. Unfortunately, it is known that there is no equally powerful rule for multi-layered perceptrons.

The simplest multi-layered perceptron implementing back-propagation is a three-layered perceptron with feed-forward connections from the input layer to the hidden layer and from the hidden layer to the output layer. This function-estimating ANN stores pattern pairs using a multi-layered gradient error correction algorithm. It achieves its internal representation of the training set by minimising a *cost function*. The most commonly used cost function is the sum squared error or the summation of the difference between the computed and desired output values for each output neuron across all patterns in the training set. Other cost functions include the Entropic cost function, Linear error and the Minkuouski-r or the rth power of the absolute value of the error.

In all cases, the changes made to the weight matrix are derived by computing the change in the cost function with respect to the change in each weight. The most basic version of the algorithm minimises the sum-squared error and is also known as the *generalised delta rule* (Simpson 1990).

Step 1: Assign small random weights to all weights on connections between all layers of the network as well as to all neuron thresholds. The activation used is logistic sigmoid function as given by (3.2) with $\lambda = 1$.

Step 2: For each pattern pair in the training set:

(a) Read the environmental stimuli into the neurons of the input layer and proceed to calculate the new activations for the neurons in the hidden layer using

$$hidden_i = f \left(\sum_{h=1}^{n} input_h \; w_{hi} + \theta_i \right) \tag{3.3}$$

where $f(.)$ is the activation function, there are n input neurons and θ_i is the threshold for the ith hidden neuron.

(b) Use these new hidden layer activations and the weights on the connections between the hidden layer and the output to calculate the new output activations using

$$output_j = f \left(\sum_{i=1}^{n} hidden_i \; w_{ij} + \Gamma_j \right) \tag{3.4}$$

where $f(.)$ is the activation function, there are n hidden neurons and Γ_j is the threshold for the jth output neuron.

(c) Determine the difference between the computed and the desired values of the output layer activations using

$$diff_j = output_j \, (1 - output_j) \, (desired_j - output_j) \tag{3.5}$$

and calculate the error between each neuron in the hidden layer relative to the $diff_j$ using

$$err_i = hidden_i \, (1 - hidden_i) \sum_{i=1}^{n} w_{ij} diff_j \tag{3.6}$$

(d) Modify each connection between the hidden and output layers, and $\Delta w_{ij} = \alpha \; hidden_i \; diff_j$, which is the amount of change to be made to the weight on the connection from the ith neuron in the hidden layer to the jth neuron in the output layer. α is a positive constant that controls the rate of modification or learning.

(e) Perform a similar modification to the weights on the input to hidden layer connections with $\Delta w_{hi} = \beta \; err_i$ for the hidden units and $\Delta\Gamma_j = \alpha \; diff_j$ for the output units.

Step 3: Repeat Step 2 until all the *diff$_j$*s are either zero or sufficiently low.

Step 4: After the BP-ANN has been trained; recall consists of two feed-forward operations, which create hidden neuron values.

$$hidden_i = f\left(\sum_{h=1}^{n} input_i \ w_{hi} + \theta_j\right) \qquad (3.7)$$

and then we use them to create new output neuron values

$$output_j = f\left(\sum_{i=1}^{n} hidden_i \ w_{ij} + \Gamma_j\right) \qquad (3.8)$$

Back-propagation is guaranteed only to find the local, not the global error minimum. And while this technique has proven extremely successful for many practical applications, it is based on gradient descent, which can proceed very slowly because it is working only with local information. Practical implementation factors that must be considered include:

- The number of units in the hidden layer.
- The value of the learning rate constants.
- The amount of data that is necessary to create the proper mapping.

Once these issues have been addressed, the power of back-propagation is realised in a system that has the ability to store many more patterns that the number of dimensions inherent in the size of its input layer. It also has the ability to acquire arbitrarily complex non-linear mappings. This is possible if the application allows for a reasonably long training time in an off-line mode.

Current research in the area of back-propagation improvements is looking at:

- Optimising the number of units in the hidden layer and the effect of the inclusion of more than one layer of hidden units.
- Improving the rate of learning by dynamic manipulation of the learning rates and by the use of techniques such as momentum.
- The effects of dynamically changing and modular connection topologies.
- Analysing the scaling and generalisation properties of this ANN model.
- Employing higher-order correlations and arbitrary threshold functions.

During training the nodes in the hidden layers organize themselves such that different nodes learn to recognize different features of the total input space.

During the recall phase of operation the network will respond to inputs that exhibit features similar to those learned during training. Incomplete or noisy inputs may be completely recovered by the network.

In its learning phase, you give it a training set of examples with known inputs and outputs.

An overview of training

The objective of training the network is to adjust the weights so that application of a set of inputs produces the desired set of outputs. For reasons of brevity, these input–output sets can be referred to as vectors. Training assumes that each input vector is paired with a target vector representing the desired output; together these are called a training pair. Usually, a network is trained over a number of training pairs. For example, the input part of a training pair might consist of a pattern of ones and zeros representing a binary image of a letter of the alphabet. A set of inputs for the letter A drawn on a grid. If a line passes through square, the corresponding neuron's input is one; otherwise, that neuron's input is zero. The output might be a number that represents the letter A, or perhaps another set of ones and zeros that could be used to produce an output pattern. If one wished to train the network to recognize all the letters of the alphabet, 26 training pairs would be required. This group of training pairs is called a training set.

Before starting the training process, the weights must be initialized to small random numbers. This ensures that the network is not saturated by large values of the weights, and prevents certain other training pathologies. For example, if the weights all start at equal values and the desired performance requires unequal values, the network will not learn.

Training the back-propogation network requires the steps that follow:

Step 1. Select the training pair from the training set; apply the input vector to the network input.

Step 2. Calculate the output of the network.

Step 3. Calculate the error between the network output and the desired output (the target vector from the training pair).

Step 4. Adjust the weights of the network in a way that minimizes error.

Step 5. Repeat steps 1 through 4 for each vector in the training set until the error for the entire set is acceptably low.

The operations required in steps 1 and 2 above are similar to the way in which the trained network will ultimately be used; that is, an input vector is applied and the resulting output is calculated. Calculations are performed on layer-by-layer basis.

In step 3, each of the network outputs is subtracted from its corresponding component of the target of the network, where the polarity and magnitude of the weight changes are determined by the training algorithm.

After enough repetitions of these four steps, the error between actual outputs and target outputs should be reduced to an acceptable value, and the network is said to be trained. At this point, the network is used for recognition and weights are not changed.

It may be seen that steps 1 and 2 constitute "forward pass" in that the signal propagates from the network input to its output. Steps 3 and 4 are a "reverse pass"; here the calculated error signal propagates backward through the network where it is used to adjust weights. These two passes are now expanded and expressed in a somewhat more mathematical form in Chap. 4.

% Matlab Program for Backpropagation for single hidden layer

```
% Input–output Pattern for EX-OR problem
clear all;
clc;
X=[0.1 0.1 0.1; 0.1 0.9 0.9; 0.9 0.1 0.9; 0.9 0.9 0.1];
% Training parameters
eta=1.0; % learning rate
alpha=0.6; % Momentum rate
err_tol=0.001; % Error tolerance
[row_x col_x]=size(X);
sum_err=0;
% ANN architecture
In=2; % number of input neurons
Hn=2; % number of hidden neurons
On=1; % number of output neurons
% Weight/delta weight Intialization
Wih=2*rand(In+1,Hn)-1;
Who=2*rand(Hn+1,On)-1;
DeltaWih=zeros(In+1,Hn);
DeltaWho=zeros(Hn+1,On);
deltaWihold=zeros(In+1,Hn);
deltaWhoold=zeros(Hn+1,On);
deltah=zeros(1,Hn+1);
deltao=zeros(1,On);
X_in=[ones(row_x,1) X(:,1:In)];
D_out=X(:,1:On);
sum_err=2*err_tol;
while (sum_err>err_tol)
 sum_err=0;
 for i=1:row_x
  sum_h=X_in(i,:)*Wih;
  out_h=[1 1./(1+exp(-sum_h))];
  sum_o=out_h* Who;
  out_o=1./(1+exp(-sum_o));
  error=D_out(i) - out_o;
  deltao=error.*out_o.*(1-out_o);
  for j=1:Hn+1
   DeltaWho(j,:)=deltao*out_h(j);
  end
  for k=2:Hn+1
   deltah(k)=(deltao*Who(k,:)')*out_h(k)*(1-out_h(k));
  end
```

```
  for l=2:In+1
    deltaWih(l,:)=deltah(2:Hn+1)*X_in(i,l);
  end
  Wih=Wih+DeltaWih+alpha*deltaWihold;
  Who=Who+DeltaWho+alpha*deltaWhoold;
  deltaWihold=DeltaWih;
  deltaWhoold=DeltaWho;
 sum_err=sum_err+sum(error.^2);
 end
 sum_err
end
```

Summary of back-propagation training:

- Objective is to find the global minimum on the error surface.
- Solution is obtained through gradient descent algorithm and ANN weights are adjusted to follow the steepest downhill slope.
- The error surface is not known in advance, so explore it in many small steps and the possibility to stuck in local minima is always there as shown in Fig. 3.16.

The algorithm finds the nearest local minimum, not always the global minimum. There can be two causes for this:

a. *Over-fitted ANN*

The overfitting of data is a common problem found in ANN during approximating a function, specially when ANN has too many weights. Too many weights (free parameters) in ANN approximate the function very accurately, but the generalization capability for unforeseen data is not so good. On the other hand, a network with too few weights will also give poor generalization capability as the ANN has very low flexibility and is

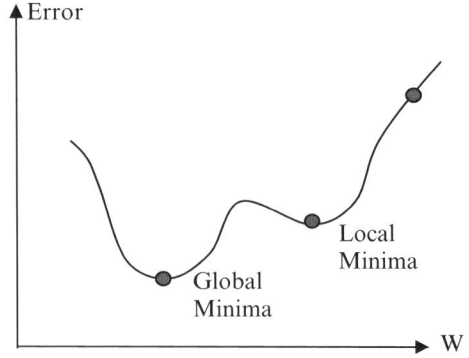

Fig. 3.16. Error curve with respect to weight

unable to approximate the function. Hence, there is a trade off between the number of training data and the size of the network.

b. *Too many hidden nodes*

- One node can model a linear function
- More nodes can model higher-order functions, or more input patterns
- Too many nodes model the training set too closely, preventing generalization.

This problem could be resolved by optimizing the ANN size.

3.7 Properties of Neural Networks

1. Neural networks are inherently parallel and implementation can be done on parallel hardware.
2. It has a capacity for adaptation.
3. In neural networks "memory" corresponds to an activation map of the neurons. Memory is thus distributed over many units giving resistance to noise. In distributed memories, such as neural networks, it is possible to start with noisy data and to recall the correct data.
4. Fault tolerant capability
   Distributed memory is also responsible for fault tolerance. In most neural networks, if some neurons are destroyed or their connections altered slightly, then the behavior of the network as a whole is only slightly degraded. The characteristic of *graceful degradation* makes neural computing systems extremely well suited for applications where failure of control equipment means disaster.
5. Capacity for generalization
   Designers of expert systems have difficulty in formulation rules which encapsulate an expert's knowledge in relation to some problem. A neural system may learn the rules simply from a set of examples. The generalization capacity of a neural network is its capacity to give a satisfactory response for an input which is not part of the set of examples on which it was trained. The capacity for generalization is an essential feature of a classification system. Certain aspects of generalization behavior are interesting because they are intuitively quite close to human generalization.
6. Ease of construction.

3.8 Limitations in the Use of Neural Networks

1. Neural systems are inherently parallel but are normally simulated on sequential machines.
   - Processing time can rise quickly as the size of the problem grows.
   - A direct hardware approach would lose the flexibility offered by a software implementation.

2. The performance of a network can be sensitive to the quality and type of preprocessing of the input data.
3. Neural networks cannot explain the results they obtain; their rules of operation are completely unknown.
4. Performance is measured by statistical methods giving rise to distrust on the part of potential users.
5. Many of the design decisions required in developing an application are not well understood.

Fruit identification problem

This is a very simple example of identification of fruits. The inputs for this problem are shape, size and colour of fruits.

Step-1 What the neural network is to learn?
Input 1 Shape = {Round, Large}
Input 2 Size = {Small, Large}
Input 3 Colour = {Red, Orange, Yellow, Green}.
Output Type of fruit = {Grape, Apple, Cherry, Orange, Banana}.
Step-2 Pre-processing of data
    a. Representation of data
       Neural Network could not work with qualitative information. Hence the input must be converted into quantitative information like 0 and 1.

            Shape = {0,1} zero stands of round and 1 for large.
            Size = {0,1} small is zero and large is 1.
            Colour = {0.0, 0.25, 0.5, 0.75, 1.0}.
    Where 0.0 - Red;   0.25 - Orange, 0.5 - Yellow, 0.75 - Green.
            Output = {0, 0.25, 0.5, 0.75, 1.0}.
    Where 0.0 Grape,   0.25 - Apple, 0.5 - Cherry, 0.75 - Orange, 1.0 Banana.
    b. Sequence of presentation of data

Input 1 shape	Input 2 size	Input 3 color	Ouptut - fruit
0 - Round	0 - Small	0.0 - Red	0 - Grape
0 - Round	0 - Small	0.25 - Orange	0.5 - Cherry
0 - Round	1 - large	0.25 - Orange	0.75 - Orange
1 - Large	1 - Large	0.5 - Yellow	1.0 - Banana
0 - Round	1 - Large	0.75 - Green	0.25 - Apple

       The inputs need not be the exact value as given in the table; we could assign some other value depending on the situation. It also helps us to incorporate the uncertainty in the model.
Step-3 Define Network Structure –
    Number of input layer neurons = 3 (Number of inputs)
    Number of output layer neuron = 1 (Number of outputs)
    Number of Hidden layer neurons = 2 (generally average of input and output neurons)

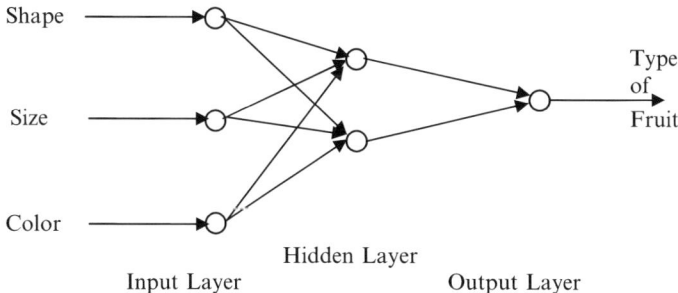

Neural Network to recognize Fruits

Step-4 Selection of Neuron Structure

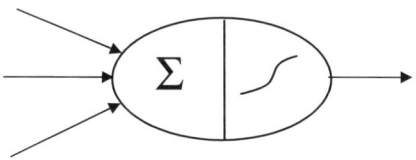

Normally the neuron structure is consisting of summation as aggregation function and sigmoid as threshold function.

Step-5 Usually the network starts with random weight in the range $(-0.1$ to $+0.1)$. Sometimes to reduce the training time, the initialization of weights is done with evolutionary algorithms.

Step-6 Training Algorithms and Error Function selection

Generally, gradient descent back-propagation training algorithm is used with or without adaptive learning and momentum factors. In this the sum squared error is fed back to modify the weight during training.

Step-7 Decision regarding selection of training parameters

Training parameters are

    a. Number of epochs = 100 (number of iterations required to reach to the desired goal)

    b. Error tolerance = 0.001 (depends on the accuracy required)

    c. Learning rate = 0.9 (near 1)

    d. Momentum facto = 0.1 (smaller)

Step-8 Training and Testing of Network

The network is trained for the above given data and then test it to check its performance. Generally, the testing data is slightly different from training data (10% new data).

Step-9 Use the trained network for prediction

% Matlab Program for identification of fruit type

```
clc;   clear all;
% Column wise input--output patterns
```

```
x=[0 0 0.0 0; 0..0 .0.25 0.5; 0..1 ..0.25 ..0.75;
    1...1..1.5 ..1.0; 0 ..1 ..0.75....0.25];
P=x(:,1:3)';
T=x(:,4)';
net=newff(minmax(P), [2 1],
          {'tansig' 'purelin'});          % defining~ANN
net.trainParam.epochs=100;                % Define number of epochs
net.trainParam.goal=0.0001;               % Define error~goal
net=init(net)                             % Initialize weights
net=train(net,P,T);                       % Training
Y=sim(net,P);                             % Testing
plot(1:5,T,'-',1:5,Y,'o')                 % Ploting the results
```

Results

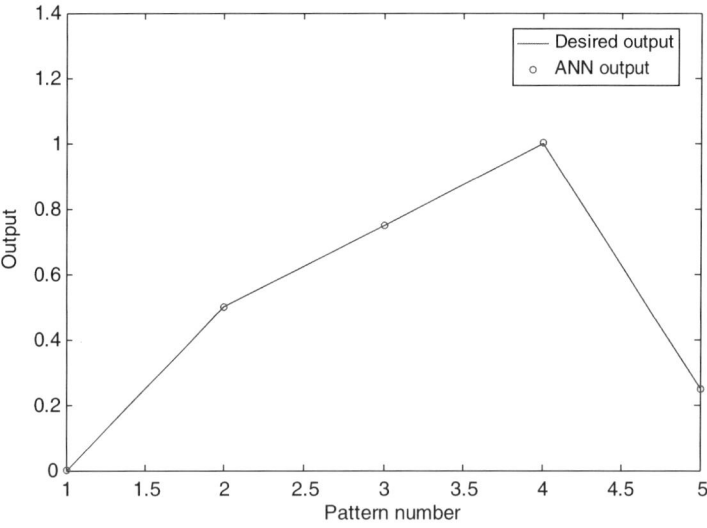

Stop Training

3.9 Summary

1. It is clear that the quantum of processing that takes place in biological neurons is far more complex. It integrates hundreds or thousands of temporal signals through their dendrites.
2. Artificial neuron is similar to biological neurons and receives weighted input, which passes through aggregation function and activation function. A hard limiter constitutes the nonlinear element of McCulloch – Pitts neuron.

3. Artificial neural network (ANN) consists of artificial neurons in two different architectures: feed forward and feedback. Feedforward networks are static and the output depends only on input, but in feedback ANN output is also feedback and therefore, it is dynamic in nature.

3.10 Bibliography and Historical Notes

The pioneer work in the area of neural network was done by McCulloch and Pitts way back in 1943. McCulloch was a psychiatrist and neuroanatomist by training. He spent nearly two decades in understanding the event in the nervous system. Pitts was a mathematical prodigy. McCulloch and Pitts developed neuron model at the University of Chicago.

The next major development in the field of neural network came in 1949, when Hebb wrote a book on *The Organization of Behavior*. He proposed that the connections strength between neurons change while learning. Rochester et al. (1956) is probably the first attempt to formulate neural learning theory based on Hebb's work.

Minsky submitted his doctorate thesis at Princeton University on the topic of *Theory of Neural – Analog Reinforcement Systems and Its Application to the Brain-Model Problem* in 1954. Then he published a paper on *Steps Toward Artificial Intelligence*. The significant contributions to the early development of associative memory papers by Taylor (1956), Anderson (1972), and Kohonen (1972).

In 1958 Rossenblatt introduced a novel method of supervised learning. In 1960 Widrow and Hoff gave least mean square (LMS) algorithm to formulate ADALINE. Later on Widrow and his students developed MADALINE. The books by Wasserman Philip (1989) and Nielsen (1990) also contain treatment of back propagation algorithms. Minsky and Papert (1969) demonstrated the fundamental limitations of perceptron.

In 1970s self-organizing maps using competitive learning was introduced (Grossberg 1967, 1972). Carpenter and Grossberg also developed adaptive resonance theory (ART) in 1980 and used it for pattern recognition (Carpenter and Grossberg 1987, 1988, 1990, 1996). Hopfield used an energy function to develop recurrent networks with symmetric synaptic connections. Rumelhart et al. (1986) developed back propagation algorithm. In early 1990s, Vapnik et al. invented a computationally powerful class of supervised learning networks called support vector machines for different applications. An excellent review article is that by Lippmann (1987). Kosko (1988) discusses on bidirectional associative memory (BAM).

3.11 Exercises

1. Explore the method of steepest descent involving a single weigh w by considering the following cost function:

   $f(w) = 0.7 A + W^*B + C^*W^2$ where A, B, and C are constants.

2. The function expressed by $f(x) = 1/x^2$
   a. Write a matlab program to generate the two sets of data:
      1. Training data
      2. Testing data.
   b. Use a three layer network and train it with back propagation learning algorithm for the data generated in part a. Consider the error tolerance 0.01.
   c. Test the network for generated testing data.
   d. Compare the network performance for
      1. Two or more hidden layers in the network.
      2. Two or more neurons in each hidden layer.
3. In question 2, study the effect of starting (initial) weights.
4. A sigmoid function is $f(x) = 1/(1 + e^{-\lambda x})$. Find its inverse function and plot both the function and its inverse for different values of λ.
5. Find the derivative of the above mentioned function with respect to x.
6. Find appropriate weights and threshold of neuron for logical AND problem.
7. Solve the 3-bit even parity problem with three layer ANN using 3-hidden neurons. Write the matlab program to solve the parity problem.

4

Factors Affecting the Performance of Artificial Neural Network Models

Artificial neural network is widely used in various fields like system's modelling, forecasting, control, image processing and recognition, and many more. The development of multi-layered ANN model for a particular application involves many issues which affect its performance. ANN performance depends mainly upon the following factors:

1. Network
2. Problem complexity
3. Learning Complexity.

4.1 Network Complexity

Network complexity broadly depends on

a. Neuron complexity
b. Number of neurons in each layer
c. Number of layers
d. Number and type of interconnecting weights.

4.1.1 Neuron Complexity

Mainly the neuron complexity could be viewed at two levels; firstly at aggregation function level and secondly at activation function level. There are two types of aggregations functions used for neuron modelling such as summation or product functions, but some researchers used combination 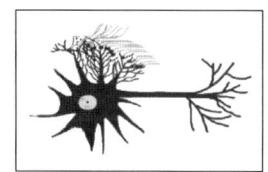 of both summation and product aggregation function such as compensatory operators (Chaturvedi et al. 1997, 1999). The threshold functions used in neuron may be discrete like hard limiter used by McCulloch and Pitts (1943)

D.K. Chaturvedi: *Soft Computing Techniques and its Applications in Electrical Engineering*, Studies in Computational Intelligence (SCI) **103**, 51 85 (2008)
www.springerlink.com   © Springer-Verlag Berlin Heidelberg 2008

in their neuron model or continuous functions like linear or non-linear function like sigmoid, Gaussian functions, etc.

The activation for a neuron can be thought of as the amount by which the neuron is affected by the input it receives. One could picture a neuron vibrating degrees depending on how excited it has become, and different neurons will be excited, or depressed the matter, by different stimuli and by differing degrees. Actually defining this state of activation for each unit within a model, and assigning a value to it, is a tricky process because the precision of the model depends on the reaction of the individual units.

Some models use a set of discrete values, that is, one of a finite set of possible values. These are often taken to be 0, 1 or -1. On the other hand, a model may take any value between two limits. This termed a continuous set of values, because for any two numbers there is always one that you can find that lies between them. In some cases, the model may have no upper or lower limit for the continuous values, but this presents problems, values can grow to an unmanageable size very quickly.

In this section, the effect of various activation functions on ANN model are considered for dc motor current prediction problem and found that the tan sigmoid function at hidden layer and pure linear function at output layer in a three layer network, where input layer is simply distributing the inputs in various hidden layer and no processing takes place there, requires least number of training epochs (i.e. 104). The comparisons of the results obtained for different activation functions are shown in bar chart, Fig. 4.1. From bar chart it is quite clear that the other functions takes more training epochs then also the model cannot be trained to the desired error level for some functions. The functions pure linear and pure linear in the model at hidden and output layers respectively also requires same number of training epochs but the results predicted for the non-linear problems are not so good. The function pair log sigmoid and log sigmoid is also able to train the model upto the desired error level but training epochs required is very large (in this case it requires 2,175). Remaining all other function pairs can not train the model up to the desired level when trained up to 2,200 epochs.

4.1.2 Number of Layers

While developing ANN model, two layers are fixed, namely input layer and output layer. Generally, at the input layer, the inputs are distributed to other neurons in the next layer and no processing takes place at this layer. Unlike the input layer, at output layer processing is done. Therefore, in a two layer network there is only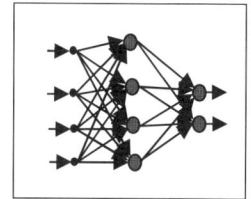
one processing layer and this type of ANN can be used for linearly separable problems. Most of the real life problems are not linearly separable in nature and hence this type of two layer network could not be used. In the literature

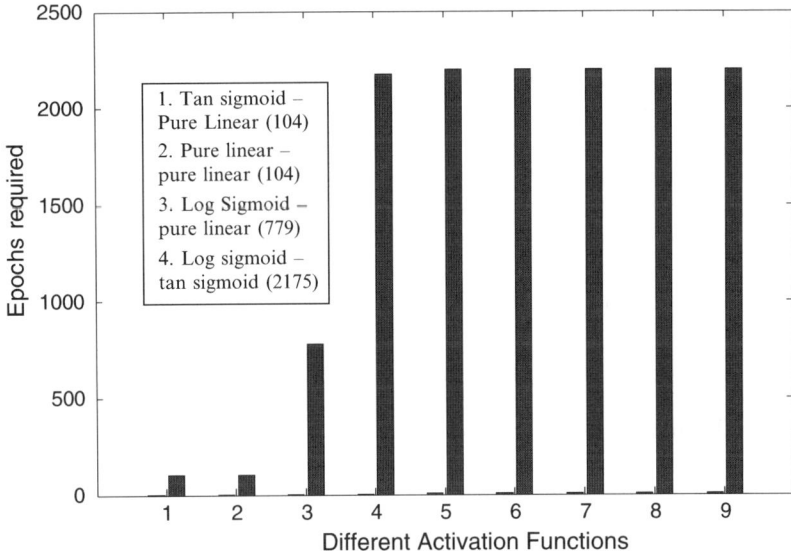

Fig. 4.1. Effect of different activation functions at different layer

it is mentioned that the three layer network is a universal approximator and could handle most of the problems. Then also for complex problems, it is difficult to train ANN with three layers network structure. Hence, most of the time the ANN developer uses trial and error method to select the number of layers in the ANN structure.

There are two ways to deal with this problem. Firstly, one can start with three layers network and then during training the number of layers and neurons may be increased till the satisfactory performance is obtained. Second method to handle this situation is, one could begin with large number of layers and then start deleting the layers and neuron, till the ANN size is optimal.

4.1.3 Number of Neurons in Each Layer

The number of neurons at input layer and output layer are equal to the number of input and output variables, but the problem lies with the number of neurons at hidden layers. It is mentioned that the number of neuron in the hidden layer is the average of number of neurons at input and output layers. But it is not hard and fast rule.

4.1.4 Type and Number of Interconnecting Weights

Generally every neuron in ANN is interconnected with its as previous layer neurons and each interconnection has some weight (signal gain), which modifies the input signal in one way or the other. The weights in the neural network could be deterministic or fuzzy in nature. Normally, ANN weights are deterministic and can be determined by some learning rule. It is well proven and logical also that it is not necessary to connect every neuron with the other neuron in the next layer. We can remove some of the connections to reduce the complexity of the ANN and ultimately the training time of it.

To select the optimal size of the network, there are two techniques generally adapted; either one could start with large number of neurons in each layer of the network and during training remove the connections till its performance is not optimal or we can start with minimal size of network and then insert the neurons and layers to achieve the optimal size of the network as mentioned earlier.

4.2 Problem Complexity

The performance of ANN models does not depend only on the size of the neural network that is chosen for the problem in hand, but it also depends on the problem complexity. The problem complexity depends on the type of functional mapping, accurate and sufficient training data acquired and their effective way of presentation to ANN during training. During the training phase of ANN, unknown neural network weights are to be determined. If the unknown network weights are more than the training data, then they could not be determined. Therefore, the training data must always be more in number than unknown weights, otherwise network will not train perfectly (means the error will never reach to global minima).

The training performance also depends on the effective way of presentation of data, in which following points have to be considered.

4.2.1 Range of Normalization of Training Data

Normalisation has a major role in the training and testing of neural networks. It is necessary to normalize the input and output in the same order of magnitude. Normalization is very critical issue in ANN. If the input and the output variables are not of the same order of magnitude, some variables may appear to have more significance than they actually do. The training algorithm has to compensate for order-of-magnitude differences by adjusting the network weights, which is not very effective in many of the training algorithms such as back propagation algorithm. For example, if one input variable has a value of thousands and other input variable has a value in tens, the assigned weight for the second variable entering a node of hidden layer 1 must be much greater than that for the first. In addition, typical transfer functions, such as a sigmoid function, or a hyperbolic tangent function, cannot distinguish between two values of xi when both are very large, because both yield identical threshold output values of 1.0.

Whenever we do normalisation of training and testing data, we need to determine minimum and maximum value of the given data. The problem is that these maximum and minimum values restrict the operating range of the network (Welstead 1994). A network that has been trained to predict a maximum change in output say 1% cannot possibly predict a change of 2%, even if the input data warrants it. This creates problems in trying to model volatile change in data. The remedy for this situation is somewhat by expanding the maximum and minimum values. First of all determine actual max-min values and then new maximum values is computed by adding 10% to the previous maximum value and a new minimum value is computed by subtracting 10% to the previous minimum value. The network can now handle values that fall within this expanded range and finally train the neural network model for these normalised data. Note that normalised data is something that is of interest only to the network. The user wants to get ANN output in the range of the actual data. For this reason, it is necessary to convert back the output of neural network into the actual range by denormalizing the ANN output.

Too large a range in relation to the actual data value has the effect of compressing the data so that it all looks the same to the network during training. If the range is too short then the neural network model could not predict the value outside that range and it will give absurd results. Hence, the selections of suitable range (i.e. max–min values) is of great importance, because it will affect the results of neural network model during testing.

The neural network is trained for different normalisation ranges and found very encouraging results. The authors have seen that if the input data of neural network model is normalised in the range of −0.9 to +0.9 and output data in the range of 0.1 to 0.9 then model took least number of epochs to train when threshold functions at hidden layer is tan sigmoid and at output layer is pure linear. The comparison of various normalisation ranges during and testing have been studied and the results given in Table 4.1, and Table 4.2 for modelling and simulation of dc motor using neural network. The ANN model was also developed for short term electrical load forecasting problem and the effect of different normalization range had been studied. The simulation results representing training and testing performance are complied in Tables 4.3–4.5 and shown in Fig. 4.2.

Generally it is found that the two layer neural network with tan sigmoid threshold functions at hidden layer and pure linear threshold function at output layer can train for any set of non-linear data and the performance will improve if the normalisation range taken between −0.9 to +0.9 for input and 0.1 to 0.9 for output.

4.2.2 Type of Functional Mapping

There are four possibilities in preparing training patterns (input and output vectors) for ANN models as shown in Fig. 4.3.

Table 4.1. DC motor current simulations with different normalisation range (Tolerable error $= 10^{-3}$, mapping actual input and actual output (X–Y) Activation functions – tan sigmoid at hidden layer and pure linear at output layer.)

Normalization	X (0.1−2.5) Y (0.1−2.5)	X (0.1−0.9) Y (0.1−2.5)	X (0.1−2.5) Y (0.1−0.9)	X (0.1−0.9) Y (0.1−0.9)	X (−0.9 to +0.9) Y (0.1−0.9)
Epochs	1500 (NT)	93	81	104	86
Test	2.1725	2.1816	2.1779	2.1951	2.2030
Results	1.9809	1.9340	1.9403	1.9518	1.9569
	1.7008	1.6976	1.7091	1.7136	1.7153
	1.4955	1.4883	1.5018	1.5014	1.5006
	1.3186	1.3096	1.3232	1.3201	1.3181
	1.6101	1.1596	1.1724	1.1682	1.1662
	1.0431	1.0349	1.0465	1.0420	1.0409
	0.9399	0.9319	0.9421	0.9379	0.9383
	0.8539	0.8478	0.8559	0.8522	0.8542
	0.5655	0.5641	0.5678	0.5666	0.5782
	0.5455	0.5446	0.5470	0.5469	0.5594

Table 4.2. DC motor speed simulations with different normalisation range (Tolerable error $= 10^{-3}$, mapping – actual input and actual output (X–Y) Activation functions – tan sigmoid at hidden layer and pure linear at output layer)

Normalization	X (−0.1 to +0.9) Y (0.1−0.9)	X (0.1−2.5) Y (0.1−0.9)	X (0.1−0.9) Y (0.1−2.5)	X (0.1−2.5) Y (0.1−0.9)	X (0.1−0.9) Y (0.1−0.9)
Epochs	53	64	71	3,000 (NT)	3,000 (NT)
Test	43.7420	40.2860	41.5429	41.0291	41.9124
Results	58.2756	57.4027	57.2390	57.3006	57.5467
	71.6113	72.3925	71.2981	71.9746	71.7368
	83.1838	84.7799	83.2701	84.3292	83.8102
	92.9339	94.7392	93.2272	94.3757	93.7644
	101.0195	102.6546	101.4171	102.4133	101.8504
	107.6690	108.9259	108.1201	108.8045	108.3780
	113.1148	113.9000	113.5955	113.8828	113.6386
	117.5664	117.8571	118.0663	117.9255	117.8803
	121.2035	121.0171	121.7179	121.1547	121.3059
	124.1752	123.5504	124.7023	123.7417	124.0782
	126.6045	125.5886	127.1431	125.8227	126.3263
	127.2335	127.2335	129.1404	127.5014	128.1526

(1) Actual input vector and actual output vector (**X-Y** mapping)
(2) Actual input vector and change in previous value of output vector (**X-ΔY** mapping)
(3) Change in input vector and actual output vector (**ΔX-Y** mapping)
(4) Change in input vector and change in output vector (**ΔX-ΔY** mapping).

Table 4.3. Electrical load forecasting with different normalisation range (Tolerable error = 1, mapping – actual input and actual output (X–Y) Activation functions – tansig at hidden layer and pure linear at output layer)

Normalization	X (±0.9) Y (0.1–0.9)	X (0.1–2.5) Y (0.1–0.9)	X (0.1–0.9) Y (0.1–0.9)
Epochs	112	404	436
Test	2,257.4	2,254.4	2,285.6
Results	2,279.6	2,251.0	2,285.4
	2,704.8	2,693.3	2,697.4
	3,043.0	3,037.8	3,028.3
	3,302.5	3,286.0	3,296.0
	3,292.2	3,285.8	3,292.4
	3,191.1	3,198.2	3,197.0
	3,161.1	3,156.7	3,164.4
	2,911.6	2,929.1	2,920.3
	2,667.3	2,680.5	2,682.7
	2,751.9	2,741.2	2,755.2
	2,921.3	2,911.1	2,915.9
	3,012.9	3,015.4	3,009.9
	2,898.1	2,918.4	2,902.6
	3,040.4	3,039.2	3,040.0
	2,904.6	2,918.0	2,906.8
	3,106.4	3,098.9	3,105.6
	2,960.8	2,971.5	2,961.4
	2,911.1	2,927.8	2,918.8

Table 4.4. Comparison of ANN training with different normalization ranges (activation function "tansig – purelin", mapping x–y)

Range	ω – characteristics of DC motor	ω – t characteristics of ind. motor	P-δ characteristics of alternator	Ia-t characteristics DC motor	STLF
X (−0.1 to 0.9) Y (0.1 to 0.9)	53	1,311	85	–	–
X (−0.1 to 2.5) Y (0.1 to 0.9)	64	726	61	–	–
X (0.1 to 0.9) Y (0.1 to 2.5)	71	1,100	151	93	–
X (0.1 to 2.5) Y (0.1 to 0.9)	3000	736	62	81	404
X (0.1 to 0.9) Y (0.1 to 0.9)	3000	1,950	162	104	436
X (−0.9 to 0.9) Y (0.1 to 0.9)	a	558	69	86	112
X (−0.9 to 0.9) Y (−0.9 to 0.9)	a	1,666	54	–	–

[a] ANN not trained

Table 4.5. ANN testing with different normalization ranges for STLF (Tansig-Purelin, X–Y mapping)

Range of Normalization	Max error	Min error	SS error
X (±0.9) − Y (0.1–0.9)	5.6491	−6.0533	11.0324
X (0.1–2.5) − Y (0.1–0.9)	5.4202	−5.8098	10.6585
X (0.1–0.9) − Y (0.1–0.9)	5.1044	−5.8446	9.2511

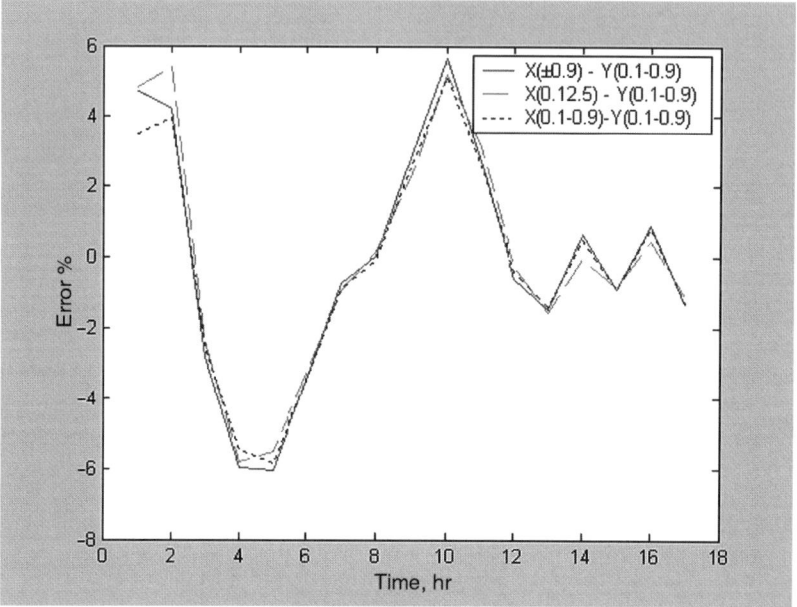

Fig. 4.2. Effect of normalization on short term load forecasting problem

There is no way of knowing a priori which of these myriad approaches is the best one. In this section the effects of all these mappings on the training and testing of following cases have been studied while

(a) Mapping of dc motor current and speed, and
(b) Predicting the electrical load demand.

The training file for dc motor consists of two inputs at adjacent time instances (say I(t-to) and I(t-2*to), where to is the sampling time) and one output O(t). Testing file contains 80% of the training file data and 20% additional data, which can test the model's performance on data from outside the training set. Similarly, for load forecasting problem we have taken data of four Mondays and predict the data of fifth Monday.

CASE – I
The dc motor data are used to train back propagation feedforward neural network for X-Y, X-ΔY, ΔX-Y, ΔX-ΔY mappings. The training algorithms

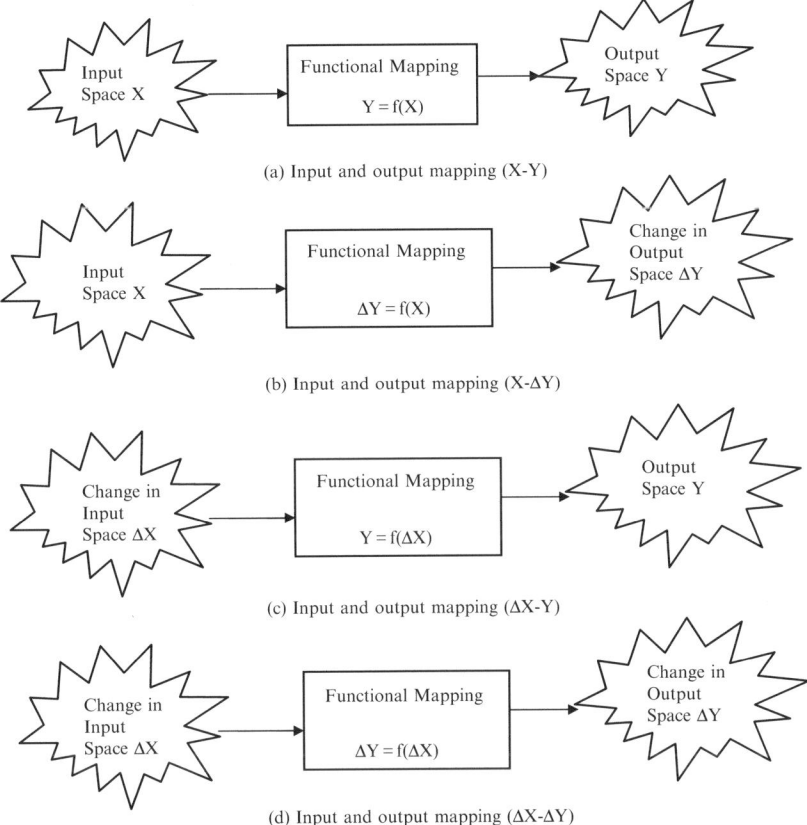

(a) Input and output mapping (X-Y)

(b) Input and output mapping (X-ΔY)

(c) Input and output mapping (ΔX-Y)

(d) Input and output mapping (ΔX-ΔY)

Fig. 4.3. Different functional mappings between input and output space

used are steepest descent based and its modifications. The modified algorithm is commonly known as Levenberg–Marquardt. These algorithms are available in **MATLAB** Tool Box on neural networks. It is found that X-Y mapping requires least number of epochs (i.e. 98) for training and X-Y mapping requires maximum number of epochs (i.e. 105). Tables 4.6 and 4.7 represent comparative analysis of the results of all these mappings and their training epochs and predicted results for dc motor current and speed prediction under starting conditions respectively. In these simulations: Tolerable error $= 10^{-3}$, Normalisation – input and output both in the range 0.1–0.9, and activation functions – Tan sigmoid at hidden layer and pure linear at output layer.

CASE – II

For electrical load forecasting problem, the comparison between all these mappings is given in Tables 4.8–4.11. Figure 4.4 shows the percentage error during forecasting of the electrical demand of the totally unforeseen data of the fifth Monday.

Table 4.6. DC motor current simulations with different mappings

Mappings	X–ΔY	ΔX–Y	X–Y	ΔX–ΔY	Actual values
Epochs	98	103	104	105	
Test	2.1921	2.1942	2.1951	2.1921	2.1905
Results	1.9394	1.9198	1.9518	1.9388	1.9463
	1.7004	1.6781	1.7136	1.6994	1.6960
	1.4909	1.4728	1.5014	1.4894	1.4860
	1.3129	1.3002	1.3201	1.3113	1.3082
	1.1641	1.1575	1.1682	1.1625	1.2305
	1.0406	1.0405	1.0420	1.0388	1.0955
	0.9384	0.9424	0.9379	0.9366	0.9839
	0.8540	0.8627	0.8522	0.8521	0.8920

Table 4.7. DC motor speed simulations with different mappings

Mappings	X–ΔY	ΔX–ΔY	ΔX–Y	Actual values
Epochs	56	140	248	
Test	48.3522	48.0219	44.6986	49.3902
Results	62.8526	62.7167	58.7081	64.2083
	75.3795	75.3599	71.9871	76.7749
	85.9263	85.9497	83.6112	87.2634
	94.7077	94.7455	93.4166	95.9506
	101.9823	102.0209	101.5404	103.1190
	107.9939	108.0276	108.2138	109.0232
	112.9552	112.9814	113.6756	113.8815
	117.0463	117.0645	118.1362	117.8775
	120.4181	120.4284	121.7800	121.1635
	123.1959	123.1988	124.7576	123.8653
	125.4836	125.4791	127.1963	126.086

4.2.3 Sequence of Presentation of Training Data

In the natural learning process of the human being, generally the simple and easy things we learn quickly. So we start our learning with simple things, which motivate and encourage us to learn more. Once we have learned simple things then more time can be spent on difficult things to learn. Hence, it is very important that how we started our learning or what is the sequence of presentation of data for learning. ANN training performance is also very much dependent on in what manner the data is to be presented to ANN. If we cluster the data and then present it to ANN, then it will learn more efficiently and quickly.

4.2.4 Repetition of Data in the Training Set

Some difficult patterns which are not remembered by ANN we have to repeat them. Now how many times that pattern is to be repeated? This is a very

Table 4.8. Short term electrical load forecasting with different mappings

Mappings	X–ΔY	ΔX–ΔY	ΔX–Y	X–Y	Actual demand
Epochs	35	87	800 (NT)*	52	
Test	2,573.4	2,187.6	2,538.0	2,456.8	2,369
results	2,568.6	2,449.7	2,744.9	2,429.7	2,380
	2,803.4	2,767.7	3,011.2	2,545.6	2,631
	2,995.7	3,034.4	3,157.9	2,738.0	2,871
	3,167.0	3,134.4	3,163.8	2,958.3	3,114
	3,161.7	3,067.1	2,950.3	3,097.5	3,182
	3,102.4	3,021.3	2,785.1	3,148.5	3,168
	3,072.9	2,816.4	2,514.0	3,162.3	3,162
	2,923.6	2,588.5	2,321.1	3,087.9	3,000
	2,784.9	2,588.5	2,367.8	2,960.0	2,827
	2,817.8	2,695.6	2,488.6	2,893.8	2,830
	2,912.4	2,796.2	2,644.3	2,901.3	2,904
	2,978.1	2,763.4	2,683.9	2,944.7	2,969
	2,919.0	2,846.2	2,735.7	2,945.9	2,917
	2,997.5	2,778.8	2,648.2	2,990.8	3,013
	2,914.2	2,878.8	2,685.6	2,974.1	2,931
	3,037.7	2,835.0	2,676.8	3,025.7	3,065

Table 4.9. ANN training performance with different functional mappings

Mapping	DC motor current	DC motor speed	Short term load forecasting
X–Y	104	107	52
ΔX–Y	103	248	800
X–ΔY	98	56	35
ΔX–ΔY	105	140	87

Table 4.10. ANN testing performance with different mappings for dc motor current

Mapping	Max error	Min error	SS error
X–Y	0.0623	−0.0176	0.0112
ΔX–Y	0.0730	−0.0037	0.0122
X–ΔY	0.0664	−0.0049	0.0111
ΔX–ΔY	0.0680	−0.0034	0.0118

important question. For example while teaching English alphabets to the students in the elementary classes; most often the students commit the mistake while writing "b" and "d". Then the teacher gives them as home assignment to repeat these alphabets 10 times, 20 times or even more depending on the students' capability. Same thing is true for ANN learning.

Table 4.11. ANN testing performance with different mappings for STLF

Mapping	Max error	Min error	MSS error
X−Y	22.6300	−5.3319	160.3354
ΔX−Y	13.7167	−5.6914	39.9838
X−ΔY	2.8178	−8.6281	13.2372
ΔX−ΔY	5.0000	−4.7046	7.3138

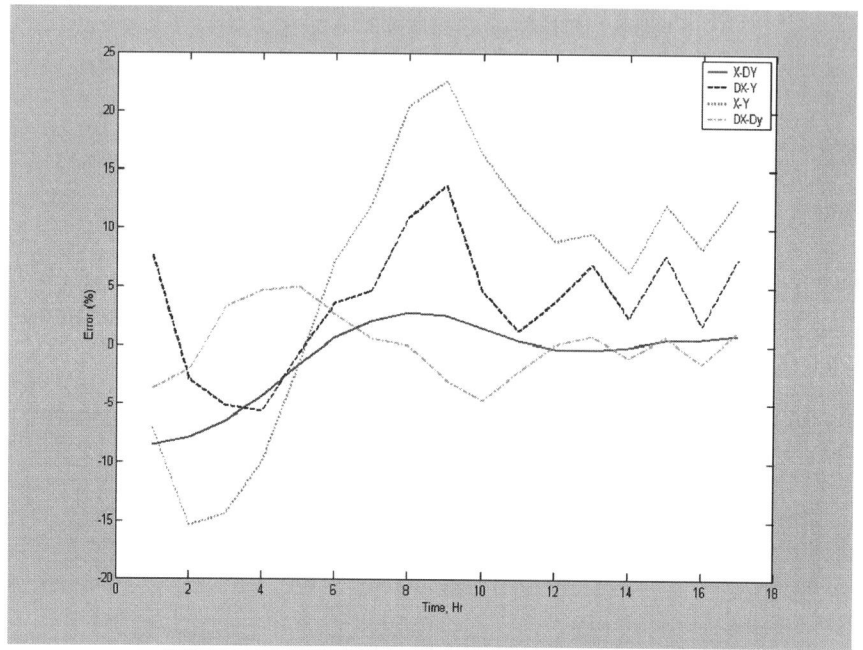

Fig. 4.4. Effect of mapping on short term load forecasting problem

4.2.5 Permissible Noise in Data

The generalization characteristics of ANN models depends on the noise included in the training data, but at the same time the accuracy reduces. Hence, we have to trade off between the generalization capability of neural networks and accuracy required in the results.

Usually when the measurements are taken by different measuring devices, are not accurate due to various reasons. Hence, the noise will be there in the measured quantities. According to the noise either in input or/and output of the training file different pattern mappings are possible. In this chapter the neural network is trained for the following mappings.

(1) Noisy input and accurate output patterns (Xnoise – Y mapping).
(2) Noisy input and noisy output patterns (Xnoise – Ynoise mapping).

Table 4.12. DC motor current simulations with noisy data (Normalisation – input 0.1 to 2.5 and output in the range 0.1–0.9)

Mappings	Xnoise–Y	X–Ynoise	Xnoise–Ynoise
Epochs	400	400	400
Error after training	0.00442398	0.00265015	0.00395185
Error during testing	1.6033 to −2.2048	−0.9237 to −3.800	2.8098 to 4.8763

Table 4.13. Training performance with noisy data for dc motor current (Ia) characteristic

Mappings	Xnoise–Y	X–Ynoise	Xnoise–Ynoise
Training error (After 400Epochs)	0.00442398	0.00265015	0.00395185
Testing error	1.6033 to −2.2048	0.9237 to −3.8	−2.8908 to 4.8763

(3) Accurate input and noisy output (X – Ynoise mapping).
(4) Accurate input and accurate output (X–Y mapping).

The training and testing statistics of the neural network model for the above combination is summarised in Tables 4.12 and 4.13 for dc motor simulation. It has been found that the X–noise mapping required least number of training epochs and also giving good results during predictions. Here the random noise of 5% is added in the training data either/both in input and output data.

4.3 Learning Complexity

Performance of supervised learning depends upon:

a. Training algorithms
b. Initialization of weights
c. Selection of error Function
d. Mode of error calculation
e. Initialization of training parameters

4.3.1 Training Algorithms of ANN

Multi-layered networks have been applied successfully to solve some difficult and diverse problems by training them in a supervised manner with a highly popular algorithm known as the *error back-propagation algorithm*. This algorithm is based on the *error-correction learning rule*.

Basically, the error back-propagation process consists of two passes through the different layers of the network; a forward pass and a backward pass. In the *forward pass*, an activity pattern (input vector) is applied to the sensory nodes of the network and its effect propagates through the network, layer-by-layer. Finally a set of outputs is produced as the actual response of the network. During the forward pass, the synaptic weights of the network are fixed. During the *backward pass*, on the other hand, the synaptic weights are adjusted in accordance with the error-correction rule. Specifically, the actual response of the network is subtracted from a desired (target) response to produce an *error signal*. This error signal is then propagated backward through the network against the direction of synaptic connections – hence the name "error back-propagation". The synaptic weights are adjusted so as to make the actual response of the network move closer to the desired response.

A multi-layered perceptron network has three distinctive characteristics:

1. The model of each neuron in the network includes a differentiable non-linearity, as opposed to the hard limiting used in McCullock and Pitt's perceptron model. A commonly used form of non-linearity that satisfies this requirement is the sigmoid non-linearity.

$$f(net) = \frac{1}{1 + \exp(-\lambda \; net)} \tag{4.1}$$

The presence of non-linearity is important to prevent reduction of the model to that of single-layered perceptron. The use of logistic function is encouraging as it is a biologically motivated function.

2. The network contains one or more hidden layers that enable the network to learn complex tasks by extracting multi-dimensional features from the input pattern vectors.

3. The network exhibits a high degree of connectivity determined by the synapses of the network. A change in the connectivity requires a change in the population of synaptic connections/weights.

All these characteristics together with the ability to learn through training that is the multi-layered perceptron derives its computing power. These same characteristics, however, are also responsible for the deficiencies in knowing the network behaviour. First, the presence of a distributed form of non-linearity and the high connectivity of the network make the theoretical analysis of a multi-layered perceptron difficult to undertake. Second, the use of hidden layers makes the learning process opaque to external environment. In an implicit sense, the learning process is rigorous enough to decide which features of the input pattern should be represented by the hidden layers and the search has to be conducted on a larger space of possible functions.

The development of the back-propagation algorithm represents a "landmark" in the field of neural networks in that it provides a *computationally efficient* method for the training of multi-layered perceptrons.

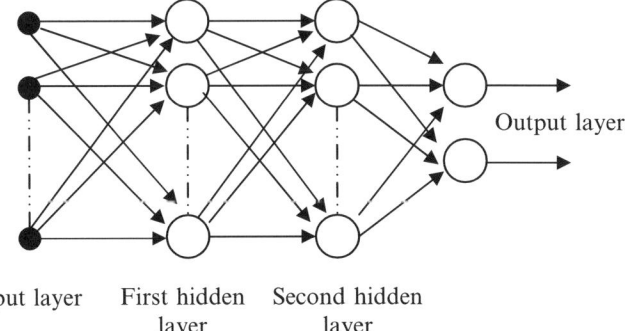

Output layer

Input layer   First hidden   Second hidden
                     layer   layer

Fig. 4.5. NN architecture with two hidden layers

4.3.1.1 Preliminary Fundamentals

The network shown in Fig. 4.5 is fully connected and the signal flows through the network in a forward direction, from left to right and on a layer-by-layer basis. The error signal flow propagates in a backward direction from right to left, again on a layer-by-layer basis.

The signals should be appropriately called function signals as they are calculated as a function of inputs and associated weights.

The error signal is so called because its computation by every neuron of the network involves an error-dependent function in one or another form.

The hidden layer(s) are not part of the input or output layers and hence designated as "hidden". Their behaviour within the architecture is totally "hidden" from analysis.

Each hidden or output neuron of a multi-layered perceptron is designated to perform two computations.

1. The computation of the function signal appearing at the output of a neuron, which is expressed as a continuous non-linear function of the input signals and synaptic weights.
2. The computation of an instantaneous estimate of the gradient, i.e. the gradient of the error surface with respect to the weights connected to the inputs of a neuron, which is needed for the backward pass through the network (Fig. 4.6).

4.3.1.2 The Back-Propagation Algorithm

Before getting into the derivation of the algorithm, we will see the notations used in the derivation.

$E(n)$ = Instantaneous sum of error squares at iteration n. The average of $E(n)$ over all values of n (i.e. the entire training set) yields the average squared error E_{av}.

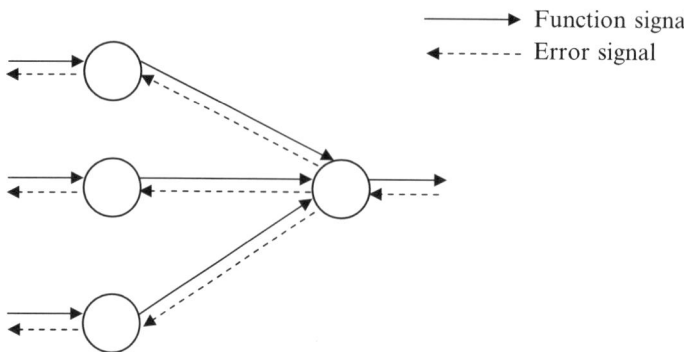

Fig. 4.6. Signal flow illustration

$e_j(n)$ = Error signal at the output of neuron j for iteration n.
$d_j(n)$ = Desired response for neuron j used to compute $e_j(n)$.
$y_j(n)$ = Function signal appearing at the output of neuron j for iteration n.
$w_{ij}(n)$ = Synaptic weight connecting neuron i to neuron j at iteration n.
$\Delta w_{ij}(n)$ = The correction applied to the synaptic weight at iteration n.
$v_j(n)$ = The net internal activity level of neuron j at iteration n.
$\varphi_j(.)$ = The activation function associated with neuron j.
θ_j = The threshold applied to neuron j which is equivalent to an extra synapse.
$x_i(n)$ = The ith element of the input vector (pattern).
$o_k(n)$ = The kth element of the overall output vector (pattern).
η = The learning-rate parameter.

The error signal at the output of neuron j at iteration n (i.e. presentation of the nth training pattern) is defined by

$$e_j(n) = d_j(n) - y_j(n), \tag{4.2}$$

neuron j is an output node.

The *instantaneous sum of squared errors* of the network at the output of neuron j can be written as

$$E(n) = \frac{1}{2} \sum_{j \in c} e_j^2(n), \tag{4.3}$$

where c is the set of all neurons in the output layer of the network.

If N is the total number of patterns in the training set, the *average squared error* over all the patterns is given by

$$E_{av} = \frac{1}{N} \sum_{n=1}^{N} E(n). \tag{4.4}$$

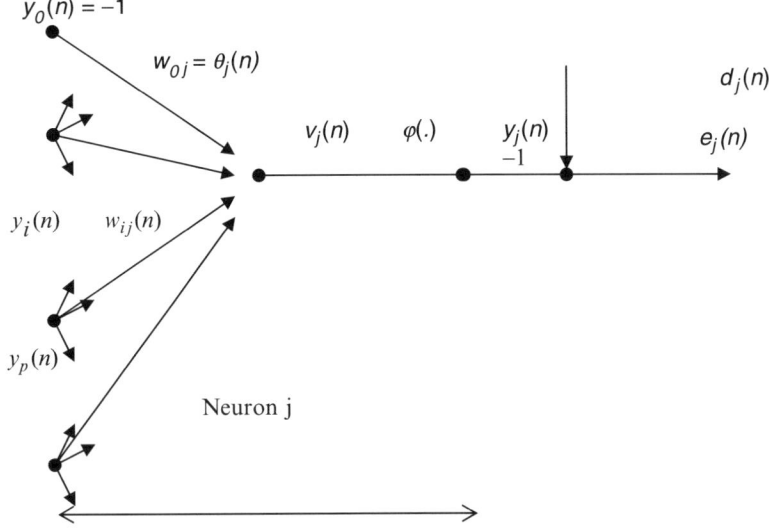

Fig. 4.7. Signal flow of output neuron j

The instantaneous sum of error squares $E(n)$, and therefore, the average squared error E_{av} is a function of the synaptic weights and thresholds. Thus E_{av} represents the *cost function* of the learning process, which adjusts the free parameters of synaptic weights and thresholds so as to minimise the *cost function*. The training is done on a *pattern-by-pattern* basis and the errors *computed* for each pattern presented to the network.

The neuron j as shown in Fig. 4.7, is fed from the layer to its left.

$$v_j(n) = \sum_{i=0}^{p} w_{ij}(n)y_i(n) \tag{4.5}$$

where p is the total number of inputs excluding the threshold applied to neuron j.

$$y_j(n) = \varphi_j(v_j(n)) \tag{4.6}$$

The back-propagation algorithm applies a correction $\Delta w_{ij}(n)$ to the synaptic weight $w_{ij}(n)$, which is proportional to the instantaneous gradient $\partial E(n)/\partial w_{ij}(n)$. According to the chain rule of partial derivatives, we may express the gradient as follows:

$$\frac{\partial E(n)}{\partial w_{ij}(n)} = \frac{\partial E(n)}{\partial e_j(n)} \times \frac{\partial e_j(n)}{\partial y_j(n)} \times \frac{\partial y_j(n)}{\partial v_j(n)} \times \frac{\partial v_j(n)}{\partial w_{ij}(n)}. \tag{4.7}$$

Now, differentiating (4.3) with respect to $e_j(n)$, we get

$$\frac{\partial E(n)}{\partial e_j(n)} = e_j(n) \tag{4.8}$$

Differentiating (4.2) with respect to $v_j(n)$, we get

$$\frac{\partial e_j(n)}{\partial y_j(n)} = -1. \tag{4.9}$$

Differentiating (4.6) with respect to $v_j(n)$ yields

$$\frac{\partial y_j(n)}{\partial v_j(n)} = \varphi'_j(v_j(n)). \tag{4.10}$$

Finally, differentiating (4.4) with respect to $w_{ij}(n)$ yields

$$\frac{\partial v_j(n)}{\partial w_{ij}(n)} = y_i(n). \tag{4.11}$$

Thus (4.7) becomes $\dfrac{\partial E(n)}{\partial w_{ij}(n)} = -e_j(n)\varphi'_j(v_j(n))y_j(n)$ \hfill (4.12)

We know by delta learning rule, the correction to weight is

$$\Delta w_{ij}(n) = -\eta\frac{\partial E(n)}{\partial w_{ij}(n)}, \tag{4.13}$$

where η is a positive constant called the *learning rate*.

From equations (4.12) & (4.13), we have

$$\Delta w_{ij}(n) = \eta\delta_j(n)y_i(n), \tag{4.14}$$

where $\delta_j(n) = e_j(n)\varphi'(v_j(n))$ is called the *local gradient* at neuron j. The *local gradient* $\delta_j(n)$ for output neuron j is equal to the product of the corresponding error signal $e_j(n)$ and the derivative $\varphi'(v_j(n))$ of the associated activation function.

We note that a key factor involved in the calculation of the weight adjustment $\Delta w_{ij}(n)$ is the error signal $e_j(n)$. There are two distinct cases of adjustment, depending on where in the network neuron j is located.

Case 1: Neuron j is an output node

When neuron j is located in the output layer of the network, the case is pretty straight forward as the neuron will be supplied with a desired response. We can use (4.2) to compute the error signal $e_j(n)$ associated with this neuron and then use (4.14) to compute the *local gradient*.

Case 2: Neuron j is a hidden node

When neuron j is located in a hidden layer of the network, there is no specific desired response for that neuron. Accordingly, the error signal for a hidden neuron would have to be determined recursively in terms of the error signals of all the neurons to which the neuron is directly connected.

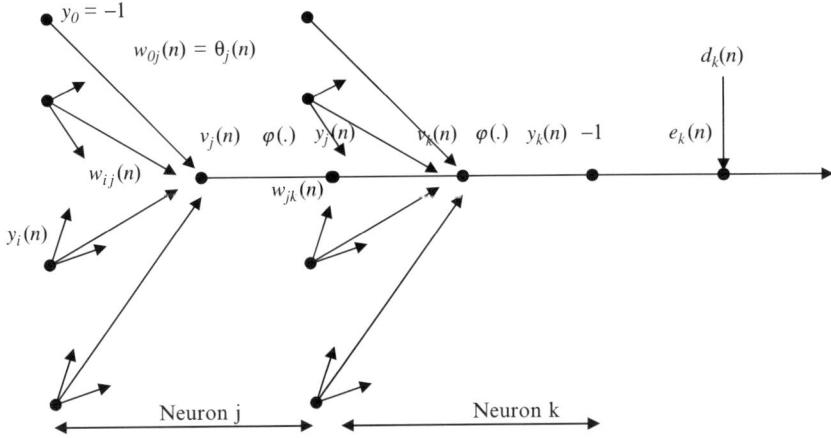

Fig. 4.8. Signal flow of hidden neuron j

Consider the case of the hidden neuron j as shown in Fig. 4.8 below. We can redefine the local gradient

$$\delta_j(n) = e_j(n)\varphi'_j(v(n)) \tag{4.15}$$

as

$$\delta_j(n) = -\frac{\partial E(n)}{\partial y_j(n)} \times \frac{\partial y_j(n)}{\partial v_j(n)} \tag{4.16}$$

$$= -\frac{\partial E(n)}{\partial y_j(n)}\varphi'_j(v_n(n)) \tag{4.17}$$

neuron j is a hidden node.

To calculate the partial derivative $\partial E(n)/\partial y_j(n)$, we may proceed as follows (see Fig. 4.4)

$$E(n) = \frac{1}{2}\sum_{k\in c} e_k^2(n) \tag{4.18}$$

neuron k is an output node

$$\frac{\partial E(n)}{\partial y_j(n)} = \sum_k e_k \frac{\partial e_k(n)}{\partial y_j(n)} \tag{4.19}$$

Using the chain rule of partial derivatives, we can write (4.19) as

$$\frac{\partial E(n)}{\partial y_j(n)} = \sum_k e_k(n)\frac{\partial e_k(n)}{\partial v_k(n)} \times \frac{\partial v_k(n)}{\partial y_j(n)} \tag{4.20}$$

However,

$$e_k(n) = d_k(n) - y_k(n)$$
$$= d_k(n) - \varphi_k(v_k(n)) \tag{4.21}$$

Hence

$$\frac{\partial e_k(n)}{\partial v_k(n)} = -\varphi'_k(v_k(n)).$$
(4.22)

Also, the net internal activity for neuron k is

$$v_k(n) = \sum_{j=0}^{q} w_{jk}(n)y_j(n),$$
(4.23)

where q is the total number of inputs (excluding the threshold) applied to neuron k.

Differentiating (4.23) with respect to $y_j(n)$ yields

$$\frac{\partial v_k(n)}{\partial y_j(n)} = w_{jk}(n)$$
(4.24)

Thus using (4.22) and (4.24), we get

$$\frac{\partial E(n)}{\partial y_j(n)} = -\sum_{k} e_k(n)\varphi'_k(v_k(n))w_{jk}(n)$$
$$= -\sum_{k} \delta_k(n)w_{jk}(n),$$
(4.25)

where we have used the definition of the *local gradient* $\delta_k(n)$ given by (4.14) with the index k substituted for j. Finally using (4.25) in (4.17), we get the local gradient $\delta_j(n)$ for the hidden neuron j as

$$\delta_j(n) = \varphi'_j(v_j(n)) \sum_{k} \delta_k(n)w_{jk}(n)$$
(4.26)

The factor $\varphi'_j(v_j(n))$ involved in the computation of the local gradient $\delta_j(n)$ depends solely on the activation function associated with the hidden neuron j. The remaining factor, namely the summation over k, depends on two sets of terms. The first set of terms, the $\delta_k(n)$, requires the knowledge of the error signals $e_k(n)$, for all those neurons that lie in the layer to the immediate right of the hidden neuron j, and that are directly connected to neuron j; the second set of terms, the $w_{jk}(n)$, consists of the synaptic weights associated with these connections.

We may summarise the relations as follows:

$$\begin{pmatrix} Weight\ correction \\ \Delta w_{ij}(n) \end{pmatrix} = \begin{pmatrix} learning\ rate\ parameter \\ \eta \end{pmatrix} \begin{pmatrix} local\ gradient \\ \delta_j(n) \end{pmatrix}$$
$$\times \begin{pmatrix} input\ signal\ of\ neuron\ j \\ y_i(n) \end{pmatrix}.$$

The local gradient $\delta_j(n)$ depends on whether neuron j is an output node or a hidden node:

1. If neuron j is an output node, $\delta_j(n)$ equals the product of the derivative $\varphi'_j(v_j(n))$ and the error signal $e_j(n)$, both of which are associated with neuron j as given by (4.14).
2. If neuron j is a hidden node, $\delta_j(n)$ equals the product of the associated derivative $\varphi'_j(v_j(n))$ and the weighted sum of the δ's computed for the neurons in the next hidden or output layers that are connected to neuron j as given by (4.26).

4.3.1.3 The Two Passes of Computation

The application of back-propagation algorithm is in two steps or two distinct passes of computation. The first pass is referred as the *forward pass* and the second pass is the *backward pass*.

In the forward pass, the synaptic weights remain unaltered throughout the network, and function signals of the network are computed on a neuron-by-neuron basis.

The function signal appearing at the output of neuron j is computed as

$$y_j = \varphi(v_j(n)), \tag{4.27}$$

where

$$v_j(n) = \sum_{i=0}^{p} w_{ij}(n) y_i(n) \tag{4.28}$$

p is the total number of inputs (excluding the threshold) applied to neuron j and $w_{ij}(n)$ is the synaptic weight connecting neuron i to j, and $y_i(n)$ is the input signal of neuron j or the function signal appearing at the output of neuron i.

If neuron j is in the first hidden layer of the network, then the index i refers to the ith input terminal of the network, for which we write

$$y_i(n) = x_i(n) \tag{4.29}$$

On the other hand, if neuron j is in the output layer of the network, the index j refers to the jth output terminal of the network, for which we can write

$$y_j(n) = o_j(n) \tag{4.30}$$

This output is compared with the desired response $d_j(n)$, obtaining the error signal $e_j(n)$ for the jth output neuron. Thus the forward phase of computation begins at the first hidden layer by presenting it with the input vector, and terminates at the output layer by computing the error signal.

In the backward pass, the error signals computed are passed leftward through the network, layer-by-layer and recursively computing the local gradient δ for each neuron. The synaptic weights are varied according to the back-propagation rule. The local gradient is computed by (4.15) or (4.26), depending on whether the neuron is in the output layer or hidden layer(s). The recursive computation is continued layer-by-layer, by propagating the changes to all synaptic weights from output layer to input layer. The computation of δ for each neuron of the multi-layered architecture requires the derivative of the activation function $\varphi(.)$ associated with that neuron. For this derivative to exist, we require the function $\varphi(.)$ to be continuous. In basic terms, *differentiability* is the only criterion that an activation function would have to satisfy. It has been observed that a non-linear activation function with maximum variation in the mid-values gives stability to the learning process. Such an activation commonly used is the sigmoid activation, whose derivative attains maximum at mid-value.

4.3.1.4 Rate of Learning and Momentum

The back-propagation algorithm provides an "approximation" to the trajectory in the error-weight space computed by the method of *steepest descent.*

According to the *method of steepest descent*, the weights are adjusted in an iterative fashion along the error surface with an aim of moving them progressively toward the optimum solution. The successive adjustments to the weights are in the direction of the *steepest descent* of the error surface.

The *rate of learning* η decides the scaling of the gradient of the error surface to be used for weight adjustment. The smaller we make the learning rate parameter, the smaller will be the changes to the synaptic weights in the network from one iteration to another and the smoother will be the trajectory in the error-weight space, this improvement being achieved at the cost of a slower learning. If we make the *rate of learning* η too large, so as to speed up the rate of learning, the resulting large changes in the synaptic weights may make the trajectory in the error-weight space oscillatory and unstable. It is better to make the learning rate adaptive, i.e. start with a larger η and progressively reduce as we move closer to the minimum. This is the implementation of back-propagation with *adaptive learning rate.*

Another simple method of increasing the rate of learning, and yet avoiding the danger of instability, is to include a *momentum* term as shown below.

$$\Delta w_{ij}(n) = \alpha \Delta w_{ij}(n-1) + \eta \delta_j(n) y_i(n), \qquad (4.31)$$

where α is usually a positive number called the *momentum constant.* The delta rule as given by (4.14) is a special case with $\alpha = 0$.

In order to see the effect of using the momentum constant α, write (4.31) as a time series with index t. The index goes from $t = 0$ to current iteration $t = n$.

$$\Delta w_{ij}(n) = \eta \sum_{t=0}^{n} \alpha^{n-t} \delta_j(t) y_i(t), \qquad (4.32)$$

$$\Delta w_{ij}(n) = -\eta \sum_{t=o}^{n} \alpha^{n-t} \frac{\partial E(t)}{\partial w_{ij}(t)}. \qquad (4.33)$$

The above equations represent a time series of length $n + 1$.

Following observations can be made:

1. The current adjustment $\Delta w_{ij}(n)$ represents the sum of an exponentially weighted time series. For the time series to be *convergent*, the momentum constant must be $0 \leq |\alpha| < 1$. The momentum constant can be positive or negative but it unlikely to use a negative α, in practice.
2. When the partial derivative $\partial E(t)/\partial w_{ij}(t)$ has the same algebraic sign on consecutive iterations, the exponentially weighted sum $\Delta w_{ij}(n)$ grows in magnitude and so the $w_{ij}(n)$ is adjusted by a large amount. Hence the inclusion of momentum in the back-propagation algorithm tends to accelerate the *descent* in steady downhill direction.
3. When the partial derivative $\partial E(t)/\partial w_{ij}(t)$ has opposite signs on consecutive iterations, the exponentially weighted sum $\Delta w_{ij}(n)$ shrinks in magnitude and so the $w_{ij}(n)$ is adjusted by a small amount. Hence the inclusion of momentum has a *stabilising effect* in the directions that oscillate in sign.

Thus the incorporation of momentum in the back-propagation algorithm represents a minor modification to the weight update and yet it can have highly beneficial effects on learning behaviour of the algorithm. The momentum term also helps in preventing the learning process from trapping in local minima. The momentum term can also be made adaptive just like the learning rate and the back-propagation implementation with adaptive η and/or α has been found to be much more efficient that the standard implementation.

4.3.1.5 The Stopping Criteria

There are several stopping criteria, each with its own practical merit, which may be used to terminate the weight adjustments. The logical thing to do is to think in terms of the unique properties of a local or global minimum of the error surface. Let the weight vector $\mathbf{w}^*$ denote a minimum, be it local or global. Various convergent criteria can be stated as follows:

- The back-propagation algorithm is considered to have converged when the Euclidean norm of the gradient vector reaches a sufficiently small gradient threshold. This means $\mathbf{g}(\mathbf{w}) \rightarrow \mathbf{0}$ at $\mathbf{w} = \mathbf{w}^*$. The drawback of this convergence criterion is that, for successful trials, learning time may be long. Also it requires the computation of the gradient vector $\mathbf{g}(\mathbf{w})$ of the error surface to the weight vector $\mathbf{w}$.
- Another unique property of a minimum that can be used is the fact that the *cost function* or error measure $E_{av}(\mathbf{w})$ is stationary at the point $\mathbf{w} = \mathbf{w}^*$. The back-propagation algorithm is considered to have converged when the absolute rate of change in the average error per epoch is sufficiently small. Typically considered ranges are from 0.01 to 1% per epoch.
- Kramer and Sangiovanni-Vincentelli (1989) suggested a hybrid criterion of convergence consisting of the former and the latter, as stated below: The back-propagation algorithm is terminated at the weight vector $\mathbf{w_{final}}$ when $\|\mathbf{g}(\mathbf{w_{final}})\| \leq \varepsilon$, where ε is sufficiently small, or $E_{av}(\mathbf{w_{final}}) \leq \tau$, where τ is also sufficiently small.

- Another useful criterion for convergence is as follows:
  After each learning iteration the network is tested for its generalisation performance. The learning is stopped when the generalisation performance is adequate, or when it is apparent that the generalisation performance has peaked.

4.3.1.6 Initialization of the Network

The first step in back-propagation is, of course, to initialise the network. A good choice for the initial values of the free parameters (i.e. adjustable synaptic weights and threshold levels) of the network can be of tremendous help in a successful network development. In cases where the prior information is available, it may be better to use the information to guess the initial values of the free parameters. But how do we initialise the network if no prior information is available? It is also important to note that if all the weights start out with equal values and the solution requires that unequal weights be developed, the system can never learn. This is because the error is propagated back through the weights in proportion to the values of the weights. This means that all hidden units connected directly to the output units will get identical error signals, and since the weight changes depend on the error signals, the weights from those units to the output units must always be the same. This problem is known as the *symmetry-breaking* problem. Internal symmetries of this kind also give the *cost function* landscape periodicities, multiple minima, (almost) flat valleys and (almost) *flat plateaus* or *temporary minima*. The last are most troublesome, because the system can get struck on such a plateau during training and take immense time to find its way down the cost function surface. Without modifications to the training set or learning algorithm, the network may escape this type of "minimum" but performance improvement in these temporary minima drops to a very low, but non-zero level because of the very low gradient of the *cost function*. In the MSE vs. training time curve, a temporary minimum can be recognised as a phase in which the MSE is virtually constant for a long time after initial learning. After a generally long training time, the approximately flat part in the energy landscape is abandoned, resulting in a significant and sudden drop in the MSE curve. The problem of unequal weights can be counteracted by starting the system with random weights. However, as learning continues, internal symmetries may develop and the network may encounter again temporary minima.

The customary practice is to set all the free parameters of the network to random numbers that are *uniformly distributed* inside a small range of values. This is because if the weights are too large, the sigmoids will saturate from the very beginning of training and the system will become struck in a kind of *saddle point* near the starting point (Haykin, 1994). This phenomenon is called *premature saturation* (Lee et al. 1991). Premature saturation is avoided by choosing the initial weights and threshold levels of the network to be uniformly distributed inside a small range of values. This is so because when the weights are small, the units operate in their *linear regions* and consequently it is impossible for the activation function to saturate. It is also maintained that premature saturation is less likely to occur when the *number of hidden neurons* is maintained *low*, and in consistent with the network requirement but the viability of this belief is under question many a times.

Gradient descent can also become struck in *local minima* of the cost function. These are isolated valleys of the cost function surface in which the system may get "stuck" before it reaches the global minimum. This is so because in these valleys, every change in the weight values causes the cost function to increase and hence the network is unable to escape. Local minima are fundamentally different from temporary minima as they cause the performance improvement of the classification to drop to zero and hence the learning process terminates even though the minimum may be located far above the global minimum. Local minima may be abandoned by including a *momentum* term in the weight updates or by adding "noise" using the on-line mode training, which is a *stochastic* learning algorithm in nature. The momentum term can also significantly accelerate the training time that is spent in a temporary minimum as it causes the weights to change at a faster rate. Other approaches include the modification of the cost function or the employment of techniques such as simulated annealing.

There are two ways of initializing the weights, from which ANN starts learning. If the initial weights are good, ANN needs less time to learn otherwise it requires more time and/or stuck in local minima

1. Random selection
2. Using evolutionary algorithm

4.3.1.7 Faster Training in Back-Propagation Learning

Plain back-propagation is terribly slow and it is desired to have faster training. There are a series of things that can be done to speed up training.

- Fudge the derivative term.
- Scale the data.
- Direct input–output connections.
- Vary the sharpness (gain) of the activation.
- Use a different activation.
- Use better algorithms.

1. *Fudge the derivative term*

   The first major improvement to back-propagation is extremely simple: fudge the derivative term in the output layer. If we are using the sigmoid function given by

$$f(net) = \frac{1}{1 + \exp(-\lambda \ net)} \tag{4.34}$$

and the derivative is

$$f'(net) = f(net)\,(1 - f(net)).\tag{4.35}$$

The derivative is largest at $net = 0.5$ and it is here that we will get the largest weight changes. Unfortunately, at values near 0 or 1, the derivative term gets close to 0 and the weight change becomes very small. In fact, if the network's response is 1 and the target is 0, the network is off by quite a lot with very small weight changes. It can take a very long time for the training process to correct this. Falhlman's solution was to add 0.1 to the derivative term making it:

$$f'_{new}(net) = 0.1 + f'(net).\tag{4.36}$$

The solution of Chen and Mass was to drop the derivative term altogether, in effect, the derivative was 1. This method passes back much larger error quotas to the lower layer, so large that a smaller η must be used there. In their experiments on 10-5-10 codec problem, they found that the best results came when η was 0.1 times the upper level η; hence they called this method the "differential step size" method. One must experiment with both upper and lower level η values to get the best results depending on the problem. Besides that, the η used for the upper layer must be much smaller that the η used without this method.

2. *Direct input–output connections*:
   Adding direct connections from the input layer to the output layer can often speed up training. It is supposed to work best when the function to be approximated is almost linear and it only needs a small amount of adjustment from non-linear hidden layer units. This method can also cut down the number of hidden layer units needed. It is not recommended when there are a large number of output units because, then there are more free parameters to the net and possibly hurt the generalisation.

3. *Adjusting the sharpness (gain) of the activation*:
   Izni and Pentland showed that training time can be decreased by increasing the sharpness or gain λ in the standard sigmoid as given in (4.34). In fact, they showed that the training time goes as $1/\lambda$ for training without momentum and $1/\sqrt{\lambda}$ for networks with momentum. This is not a perfect speed-up scheme since when λ is too large, we run the risk of becoming trapped in a local minimum. Sometimes the best value for λ is less that 1.

4.3.1.8 Better Algorithms

Everyone wants faster training and there are many variations on back-propagation that will speed up the training time enormously, but the credibility of these variations are at question, at times. Very slow online update methods will sometimes give the best results when compared to these acceleration algorithms. People have observed this with sonar data: the best results come from one pattern at a time updates. Having said this, in most cases, the acceleration algorithms work much faster than either online or batch training that they should be used first and then if better results are wanted, one can try slower online methods.

In Sect. 5.4, we have already discussed the effect of having adaptive learning rate and momentum. As the training proceeds, increase η and α adaptively if we keep going downhill, in terms of error. When the weight change gets too large, we end up on the other side of the valley and for this, we must decrease the learning rate and momentum in some way. These are basically first-order algorithms, where we use only the first order information of the error gradient in weight updating. Then there is a set of algorithms known as conjugate gradient methods, which use second order information also for faster training. We will discuss a few such algorithms.

- The Resilient propagation Algorithm.
- The Delta-Bar-Delta Algorithm.
- The Quick-propagation Algorithm.
- The Conjugate Gradient methods.

(a) *The Resilient propagation algorithm:*

The Resilient propagation is a first-order algorithm performing supervised batch learning in multi-layered perceptrons. The basic principle of Rprop is to eliminate the harmful influence of the size of the partial error derivative on the weight step. As a consequence, only the sign of the derivative is considered to indicate the *direction* of weight update. The *size* of the weight change is exclusively determined by a weight-specific, so called "update-value" $\Delta_{ij}^{(t)}$:

$$\Delta w_{ij}^{(t)} = \begin{cases} -\Delta_{ij}^{(t)}; \ if \dfrac{\partial E}{\partial w_{ij}}^{(t)} > 0 \\[2mm] +\Delta_{ij}^{(t)}; \ if \dfrac{\partial E}{\partial w_{ij}}^{(t)} < 0 \\[2mm] 0; \ otherwise \end{cases} \tag{4.36}$$

where $\dfrac{\partial E}{\partial w_{ij}}^{(t)}$ denotes the summed gradient information over the patterns of the pattern set ("batch learning").

It should be note that, by replacing the $\Delta_{ij}^{(t)}$ by a constant update-value Δ, (4.36) yields the so-called "Manhattan" Algorithm.

The second step of Rprop learning is to determine the new update values $\Delta_{ij}^{(t)}$. This is based on a sign-dependent adaptation process.

$$\Delta_{ij}^{(t)} = \begin{cases} \eta^+ \times \Delta_{ij}^{(t)}; \ if \dfrac{\partial E}{\partial w_{ij}}^{(t-1)} \times \dfrac{\partial E}{\partial w_{ij}}^{(t)} > 0 \\[2mm] \eta^- \times \Delta_{ij}^{(t)}; \ if \dfrac{\partial E}{\partial w_{ij}}^{(t-1)} \times \dfrac{\partial E}{\partial w_{ij}}^{(t)} < 0 \\[2mm] \Delta_{ij}^{(t)}; \ otherwise \end{cases} \tag{4.37}$$

where $0 < \eta^- < 1 < \eta^+$.

In other words, the adaptation rule works as follows: Every time, the partial derivative of the corresponding weight w_{ij} changes its sign, which indicates that the last update was too big and the algorithm has jumped over local minimum, the update value $\Delta_{ij}^{(t)}$ is decreased by the factor η^-. If the derivative retains the sign, the update value is slightly increased in order to accelerate the convergence in shallow regions. Additionally, in case of a change in sign, there should be no adaptation in the succeeding learning step. In practice, this can be achieved by setting $\frac{\partial E}{\partial w_{ij}}^{(t)} = 0$.

In order to reduce the number of freely adjustable parameters, often leading to a tedious search in parameter space, the increase and decrease factor are set to fixed values. The choice of decrease factor η^- was lead by the following considerations. If a jump over a minimum occurred, the previous update value was too large, for, it cannot be derived from gradient information how much the minimum was missed. We have to estimate the correct value. It will be a good guess to halve the update value (maximum likelihood estimator), so we choose $\eta^- = 0.5$. The increase factor η^+, on the other hand, has to be large enough to allow fast growth of the update values in shallow regions of error function, but, on the other hand, the learning process can be considerably disturbed if a too large increase factor leads to persistent changes of the direction of the weight-step. In several experiments, the choice of $\eta^+ = 1.2$ gave very good results, independent of examined problems. Slight variations of this value neither improve nor deteriorate convergence time.

(b) *Delta-Bar-Delta Algorithm*:

The Delta-Bar-Delta is a method that implements four heuristics regarding gradient descent. It was developed by Jacobs (1988). The method consists of a weight update rule and learning update rule. The weight update rule is applied to each weight $w_{ij}(n)$ at iteration n through the relationship given by

$$w_{ij}(n+1) = w_{ij}(n) - \eta_{ij}(n+1)\frac{\partial E(n)}{\partial w_{ij}(n)}, \qquad (4.38)$$

where $\eta(n)$ is the learning rate for the weight $w_{ij}(n)$ at update iteration n. The learning rate update rule for a given weight $w_{ij}(n)$ is defined as

$$\Delta\eta_{ij}(n) = \begin{cases} k & ; if\ \bar{\delta}_{ij}(n-1) \times \delta_{ij}(n) > 0 \\ -\phi\eta_{ij}(n) & ; if\ \bar{\delta}_{ij}(n-1) \times \delta_{ij}(n) < 0 \\ 0 & ; otherwise \end{cases} \qquad (4.38)$$

where

$$\delta_{ij}(n) = \frac{\partial E(n)}{\partial w_{ij}(n)} \qquad (4.39)$$

the partial derivative of the error with respect to $w_{ij}(n)$ at iteration n, and

$$\bar{\delta}_{ij}(n) = (1 - \theta)\delta_{ij}(n) + \theta\bar{\delta}_{ij}(n-1) \qquad (4.40)$$

where k and ϕ are constants used increment or decrement the learning rate respectively, and $0 < \theta < 1$ is an exponential "smoothing" base constant for the nth iteration.

The heuristics implemented are as follows:

1. Every parameter (weight) has its own individual learning rate.
2. Every learning rate is allowed to vary over time to adjust to changes in the error surface.
3. When the error derivative for a weight has the same sign for several consecutive update steps, the learning rate for that weight should be increased. This is because the error surface has a small curvature at such points and will continue to slope at the same rate for some distance. Therefore, the step-size should be increased to speed up the downhill movement.
4. When the sign of the derivative of a weight alternates for several consecutive steps, the learning rate for that parameter should be decreased. This is because the error surface has a high curvature at that point and the slope may quickly change sign. Thus, to prevent oscillation, the value of the step-size should be adjusted downward.

There are a few drawbacks of this algorithm. Using momentum along with the algorithm can enhance the performance; however, it can also make the search diverge wildly – especially if k is even moderately large. The reason is that momentum "magnifies" learning rate increments and quickly leads to inordinately large learning steps. One possible solution is to keep the k factor very small, but this can lead to slow increase in η and little speedup.

Another related problem is that, even with a small k, the learning rate can sometimes increase so much that the small exponential decrease is not sufficient to prevent wild jumps. Increasing ϕ exacerbates the problem instead of solving it because it causes drastic reduction of learning rate at inopportune moments, leaving the search stranded at points of high error. Thus the algorithm is very sensitive to small variations in the value of its parameters – especially k.

(c) *Quick-propagation algorithm*:
Standard back-propagation calculates the weight change based upon the first derivative of the error with respect to the weight. If the second derivative information is also available, then better step-size and optimum search direction can be found out. Back-propagation networks are also slow to train. Quick-propagation is a variation of standard back-propagation to speed up training.

The quickprop modification is an attempt to estimate and utilise the second derivative information (Fahlman 1988). This algorithm requires saving the previous gradient vector as well as previous weight change. The calculation of weight change uses only the information associated with the weight being updated.

$$\Delta w_{ij}(n) = \frac{\nabla w_{ij}(n)}{\nabla w_{ij}(n-1) - \nabla w_{ij}(n)} \times \Delta w_{ij}(n-1), \qquad (4.41)$$

where $\nabla w_{ij}(n)$ is the gradient vector component associated with weight vector w_{ij} in step n, $\nabla w_{ij}(n-1)$ is the gradient vector component associated with weight w_{ij} in the previous step and $\Delta w_{ij}(n-1)$ is the weight change in step $n-1$.

A maximum growth factor μ is used to limit the rate of increase of step-size like

If $\Delta w_{ij}(n) > \mu \Delta w_{ij}(n-1)$, then $\Delta w_{ij}(n) = \mu \Delta w_{ij}(n-1)$

Fahlman suggested an empirical value 1.75 for μ.

There are some complications in this method. First is the step-size calculation that requires the previous value, which is not available at the time of starting. This is overcome by using the standard back-propagation method for weight adjustment. The gradient descent weight change is given by

$$w_{ij}(n+1) = w_{ij}(n) - \eta \nabla w_{ij} \qquad (4.42)$$

Value of η is taken suitably small.

Second problem is that the weight values are unbounded. They become so large that they may cause an overflow. Suitable scaling of slope by a factor less than 1 reduces the rate of increase of the weights.

(d) *The conjugate gradient (CG) methods*:
The conjugate gradient algorithms have become very popular for training back-propagation networks. Just like all the second order methods, the CG algorithm is implemented in batch-mode. The CG algorithm can search the minimum of a multivariate function faster than the conventional gradient descent procedure for BP networks. Each conjugate gradient step is, at least, as good as the steepest descent method from the same point. The formula is simple and the memory usage is in the same order as the number of weights. Most important, the CG technique obviates the tedious tasks of determining optimal learning parameters. Moreover, the CG technique has very reliable convergence behaviour as compared with the first-order gradient methods.

The basic back-propagation algorithm adjusts the weights in the steepest descent direction, i.e. negative of the gradient. This is the direction in which the performance function is decreasing most rapidly. It turns out that although the function decreases most rapidly along the negative of the gradient, this does not necessarily produce faster convergence. In the conjugate gradient algorithms, a search is performed along the conjugate directions, which produces generally faster convergence (conjugate directions means at orthogonal directions), than steepest descent directions.

In most training algorithms that we have discussed up to this point, a *learning rate* is used to determine the length of the weight update (step-size). In CG methods, the step-size is adjusted at every iteration. A search is made along conjugate gradient directions to determine the step-size, which will minimise the performance function along that line.

The basic attempt in all second order enhancement methods is that the current search direction $\mathbf{d}(n)$ to be a compromise between the *exact* gradient $\nabla E(n)$ and the previous search direction $\mathbf{d}(n-1)$, i.e. $\mathbf{d}(n) = -\nabla E(n) + \beta \mathbf{d}(n-1)$ with $\mathbf{d}(0) = -\nabla E(0)$.

The search direction is chosen (by appropriately setting β so that it distorts as little as possible the minimisation achieved by the previous step. In conjugate gradient methods, the current search is chosen to be conjugate to the previous search direction. Analytically, we require

$$\bar{d}(n-1)^t H(n-1)\bar{d}(n) = 0 \qquad (4.43)$$

where the Hessian $H(n-1)$ is assumed to be positive definite (H is the Hessian matrix with components $H_{ij} = \frac{\partial^2 E}{\partial w_i \partial w_j}$).

β plays the role of an adaptive momentum and chosen according to the Polack–Ribiere rule

$$\beta = \beta(n) = \frac{[\nabla E(n) - \nabla E(n-1)]^t \nabla E(n)}{\|\nabla E(n-1)\|^2} \qquad (4.44)$$

Thus the search direction in the conjugate gradient methods at iteration n is given by

$$\bar{d}(n) = -\nabla E(n) + \beta \bar{d}(n-1)$$
$$= -\nabla E(n) + \frac{[\nabla E(n) - \nabla E(n-1)]^t \nabla E(n)}{\|\nabla E(n-1)\|^2} \bar{d}(n-1) \qquad (4.45)$$

Now using $\mathbf{d}(n-1) = (1/\rho)\Delta\mathbf{w}(n-1)$ and substituting the preceding expression for $\mathbf{d}(n)$ in $\Delta\mathbf{w}(n)$ leads to the weight update rule:

$$\Delta\bar{w}(n) = -\rho \nabla E(n) + \beta(n)\Delta\bar{w}(n-1) \qquad (4.46)$$

When E is quadratic, the conjugate methods theoretically converge in N or fewer iterations. In general, E is not quadratic, and therefore, this method would be slower than what theory predicts. However, it is reasonable to assume that E is approximately quadratic near a local minimum. Therefore, conjugate gradient descent is expected to accelerate the convergence of back-propagation once the search enters a small neighbourhood of a local minimum.

4.3.2 Selection of Error Functions

Normally in the supervised learning of multi-layer neural networks, sum squared error is used. There are many other error functions which may used for ANN training as given in Table 4.14.

These error functions, their derivates and delta functions have been plotted against error as shown in Fig. 4.9. The three-dimensional surfaces for some error function are shown in Fig. 4.10.

Table 4.14. Different error functions for ANN learning

1.	Sum square error	$\frac{1}{2}\sum e_i{}^2$
2.	Logarithmic error	$\sum[(1+y_pk)\text{In}\{(1+y_pk)/(1+O_pk)\}]+$ $\sum[(1-y_pk)\text{In}\{(1-y_pk)/(1-O_pk)\}]$
3.	Mean fourth power error	$\sum e_i{}^4/p$
4.	Hyperbolic square error	$\sum\text{In}\{(1-e_i{}^2)/(1+e_i{}^2)\}$
5.	Hubber's error	$\sum e_i{}^2/2$ if $|e_i|<c$ $\sum c(e_i-c/2)$ if $|e_i|>=c$
6.	Cauchy's error	$\sum c^2[\text{In}\{1+(e_i/c)^2\}]/2$
7.	Geman–McClure error	$\sum e_i{}^2/\{2(1+e_i{}^2)\}$
8.	Welsch error	$\sum c^2[1-e_i-(e_i/c)^2]/2$
9.	fair's error	$\sum c^2[e_i/c-\text{In}(1-e_i/c)]$
10.	Mean median error	$\sum 2[(1+e_i/2)1/2-1]$
11.	Log-cos-hyperbolic error (Tasos Falas 1999)	$\sum\text{In}[\cos\ h(e_i{}^2)]$
12.	Andrew error	$\sum\cos(\pi^*e_i)/\pi 2$ If $e_i<=1$ $\sum e_i$ If $e_i>1$
13.	Entropy error	$-\text{In}(1-e_i)$
14.	Hamlet error	$e_i-\text{In}(1-e_i)$
15.	Fahlman error	$\sum[(1+e_i)\text{In}(1+e_i)+(1-e_i)\text{In}(1-e_i)]$

4.3.3 Mode of Error Calculation

The error can be calculated in the pattern mode or in batch mode. In pattern mode of error calculation the error is calculated after present each pattern, i.e. one set of training inputs, which is used to modify the weights. In batch mode all the patterns are presented and the errors are calculated for each pattern and then sum square error is used to modify the weights.

4.4 Summary

Finally it could be concluded that the back propagation feedforward neural networks training and testing performance is dependent on network complexity, problem complexity and complexity of learning algorithm. In this chapter it has been found that:

1. Tan sigmoid activation function at hidden layer and pure linear at output layer is taking less training epochs and also giving very good results during testing. Pure linear–pure linear combination is also taking same training time but the results predicted with this pair is not very encouraging.
2. Prediction accuracy is comparable for different mappings. However training time is minimum for X-Y type of mapping (it requires only 98, 56, and 35 epochs for dc machine current, speed predictions and electrical load forecasting problem, respectively. The other mappings require significantly large training epochs.

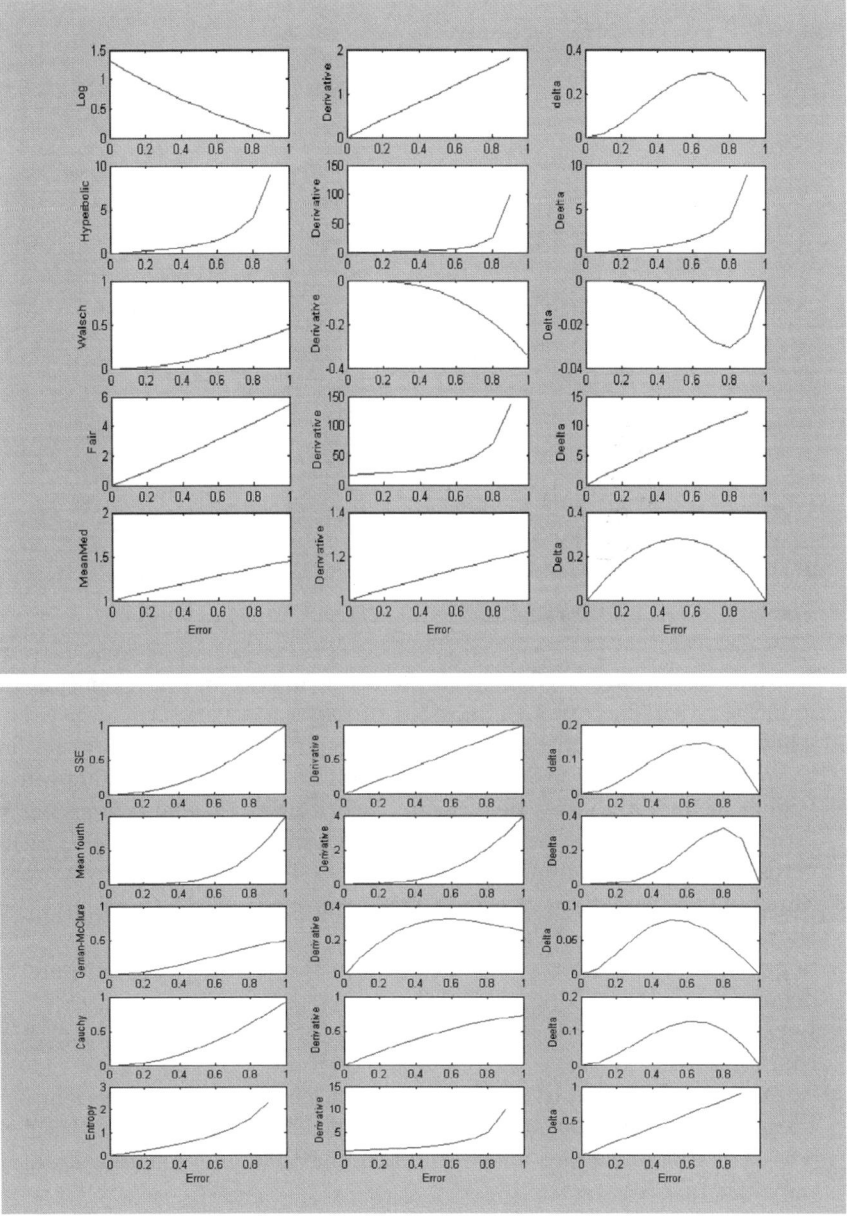

Fig. 4.9. Error functions, their derivates and delta functions for different error value

3. Xnoise–Y mapping is able to train up to error level 0.00265015 in 400 training epochs and the error during testing is also low as compared to the other noisy mappings as shown in Table 2.4.

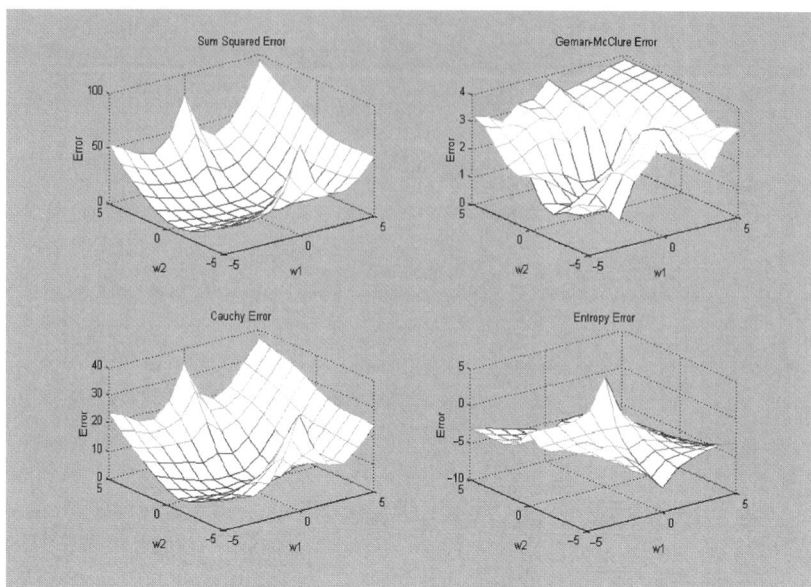

Fig. 4.10. Error surfaces for different error functions

4. Normalization ranges −0.9 to 0.9 for input and 0.1 to 0.9 for output are found very satisfactory for almost all problems.
5. They change the weights each time by some fraction of the change needed to completely correct the error. This fraction, ß, is called learning rate.
6. High learning rates cause the learning algorithm to take large steps on the error surface, with the risk of missing a minimum, or unstably oscillating across the error minimum.
7. Small steps, from a low learning rate, eventually find a minimum, but they take a long time to get there.
8. Some NN simulators can be set to reduce the learning rate as the error decreases.
9. Local minima problem can be avoided by introducing the momentum term.
10. To increase the speed of back propagation learning algorithms, adaptive learning rate and momentum factor is considered.
11. The error tolerance also affects the training time and generalization capabilities of ANN.

4.5 Bibliography and Historical Notes

In multilayer ANN, it is reported that the training is very time-consuming phase. Among the different approaches suggested to ease the back-propagation training process, input data pre-treatment has been pointed out, although no specific procedure has been proposed. We have found that input data normalization with certain criteria, prior to a training process, is crucial to obtain good results as well as to fasten significantly the calculations. There are some researchers (Sevilla Sola 1997; Kartam 1997) reported that how data normalization affects the training performance of ANN.

4.6 Exercises

1. Consider a neuron whose activation function is sigmoid $f(x) = \frac{1}{1+e^{-\lambda x}}$
   a. Prove that the derivative of f(x) with respect to x is given as $f'(x) = \lambda$. $f(x)(1 - f(x))$.
   b. Write a MATLAB program for plotting the activation function and its derivative for different values of x for $\lambda = 0.1, 0.5, 1.0$.
2. Repeat (a) and (b) parts of questions 1 with hyperbolic activation function.
3. Implement the following logic gates using feedforward bckpropagation ANN with one, two and three hidden layers:
   a. NAND gate
   b. NOR gate
   c. EX-OR gate.
4. Repeat question 3 for product aggregation neuron and compare the training and testing performance of above ANN and Product ANN.
5. Write a step by step procedure for backpropagation algorithms.
6. Study the effect of different intial weights on the training performance on backpropagation learning algorithms.
7. Write a MATLAB program to generate at least 100 training and 25 testing data for the following function

$$f(x) = x^* e^x.$$

   a. Develop ANN model to map this function and compare the results for fixed parameter backpropagation and adaptive backpropagation learning.
   b. Also train the ANN model using 5% noise in the training data and test with actual data.

5

Development of Generalized Neuron and Its Validation

More recently, ANNs and fuzzy set theoretic approach have been proposed for many different industrial applications. A number of papers have been published in the last two decades. An illustrative list is given in bibliography. Both techniques have their own advantages and disadvantages. The integration of these approaches can give improved results.

In the previous chapter, the performance aspect of ANN has been discussed in detail. To overcome some of the problems of ANN and improve its training and testing performance, the simple neuron is modified and a generalized neuron is developed in this chapter.

In the common neuron model generally the aggregation function is summation, which has been modified to obtain a generalized neuron (GN) model using fuzzy compensatory operators as aggregation operators to overcome the problems such as large number of neurons and layers required for complex function approximation, which not only affect the training time but also the fault tolerant capabilities of the artificial neural network (ANN) (Chaturvedi 1997).

5.1 Existing Neuron Model

The general structure of the common neuron is an aggregation function and its transformation through a filter. It is shown in the literature (Widrow and Lehr 1990) that the ANNs can be universal function approximators for given input–output data. The common neuron structure has summation or product as the aggregation function with linear or nonlinear (sigmoid, radial basis, tangent hyperbolic, etc.) as the threshold function as shown in Fig. 5.1.

If variation at aggregation is only considered at the neuron level, two types of neurons are possible:

1. Summation type neuron ($\sum$ - Neuron)

   In summation type neuron summation function at aggregation level and sigmoid function at activation level is considered as shown in Fig. 5.1a.

D.K. Chaturvedi: *Soft Computing Techniques and its Applications in Electrical Engineering*, Studies in Computational Intelligence (SCI) **103**, 87–122 (2008)
www.springerlink.com   © Springer-Verlag Berlin Heidelberg 2008

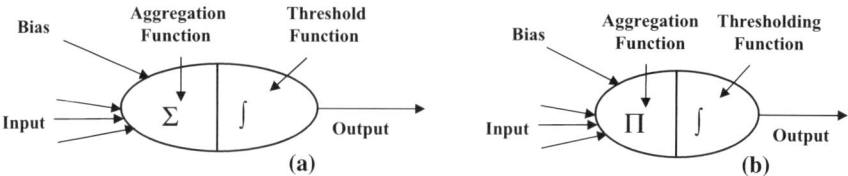

Fig. 5.1. (a) Simple summation neuron model. (b) Simple product neuron model

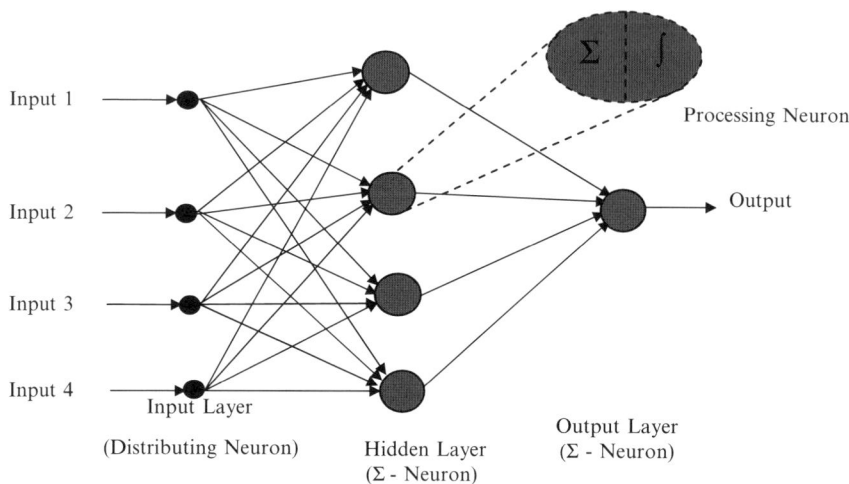

Fig. 5.2. Summation type neural network ($\sum$ - ANN)

2. Product type neuron (Π – Neuron)

It consists of product function at aggregation level and sigmoid function at activation level as shown in Fig. 5.1b.

Now using these neuron models four type of neural network could be developed:

a. Summation type neural network ($\sum$ - ANN)

It contains all summation neuron at hidden layer as well as output layer as shown in Fig. 5.2.

b. Product type neural network (Π – ANN)

It is made up of all product type neurons at both hidden layer and output layer as shown in Fig. 5.3.

c. Mixed type neural network

In mixed type neural networks both summation and product type neurons could be kept in two ways in the network as mentioned below:

1. Summation – product type neural network ($\sum -\Pi$ – ANN)

Here summation type ($\sum$) neuron is considered at the hidden layer and product type (Π) neuron is considered at the output layer as shown in Fig. 5.4.

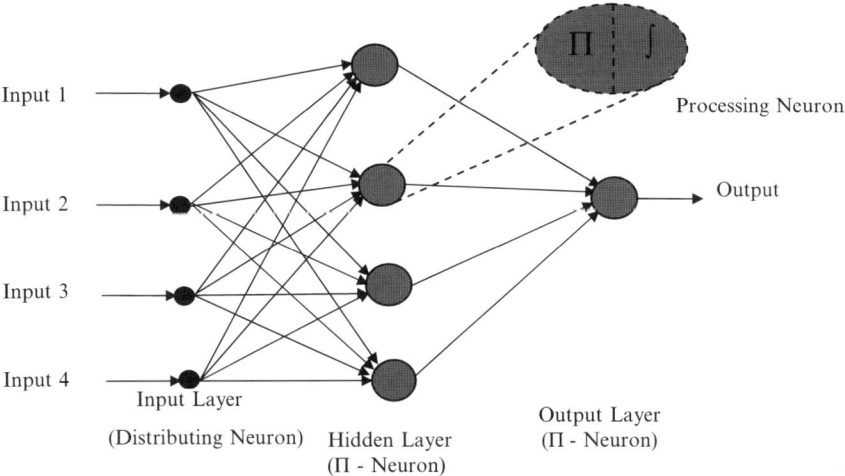

Fig. 5.3. Product type neural network (Π - ANN)

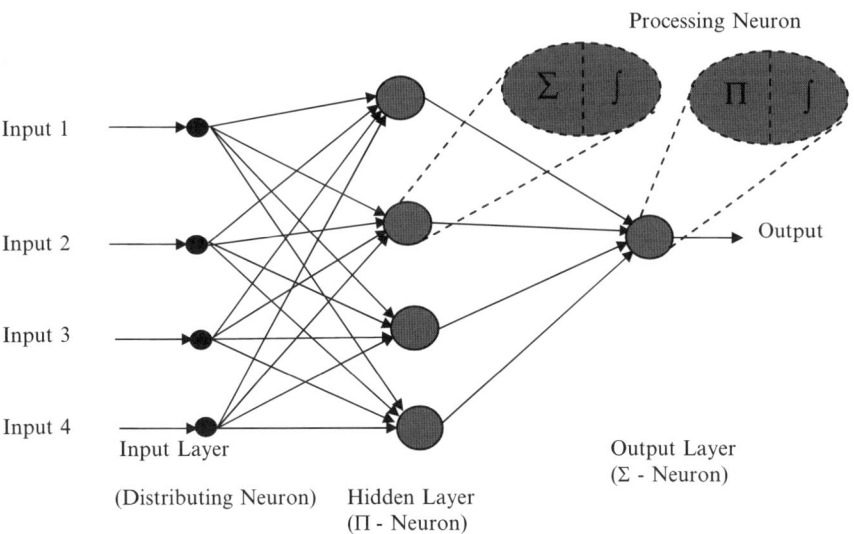

Fig. 5.4. Summation - Product type neural network ($\sum -\Pi$ - ANN)

2. Product – summation type neural network (Π-$\sum$ – ANN)
   This is a network in which Π – neurons are taken at the hidden layer
   and $\sum$ – neurons are at output layer as shown in Fig. 5.5.

Then all these four types of networks shown in Figs. 5.2–5.5 are used to
model the non-linear starting speed – torque characteristic of induction motor
and their performance have been compared for same initial weights and same
ANN learning parameters as shown in Table 5.1.

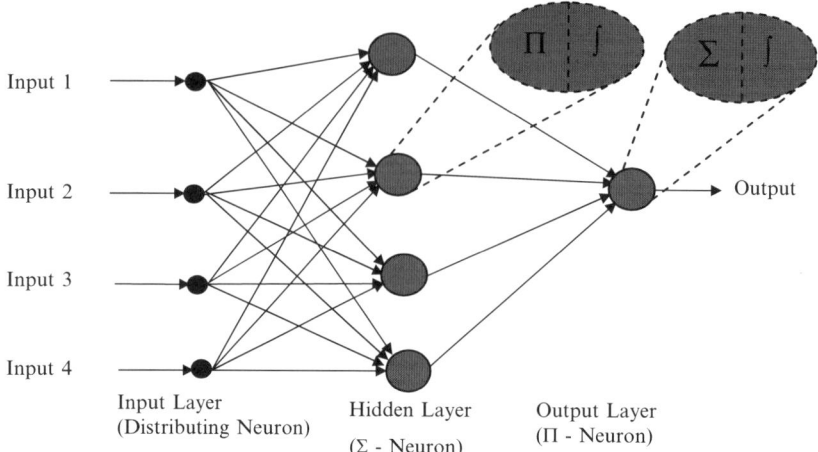

Fig. 5.5. Product-summation type neural network ($\Pi - \sum$ -ANN)

Table 5.1. ANN learning parameters

Learning rate	η	0.4
Momentum factor	α	0.6
Gain scale factor	λ	1.0
Error tolerance	E	0.005

Table 5.2. Performance of different ANN models for mapping induction motor characteristics

Models	Training performance	Testing performance		
		RMS error	Min error	Max error
$\sum$ - **ANN**	1150	0.02568	0.000376	0.888082
Π – **ANN**	4090	0.14097	0.000632	0.287738
$\sum$ –Π – **ANN**	50	0.01661	0.000430	0.250279
Π – $\sum$ - **ANN**	50	0.01358	0.000614	0.297254

The training and testing performance of all four type neural network is given in Table 5.2. It is quite clear that the combination of these different types of neuron layers in the network gave very interesting results. The mixed type neural network only needed 50 iterations (epochs) during training and testing results are also quite good for these type of networks.

5.2 Development of a Generalized Neuron (GN) Model

It is very clear from the above discussion that the combinations of summation ($\sum$) neurons and product (Π) neurons at different layers are giving quite good results as compared to only summation neuron or product neuron in the whole network; which motivated to explore the possibilities of different combinations. Thus, a generalized neuron model has been developed that uses the fuzzy compensatory operators (listed in Table 5.3) that are partly union and partly intersection given by Mizumoto in his paper on pictorial representation of fuzzy connectives II in 1989.

Use of the sigmoid threshold function and ordinary summation or product as aggregation functions in the existing models fails to cope with the

Table 5.3. Compensatory operators suggested by Mizumoto (1989)

S. No.	Summation type operator	Product type operator
1.	$[X_1 \cap X_2]*W + [X_1 \cup X_2]*(1-W)$	$[X_1 \cap X_2]^W*[X_1 \cup X_2]^{(1-W)}$
2.	$(X_1*X_2)*W + (X_1 + X_2 - X_1*X_2)*(1-W)$	$(X_1*X_2)^W*(X_1 + X_2 - X_1*X_2)^{(1-W)}$
3.	$[0 \cup (X_1*X_2)]*W] + [1 \cap (X_1 + X_2)]*(1-W)$	$[0\cup(X_1*X_2)]^W + [1\cap(X_1+X_2)]^{(1-W)}$
4.	$[X_1 \cap X_2]*W + (X_1 + X_2 - X_1*X_2)*(1-W)$	$[X_1 \cap X_2]^W*(X_1 + X_2 - X_1X_2)^{(1-W)}$
5.	$[X_1 \cup X_2]*W + (X_1*X_2)*(1-W)$	$[X_1 \cup X_2]^W*(X_1*X_2)^{(1-W)}$
6.	$[X_1 \cap X_2]*W + [1 \cap (X_1 + X_2)]*(1-W)$	$[X_1 \cap X_2]^W*[1 \cap (X_1 + X_2)]^{(1-W)}$
7.	$[X_1 \cup X_2]*W+[0\cup(X_1+X_2-1)]*(1-W)$	$[X_1 \cup X_2]^W*[0\cup(X_1 + X_2 - 1)]^{(1-W)}$
8.	$[X_1*X_2]*W + [1 \cap (X_1 + X_2)]*(1-W)$	$[X_1*X_2]^W*[1 \cap (X_1 + X_2)]^{(1-W)}$
9.	$[X_1 + X_2 - X_1*X_2]*W + [0\cup(X_1+X_2-1)]*(1-W)$	$[X_1 + X_2 - X_1*X_2]^W*[0 \cup (X_1 + X_2 - 1)]^{(1-W)}$
10.	$[X_1 \cap X_2]*W+[0\cup(X_1+X_2-1)]*(1-W)$	$[X_1 \cap X_2]^W*[0 \cup (X_1 + X_2 - 1)]^{(1-W)}$
11.	$[X_1 \cup X_2]*W + [1 \cap (X_1 + X_2)]*(1-W)$	$[X_1 \cup X_2]^W*[1 \cap (X_1 + X_2)]^{(1-W)}$
12.	$[X_1*X_2]*W+[0\cup(X_1+X_2-1)]*(1-W)$	$[X_1*X_2]^W*[0 \cup (X_1 + X_2 - 1)]^{(1-W)}$
13.	$(X_1 + X_2 - X_1*X_2)*W + [1 \cap (X_1 + X_2)]*(1-W)$	$(X_1 + X_2 - X_1*X_2)^W*[1 \cap (X_1 + X_2)]^{(1-W)}$
14.	$[X_1 \cap X_2]*W + [X_1*X_2]*(1-W)$	$[X_1 \cap X_2]^W*[X_1*X_2]^{(1-W)}$
15.	$[X_1 \cup X_2]*W + (X_1 + X_2 - X_1*X_2)*(1-W)$	$[X_1 \cup X_2]^W*(X_1 + X_2 - X_1X_2)^{(1-W)}$
16.	$[X_1 \cap X_2]*W + [(X_1 + X_2)/2]*(1-W)$	$[X_1 \cap X_2]^W*[(X_1 + X_2)/2]^{(1-W)}$
17.	$[X_1 \cup X_2]*W + [(X_1 + X_2)/2]*(1-W)$	$[X_1 \cup X_2]^W*[(X_1 + X_2)/2]^{(1-W)}$
18.	$[X_1*X_2]*W + [(X_1 + X_2)/2]*(1-W)$	$[X_1*X_2]^W*[(X_1 + X_2)/2]^{(1-W)}$
19.	$[X_1 + X_2 - X_1*X_2]*W + [(X_1 + X_2)/2]*(1-W)$	$[X_1 + X_2 - X_1*X_2]^W*[(X_1 + X_2)/2]^{(1-W)}$
20.	$[0 \cup (X_1 + X_2 - 1)]*W + [(X_1 + X_2)/2]*(1-W)$	$[0\cup(X_1+X_2-1)]^W*[(X_1+X_2)/2]^{(1-W)}$
21.	$[1 \cap (X_1 + X_2)]*W + [(X_1 + X_2)/2]*(1-W)$	$[1 \cap (X_1 + X_2)]^W*[(X_1 + X_2)/2]^{(1-W)}$
22.	$[(X_1 \cap X_2)]*W+[(X_1+X_2)/2]*(1-W)$	$[(X_1 \cap X_2)]^W*[(X_1 + X_2)/2]^{(1-W)}$
23.	$[X_1 \cup X_2]*W + [1 - \sqrt{(1 - X_1)(1 - X_2)}]*(1-W)$	$[X_1 \cup X_2]^W*[1 - \sqrt{(1 - X_1)(1 - X_2)}]^{(1-W)}$
24.	$[X_1 \cap X_2]*W + [1 - \sqrt{(1 - X_1)(1 - X_2)}]*(1-W)$	$[X_1 \cap X_2]^W*[1 - \sqrt{(1 - X_1)(1 - X_2)}]^{(1-W)}$
25.	$[X_1 \cup X_2]*W + [\sqrt{(X_1*X_2)}]*(1-W)$	$[X_1 \cup X_2]^W*[\sqrt{(X_1*X_2)}]^{(1-W)}$
26.	$\sqrt{(X_1*X_2)}*W + [1 - \sqrt{(1 - X_1)(1 - X_2)}]*(1-W)$	$\sqrt{(X_1*X_2)}^W*[1 - \sqrt{(1 - X_1)(1 - X_2)}]^{(1-W)}$
27.	$[2X_1X_2/(X_1 + X_2)]*W + [(X_1 + X_2 - 2X_1X_2)/(2 - X_1 - X_2)]*(1-W)$	$[2X_1X_2/(X_1 + X_2)]^W*[(X_1 + X_2 - 2X_1X_2)/(2 - X_1 - X_2)]^{(1-W)}$

Table 5.3. (*Continued*)

S. No.	Summation type operator	Product type operator
28.	$[(X_1 + X_2)/2]*W + \surd(X_1X_2)]*(1 - W)$	$[(X_1 + X_2)/2]^W*\surd(X_1X_2)^{(1-W)}$
29.	$[(X_1 + X_2)/2]*W + [1 - \surd(1 - X_1)(1 - X_2)]*(1 - W)$	$[(X_1 + X_2)/2]^W*[1 - \surd(1 - X_1)(1 - X_2)]^{(1-W)}$
30.	$[(X_1 + X_2)/2]*W + [2X_1X_2/(X_1 + X_2)]*(1 - W)$	$[(X_1 + X_2)/2]^W*[2X_1X_2/(X_1 + X_2)]^{(1-W)}$
31.	$[(X_1 + X_2)/2]*W + [(X_1 + X_2 - 2X_1X_2)/(2 - X_1 - X_2)]*(1 - W)$	$[(X_1 + X_2)/2]^W*[(X_1 + X_2 - 2X_1X_2)/(2 - X_1 - X_2)]^{(1-W)}$
32.	$[(X_1X_2)(X_1 + X_2 - 2X_1X_2)]*W + [X_1 + X_2 - X_1X_2(X_1 + X_2 - 2X_1X_2)]*(1 - W)$	$[(X_1X_2)(X_1 + X_2 - 2X_1X_2)]^W*[X_1 + X_2 - X_1X_2(X_1 + X_2 - 2X_1X_2)]^{(1-W)}$
33.	$[(X_1X_2)(X_1 \cap X_2)]*W + [X_1 + X_2 - X_1X_2 + X_1 \cup X_2 - (X_1 + X_2 - X_1X_2)(X_1 \cap X_2)]*(1 - W)$	$[(X_1X_2)(X_1 \cap X_2)]^W*[X_1 + X_2 - X_1X_2 + X_1 \cup X_2 - (X_1 + X_2 - X_1X_2)(X_1 \cap X_2)]^{(1-W)}$
34.	$[(X_1X_2) + (X_1 \cap X_2) - (X_1X_2)(X_1 \cap X_2)]*W + [(X_1 + X_2 - X_1X_2)(X_1 \cup X_2)]*(1 - W)$	$[(X_1X_2) + (X_1 \cap X_2) - (X_1X_2)(X_1 \cap X_2)]^W*[(X_1 + X_2 - X_1X_2)(X_1 \cup X_2)]^{(1-W)}$

X_1 - Input # 1 for Σ – aggregation and
X_2 – Input # 2 for Π - aggregation
W - Weight or parameter of the operator varies between 0 and 1

Note: Output of Σ – part of neuron may be considered as union operator of fuzzy and Output of Π – part of neuron may be considered as intersection operator of fuzzy system

non-linearities involved in real life problems. To deal with these, the proposed model has both sigmoid and Gaussian functions with weight sharing. The generalized neuron model has flexibility at both the aggregation and threshold function level to cope with the non-linearity involved in the type of applications dealt with. The neuron has both Σ and π aggregation functions. The Σ aggregation function has been used with the sigmoid characteristic function while the π aggregation function has been used with the Gaussian function as a characteristic function. The final output of the neuron is a function of the two outputs O_Σ and O_π with the weights W and (1–W) respectively as shown in Figs. 5.6 and 5.7. Mathematically the output of summation type generalized neuron (GN) may be written as

$$\text{GN output} = O_\Sigma * W + O_\Pi * (1 - W),$$

where

O_Σ – output of the summation part of the neuron Σ_1
W – weight associated with O_Σ
O_Π – output of the product part of the neuron (π).

The neuron model described above is known as the summation type compensatory neuron model, since the outputs of the sigmoidal and Gaussian functions are summed up. Similarly, the product type compensatory neuron models may also be developed. It is found that in most of the applications

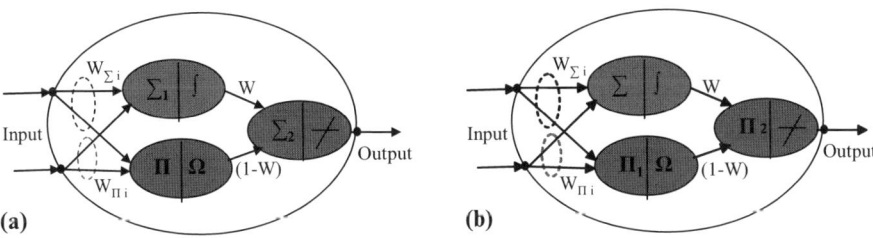

Fig. 5.6. (a) Internal structure of summation type. **(b)** Internal structure of product type generalized neuron

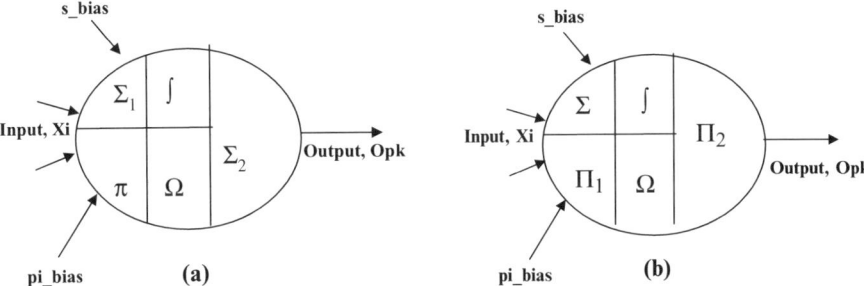

Fig. 5.7. (a) Symbolic representation of summation type generalized neuron model. **(b)** Symbolic representation of product type generalized neuron model

summation type compensatory neuron model works well (Chaturvedi 2002). Mathematically the output of product type generalized neuron may be written as –

$$\text{GN Output} = O_{\Sigma}{}^{W} * O_{\Pi}^{(1-W)}$$

5.3 Advantages of GN

1. Less number of unknown weights
   The number of weights in the case of a GN is equal to twice the number of inputs plus one, which is very low in comparison to a multi-layer feed-forward ANN.
2. Less training time
   The weights are determined through training. Hence, by reducing the number of unknown weights, training time can be reduced.
3. Less number of training patterns
   The number of training patterns required for GN training is dependent on the number of unknown weights. The number of training patterns must be greater or equal to number of GN weights. As mentioned above the number of GN weights are lesser than multi layered ANN, hence the number of training patterns required is also lesser.

4. Size of hidden layers
   There is no hidden layer required in case of GN and single neuron is capable to solve most of the problems.
5. Complexity of GN
   GN model is less complex as compared to multilayered ANN models.
6. Structural level flexibility
   GN models are more flexible at structural level. The aggregation and activations functions could be chosen depending on the problem in hand.

5.4 Learning Algorithm of a Summation Type Generalized Neuron

The following steps are involved in the training of a summation type generalized neuron:

1. *Foreward calculations*
   Step-1: The output of the Σ_1 part of the summation type generalized neuron is

$$O_\Sigma = \frac{1}{1 + e^{-\lambda s * s\_net}} \tag{5.1}$$

where $s\_net = \sum W_{\Sigma i} X_i + X_{o\Sigma}$.
Step-2: The output of the π part of the summation type generalized neuron is

$$O_\Pi = e^{-\lambda p * pi\_net^2} \tag{5.2}$$

where $pi\_net = \prod W_{\Pi i} X_i * X_{o\Pi}$.
Step-3: The output of the summation type generalized neuron can be written as

$$O_{pk} = O_\Pi * (1 - W) + O_\Sigma * W \tag{5.3}$$

2. *Reverse calculation*
   Step-4: After calculating the output of the summation type generalized neuron in the forward pass, as in the feed-forward neural network, it is compared with the desired output to find the error. Using back-propagation algorithm the summation type GN is trained to minimize the error. In this step, the output of the single flexible summation type generalized neuron is compared with the desired output to get error for the ith set of inputs:

$$\text{Error } Ei = (Yi - Oi) \tag{5.4}$$

Then, the sum-squared error for convergence of all the patterns is

$$Ep = 0.5 \sum Ei^2 \tag{5.5}$$

A multiplication factor of 0.5 has been taken to simplify the calculations.
Step-5: Reverse pass for modifying the connection strength.

(a) Weight associated with the Σ_1 and Σ_2 part of the summation type generalized neuron is:

$$W(k) = W(k-1) + \Delta W \tag{5.6}$$

where $\Delta W = \eta \delta_k (O_\Sigma - O_\Pi) Xi + \alpha W(k-1)$
and $\delta_k = \sum (Yi - Oi)$

(b) Weights associated with the inputs of the Σ_1 part of the summation type generalized neuron are:

$$W_{\Sigma i}(k) = W_{\Sigma i}(k-1) + \Delta W_{\Sigma i} \tag{5.7}$$

where $\Delta W_{\Sigma i} = \eta \delta_{\Sigma j} Xi + \alpha W_{\Sigma i}(k-1)$
and $\delta_{\Sigma j} = \sum \delta_k W(1 - O_\Sigma) * O_\Sigma$

(c) Weights associated with the input of the π- part of the summation type generalized neuron are:

$$W_{\Pi i}(k) = W_{\Pi i}(k-1) + \Delta W_{\Pi i} \tag{5.8}$$

where $\Delta W_{\Pi i} = \eta \delta_{\Pi j} Xi + \alpha W_{\Pi i}(k-1)$
and $\delta_{\Pi j} = \sum \delta_k (1 - W) * (-2 * pi\_net) * O_\Pi$
α – momentum factor for better convergence
η – learning rate

Range of these factors is from 0 to 1 and is determined by experience.

Matlab Program for Summation type GN model

```
% Main Programm for Summation type Generalized neuron (GN)
clear all;
clc;
tr_exor;      % training file name (tr_pat.m)
[i_row i_col]=size(x_tr);
patterns=i_row;
% Initialization of GN model
% weight Initialization
w=randn(in,on)*0.1;               % Weight of sum part (size [in x on])
wpi=ones(in,on)+randn(in,on)*0.1;   % Weight of product part
                                    (size [in x on])
w1=0.6;                           % weight of sum-sum part
pi_bais=0.05;                     % bais of product part
s_bais=0.05;                      % bais of sum part
delta_w1=0.0;                     % Change in weights for sum-sum part
delta_w=zeros(in,on)              % Change in weights for sum part
delta_wpi=zeros(in,o              % Change in weights for product part
delta_s_bais=0.0;                 % Change in bais for sum part
delta_pi_bais=0.0;                  % Change in bais for sum part
ss_err=0;                         % sum quared error
```

```
% Network parameters
eta=input('Value of Learning rate='); %learning rate
alpha=input('Value of momentum factor='); % momentum factor
lemda_s=1;                            % gain scale factor of sigma part
lemda_pi=1;                           % gain scale factor of pi part
err_tol=0.001;                        % error tolrence
max_epoch=20000;                      % maximum number of iterations
disp_itr=max_epoch/100;               % Display training results after
                                        so many epochs
x_in_tr=x_tr(:,1:in);                 % input pattern for training
y_desired=x_tr(:,(in+1):i_col);       % Desired output
count1=0;
for epoch=1:max_epoch                 % Loop or cycle (both forward and
                                        reverse calculation)
for i=1:patterns                      % Loop for calculating output
                                        for GN model
s_net(i,:)=x_in_tr(i,:)*w+s_bais;     % sigma of (xi*wi)
x_wpi=x_in_tr(i,:).*wpi';             % product of (xi*wi)
pi_net(i,:)=pi_bais+prod(x_wpi);
end
s_out=1./(1+exp(-lemda_s*s_net));          % output of sigma part
pi_out=exp(-lemda_pi*(pi_net.^2));         % output of pi part


y_cnn=(w1*s_out+(1-w1)*pi_out);    % Final output

% Reverse calculation for adjusting weights and train GN model
% Network error
error=y_desired-y_cnn;                % Error of GN model
s_err=(error.^2)./2;                       % square error
ss_err1=ss_err;
ss_err=sum(s_err);                         % sum square error
err_dot=ss_err1-ss_err;               % change in sum square error
tr_res(epoch,:)=[epoch ss_err err_dot];    % training results

if ss_err<=err_tol; break; end
if count1==disp_itr
ss_err
count1=0;
end
count1=count1+1;

% weight adjustment of sigma-sigma part
delta_w1=eta*error'*(s_out-pi_out)+alpha*delta_w1;
w1=w1+delta_w1;

% weight adjustment of input-sigma part
f_desh=s_out.*(1-s_out);
delta_w=(lemda_s*eta*w1*(error.*f_desh)'*x_in_tr)'+alpha*delta_w;
w=w+delta_w;                          % New weights for sum part
```

```
% weight adjustment of input-pi part
fdesh_pi=-2*pi_out.*pi_net;
delta_wpi=lemda_pi*eta*(1-w1)*sum(error.*fdesh_pi.*(pi_net-pi_bais))./
        wpi+alpha*delta_wpi;
wpi=wpi+delta_wpi;              % New weights for product part

% modify bais of first sigma part
delta_s_bais=lemda_s*eta*w1*error'*f_desh+alpha*delta_s_bais;
s_bais=s_bais+delta_s_bais;

% modify bais of pi part
delta_pi_bais=lemda_pi*eta*(1-w1)*error'*fdesh_pi+alpha*delta_pi_bais;
pi_bais=pi_bais+delta_pi_bais;
end

% Testing of GN model
tst_exor;                      % Test File name (tst_pat.m)
[x_R x_c]=size(x_tst);
patterns1=x_R;

x_in1=x_tst(:,1:in);
y_desired1=x_tst(:,(in+1):i_col);

for i=1:patterns1
  s_net1(i,:)=x_in1(i,:)*w+s_bais;      % sigma of (xi*wi)
  x_wpi1=x_in1(i,:).*wpi';              % product of (xi*wi)
  pi_net1(i,:)=pi_bais+prod(x_wpi1);
end

s_out1=1./(1+exp(-lemda_s*s_net1));    % output of sigma part
pi_out1=exp(-lemda_pi*(pi_net1.^2));   % output of pi part

y_cnn1=(w1*s_out1+(1-w1)*pi_out1);     % GN output
err_tst=(y_desired1-y_cnn1);           % Error during testing

% plotting of training results
subplot(1,2,1);                        % Divide the display screen
                                       % in 1 row and 2 columns
plot(tr_res(:,1),tr_res(:,2),'k-');    % plot (x,y)
xlabel('Number of Epochs') ;           % Label x-axis
ylabel('training ss_err');             % Label y-axis
title('Error');                        % Title for the graph
subplot(2,2,2)
plot(tr_res(:,1),tr_res(:,3),'k-');
axis([0 20000 -0.001 0.001]);
xlabel('Number of Epochs')
ylabel('Err_dot')
title('Derivative of Error')

% plotting of test results
subplot(1,2,2)
```

```
plot(y_desired1,'k--');hold on
plot(y_cnn1,'k-')
xlabel('Number of Output')
ylabel('Generalized Neuron output')
title('GN output during testing')
subplot(2,2,4)
plot(1:x_R, err_tst,'k-*')
xlabel('Number of Output')
ylabel('Error during testing')
title('testing Error')
```

5.5 Benchmark Testing of Generalized Neuron Model

The generalized neuron model developed must be verified on Benchmark problems and compared with feed-forward multi-layered ANN under same training conditions such as same gain scale factor, learning rate, momentum, initial weights and error function used in back-propagation learning algorithm.

5.5.1 Ex-OR Problem

The multi-layered feed-forward ANNs are trained to produce an output of one (zero) when binary input has an odd (even) number of bits. The Ex-OR problem is a classification problem, which is linearly non-separable. It requires minimum one hidden layer having two neurons for its solution. The input–output pattern of Ex-OR problem is given in Table 5.4.

It arises in the case of XOR problem, which may be viewed as a special case of points in the unit hypercube. Each point in the hypercube is class 0 or class 1. However, in the special case of the XOR problem, we need only the four corners of the unit square that corresponds to the input patterns (0,0), (0,1), (1,0) and (1,1). The first and third patterns are in class 0 and the input patterns (0,1) and (1,0) are also at the opposite corners of the square, but are classified together as output 1.

The use of a single neuron with two inputs results in a straight line for decision boundary in the input space. For all points on one side of the line, the neuron outputs 1; for all points on the other side of the line, it outputs 0. The position and orientation of the line in the input space are determined by the synaptic weights of the neuron connected to the input nodes, and the

Table 5.4. Input–output patterns for Ex-OR problem

Inputs		Output
0	0	0
0	1	1
1	0	1
1	1	0

threshold applied to the neuron. With the input patterns (0,0) and (1,1) located on opposite corners of the unit square and likewise for the other two input patterns (0,1) and (1,0), it is clear that we cannot construct a straight line for a decision boundary so that (0,0) and (1,1) lie in one decision region and (0,1) and (1,0) lie in the other decision region. In other words, an elementary perceptron cannot solve the XOR problem (Minsky and Papert 1969).

We may solve the XOR problem by using a single hidden layer with two neurons. The signal flow graph is shown in Fig. 5.8 and the decision boundaries formed in Fig. 5.9. The following assumptions are made here:

- Each neuron is represented by a McCulloch–Pitts model.
- Bits 0 and 1 are represented by 0 and +1.

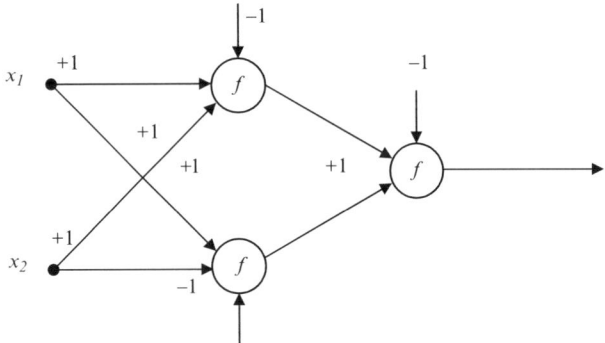

Fig. 5.8. XOR problem network

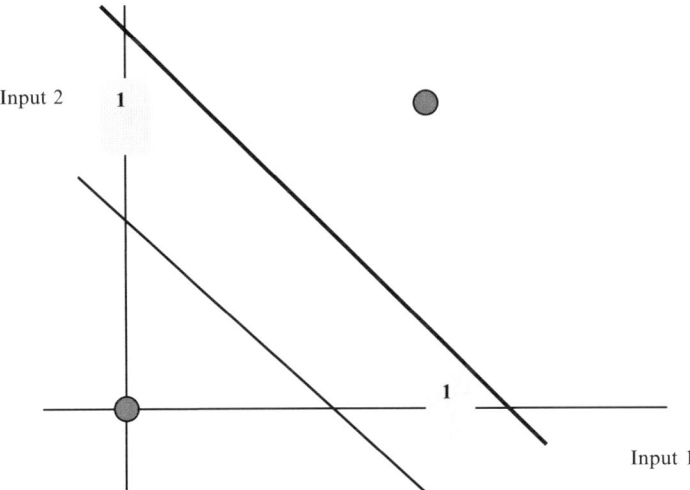

Fig. 5.9. Boundaries for Ex-OR problem

The top neuron, labelled 1 in the hidden layer is characterised as follows:

$$w_{11} = w_{12} = +1$$
$$\theta_1 = +1.5$$

The bottom neuron, labelled 2 in the hidden layer is characterised as follows:

$$w_{21} = w_{22} = +1$$
$$\theta_2 = +0.5$$

The output neuron, labelled 3 is characterised by:

$$w_{13} = -2$$
$$w_{23} = +1$$
$$\theta_3 = +0.5$$

ANN and GN model both have been trained using gradient descent back-propagation learning algorithm for 0.001 error tolerance with same training parameters. The reduction in error during training of ANN and GN Model is shown in Fig. 5.10 for Ex-or problem. Reduction in error during training is faster for GNM in comparison to ANN as given in Table 5.5.

The training and testing performance of GNM are shown in Figs. 5.11 and 5.12 for all four input patterns for Ex-or problem. The testing results in terms of RMS, max and min error of ANN and GNM during testing are presented in Table 5.6. It is found that during testing GN model is giving very good performance than ANN as the errors in output shown by them are considerably less than ANN.

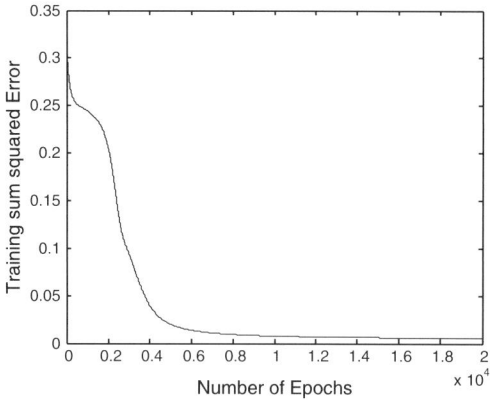

Fig. 5.10. Training performance of GN model for Ex-OR problem

Table 5.5. Training performance of ANN and GN Model (GNM) for EX-OR problem

Models	Structure	Training epochs
ANN	2-2-1	60,680
GN Model	Single neuron	20,435

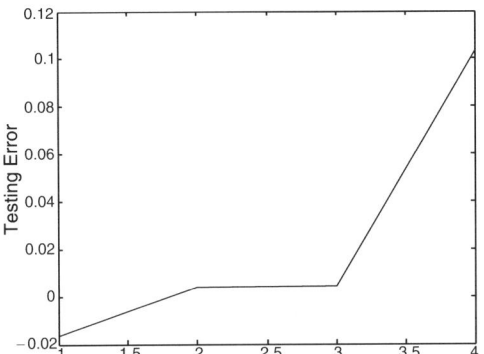

Fig. 5.11. Testing performance of GN model for Ex-OR problem

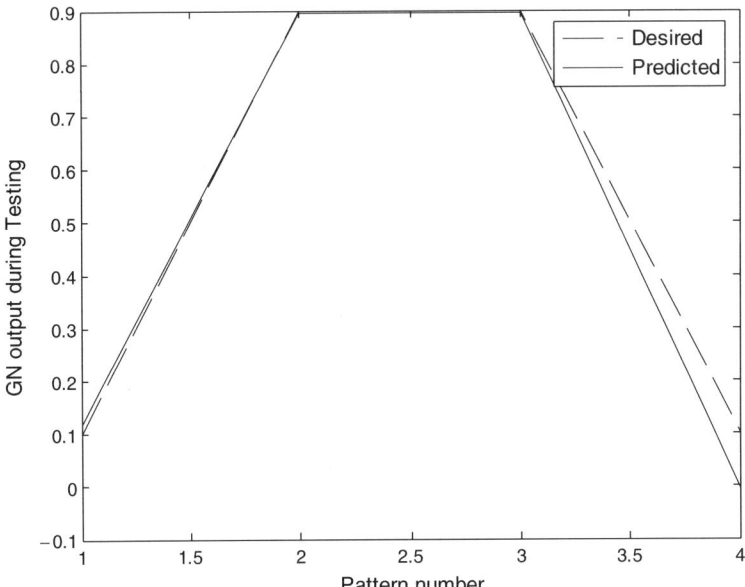

Fig. 5.12. Performance of generalized neuron output during testing

Table 5.6. Testing performance of ANN and GN model For Ex-OR problem with 5% noise

Models	RMS error	MAX error	MIN error
ANN	0.33673	0.21758	−0.11341
GN Model	0.11671	0.11201	−0.01832

```
% Training patterns for Ex-OR problem (train_pat.m)
% Normalized input output patterns
in=2; % number of inputs (X1 and X2)
on=1; % number of outputs (Y)
```

%	X1	X2	Y
x=[	0.9	0.9	0.1
	0.1	0.9	0.9
	0.9	0.1	0.9
	0.1	0.1	0.1];

% Normalized testing file (tst_pat.m)

%	X1	X2	Y
x=[	0.89	0.9	0.1
	0.1	0.9	0.9
	0.9	0.1	0.9
	0.12	0.1	0.1];

5.5.2 The Mackey-Glass Time Series

The Mackey-Glass (MG) time series is the most common problem to evaluate a network for its prediction capabilities. The MG series is a model of chaotic series. The Mackey-Glass equation represents a model for white blood cells production in leukaemia patients. It mimics the non-linear oscillations in the physiological processes involved. The Mackey–Glass delay difference equation is given below

$$x(t + 1) = 0.9x(t) + [0.2 \times (t - \tau)/(1 + x^{10}(t - \tau))].$$

The function plot is shown in Fig. 5.13.

The model is complicated due to the addition of a time delay τ in the non-linear equations. The objective of this analysis is to evaluate the efficiency of networks to predict future values using a set of past values. The above M-G equation is implemented with $\tau = 1.7$, $x(0) = 1.2$, $x(t) = 0$ for $t < 0$. A total of 301 points have been generated from $t = 0$ to $t = 300$, all points have been used for training. The 0th, 6th, 12th, and 18th points have been used to predict 19th point and so on.

The training results of ANN and GNM are shown in Table 5.7 and Fig. 5.14 for Mackey–Glass problem for error level 0.002. Reduction in error during training is faster for GN Model in comparison to ANN. GN model is consistently giving good results in training of this time series problem.

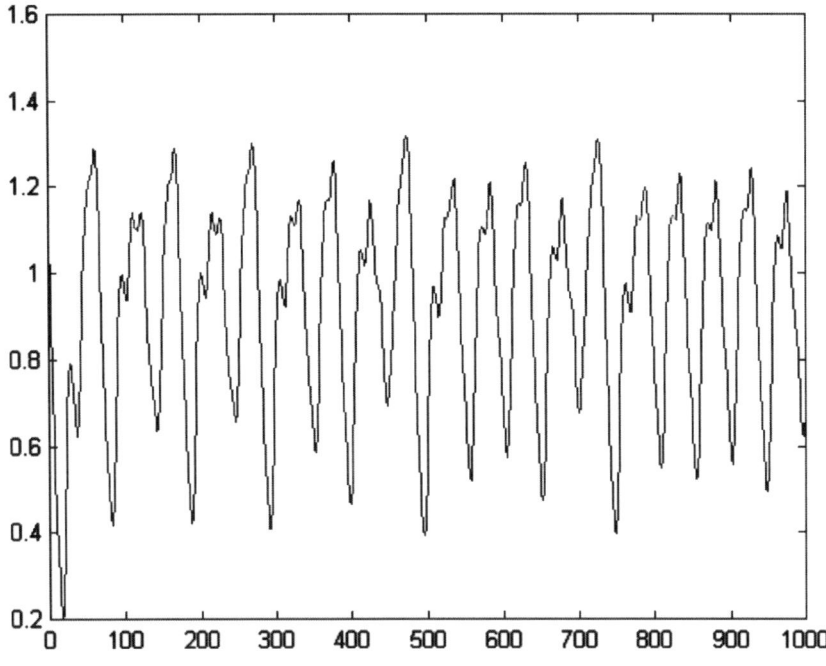

Fig. 5.13. The Mackey-Glass time series

Table 5.7. Training performance of ANN and GN model for Mackey–Glass problem

Models	Structure	Training epochs
ANN	4-4-1	13,340
GN model	Single neuron	20,083

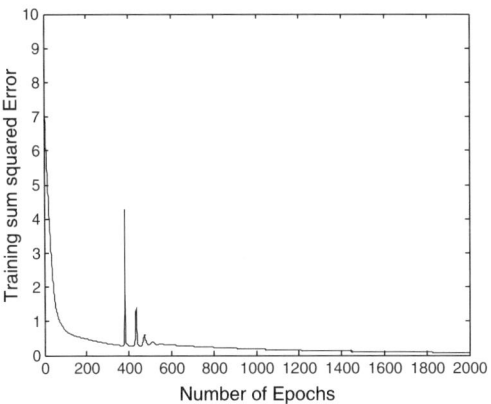

Fig. 5.14. Error during training

Once ANN and GNM are carefully trained for Mackey–Glass Problem, it has been used for testing. The training and testing performance of ANN and GNM are shown in Figs. 5.15 and 5.16. The test output of GN model is nearly coinciding with the actual data for all test patterns. The results in terms of RMS, max and min errors of ANN and GN Model during testing are presented in Table 5.8.

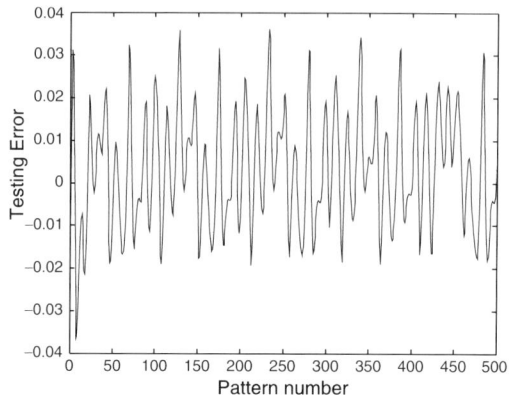

Fig. 5.15. Error during testing

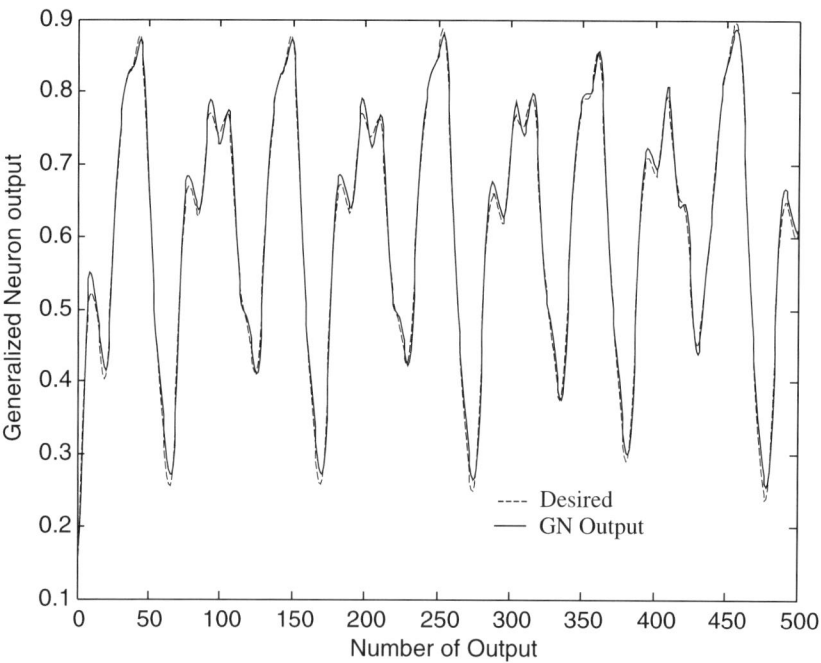

Fig. 5.16. Test results of Mackey–Glass problem during testing

Table 5.8. Testing performance of ANN and GN model with MACKEY-GLASS problem

Models	Errors (after 200 training epochs)		
	RMS error	MAX error	MIN error
ANN	0.17301	0.32675	−0.20943
GN model	0.02101	0.03502	−0.03730

5.5.3 Character Recognition Problem

The GN model is used to distinguish five different characters, A, X, H, B, I. Each character is represented by 5×7 dots. Hence, there are 35 inputs for each character as shown below:

```
input =[0 0 1 0 0   0 1 0 1 0   1 0 0 0 1   1 0 0 0 1   1 1 1 1 1   1 0 0 0 1   1 0 0 0 1
        1 0 0 0 1   0 1 0 1 0   0 0 1 0 0   0 0 1 0 0   0 0 1 0 0   0 1 0 1 0   1 0 0 0 1
        1 0 0 0 1   1 0 0 0 1   1 0 0 0 1   1 1 1 1 1   1 0 0 0 1   1 0 0 0 1   1 0 0 0 1
        1 1 1 1 1   1 0 0 0 1   1 0 0 0 1   1 1 1 1 1   1 0 0 0 1   1 0 0 0 1   1 1 1 1 1
        0 0 1 0 0   0 0 1 0 0   0 0 1 0 0   0 0 1 0 0   0 0 1 0 0   0 0 1 0 0   0 0 1 0 0];
```

The output of GN model is these characters. We can assign some values to these characters, because Neural system can give only numerical outputs.

$$
\begin{array}{ll}
\text{A -} & 0.1 \\
\text{X -} & 0.3 \\
\text{H -} & 0.5 \\
\text{B -} & 0.7 \\
\text{I -} & 0.9
\end{array}
$$

Let's train the GN model to recognize these characters. The training file consists of the set of inputs–ouput data. Here the input has value 0 or 1. For 0 input product part of GN model does not work, therefore it is necessary to normalize the data in 0.1–0.9 range and present them to GN model for training. GN model uses the following parameters for training:

Learning rate = 0.1, momentum factor = 0.5, error tolerance = 0.0001 and maximum epochs = 1,000. The following results are obtained – given error tolerance level is achieved in 210 epochs (cycles) and weights of size $w1(1 \times 1)$, $w(1 \times 35)$, $wpi(1 \times 35)$ are as follows:

$w1 = [1.1256];$

$w = [0.0024 \ − 0.1337 \ 0.1996 \ − 0.0314 \ 0.0425 \ 0.1887 \ − 0.2568 \ 0.3413$

$\qquad −0.5350 \ 0.1317 \ − 0.1299 \ − 0.0076 \ldots$

$\qquad 0.1993 \ − 0.0968 \ 0.0157 \ − 0.0183 \ 0.0938 \ 0.4312 \ 0.2867 \ − 0.1531$

$\qquad −0.0449 \ − 0.2633 \ 0.1360 \ − 0.1022 \ \ldots$

$\qquad −0.0940 \ − 0.0982 \ − 0.1757 \ 0.4440 \ − 0.1252 \ − 0.0573 \ − 0.2875$

$\qquad 0.1390 \ 0.7092 \ 0.1367 \ − 0.3039];$

wpi = [1.0247 0.8564 1.0149 0.8307 1.0719 1.1142 1.1552 1.1384 0.9242
    1.0443 1.0911 0.8926 ...
    1.0202 1.0763 0.8712 0.9047 1.0778 0.9994 1.0524 1.1364 1.0482 0.9213
    1.0752 0.9833 ...
    0.9184 1.2094 1.0080 0.9063 1.0636 1.1682 1.0594 1.0790 1.0105 0.9841
    1.0871];

The training and testing performance is graphically shown in Figs. 5.17 and 5.18.

For the same given inputs GN output is [0.1005 0.3021 0.4968 0.7040 0.8870].

Now, one bit is changed in every character and then input data is prepared for testing of GN performance. The GN output for

these set of testing data is   [0.0869 0.3221 0.4999 0.7040 0.8737];
and expected output vector is   [0.1   0.3   0.5   0.7   0.9].

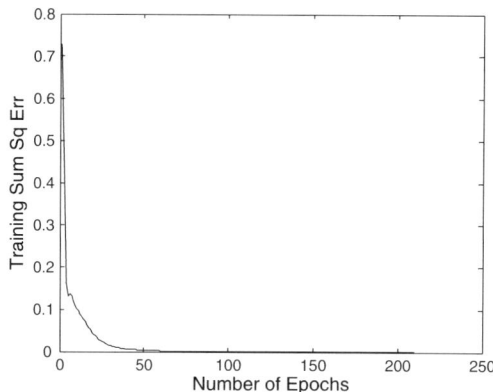

Fig. 5.17. Training performance of GN model for character recognition problem

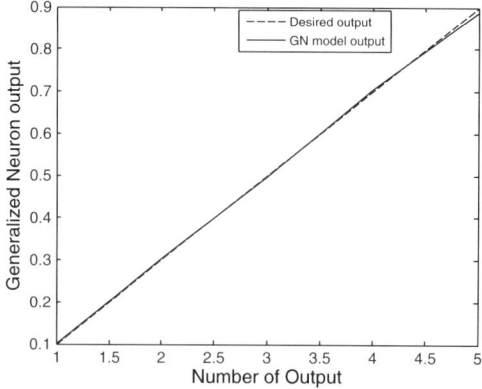

Fig. 5.18. Testing performance of GN model for character recognition problem

Table 5.9. The performance comparison of GN model and ANN model

	ANN	GNN
Network size	35-5-1 (three layer network)	35-1 (Single neuron)
Training cycles	1000	210
Error achieved in training	0.008223	0.0001
Change one bit in each character	0.0120	0.0028

It shows that the GN performance does not detoriate too much if some noise is present in the testing data set. It means the training patterns are learned well. What happens if, we present foreign characters to the GN model? Let us consider the letter M and J, as follows:

Testing = [1 0 0 0 1   1 1 0 1 1   1 0 1 0 1   1 0 0 0 1   1 0 0 0 1   1 0 0 0 1   1 0 0 0 1
pattern   0 0 1 0 0   0 0 1 0 0   0 0 1 0 0   0 0 1 0 0   0 0 1 0 0   0 0 1 0 0   0 1 1 1 0];

The results should show each foreign character in the category closest to it. The results obtained from the model is [0.1977 0.8741]. In the first pattern, M is categorized as A and J is categorized as I as expected.

The performance of GN model is also compared with ANN model and results are given in Table 5.9.

5.5.4 Sin(X1) * Sin(X2) Problem

This is a functional mapping problem to test the capabilities of GN model for various types of functions. The mapping of two functions sin(x1) and sin(x2) on to their product is used here. This is a popular functional mapping problem Y = sin(x1)* sin(x2) as shown in Fig. 5.19. The training data for sin(x1)* sin(x2) problem of the GNM and ANN is given in Table 5.10.

The ANN and GN model have been trained for Sin(x1) * Sin(x2) problem and training performance of both types of architectures is compared and given in Table 5.9 and Fig. 5.20.

The testing performance of ANN and GN Model are shown in Fig. 5.21 with 5% noise in the testing data. The test output given by GN Model nearly coincides with the actual data for all test patterns; however the test output given by ANN is far away to coincide with actual data. The results in terms of RMS, max and min errors of ANN and GN Model during testing are presented in Table 5.11. It is found that during testing GN Model gives very good performance than ANN as the errors in output shown by it is considerably less than ANN.

5.5.5 Coding Problem

In this problem the network is presented with n distinct binary input patterns, each with different bit positions and the network is trained to produce a

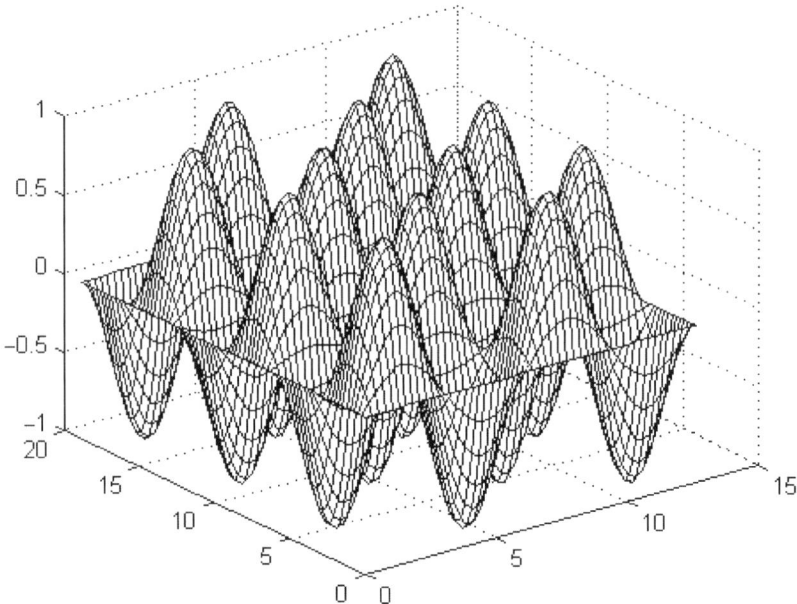

Fig. 5.19. The Sin(x1) Sin(x2) problem

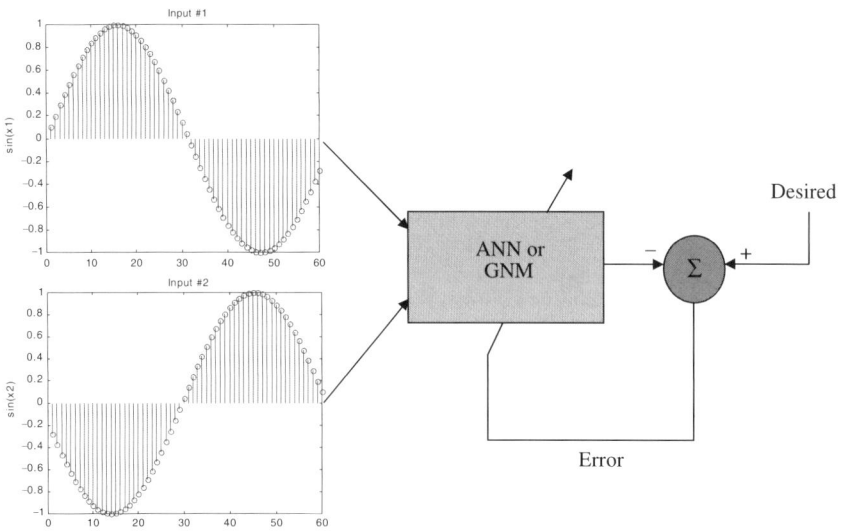

Table 5.10. Training performance of ANN AND GN model for SIN (x1) * SIN (x2) problem

Models	Structure	Training epochs
ANN	4-4-4-1	50,870
GN model	Single neuron	764

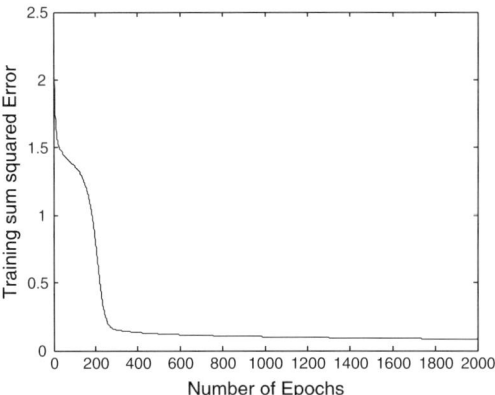

Fig. 5.20. Training results of GN model

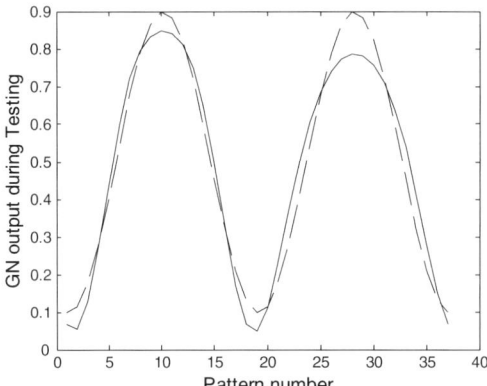

Fig. 5.21. Testing results of GN model

Table 5.11. Testing performance of ANN and GNM SIN (X1) * SIN (X2) problem

Models	Errors (After 10,000 training epochs)		
	RMS error	MAX error	MIN error
ANN	0.36632	0.71788	−0.85541
GN model	0.04011	0.04442	−0.06601

particular output value corresponding to each set of bit. The training set for coding problem is given below in Table 5.12. The training patterns are normalized in the range 0.1–0.9 and then presented to ANN and GN model for training. Once the model is trained then it can be used for testing. The training and testing results are shown in Figs. 5.22–5.23 and Tables 5.13 and 5.14.

Table 5.12. Training patterns for coding problem

Input pattern	Output
1 1 1 1	1.5
1 1 1 0	1.4
1 1 0 1	1.3
1 1 0 0	1.2
1 0 1 1	1.1
1 0 1 0	1.0
1 0 0 1	0.9
1 0 0 0	0.8
0 1 1 1	0.7
0 1 1 0	0.6
0 1 0 1	0.5
0 1 0 0	0.4
0 0 1 1	0.3
0 0 1 0	0.2
0 0 0 1	0.1

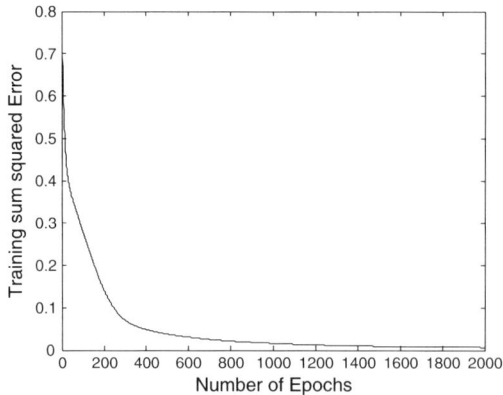

Fig. 5.22. Performance of GN model during training

For the GN model following points are important to discuss:

1. The GN input must be normalized in the appropriate range. Most of the time 0.1–0.9 normalization range works very well. If the input is slightly outside to the normalization range then there is a margin of 0.1 on both sides, so the GN model could give appropriate results. If normalization range is between 0 and 1, then for 0 inputs the output of product part of GN model is zero. Hence this normalization range in not suitable for GN model.

2. Learning rate and momentum factor should be decided in such a way that GN model learns faster and give stable response. One can start with very low value of learning rate and higher value of momentum factor. The

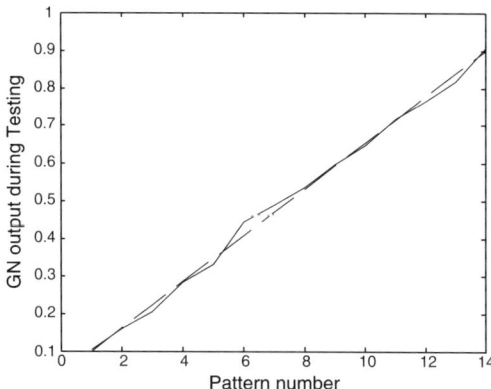

Fig. 5.23. Performance of GN model during testing

Table 5.13. Training performance of ANN and GN model for coding problem

Models	Structure	Epochs
ANN	4-4-1	38,430
GN model	Single neuron	689

Table 5.14. Testing performance of ANN and GN model for coding problem

Model	RMS error	MAX error	MIN error
ANN	0.36543	0.79301	−0.55296
GN model	0.00079	0.0018	−0.00111

typical values are – Learning rate = 0.001 and momentum factor = 0.1 to start with GN model. Then it could be increased.

3. For better generalization capability of GN model little amount of noise may be included in the training data (e.g. 0–5%).

D. 3-D Surfaces for different types of neurons

A simple matlab program is written to draw the 3-D error surfaces for single input single output system for all four type of neurons (i.e. Σ – neuron, Π – neuron, summation type generalized neuron and product type generalized neuron) as shown in Fig. 5.24–5.27. It is seen that for the simple Σ – neuron the error surface is just a inclined plane. Hence, it can never map the non-linear function. On the other hand, in Π – neuron the 3-D surface is a curved one and it can handle non-linear problems, but the error surface suddenly changes. In case for summation or product type generalized neuron the error surfaces are curved and learn quickly.

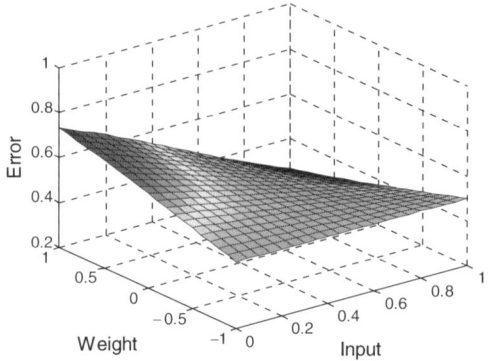

Fig. 5.24. 3-D surface for conventional Σ – neuron

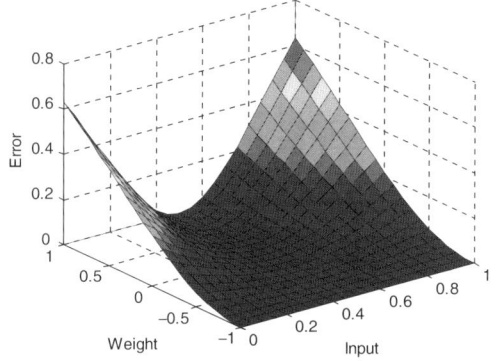

Fig. 5.25. 3-D surface for conventional Π – neuron

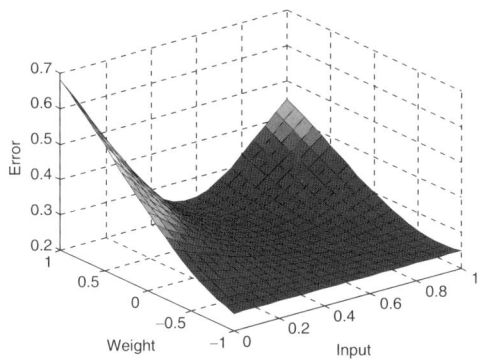

Fig. 5.26. 3-D surface for summation type GN model

% Matlab program for surface generation for different types of Neuron Models

```
clear all;
```

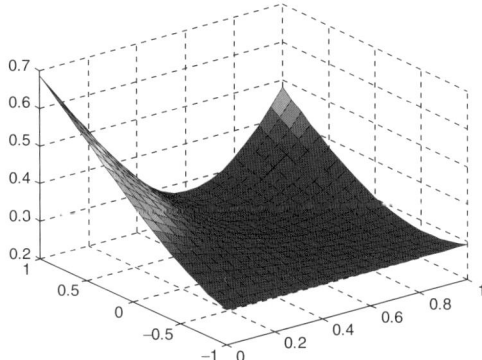

Fig. 5.27. 3-D surface for product type GN model

```
w1=[-1:.1:1];                    % Weight varaition in the range -1 to+1.
x=[0:.05:1];                     % Input variation from 0 to 1
 for i=1:21                      % Loop
 net_s=x(i)*w1;   % Calculating weighted sum for the neuron output
 net_pi=x(i)*w1;                 % Calculating the product of input
                                     and weight
 sumout=1./(1+exp(-net_s));      % Output of summation part
 piout=exp(-net_pi.^2);          % Output of product part
 ysum(i,:)=sumout;               % Output of summation neuron
 ypi(i,:)=piout;                 % output of product neuron
 ySGNM(i,:)=((sumout+piout)./2);    % output of summation type
                                       generalized neuron
 yPGNM(i,:)=(sqrt(sumout.*piout));  % output of product type
                                       generalized neuron
end
figure(1)
surf(x,w1,1-ysum);                      % Ploting the 3-D surface
xlabel('Input'); ylabel('Weight'); zlabel('Error');
title('Standard Summation Neuron');
figure(2)
surf(x,w1,1-ypi); xlabel('Input');ylabel('Weight'); zlabel('Error');
title('Standard Product Neuron');
figure(3)
surf(x,w1,1-ySGNM); xlabel('Input');ylabel('Weight'); zlabel('Error');
title('Summation type Generalized Neuron');
figure(4)
surf(x,w1,1-yPGNM);
surf(x,w1,1-ySGNM); xlabel('Input');ylabel('Weight'); zlabel('Error');
title('Product type Generalized Neuron');
```

5.6 Generalization of GN model

There are many GN models proposed based on the flexibility at both the aggregation and activation function level to cope with the non-linearity involved in the type of applications dealt with. The neuron can use "n" number of aggregation and "m" number of activation functions. The final output of the neuron is a function of output of all activation functions as shown in Fig. 5.28–5.29.

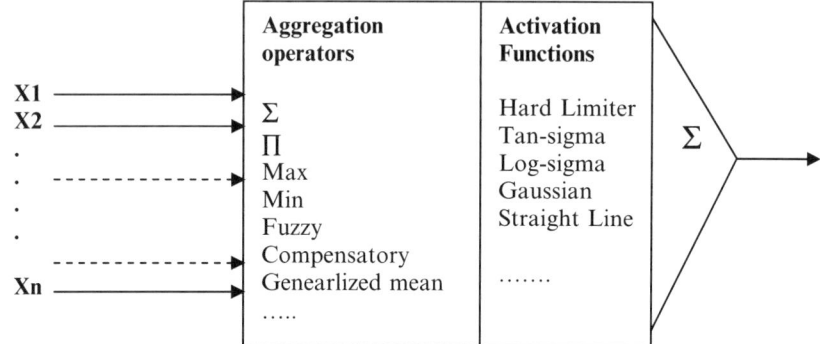

Fig. 5.28. Generalization of GN model

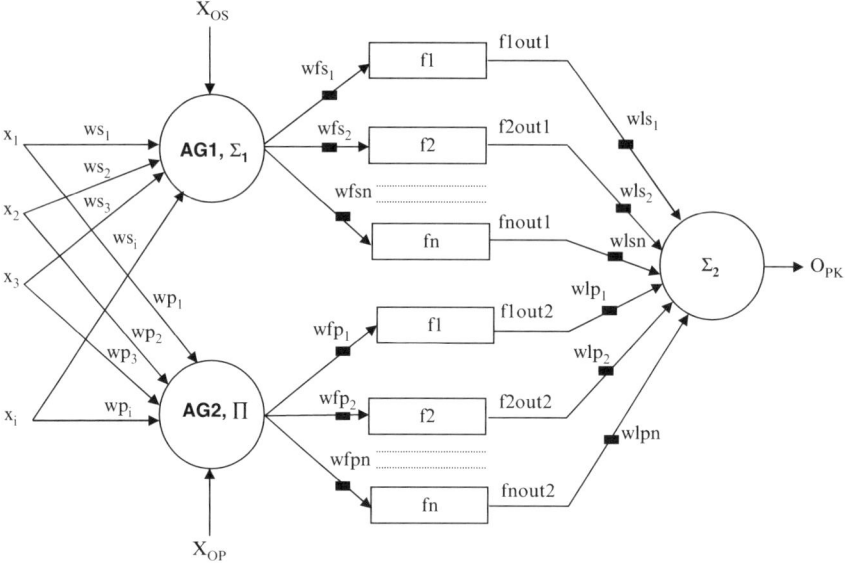

Fig. 5.29. Detailed diagram of generalization of GN model

5.6.1 GN Model-1

In this model of generalized neuron two aggregation functions (Σ and Π) and two aggregation functions (Sigmoidal and Gaussian) have been considered. Finally, the outputs are summed up to get the neuron output. The output of new neuron can be mathematically written as:

$$Opk = f1out1 * W1s1 + f1out2 * W1p1 + f2out1 * W1s2 + f2out2 * W1p2 \tag{2.22}$$

where,

$$W1p1 = (1 - W1s1),$$
$$W1p2 = (1 - W1s2)$$

Outputs of sigmoid activation functions are

$$f1out1 = \frac{1}{1 + e^{(-sumsigma^* Wfs1)}}$$
$$f1out2 = \frac{1}{1 + e^{(-product^* Wfp1)}}$$

Outputs of Gaussian activation functions are

$$f2out1 = e^{-(sumsigma^* Wfs2)^2}$$
$$f2out2 = e^{-(product^* Wfp2)^2}$$

where

$$Wfs2 = (1 - Wfs1), \;\; Wfp2 = (1 - Wfp1)$$

5.6.2 GN Model-2

In GN model-2 three activation functions sigmoid, Gaussian and straight line have been tried with two aggregation functions "Σ" and "Π". The outputs of functions used in this case of the model are given below.

Outputs of activation function for Σ part are f1out1, f2out1 are same as in Case-1.

$$f3out1 = K^* sumsigma, \quad \text{"straight line function"}$$

Output of activation functions for Π part are f1out2, f2out2 are same as in Case-1.

$$f3out2 = K^* product \quad \text{'straight line function'}$$
$$K = \text{slope of straight line}$$

Output of the neuron is

$$Opk = f1out1 * W1s1 + f1out2 * W1p1 + f2out1 * W1s2 + f2out2 * W1p2$$
$$+ f3out1 * W1s3 + f3out2 * W1p3 \tag{2.23}$$

5.6.3 GN Model-3

In this case of new GN model-3 also three activation functions (sigmoid, Gaussian and straight line) have been tried with two aggregation functions Σ and Π. Three output weights (wl1, wl2, wl3) are independent for activation functions, the other three dependent output weights are taken as given below:

$$wlp1 = 1 - wls1, \quad wlp2 = 1 - wls2, \quad wlp3 = 1 - wls3.$$

Output of the neuron is

$$
\begin{aligned}
Opk = {} & f1out1 * W1s1 + f1out2 * (1 - W1s1) + f2out1 * W1s2 \\
& + f2out2 * (1 - W1s2) + f3out1 * W1s3 + f3out2 * (1 - W1s3)
\end{aligned}
\tag{2.24}
$$

5.6.4 GN Model-4

In this case of new GN model-4 four activation functions namely sigmoid, Gaussian, straight line and sinusoidal have been tried with two aggregation functions Σ and Π. Output of the neuron is

$$
\begin{aligned}
Opk = {} & f1out1 * W1s1 + f1out2 * Wlp1 + f2out1 * W1s2 + f2out2 * Wlp2 \\
& + f3out1 * W1s3 + f3out2 * Wlp3 + f4out1 * W1s4 + f4out2 * Wlp4
\end{aligned}
\tag{2.25}
$$

where $W1p1 = (1 - W1s1)$, $W1p2 = (1 - W1s2)$, $W1p3 = (1 - W1s3)$, $W1p4 = (1 - W1s4)$

The above mentioned GN models have been tested for different benchmarks problems and compared with ANN with the parameters shown in Table 5.15. The ANN (4-4-4-1) and GNM both have been trained using gradient descent back-propagation learning algorithm for 0.002 error tolerance with same training parameters. The results are as given in Tables 5.16 and 5.17.

3-D surfaces for GN models 1–4 are given in Figs. 5.30–5.33.

Table 5.15. Neural network parameters for ANN and GNM

Learning rate	–	0.0001
Momentum	–	0.9
Gain scale factor	–	1.0
Tolerance	–	0.002
All initial weights	–	0.95

Table 5.16. Training epochs of ANN AND GN models when 5% noise is included in data

Problems	ANN	GNM-1	GNM-2	GNM-3	GNM-4
Ex-OR	150,680	50,435	3,452	7,967	3,470
4-bit parity	255,430	120,423	9,998	26,335	10,122
Mackey-Glass time series	30,340	183	131	150	162
Character recognition	70,945	3,037	2,358	268	241
Sin(x1) Sin(x2)	50,870	764	242	315	229
Coding problem	38,-30	689	7,478	421	223

Table 5.17. Testing performance (rms error) of ANN and GNM For benchmark problem with 5% noise in testing data

Problems	ANN	GNM-1	GNM-2	GNM-3	GNM-4
Ex-OR	0.63673	0.48671	0.00261	0.00627	0.00292
4-bit parity	0.46543	0.12257	3.02×10^{-10}	1.18×10^{-8}	1.09×10^{-9}
Mackey-Glass time series	0.27301	0.24426	0.02058	0.02654	0.02426
Character recognition	0.39375	0.13432	0.15084	0.03376	0.03201
Sin(x1) Sin(x2)	0.36632	0.04011	0.01841	0.03941	0.00613
Coding problem	0.36543	0.00079	3.74×10^{-6}	9.91×10^{-2}	2.21×10^{-12}

5.7 Discussion on Benchmark Testing

Training performances of ANN and GN Models during training on various benchmark problems are discussed in this chapter. Convergence in GN model is much faster as compared to ANN. For Ex-OR, 4-bit Parity and Mackey–Glass problems GN Model-2 requires only 3,452, 9,998 and 131 epochs, however, ANN requires 150,680, 255,430 and 30,340 epochs, respectively, to achieve same tolerance (0.002) in output. In character recognition, sin (x1)∗ sin (x2) and coding problems GN Model - 4 requires 241, 229 and 223 epochs only, which are much less as compared to epochs 70,945, 50,870 and 38,430 required by ANN to achieve same tolerance level in output.

This shows that GN models have better training performance than feed-forward commonly used ANN. It is also observed that the performance of GN model not only depends on type of problem but also depends on type and number of activation and aggregation functions used. Apart from this, the above model gives good performance over 5% noise in testing input data as shown earlier.

In the GN model structural complexity is very less as compared to ANN. Comparison of structural complexity associated with ANN and GN Model is represented in Table 5.18. Four layered ANN with 13 neurons and 52 interconnections with 13 number of biases uses for modeling of benchmark problems, however GN Model - 4 uses only 1 neuron with 24 number of interconnections and 2 biases. Further, ANN uses 13 activation functions for all its neurons;

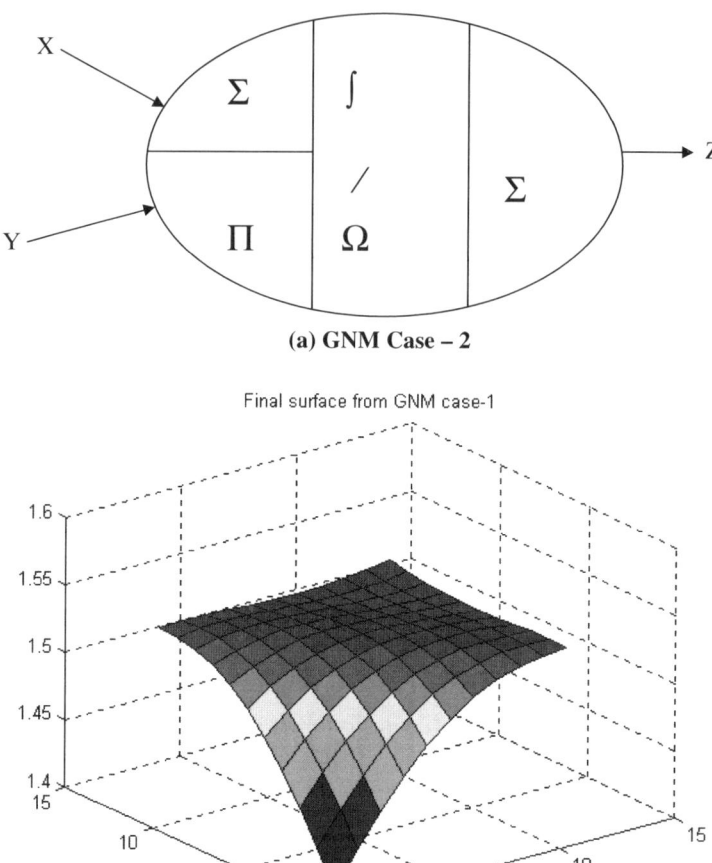

Fig. 5.30. (a) GNM Case-2. (b) Final output surface obtained from GNM-1

however, GN model -4 uses maximum eight activation functions. This shows that ANN requires more complex structure as compared to GN model to model a problem.

The computational complexity of ANN and GN models are also represented in Table 5.19. In one stroke, ANN requires 91 total number of operations, however GN Model requires only 31 operations as in the GNM case-4. It means structural complexity as well as computational complexity involved both are reduced in GN model as compared to ANN. Further, the computation time required on PC – Pentium III in one stroke is also less, i.e. 125.5 ms in case of GN model; however it is 808 ms for ANN.

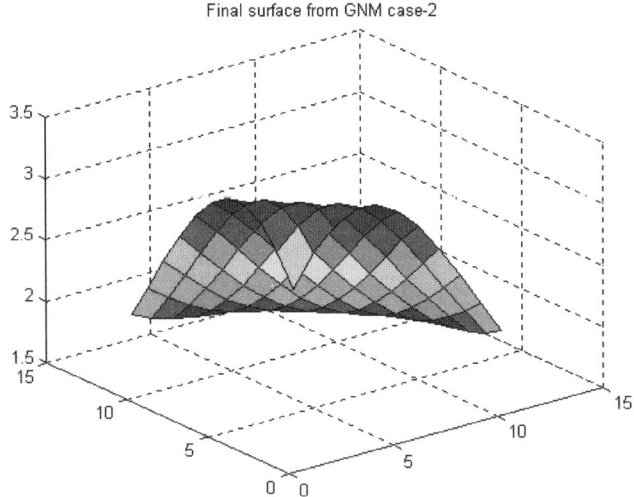

Fig. 5.31. Final output surface obtained from GNM-2

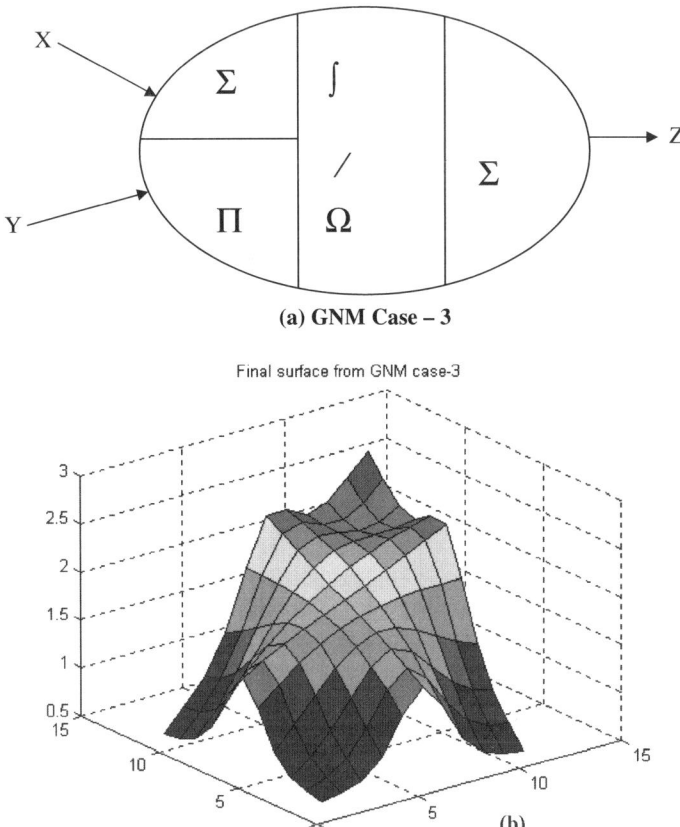

Fig. 5.32. (**a**) GNM Case-3 (**b**) Final output surface obtained from GNM-3

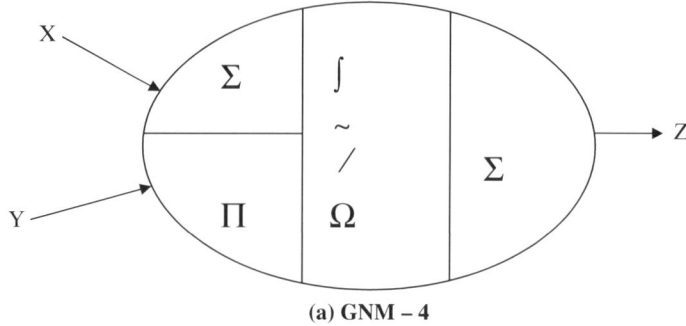

(a) GNM – 4

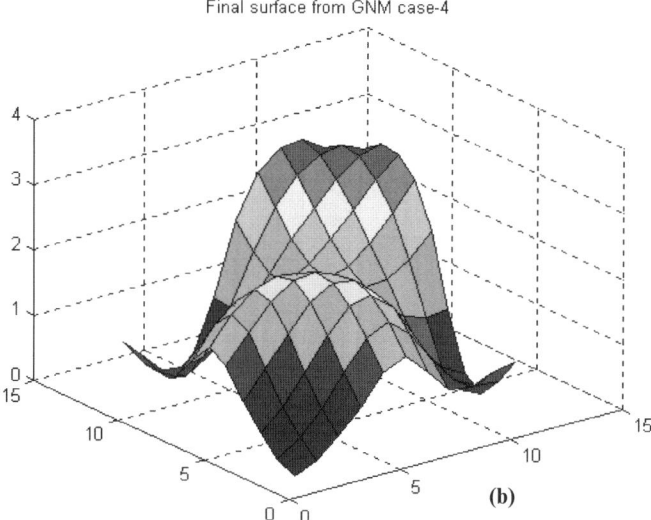

Fig. 5.33. (a) GNM-4 (b) Final output surface obtained from GNM-4

Table 5.18. Comparison of network complexity involved in ANN and GNM

Components	ANN	GN model-1	GN model-2	GN model-3	GN model-4
Number of neurons used	13	1	1	1	1
Number of layers in network	4	1	1	1	1
Number of interconnections in network	52	16	20	20	24
Number of biases	13	02	02	02	02
Number of aggregation functions	13	03	03	03	03
Number of activation functions used	13	4	6	6	8

Table 5.19. Comparison of ANN and GN model in one stroke for Mackey–Glass problem

Operations	ANN	GN Model-1	GN Model-2	GN Model-3	GN Model-4
No. of summations	13	02	02	02	02
No. of product	52	17	21	21	25
No. of divisions	13	02	02	02	02
No. of exponential functions computed	13	04	04	04	04
No. of sin functions computed	–	–	–	–	02
Total number of operations	91	25	29	29	31
Time consumed in one stroke (ms)	808	116.3	120.3	121.5	125.5

5.8 Summary

1. The training time of neural network models is a function of type and number of aggregation activation functions used and configuration among different functions. More number of configurations is possible in GN model as compared to ANN because of large number of aggregation and activation functions used, which is helpful to reduce the training time.
2. Among all models GN Model-4 suits best for character recognition, $Sin(x1) * Sin(x2)$ and coding problems taking minimum training epochs, however for Ex-OR, Parity–4 and Mackey-Glass time series problem GN Model-2 requires minimum training epochs to reach same tolerance level. It shows that GN Model has flexibility to select proper configuration of itself according to problem in hand.
3. The results reveal that convergence capability of GN Model is very good for all benchmark problems.
4. The requirement of the total number of neurons and hidden layers is reduced drastically in case of the GN models.
5. The GN model exhibits much superior property both in terms of convergence time during training as well as prediction error during testing.
6. The performance of generalized neuron model is better compared to ANN with noisy data also.
7. The structural complexity as well as computational complexity in GN Model is reduced as compared to ANN.
8. The computation time in seconds has also been reduced in GN Model as compared to ANN.

5.9 Exercises

1. Explain various factors on which ANN performance depends
2. Write a matlab program for one hidden layer ANN and study the effect of variation of hidden neurons.
3. What do you mean by over fitting of neural network? How this problem may be overcome?
4. Use generalized mean as aggregation function of GN and sigmoidal as activation function. Write a Matlab programfor this GN and train it with back-propagation algorithm.
5. Test the above developed GN for benchmark problems.
6. Write a Matlab program for GN training with adaptive backpropagation learning by varying learning rate and momentum factor.
7. Study the effect of noise on training and testing performance of GN.
8. Write step by step solution for training of GN with Gaussian as aggregation function.
9. Let us consider a function $f(x) = x^{2*}e^{-5X}$

Deteremine 100 training and 25 testing patterns for ANN and GN. Compare the performance of ANN and GN while training and testing. Also compare the performance of both if 5% random noise is added in training.

6

Applications of Generalized Neuron Models

In the earlier chapter, the development details of GN models have been studied. GN models have also been tested on benchmark problems. It is found that the GN models are much better than multilayered ANN in all the benchmark problems, which encouraged to use for different problems like modeling and simulation of electrical machines, short term electrical load forecasting and various control applications.

6.1 Application of GN Models to Electrical Machine Modeling

Conventional models of rotating electrical machines give satisfactory results over only certain ranges, as they fail to deal with the non-linear behavior of the components involved in such systems. The neural network models of these machines can deal with such problems.

The modeling of DC motor, induction motor and synchronous machine have been done using artificial neural network with the new neuron models and the results have been compared with the existing back-propagation multilayered ANN model. The training data for DC motor, induction motor and synchronous generator have been generated using system dynamic models (Chaturvedi 1992, 1994; Chaturvedi and Satsangi 1993) of these machines using DYNAMO software. The ANN models were trained and tested over a wide range of data. The models have been developed and implemented. The results and inferences presented for the following GN models.

6.1.1 GN Models

1. *Model-0*

   In this model both $\sum$ as well as Π have been taken as the aggregation functions and the output of these aggregation functions have been passed

D.K. Chaturvedi: *Soft Computing Techniques and its Applications in Electrical Engineering*, Studies in Computational Intelligence (SCI) **103**, 123–221 (2008)
www.springerlink.com   © Springer-Verlag Berlin Heidelberg 2008

through the sigmoid and Gaussian functions, respectively. Finally, the outputs are summed up to get the neuron output. This type of summation type GN model structure has been shown in Fig. 5.7a. The output of the neuron can be mathematically written as:

$$O_{pk} = O_{\Sigma}{}^{*}W_{\Sigma} + O_{\Pi}W_{\Pi} \tag{6.1}$$

2. *Model-1*

This model is similar to the above mentioned model-0. The only difference is that in this model the weight associated with the output of the product aggregation function when passed through the Gaussian function is $(1-Ws)$. Hence the output of the neuron becomes:

$$O_{pk} = O_{\Sigma}{}^{*}W_{\Sigma} + O_{\Pi}(1 - W_{\Sigma}) \tag{6.2}$$

The above mentioned neuron model is known as summation type compensatory neurons model, since the outputs of the sigmoid and Gaussian functions have been added up.

3. *Model-2*

The neuron model-2 is not a summation neuron model but a product type compensatory neuron model. The output of the sigmoid and Gaussian functions have been multiplied after being exponentiated to the powers W_{Σ} and $(1-W_{\Sigma})$. The output of this product type GN model is in the form of product as given below and shown in Fig. 5.7b:

$$O_{pk} = O_{\Sigma}{}^{W\Sigma}{}^{*}O_{\Pi}{}^{(1-W\Sigma)} \tag{6.4}$$

4. *Model-3*

This neuron model has a complicated aggregation function which is neither a summation function nor a product function alone but a summation type compensatory aggregation function. The output of this neuron model is:

$$O_{pk} = (O_{\Sigma} - O_{\Pi} - O_{\Sigma}O_{\Pi})^{*}W_{\Sigma} - O_{\Sigma}W_{\Pi} \tag{6.5}$$

5. *Model-4*

This model is similar to the above model but the output of the compensatory neuron is in the product form. As given below

$$O_{pk} = (O_{\Sigma} - O_{\Pi} - O_{\Sigma}O_{\Pi})^{*W\Sigma}{}^{*}O_{\Sigma}{}^{W\Pi} \tag{6.6}$$

6. *Model-5*

This is also a summation type compensatory neuron model, however it uses the arithmetic and geometric means of the output of sigma and the product aggregation functions. The neuron output is as follows:

$$O_{pk} = (O_{\Sigma} + O_{\Pi})(1 - W_{\Sigma})/2 + \sqrt{(O_{\Sigma}O_{\Pi})}W_{\Sigma} \tag{6.7}$$

7. *Model-6*

This model is similar to the above mentioned as follows:

$$O_{pk} = (1/2)(O_{\Sigma} + O_{\Pi})^{(1-W\Sigma)*}\sqrt{(O_{\Sigma}O_{\Pi})}{}^{W}{}_{\Sigma} \tag{6.8}$$

6.1.2 Results

The results for modeling and simulation of induction motor, synchronous and DC machines using ANN model and GN model-0 to GN model-6 are shown in Figs. 6.1–6.3. The figures clearly show that the mappings using the new models are far closer to the actual values as compared to the ordinary back propagation model. The performance of the proposed neuron models and the existing model for the simulation of aforesaid rotating electrical machine characteristics with regard to the number of epochs, rms, min and max errors are summarized in Tables 6.1–6.3, respectively. These simulation results related with the precise input data (i.e. there is no noise in the input data). The following inferences can be drawn from these results.

6.1.3 Discussions

The mapping is much closer to the actual values with the GN models as seen in Figs. 6.1–6.3 which show the speed torque characteristics of induction motors and armature current characteristics of D.C. Motor. The existing model deviates from actual values much more than the proposed models. From the tables it is quite clear that the rms error in GN model mode is minimum as compared to other models as shown in Tables 6.1–6.3. But for synchronous machine power angle characteristics the neuron model-4 is best to predict the results for unforeseen data. For all the above simulation learning rate was 0.4, momentum scale factor 0.6 and error tolerance was 0.005.

6.1.4 Training Time and Data Required

The training time needed reduces drastically with reduction in the number of neurons and hidden layers in the proposed models. The time for convergence is lesser for the new models as given in Table 4.3, the number of epochs required to train an existing summation neuron model for induction motor characteristics is 1,150 whereas for model-1 number of epochs required are 60. Model-1 gives consistently good results with respect to the epochs required in training for DC motor, induction motor, and synchronous machine characteristics. The data required for the training of networks using new neuron models is nearly half as compared to the existing model.

6.1.5 Fault Tolerant Capabilities

The proposed neuron models are better in terms of fault tolerant capabilities. The proposed neuron models have been tested with the noisy data (5% noise is added in the data) and compared with the existing neuron model. The results obtained for induction motor, synchronous machine, and DC machine are summarized in Tables 6.4–6.6, respectively. The results obtained for various rotating electrical machines show that the proposed generalized neuron

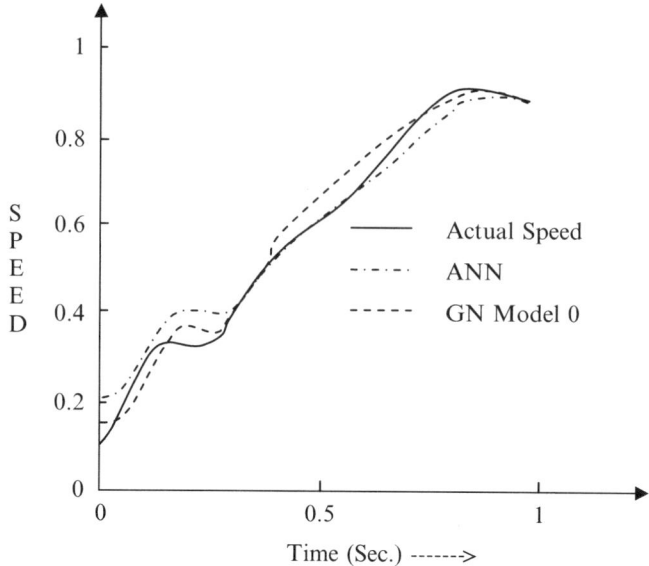

(a) Comparison of GN model 0 and ANN model with actual speed

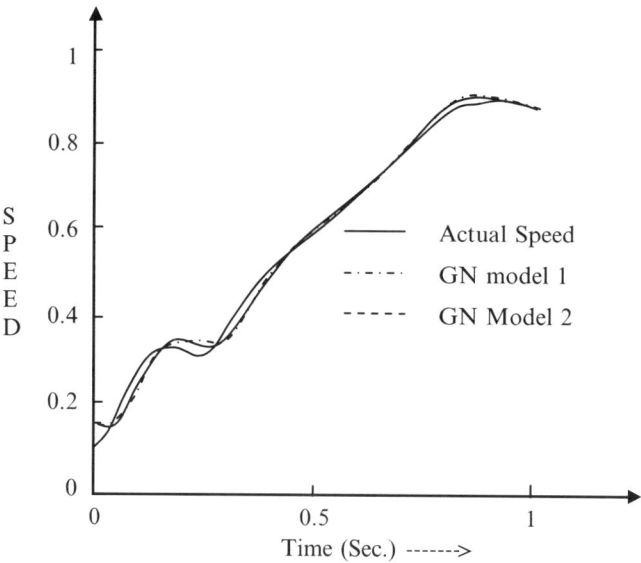

(b) Comparison of GN model 1 and model 2 with actual speed

Fig. 6.1. Simulation results of induction motor using ANN and GN models 0–6

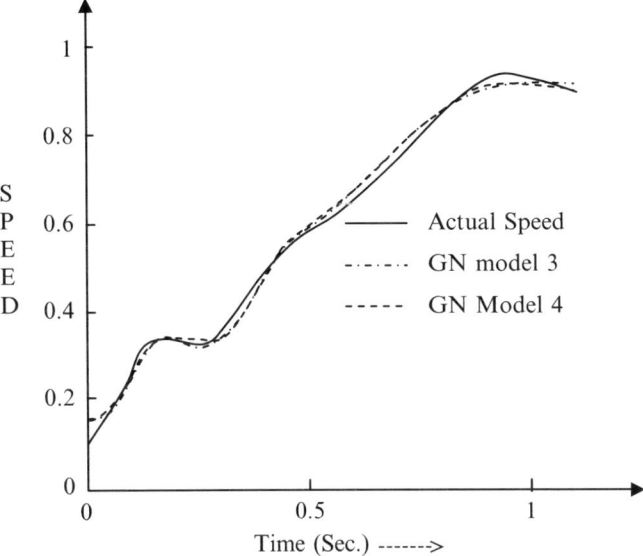

(c) Comparison of GN model 3 and model 4 with actual speed

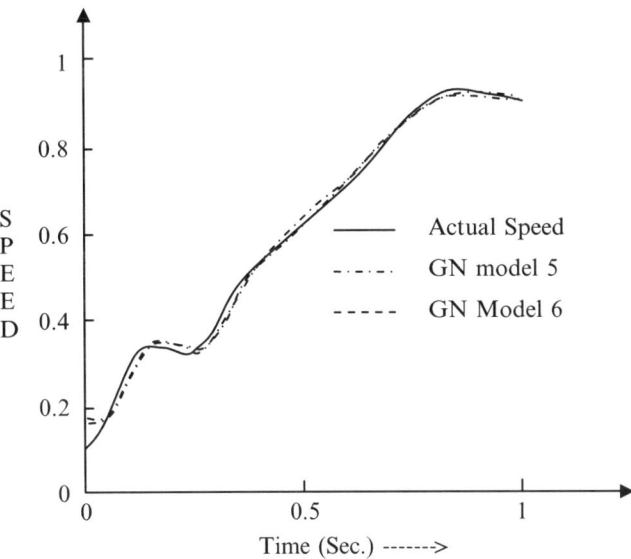

(d) Comparison of GN model 5 and model 6 with actual speed

Fig. 6.1. (*Continued*)

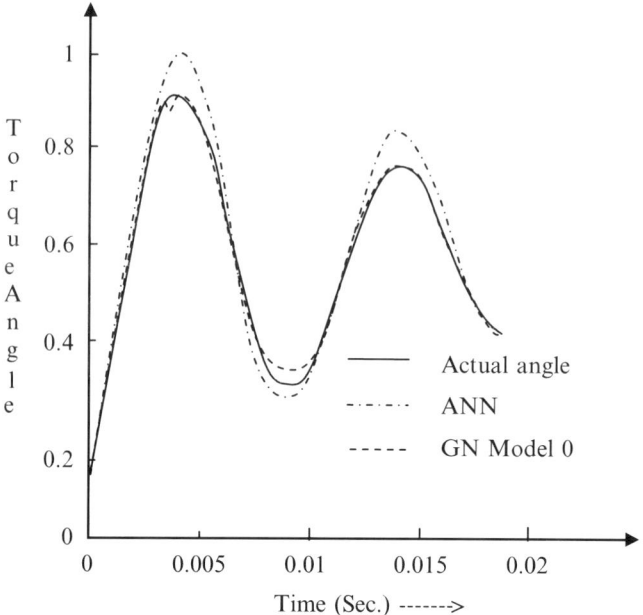

(a) Comparison of ANN model and GN model 0 with actual torque angle

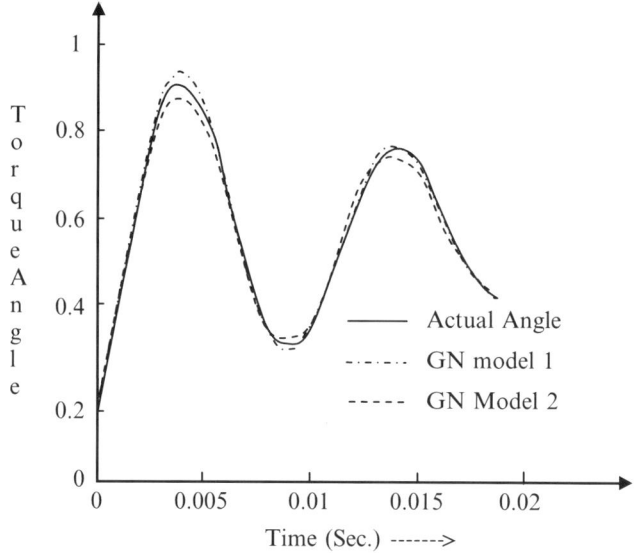

(b) Comparison of GN model 1 and 2 with actual torque angle

Fig. 6.2. Simulation results of synchronous machine using ANN and GN models 0–6

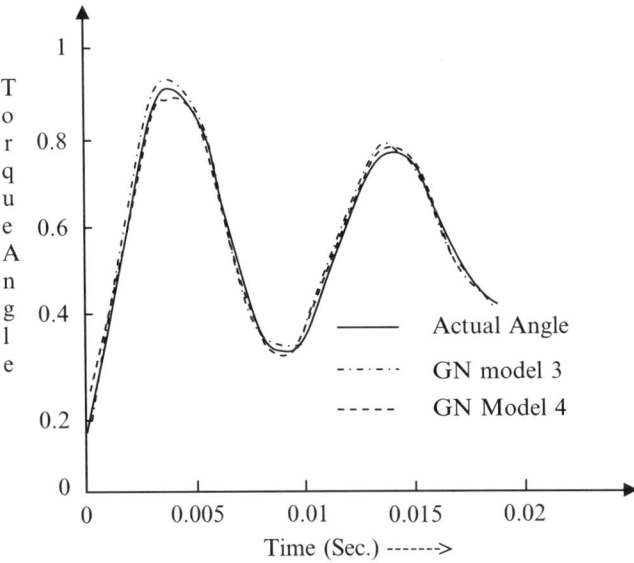

(c) Comparison of GN models 3 and 4 with actual torque angle

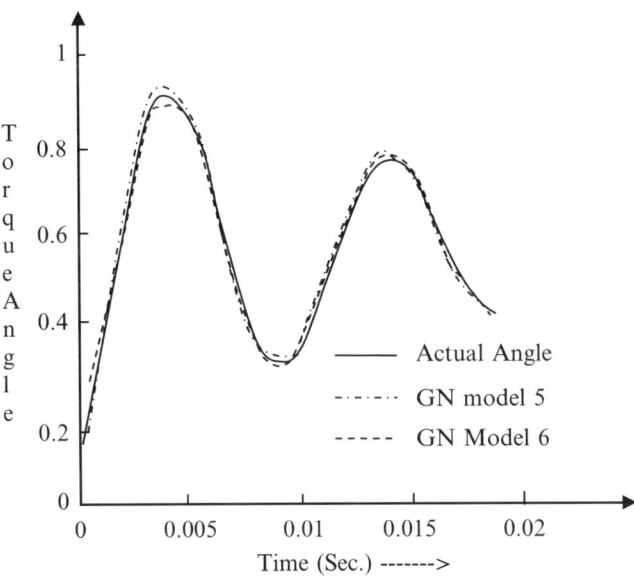

(d) Comparison of GN model 5 and 6 with actual torque angle

Fig. 6.2. (*Continued*)

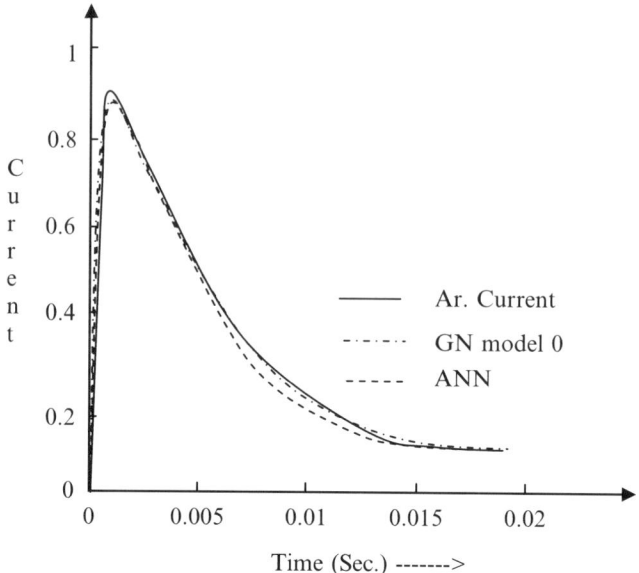

(a) Comparison of GN model 0 and ANN model with armature current

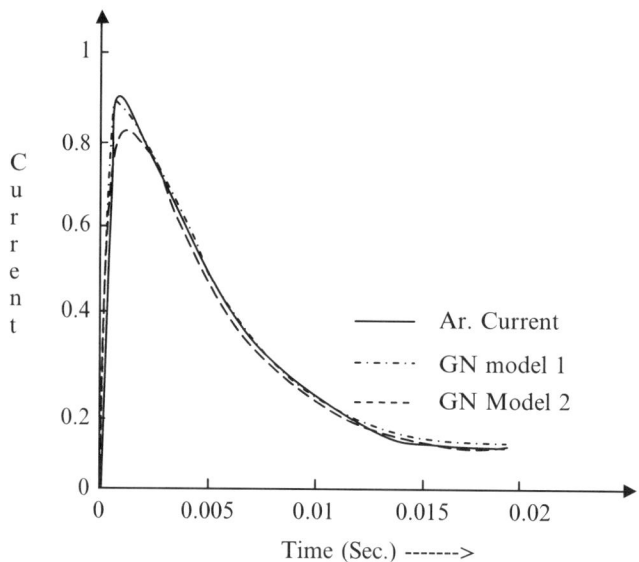

(b) Comparison of GN model 1 and 2 output with motor armature current

Fig. 6.3. Simulation results of DC motor armature current using ANN and GN models 0–6

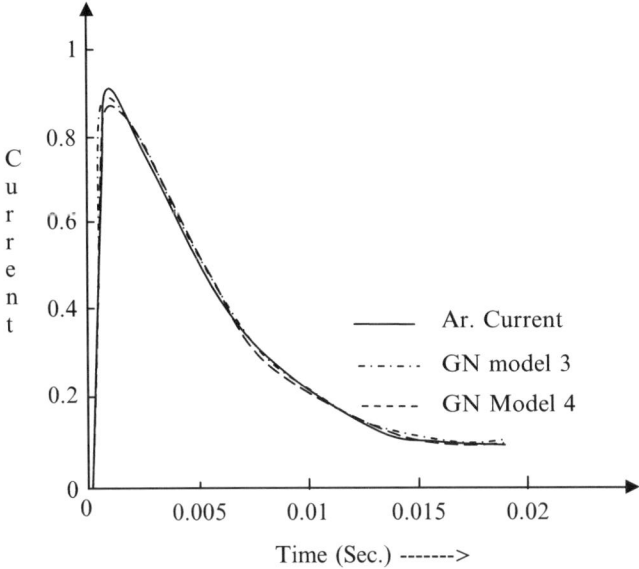

(c) Comparison of GN model 3 and 4 output with motor armature current

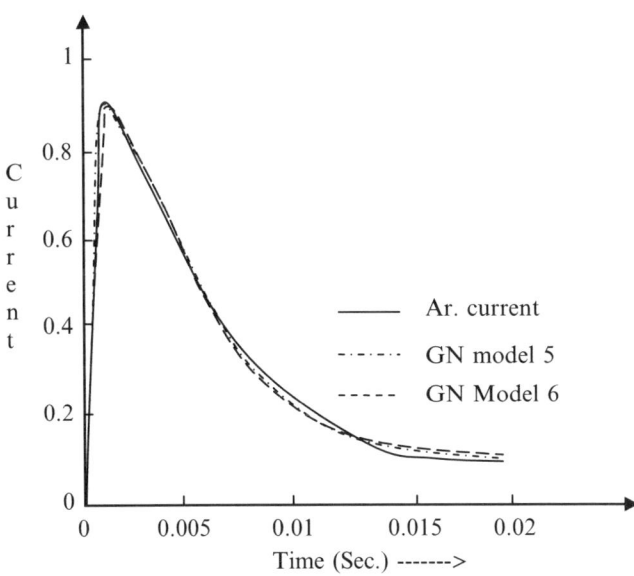

(d) Comparison of GN model 5 and 6 output with motor armature current

Fig. 6.3. (*Continued*)

Table 6.1. Comparison of different GN models for modeling of induction motor starting characteristics

Models	Training epochs	Testing error		
		Rms	Min	Max
ANN	1150	0.025682	0.000376	0.888082
GN model-0	220	0.019238	0.000222	0.747365
GN model-1	60	0.023029	0.000291	0.952220
GN model-2	117	0.021297	0.001609	0.891138
GN model-3	850	0.012626	0.000543	0.614925
GN model-4	900	0.014925	0.000364	0.630239
GN model-5	200	0.020876	0.000156	0.871928
GN model-6	300	0.019674	0.000086	0.817356

Table 6.2. Comparison of different GN models for modeling of synchronous machine P-δ characteristics

Models	Training epochs	Testing error		
		Rms	Min	Max
ANN	1800	0.010960	0.000751	0.096844
GN model-0	60	0.003394	0.000311	0.042720
GN model-1	45	0.007196	0.000851	0.096744
GN model-2	50	0.003355	0.000163	0.068143
GN model-3	250	0.002156	0.000052	0.050060
GN model-4	1585	0.002116	0.000146	0.046841
GN model-5	80	0.002814	0.000003	0.060138
GN model-6	295	0.002146	0.000150	0.047878

Table 6.3. Comparison of different GN models for modeling of D.C. Motor Ia – Time characteristics

Models	Training epochs	Testing error		
		Rms	Min	Max
ANN	515	0.03830	0.031818	0.813412
GN model-0	490	0.066047	0.034573	0.837133
GN model-1	55	0.059934	0.015513	0.712920
GN model-2	90	0.050138	0.030655	0.666004
GN model-3	320	0.047194	0.027005	0.687586
GN model-4	220	0.049158	0.028547	0.708765
GN model-5	170	0.053734	0.034077	0.671988
GN model-6	325	0.062358	0.062358	0.693487

models are also fault tolerant. In case of induction machine the model-3 gives least rms error during testing when the input set is noisy. Similarly, for synchronous machine and DC machine model-3 and model-5 give least rms error, respectively.

Table 6.4. Comparison of different neuron models for predicting induction motor characteristics (with 5% noise in input data)

Model	RMS error	MIN. error	MAX. error
ANN	0.050128	0.003176	0.088808
GN model-0	0.049461	0.008725	0.075220
GN model-1	0.045468	0.013412	0.071863
GN model-2	0.045618	0.010788	0.078343
GN model-3	0.043670	0.003191	0.070937
GN model-4	0.047334	0.010546	0.097423
GN model-5	0.048881	0.000702	0.106806
GN model-6	0.049498	0.000684	0.098497

Table 6.5. Comparison of different neuron models for predicting synchronous machine characteristics (with 5% noise in input data)

Model	RMS error	MIN. error	MAX. error
ANN	0.048658	0.003854	0.08474
GN model-0	0.038658	0.002854	0.08474
GN model-1	0.037912	0.001002	0.07591
GN model-2	0.029735	0.001420	0.07381
GN model-3	0.023402	0.004211	0.037016
GN model-4	0.025421	0.001427	0.039361
GN model-5	0.024175	0.000367	0.065905
GN model-6	0.024128	0.002098	0.038062

Table 6.6. Comparison of different neuron models for predicting dc machine characteristics (with 5% noise in input data)

Model	RMS error	MIN. error	MAX. error
ANN	0.087643	0.002818	0.0251999
GN model-0	0.082534	0.001813	0.0251999
GN model-1	0.084161	0.002746	0.0251999
GN model-2	0.082962	0.003718	0.0251999
GN model-3	0.085755	0.002416	0.0251999
GN model-4	0.086641	0.000712	0.0251999
GN model-5	0.082125	0.002726	0.0251999
GN model-6	0.085456	0.000464	0.0251999

6.1.6 Effect of Different Mappings on GN Models

There are four types of mappings generally used as explained in Chap. 4. The mapping with accurate data and noisy data in training are studied. Both the mappings are classified as input–output mapping, change in input–output, input-change in output and change in input and change in output. It is not possible to predict which mapping will be the best for the given problem. Suitability of the mapping for a problem in hand cannot be guaranteed until

Table 6.7. Effects of different mappings on results of DC machine predicted using GN models and ANN model

Models	Epochs for X-Y (error 0.01)	Epochs for DX-Y (error 0.01)	Error for X-DY	Epochs for X-DY	Error for DX-DY	Epochs for DX-DY
GN model-0	380	7270	0.001	1,090	0.001128	144,698
GN model-1	140	1970	0.001	10	0.001	1,790
GN model-2	159	2333	0.02921	13,605	0.029932	8,644
GN model-3	700	16,950	0.030052	21,600	0.03	29,000
GN model-4	550	8,440	0.029931	7,350	0.0299	10,000
GN model-5	300	3,250	0.001	350	0.001	17,100
GN model-6	1050	2,750	0.029935	3,800	0.02991	6,450
ANN	950	3,600	0.0299932	52,114	0.02966	25,900

Table 6.8. Effects of different mappings on results predicted using GN models and existing model of induction machine

Models	Epochs for X-Y (error 0.01)	Error for DX-Y	Epochs for DX-Y	Error for X-DY	Epochs for X-DY	Error for DX-DY	Epochs for DX-DY
GN model-0	130	0.46	6,590	0.09487	62,278	0.048415	52,734
GN model-1	710	0.381368	69,244	0.094923	58,628	0.027967	20,380
GN model-2	429	0.365593	109,461	0.131552	203,299	0.05039	13,728
GN model-3	800	0.50	21,500	0.082394	154,666	0.07	61,964
GN model-4	1,810	0.27948	149,698	0.341024	152,790	0.049549	39,594
GN model-5	400	0.37558	50,014	0.202034	34,614	0.04796	32,400
GN model-6	850	0.201545	56,444	0.084450	59,264	0.050334	22,550
ANN	400	0.097976	46,124	0.080189	69,286	0.034374	151,026

and unless it is trained with different mappings or some sound mathematical basis is available. Unfortunately the second option is not so easy as it requires sophisticated mathematical background. In the present work, the effect of different mappings (without and with noise in data) has been studied for different neuron models and the results are summerised in Tables 6.7–6.12. From Table 6.7, it is seen that the least epochs are required (i.e. 10) in training of d.c. machine characteristics upto error level 0.001 using X-Y mapping with generalized neuron model-1. But at the time of testing of these models for dc machine data, it was found that model-0 was the best one. Similarly in the mapping of induction motor and synchronous motor characteristics without noise, X–Y mapping using generalized neuron model-0 and mapping using generalized neuron model-2, respectively, require least training time, to reach upto the same error level.

In the real life situations it is not possible to get noiseless inputs, but always the inputs have some noise and that noise may change with time or the

Table 6.9. Effect of different mappings on training time of synchronous machine models

Models	Epochs for X-Y (error 0.01)	Epochs for X-DY	Error for X-DY	Epochs for DX-DY	Error for DX-DY
GN model-0	260	660	0.01	87,214	0.031559
GN model-1	120	145,336	0.025023	62874	0.001
GN model-2	117	168,648	0.026861	65350	0.026579
GN model-3	600	149,376	0.027403	104328	0.031093
GN model-4	540	100,038	0.024880	47294	0.026718
GN model-5	200	114,178	0.025319	53114	0.031767
GN model-6	650	63,614	0.026540	58064	0.030166
ANN	550	35,464	0.025341	33216	0.032361

Table 6.10. Effect of noisy mappings on results predicted using GN models and existing model of DC machine (with 5% random noise in training data and error level = 0.01)

Models	Epochs for X-Y	Epochs for Xnoise-Y	Epochs for X-Ynoise	Epochs for Xnoise-Ynoise
GN model-0	380	380	390	380
GN model-1	140	130	130	130
GN model-2	159	163	158	158
GN model-3	700	700	750	700
GN model-4	550	550	570	560
GN model-5	300	300	300	300
GN model-6	1,050	1,000	1,100	1,000
ANN	950	950	1,000	950

Table 6.11. Effect of noisy mappings on results predicted using GN models and existing model of induction machine (with 5% random noise in training data)

Models	Epochs for X-Y (error 0.01)	Epochs for Xnoise-Y (error 0.01)	Epochs for X-Ynoise (error 0.01)	Epochs for Xnoise-Ynoise (error 0.01)
GN model-0	130	140	810	790
GN model-1	710	980	810	730
GN model-2	429	755	626	500
GN model-3	800	1,250	1,400	1,200
GN model-4	1810	2,380	2700	2,420
GN model-5	400	800	650	500
GN model-6	850	1,050	1,150	1,050
ANN	400	150	950	900

Table 6.12. Effect of noisy mappings on results predicted using GN models and existing model of synchronous machine (with 5% random noise in training data and error level = 0.01)

Models	Epochs for X-Y	Epochs for Xnoise-Y	Epochs for X-Ynoise	Epochs for Xnoise-Ynoise
GN model-0	260	210	250	230
GN model-1	120	120	120	110
GN model-2	117	120	119	110
GN model-3	600	600	800	650
GN model-4	540	520	2,540	1,430
GN model-5	200	200	200	200
GN model-6	650	650	900	750
ANN	550	550	550	550

magnitude of the input. Therefore these noisy inputs in real systems will give outputs which are different from the output predicted by generalized neuron networks. Therefore, it is necessary to study the effect of noise in the input and the output on the training time and the accuracy of the results predicted by the different generalized neuron models. In this regard, all the aforesaid mappings are tried with 5% random noise in the input output and in the both input and output and the results are given in Tables 6.10–6.12. From Table 6.10 it can be inferred that the generalized neuron model-1 requires minimum number of epochs (i.e. 130) for Xnoise-Y, X-Ynoise and Xnoise-Ynoise mappings for the dc machine problem. For induction motor the model-0 with X-Y mapping and for synchronous machine Xnoise-Ynoise mapping with models 1 and 2 need least time for training when 5% random noise is mixed in both input and output, respectively.

6.1.7 Effect of Different Normalizations on GNN Models

As in the earlier Chap. 4, it is seen that the training time and the accuracy of the predicted result of existing neural networks depend upon the normalization ranges which are used for the inputs. If the normalization ranges are small then all the inputs are crowded in that range and the output of the neural network is not significantly different. On the other hand, if the normalization range is very wide, then neural network will clip the output. Hence it is very essential to study the effect of different normalization ranges on generalized neural network models. The effect different normalization ranges has been tried for the GNN models developed for the rotating electrical machines and the results are tabulated in Tables 6.13–6.15. In case of dc machine −0.9 to 0.9 normalization range needs the least training epochs (40) as given in Table 6.13. From Table 6.14, it is quite evident that the normalization range −0.1 to 0.9 with model-5 requires 300 epochs and from Table 6.15, in synchronous machine −0.9 to 0.9 normalization range with model-1 requires least epochs(only 20 epochs).

Table 6.13. Effect of normalization on training time of DC machine models

	0.1 to 1.8		0.1 to 2.7		−0.9 to −0.1		−0.9 to 0.9		−1.8 to 1.8		−2.7 to 2.7	
	Epoch	Error	Epoch	Error	Epoch	Error	Epoch	Error	Epoch	Error	Epoch	Error
GNM0	32,750	.0220	32750	.0220	490	0.01	32,750	0.8593	32,750	3.437	32,750	7.4569
GNM1	50	0.01	50	0.01	90	0.01	40	0.01	32750	0.004	2,450	2.4558
GNM2	–	–	–	–	12,801	3.69	16,893	2.563	4078	10.26	–	–
GNM3	32,750	0.73	1,300	3.69	32,750	3.59	9,750	2.572	21300	10.69	32,750	15
GNM4	24,120	0.648	16,220	3.607	8,500	3.69	32,750	2.562	23800	10.69	13,480	15.015
GNM5	–	–	1,800	0.628	450	0.01	32,750	2.570	24800	10.58	7,050	12.721
GNM6	19800	0.655	11,850	3.612	8,350	3.69	5,000	3.003	32750	12.25	32,750	15.018
ANN	65500	0.647	32,750	3.609	13,700	3.69	65,500	2.560	65500	10.68	65,500	15.014

Table 6.14. Effect of normalization on training time of induction machine models

	0.1 to 1.8		0.1 to 2.7		−0.9 to −0.1		−0.9 to 0.9		−1.8 to 1.8		−2.7 to 2.7	
	Epoch	Error	Epoch	Error	Epoch	Error	Epoch	Error	Epoch	Error	Epoch	Error
GNM0	32,750	0.01	32,750	0.02	490	0.01	32,750	0.690	32,750	2.717	32,750	6.226
GNM1	2,450	0.01	2,010	0.01	330	0.01	1,670	0.01	32,750	0.039	32,750	6.226
GNM2	14,557	0.01	–	–	32,750	2.465	32,750	0.672	32,750	2.730	–	–
GNM3	32,750	2.624	32,750	13.38	32,750	2.465	32,750	0.703	32,750	16.16	32,750	4.583
GNM4	20,850	0.01	32,750	13.37	32,750	2.465	32,750	0.673	32,750	4.582	32,750	15.73
GNM5	1,100	0.01	32,750	13.58	300	0.01	32,750	0.688	32,750	2.745	32,750	–
GNM6	32,750	2.534	32,750	13.36	32,750	2.465	32,750	0.698	32,750	4.530	32,750	15.72
ANN	32,750	2.523	32,750	13.34	32,750	2.465	32,750	0.670	32,750	4.58	32,750	15.72

Table 6.15. Effect of normalization on training time of synchronous machine models

	0.1 to 1.8		0.1 to 2.7		−0.9 to −0.1		−0.9 to 0.9		−1.8 to 1.8		−2.7 to 2.7	
	Epoch	Error	Epoch	Error	Epoch	Error	Epoch	Error	Epoch	Error	Epoch	Error
GNM0	70	0.01	32,750	0.138	32,750	0.030	32,750	0.041	32,750	3.372	32,750	8.160
GNM1	150	0.01	32,750	0.089	100	0.01	20	0.01	32,750	0.118	32,750	0.869
GNM2	101	0.01	–	–	32,750	4.985	32,750	0.853	32,750	3.561	–	–
GNM3	32,750	3.032	32,750	18.09	32,750	4.985	32,750	0.907	32,750	4.806	32,750	16.14
GNM4	32,750	2.90	32750	18.12	32750	4.985	32750	0.853	32,750	4.798	32,750	16.12
GNM5	–	–	32750	18.24	950	0.01	32750	0.889	32,750	4.545	32,750	8.04
GNM6	32750	2.876	32750	18.03	32750	4.985	32750	0.905	32,750	4.805	32,750	16.13
ANN	32750	2.859	32750	18.24	32750	4.985	32750	0.835	32,750	4.817	32,750	16.13

6.1.8 Conclusions

It is observe that the generalized neuron models are superior to the existing neuron model, in all respects, like lesser number of neurons, lesser number of weights and hidden layers, lesser training time and fewer training data required to map complicated functions using artificial neural networks. The models are tested for rotating electrical machines like dc machines, induction machines and synchronous machines and the results obtained can be compared as shown in Tables 6.16–6.18. From these tables, it can be seen that the new neuron

Table 6.16. RMS error while predicting the results of D.C. machine with different mappings

Model	X-Y	DX-Y	DX-DY	X-DY
GN model-0	0.034263	0.036162	0.012260	0.01
GN model-1	0.033124	0.052313	0.015183	0.01133
GN model-2	0.029221	0.054379	0.063174	0.063162
GN model-3	0.035681	0.036397	0.063219	0.063300
GN model-4	0.034431	0.036333	0.063170	0.063177
GN model-5	0.033932	0.045148	0.014938	0.011184
GN model-6	0.034800	0.087006	0.063157	0.063166
ANN	0.034882	0.037734	0.063209	0.06

Table 6.17. RMS error while predicting the results of induction machine with different mappings

Model	X-Y	DX-Y	DX-DY	X-DY
GN model-0	0.073581	0.212051	0.064401	0.084923
GN model-1	0.027614	0.191447	0.047364	0.084945
GN model-2	0.028087	0.202443	0.065865	0.102790
GN model-3	0.028618	0.207573	0.078158	0.170261
GN model-4	0.028348	0.166662	0.065298	0.084272
GN model-5	0.028101	0.187399	0.064304	0.143647
GN model-6	0.028425	0.133099	0.065570	0.08345
ANN	0.028411	0.190616	0.065451	0.101044

Table 6.18. RMS error while predicting the results of synchronous machine with different mappings

Model	X-Y	DX-Y	DX-DY	X-DY
GN model-0	0.049373	0.025401	0.037032	0.092530
GN model-1	0.033758	0.032966	0.06733	0.078089
GN model-2	0.028179	0.034126	0.033965	0.213363
GN model-3	0.020401	0.034515	0.036766	0.020785
GN model-4	0.020595	0.032885	0.034075	0.020839
GN model-5	0.020696	0.033167	0.037167	0.034974
GN model-6	0.020411	0.033906	0.036147	0.020916
ANN	0.043384	0.034190	0.037501	0.032162

models are very efficient and accurate. The proposed neuron models are also tested when 5% noise is there in the input data and it is found that the new neuron models are better in terms of fault tolerant capabilities. The successful results validate the approach of applying compensatory operators to neural networks. This approach may be extended to other applications of artificial intelligence like pattern recognition, financial forecasting, classification tasks, speech synthesis, and adaptive robotics control and data compression.

6.2 Electrical Load Forecasting Problem

The increase in demand of electrical energy has drawn the attention of power system engineers towards the reliable operation of power systems. For reliable operation of integrated power supply systems, a close tracking of electrical load is required. For the economy of operations, this must be accomplished over a broad spectrum of time intervals. While for a short range of seconds or even minutes, the automatic generation control function involving economic dispatch is used to ensure the matching of the load with economic allocation among the committed generation sources, the security of supply still depends on the availability of hot and cold reserves which in turn depend on the total load demand at any time. Specially for periods of hours and day where wider variation of loads occur, meeting the demand entails the start up and shut down of the entire generating unit or interchange of the power with neighbouring systems. For the preparation of the maintenance schedule of the different units and auxiliaries, it is desirable to know prior to the demand profile of important nodes of the system for wider length of time. All these necessitate an accurate forecasting of the load with reasonable degree of accuracy.

The time range, ahead of which the forecast is required, has to be viewed from the functional areas of planning, operation and management. Depending on the time range, there are three types of forecasting, e.g. Long term (a few months to a few years), medium term (one week to few months) and short term forecasting (a few minutes to a few hours). It can also be classified depending on the specific need and applicability.

Long term load forecasting (Brown 1983; Gupta 1994) is mainly concern with the generation expansion planning, transmission and distribution planning, financial planning, energy exchange policy between organizations, planning for peaking capacity and maintenance of plants. Medium term forecasting refers to economic scheduling of various energy sources, inventory control of coal and liquid fuels, reservoir utilization and water management for irrigation. It also helps in maintaining security constraints and proper planning of load shedding. Short term load forecasting helps in load management with on-line dynamic voltage control, load flow studies and exchange of power as requirement for load frequency control.

There are various methods of electrical load forecasting. Methods of forecasting vary from simplest, intuitive and naïve ones to be most sophisticated learning system models. Selection of forecasting methods is guided by the following factors (Basu 1993):

1. *Accuracy of forecast.* This is a major criterion in selecting a model for forecasting and affects the operation cost of a system.
2. *Data.* Pattern, type and length of the data decide the nature of model.
3. *Cost of computation.* This is very important for short-term load forecasting.
4. *Ease of applicability.* User must feel at ease to handle the model.
5. *Interactive facility.* Forecast procedure should be normally automatic and provision must be kept for intervention through external control.
6. *Constant monitoring facility.* for adjustments during abnormal load behaviour.
7. *Risk due to load forecasting uncertainty:* Operational risk independent of the load forecasting such as lead-time (Douglas et al. 1998a,b).

6.2.1 Litreture Review

The variation in electrical load of a power system is inherently a stochastic process. It is influenced by a number of factors. These factors are:

1. Economic factors:
   Economic factors such as changes in the farming sector, levels of industrial activity and economic trends have significant impact on the system load growth.
2. Time:
   Three principal time factors namely seasonal effects, weekly-daily cycle, legal and religious holidays play an important role in influencing load patterns.
3. Weather:
   Meteorological conditions like temperature, wind, cloud cover and humidity are responsible for significant variations in the load pattern. The weather sensitive loads are space heaters, air-conditioners and agricultural motors.
4. Random:
   The IEEE load forecasting working group (IEEE committee report 1980, 1981) has published a general philosophy of load forecasting on the economic issues in 1980–1981. Some of the older techniques stated in the literature are general exponential smoothing (Christiaanse 1971), State Space and Kalman filter (IEEE report 1981), and multiple regression (Mathewman 1968). Kalman filtering is a state space method. The difficult aspect of Kalman filtering is the selection of the process and observation noise. The stochastic time series model was also used by Hagan (1987) for short term load forecasting. Auto regressive moving average (ARMA) models have received a great deal of attention in the literature (Galiana et al. 1974). ARMA models can be used to model stationary processes

with finite variances. Non-stationary processes can be modeled by differencing the original processes. The differencing operation produces an auto regressive integrated moving average (ARIMA) models. ARMA models fall into the time series category and can be implemented in state space formulation. Generally, time series approaches assume that the load can be decomposed into two components. One is weather dependent and the other weather independent. Each component is modeled separately and the sum of these two gives the total load forecast. The behaviour of weather independent load is mostly represented as a function of time. It has been observed that above models are acceptable during normal operating conditions. However, improvement is needed in the forecast during vast and rapidly changing weather conditions (Sharma and Mahalanabis 1974; Peng et al. 1992; Douglas et al. 1998a). Rajurkar and Nissen (1985) introduced stochastic modelling and analysis methodology called data-dependent systems (DDS) for short time load forecasting (STLF), while Goh and Ong refined the approach through stochastic time series analysis so that with routinely available data from a number of key substations, the substation demand patterns are separately characterized. Jenq-Neng Hwang and Seokyong Moon (1991) discussed a power load forecasting system based on a temporal difference (TD) method.

On the other hand, heuristic approaches like expert Systems, artificial neural networks and fuzzy systems are also being used for load forecasting purpose. Also Rahaman and Batnagar (1988), Ho (1990), and Rahaman and Hazim (1993), Rahaman and Shrestha (1993) have proposed knowledge based expert system. Further, in 1991–192 Park (1991) and Peng (1992) used artificial neural network (ANN) for short term load forecasting model. The model suggested by Peng does not consider the weather dependency of the load. Kalra (1995) improved the ANN model by incorporating the weather dependent variables to predict the better results. On-line ANN model for short term load forecasting for a feeder load was suggested by Khincha and Krishnan (1996). One drawback of the ANN model is the large training time required for model development. To reduce the training time Drezga and Rahaman (1997) proposed input variable selection technique for ANN based load-forecasting model. It was also felt that in the model development using ANN, one needs accurate and sufficient data, but in short term load forecasting it is very difficult, or sometimes impossible, to obtain accurate and sufficient data, which are related with weather. Hence, it was proposed by Dash et al. (1997) a real time load forecaster using functional link neural network incorporating the non-linearity due to temperature variation.

Over the past few years, the artificial neural networks (ANN) have received a great deal of attention and are now being proposed as a powerful computational tool (Patterson 1995; Al-Shakararchi and Ghulaim 2000; Krunic and Rajakovik 2000; Sinha 2000; Shantiswarup and Satish 2002; Osowski and Siwek 2002). They have been successfully applied in pattern recognition classification and non-linear control problems. It has been demonstrated that multi-layered feed-forward back-propagation ANNs are *universal approximators* and they are able to approximate any nonlinear continuous function upto the desired level of accuracy. Efforts have also been made to improve accuracy of short-term load forecasting using the ANN technique by introducing the fuzzy concepts by Dash et al. (1993). Mandal and Agarwal (1997) has developed a Fuzzy – Neural Network for Short Term Load Forecasting considering the Network Security. In that paper, ANN creates non-linear relationship between fuzzy inputs and outputs. Daneshdoost et al. (1998) used fuzzy set technique for hourly data classification into various classes of weather conditions and then ANN model was developed. Hong Tzer Yang and Huang (1998) developed self-organizing fuzzy ARMAX model for forecasting hourly load. Chow et al. (1998) used fuzzy multi-objective decision-making approach for land use based spatial load forecasting. Further, Douglas et al. (1998a,b) considered the effect of weather forecast uncertainty in the short term load forecasting. Papadakis et al. (1998) forecasted the demand using fuzzy neural network approach on the basis of maximum and minimum load of the day. Drezga and Rahaman (1999) developed a local ANN load predictor by active selection of training data employing K-nearest neighbour concept. Nazarko and Zalewski (1999) used fuzzy regression approach for forecasting the peak load for 15 min ahead in distribution system.

Dillon et al. (1991), Ishibashi et al. (1992) and T. Matsumoto et al. (1993) presented a method of short term load forecasting using artificial neural networks. *Azzam-ul-Asar, McDonald and Khan* investigated in 1992 the effectiveness of an ANN approach to short-term load forecasting in power systems. Examples demonstrate the learning ability of an ANN in predicting the peak load of the day by using different preprocessing approaches and by exploiting different input patterns to observe the possible correlation between historical load and temperatures. In 1993, *Li Guangxi and Xiong Manli* presented a method of changing a topological ANN to forecast the load of a power system. The model is almost an all-round reflection of various factors, which affect the changing of load. *Papalexopoulos et al. (1994)* presented an ANN based model for the calculation of next day's load forecasts. The most significant aspects of the model fall into the following two areas: training process and selection of the input variables. At the same time *Lee et al. (1993)* presented a diagonal recurrent ANN with an adaptive learning rate. In 1993, *S.D. Chaudhary, P.K. Kalra, S.C. Srivastava & D.M.V. Kumar (1993)* presented a fast and accurate method of STLF using combinations of self-organising maps (SOM) and multi-layer perception model. The SOM recognizes the type of day examining the variation of load which, along with the past load, temperature, humidity, etc. *Peng et al. (1993)* used a linear adaptive neuron or adaptive linear combiner called Adaline for STLF. Hence, it is very clear that ANN is gaining momentum in load forecasting due to various reasons like ability to cope up with non-linearity, adaptivity, intelligent and simplicity. *Chaturvedi et al. (2001)* used the generalized neural network (GNN) approach for electrical STLF problem to overcome the problem of ANN. The performance of

Table 6.19. Features of conventional and ANN approach for short-term load forecasting

Features	Time-series method	Regression analysis	ANN
Load information	Required	Required	Required
Weather information	Not required	Required	Not necessarily required
Functional relationship between load and weather information	Required	Required	Not required
Complex mathematical calculations	Required	Required	Not required
Time required in prediction	More	More	Less
Adaptability	Less	Less	More

GN model has been again improved using $(Xi + Wi)^n$ instead of weighted input (Xi^*Wi), so that a closed surface may be generated depending on the requirements, by selecting the proper value of n.

The above discussion shows that artificial neural networks (ANN) offer a reasonable alternative to the classical methods of load forecasting. As far as the question of load forecasting is concerned, the main concern is to improve the accuracy of forecasting as given in Table 6.19.

6.2.2 Short Term Load Forecasting Using Generalized Neuron Model

The implementation of GN model based forecasting consists of two phases:

i. *Model development phase.* Development of the GN Model with its learning and testing – this is an offline process; and
ii. *Model testing phase.* The online process in which the trained GN Model developed in phase (1) is used for forecasting the load in daily operations. The suitable number of past data (load history) is provided to train the GN model for better predictions. Once the training is over then GN model is validated and tested for little new data.

6.2.2.1 Model Development Phase

The model development using artificial neural network consists of the following important considerations as shown in Fig. 6.4:

a. Data preparation
b. Selection of neural network structure
c. Selection of proper training algorithm

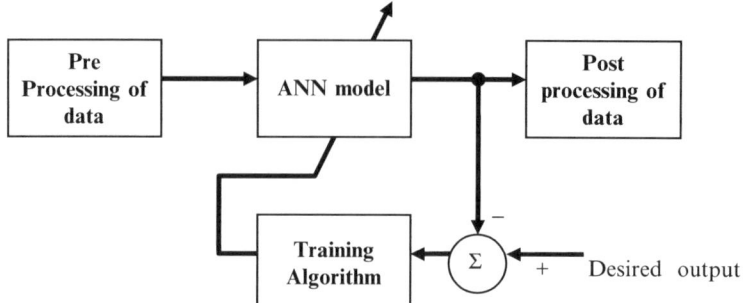

Fig. 6.4. Block diagram for model development using ANN

The data preparation consists of selection of input and output variables, collection of accurate and sufficient training data, and decision of proper normalization range for input, output data.

Selection of Input Variables

The most important work in developing an ANN or GN model based load forecasting model is the selection of input variables. The system load depends on several factors, such as weather, type of day, load of previous day at same hour, social and other activities, etc. The objective of neural network based forecaster is to recognize these factors and forecast the load accordingly. There is no general rule that can be followed in the selection of input variables. It largely depends on engineering judgment and experience. For solving a short-term load-forecasting problem, all of these inputs are not needed at the same time. Depending on the forecast to be made whether daily or hourly, the choice of input variable changes. For daily load forecasting, the time input variables like temperature, humidity, wind speed, etc. are required along with past load data. However, for hourly load forecasting, the past load data is generally sufficient as input variable. Using auto-correlation factor it has been found in the literature [56] that the hourly load has a very high correlation with the load of previous same day type at same hour and the load of previous weekdays at same hour. Therefore, for modeling and simulation of short-term load forecast problem using ANN and GN model the following four inputs are considered to get one output of next hour load as given in Table 6.20.

a. Load of previous day (of the same day type) at same hour and
b. Same hour load of three previous weeks on the same day are used.

Normalization of Input and Output Data

The input and output variables for the neural network may have very different ranges if actual hourly load data is directly used. This may cause convergence problem during the learning process. To avoid this, the input and output load

Table 6.20. Number of inputs for ANN/GN model

Input 1	–	Electrical demand of first last Monday (say), at time (t)
Input 2	–	Electrical demand of second last Monday at time (t)
Input 3	–	Electrical demand of third last Monday at time (t)
Input 4	–	Electrical demand of previous day at time (t)
Output	–	Electrical demand of fourth Monday at time (t)
		Where (t) is the time for forecasting

data are scaled such that they remain within the range (0.1–0.9). The lower limit is 0.1, so that during testing it could not go far beyond lower extreme limit, which is 0. Similarly, the upper limit is taken as 0.9, so that the data could go upto upper extreme limit, which is 1.0, in testing. These margins of 0.1 on both sides (i.e. upper and lower) are called safe margins. The actual load is scaled using the following expression for short-term load forecasting problem:

$$Ls = \frac{(Y\max - Y\min)}{(L\max - L\min)} * (L - L\min) + Y\min \qquad (6.9)$$

where

L = The actual load, MW
Ls = The scaled load which is used as input to the net
$L\max$ = the maximum load
$L\min$ = the minimum load
$Y\max$ = Upper limit (0.9) of normalization range
$Y\min$ = Lower limit (0.1) of normalization range

Selection of Neural Network Structure

The structure of artificial neural network is decided by selecting the number of hidden layers and number of neurons in hidden layer, input neurons, output neurons. Hidden layers in ANN structure create decision boundaries for the outputs. Larger the number of hidden layers (upto three), larger is the capability to create complex decision boundaries for the outputs of non-linear problems (Lee et al. 1993). Here, structure of ANN having three hidden layers is selected for the short-term load-forecasting problem. The number of input neurons is fixed with the number of input scalars in the input vector. Since, for this short-term load forecasting problem, four input vectors (variables) have been selected, therefore number of input neurons is taken four. Similarly, number of neurons in the output layers is equal to number of output scalars, so number of neuron selected in output layer is one for this problem. Regarding selection of number of neurons in a hidden layer, it has been found that with too few neurons in a hidden layer, the network is unable to create complex decision boundaries and so it will create difficulties in convergence during training (Sinha et al. 2000). Therefore, four neurons in each hidden layer have been taken, which are equal to input neurons. The selected structure of ANN is shown in Fig. 6.5.

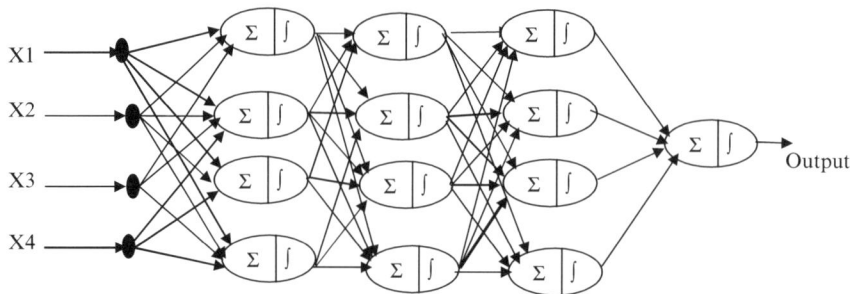

Fig. 6.5. Four-layer structure of ANN used in short-term load forecasting

Selection of Proper Training Algorithm

The training or learning algorithm, which is commonly used for feed-forward ANN, is gradient descent back-propagation algorithm. It is a popular learning algorithm for multi-layered ANN mainly because of its computation simplicity, ease in implementation and good results generally obtained for large number of problems in many different areas of application. The gradient descent back-propagation algorithm adjusts the connection weights between different neurons in proportion to the difference between the desired and computed values of each layer in ANN structure. Addition of momentum term improves the stability of the process. The weight adjustment equation of back propagation algorithm is given by

$$Wnew = Wold + \Delta W, \tag{6.10}$$

where,

$$\Delta W(k) = -\eta * \frac{\partial Ess}{\partial W} + \alpha * \Delta W(k-1)$$

η = learning rate,
α = momentum coefficient and
Ess = error function used.

The learning rate η and momentum α have very significant effect on learning speed of the back propagation algorithm. Large value of η results in faster convergence but subjects to network oscillations. Where a small value of η stabilizes the process but results in slower convergence. Similarly for higher values of momentum coefficient α connection weights are updated in correct direction and improve the convergence.

The performance of gradient descent back propagation algorithm also depends on error function used. The sum-squared error function is the most popular error function used in back propagation learning algorithm because of its computation simplicity, it is being used here for the short-term load-forecasting problem.

6.2.3 Training of ANN and GN Model

The short-term demand of Gujarat State Electricity Board has been collected and arranged in a proper format and normalized in the range of 0.1–0.9 as shown in Table 6.21.

The parameters have been taken for GN model and commonly used ANN are given in Table 6.22.

The proposed GN model and conventional ANN, with three hidden layers, have been trained using the back propagation learning algorithm for the data given in Table 6.22 with the goodness of fit represented by tolerable mean squared error equal to 0.002. Training performance of ANN and GN model are shown in Figs. 6.6a–6.11a graphically. The comparison of training and testing

Table 6.21. Normalized training data for ANN And GN model

Input 1	Input 2	Input 3	Input 4	Desired output
0.2690	0.1389	0.1461	0.1000	0.2201
0.1883	0.1768	0.1461	0.1068	0.1745
0.1526	0.1676	0.1154	0.1192	0.1248
0.1000	0.1000	0.1398	0.1733	0.1580
0.1019	0.1891	0.1000	0.1372	0.1000
0.2512	0.3622	0.2600	0.3133	0.3567
0.5845	0.6982	0.5178	0.7172	0.6320
0.8117	0.8068	0.4431	0.8425	0.6889
0.8343	0.8580	0.6142	0.8312	0.7965
0.8624	0.8283	0.7484	0.8300	0.8617
0.8709	0.7720	0.7603	0.7804	0.8700
0.7657	0.6460	0.6743	0.7375	0.7375
0.5620	0.4544	0.5576	0.5310	0.4829
0.7000	0.4636	0.6610	0.7398	0.5067
0.7150	0.5446	0.7016	0.7262	0.6019
0.7310	0.6091	0.7672	0.7702	0.6692
0.7451	0.6265	0.7421	0.7003	0.6485
0.8042	0.6449	0.7400	0.7567	0.7799
0.9000	0.8416	0.8420	0.8571	0.7965
0.8606	0.9000	0.9000	0.9000	0.8876
0.8192	0.7709	0.8497	0.8898	0.9000
0.6277	0.6306	0.7728	0.6924	0.7189
0.4962	0.5425	0.6890	0.6958	0.7220
0.3488	0.4370	0.5493	0.4870	0.5305

Table 6.22. Learning parameters

Learning rate	–	0.0001
Momentum	–	0.9
Gain scale factor	–	1.0
Tolerance	–	0.002
All initial weights	–	0.95

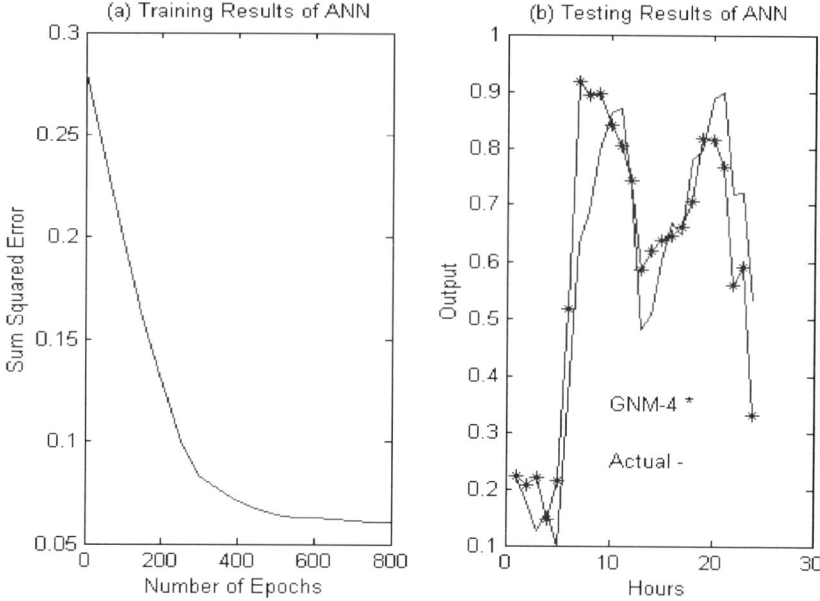

Fig. 6.6. Training and testing performance curves of ANN

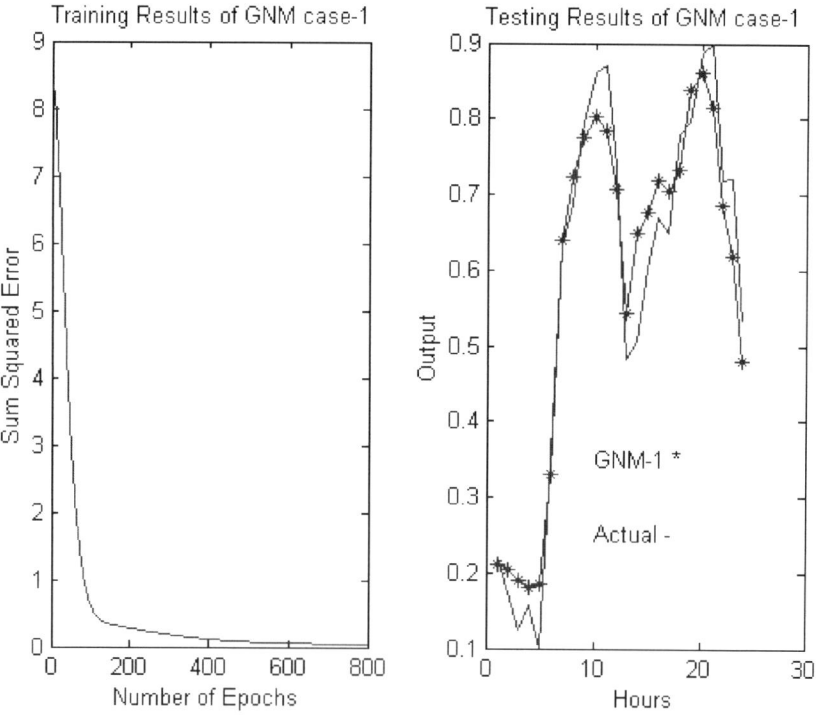

Fig. 6.7. Training and testing performance curves of GN Model-1

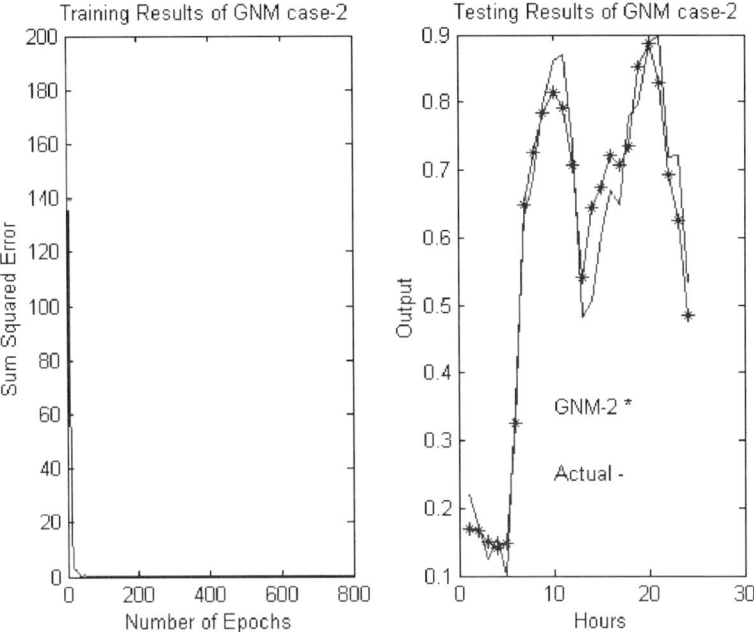

Fig. 6.8. Training and testing performance curves of GN Model-2

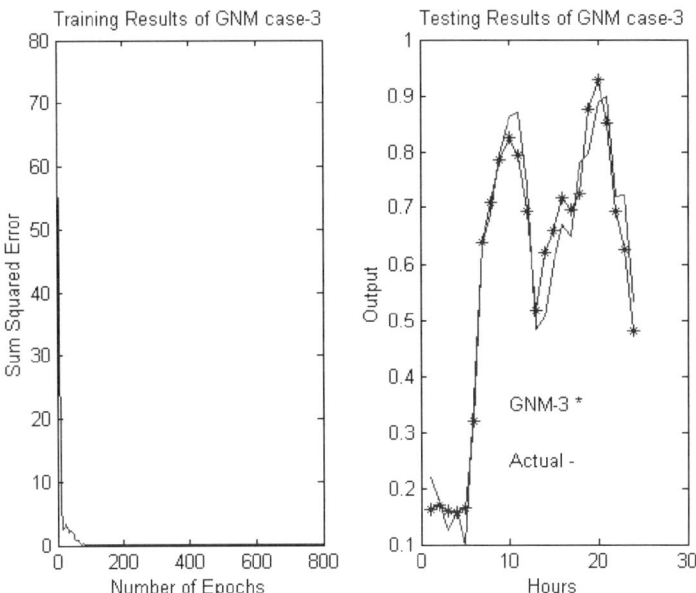

Fig. 6.9. Training and testing performance curves of GN Model-3

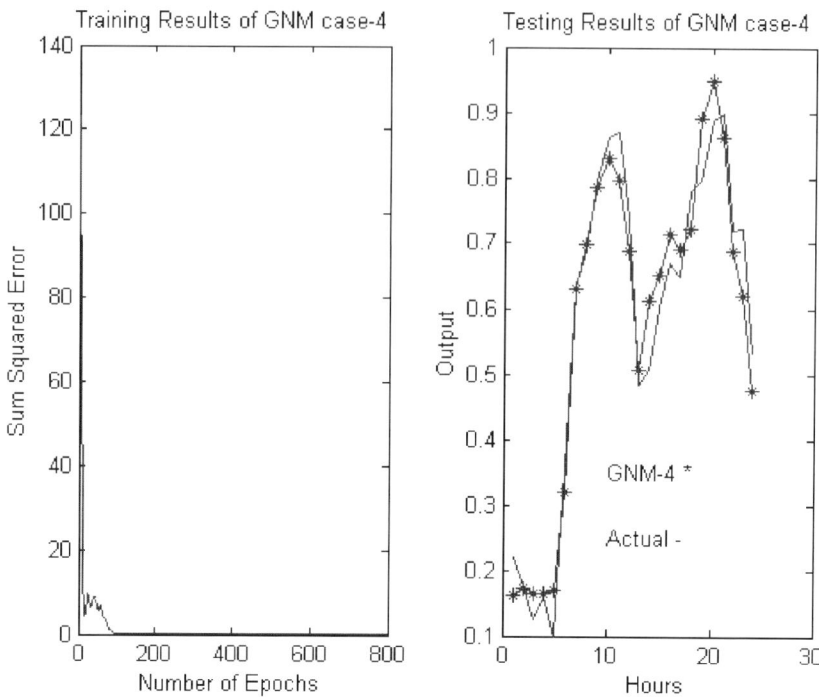

Fig. 6.10. Training and testing performance curves of GN Model-4

Table 6.23. Comparison of training performance of ANN and GN model for short term load forecasting problem (mean squared error tolerance = 0.002)

Models	Structure	Number of epochs
Existing ANN	4-4-4-1	Above 50,000
GNM-1	Single neuron	722
GNM-2	Single neuron	**152**
GNM-3	Single neuron	168
GNM-4	Single neuron	202

performance of GN Model and commonly used ANN in terms of training epochs required to train the models upto desired tolerable error is given in Table 6.23.

6.2.4 Testing of ANN and GNM

The test results of ANN and GN model are shown graphically in Figs. 6.6b–6.11b. Table 6.24 represents actual output, and predicted output of ANN and GN model. Prediction errors are shown in Table 6.25. The comparison of testing performance of ANN and GN Models in terms of RMS, maximum and minimum errors are shown in Table 6.26.

Table 6.24. Actual and predicted outputs by ANN and GN models (epochs = 5,000, learning rate = 0.0001, momentum = 0.9, for all models)

Actual output	ANN (4-4-4-1)[a]	GNM case-1 (1)[a]	GNM case-2 (1)[a]	GNM case-3 (1)[a]	GNM case-4 (1)[a]
0.2201	0.2231	0.1721	0.1667	0.1572	0.1490
0.1745	0.2063	0.1687	0.1574	0.1702	0.1703
0.1248	0.2203	0.1567	0.1355	0.1607	0.1560
0.1580	0.1448	0.1496	0.1140	0.1554	0.1377
0.1000	0.2155	0.1539	0.1281	0.1667	0.1643
0.3567	0.5183	0.3042	0.3452	0.3177	0.3367
0.6320	0.9184	0.6444	0.6593	0.6498	0.6598
0.6889	0.8928	0.7321	0.7356	0.7080	0.7030
0.7965	0.8962	0.7876	0.7817	0.7887	0.7900
0.8617	0.8398	0.8159	0.8054	0.8240	0.8232
0.8700	0.8033	0.7945	0.7862	0.7907	0.7927
0.7375	0.7447	0.7109	0.7128	0.6919	0.6923
0.4829	0.5860	0.5311	0.5685	0.5113	0.5235
0.5067	0.6200	0.6454	0.6560	0.6084	0.5939
0.6019	0.6392	0.6774	0.6826	0.6540	0.6490
0.6692	0.6448	0.7254	0.7201	0.7172	0.7121
0.6485	0.6611	0.7089	0.7098	0.6972	0.7024
0.7799	0.7070	0.7389	0.7350	0.7208	0.7182
0.7965	0.8161	0.8528	0.8401	0.8710	0.8667
0.8876	0.8155	0.8816	0.8673	0.9268	0.9214
0.9000	0.7677	0.8310	0.8134	0.8505	0.8389
0.7189	0.5581	0.6926	0.6928	0.7054	0.7154
0.7220	0.5900	0.6215	0.6329	0.6362	0.6407
0.5305	0.3291	0.4672	0.5067	0.4861	0.5071

[a]Structure of different models.

The fault tolerant capabilities of neural networks are generally tested with noisy data. The performance of ANN and GN model are tested with noisy data and the results are as shown in Table 6.27.

6.2.5 Discussion on Training and Testing Results

The training results of ANN and GN Models have been shown in Figs. 6.6a–6.10a graphically, for short-term load forecasting problem. As shown in Fig. 6.6a, during training of ANN, the reduction in error upto 400 epochs is faster, but after 600 epochs the reduction in the error is too slow that it could not achieve the required error (tolerance) level of 0.002 even in 50,000 training epochs as indicated in Table 6.23. The training epochs taken by various cases of GNM to achieve output error (tolerance) level of 0.002 are shown in Table 6.23. The required training epochs (722, 152, 168, 202) to achieve the above tolerance level are much less as in all cases of GNM as compared

Table 6.25. Prediction error of ANN And GNM

ANN	GNM case-1	GNM case-2	GNM case-3	GNM case-4
−0.0031	0.0480	0.0534	0.0629	0.0710
−0.0318	0.0058	0.0171	0.0043	0.0043
−0.0955	−0.0319	−0.0106	−0.0359	−0.0312
0.0132	0.0084	0.0439	0.0026	0.0202
−0.1155	−0.0539	−0.0281	−0.0667	−0.0643
−0.1616	0.0525	0.0115	0.0389	0.0199
−0.2865	−0.0124	−0.0273	−0.0179	−0.0279
−0.2039	−0.0432	−0.0467	−0.0191	−0.0141
−0.0996	0.0089	0.0148	0.0078	0.0065
0.0219	0.0458	0.0563	0.0377	0.0385
0.0667	0.0755	0.0838	0.0793	0.0773
−0.0072	0.0267	0.0247	0.0456	0.0453
−0.1031	−0.0482	−0.0856	−0.0284	−0.0406
−0.1132	−0.1387	−0.0963	−0.0988	−0.0992
−0.0373	−0.0755	−0.0806	−0.0520	−0.0470
0.0244	−0.0562	−0.0509	−0.0480	−0.0429
−0.0126	−0.0604	−0.0613	−0.0487	−0.0539
0.0729	0.0410	0.0450	0.0591	0.0618
−0.0196	−0.0563	−0.0436	−0.0745	−0.0701
0.0720	0.0060	0.0203	−0.0393	−0.0338
0.1323	0.0690	0.0866	0.0495	0.0611
0.1608	0.0263	0.0261	0.0135	0.0035
0.1320	0.1009	0.0891	0.0858	0.0903
0.2014	0.0633	0.0238	0.0444	0.0235

Table 6.26. Testing performance of ANN and GNM For short term load forecasting problem without noise in testing data (after 5,000 epochs)

Model	RMS error	MAX error	MIN error
ANN	0.1166	0.2014	−0.2865
GNM case-1	0.0671	0.1009	−0.1387
GNM case-2	0.0426	0.0891	−0.0963
GNM case-3	0.0494	0.0858	−0.0988
GNM case-4	0.0525	0.0903	−0.0992

Table 6.27. Testing performance of ANN and GNM for short term load forecasting problem with 5% noise in testing data (after 5,000 epochs)

Model	RMS error	MAX error	MIN error
Existing ANN	0.1223	0.2490	−0.2756
GNM case-1	0.0696	0.1519	−0.1545
GNM case-2	0.0448	0.0819	−0.0923
GNM case-3	0.0500	0.0901	−0.0997
GNM case-4	0.0596	0.0940	−0.1052

to ANN. Among all GN Models, the GN Model-2 shows fastest convergence of error requiring only 152 training epochs to achieve same level of tolerance 0.002. It shows much faster convergence of GN Model for the short-term load forecasting problem as compared to ANN.

The testing results of ANN and GN Model have been shown in Figs. 6.6b–6.10b graphically, for the short-term load forecasting problem. The closeness between actual and predicted output (load) curves in all cases of GN Model is more as compared to ANN. Among all cases of GN Model, the closeness between actual and predicted output curves is maximum in GN Model-2, this fact is supported by outputs predicted by ANN and GN Model given in Table 6.24. The error in output for ANN and GN Model is shown in Table 6.25. The rms, maximum and minimum errors during testing (without noise) of ANN and GN Model for the load forecasting problem are shown in Table 6.26. These errors are less in case of GN Model as compared to ANN. Among all cases of GN Model, the GN Model-2 shows minimum testing errors. The ANN and GN Model are also tested for 5% noisy data, the testing errors for noisy data are given in Table 6.27. GN Model performs better on noisy data also as the errors shown in the Table 6.27 are less than the errors given by ANN. Here also, the GN Model-2 performs best among all cases showing minimum testing errors for noisy data.

6.3 Load Frequency Control Problem

In power system both active and reactive power are never steady, and they continuously change with the rising and falling trends. Steam input to steam turbine of turbo generator (or water input to hydro turbine of hydro generators) must therefore be continuously regulated to match the active power demand, failing which machine speed will vary with consequent change in frequency, which may be highly undesirable. Change in frequency causes change in speed of the consumer's plant affecting production process. Maximum permissible change in frequency is ± 0.5 Hz. Also the excitation of generators must be continuously regulated to match the reactive power demand with the reactive generation, otherwise the voltages at various systems buses may go beyond the prescribed limits. In modern age large interconnected systems, manual regulations are not feasible and therefore automatic generation and voltage regulation equipment is installed on each generator. The controllers are set for particular operating conditions and take care of small changes in load demand without frequency and voltage exceeding the prescribed limits. With the passage of time, as the change in load demand becomes large, the controllers must be reset either manually or automatically.

It is known that for small changes active power is dependent on internal machine angles (power angle) δ and is independent of bus voltage V: while bus voltages are dependent on machine excitation (i.e. on reactive generation Q) and is independent of machine angle δ. Change in δ is caused by momentary change in generators speed. Therefore, load frequency and excitation voltage controls are non-interactive for small changes and can be modelled and analyzed independently. Further more, excitation voltage control is fast acting in which the major time constant encountered is that of the generator field; while the power frequency control is slow acting with the major time constant contributed by the turbine and the generator moment of inertia – this time constant is much larger than that of generator field. Thus, the transients in excitation voltage control vanish much faster and do not affect the dynamics of power frequency control.

Changes in load demand can be identified as:

1. Slow varying changes in mean demand, and
2. Fast random variations around the mean.

The regulators must be designed to be insensitive to fast random changes; otherwise the system will be prone to hunting, resulting in excessive wear and tear of rotating machines and control equipment.

6.3.1 Need of Load Frequency Control

It is important to maintain (Zeynelgil et al. 2002) frequency constant in the integrating power systems. The variations on the frequency cause some problems as mentioned below:

1. Most of a.c. motors run at speeds that are directly related to the frequency. The speed, induced electromotive force may vary because of the change in frequency of the power system.
2. When a system operates at frequencies below 49.5 Hz, turbines/rotors undergo excessive vibration in certain turbine rotors states which results in metal fatigue and blade failures.
3. A large number of electrically operated clocks are used. If frequency changes, then operation of these clocks is affected.
4. The turbine regulated devices fully open when the frequency falls below 49 Hz; this situation causes the extra loadings on the generators. The decrease in frequency may cause a reduction in the equipment efficiency.
5. The change in frequency can cause maloperation of power converters by producing harmonics.

To study the load frequency control the excitation control influence on the system performance can be ignored for the following reasons:

1. For small change it may safely be assumed that change in real power only causes change in load angle. That's why the vibrations in the load angle cause momentarily change in generation speed. This means that the generator has been supplied with sufficient reactive power to maintain the voltage constant at the terminals.

2. The time constants involved in the load frequency are turbines inertia of the generators. Hence these time constants are much larger as compared to the excitation system time constants. Therefore, it is fair to assume that the transient of the excitation system will vanish much faster than the transient of load frequency control system and doesn't affect the response of the load frequency control systems.

The load frequency control is based on an error signal called area controlled error (ACE) which is linear combination of net interchange and frequency error. The conventional controls strategy used in industries is to take the integral of ACE as the control signal. It has been found that the use of ACE is to calculate the control signal reduces the frequency and tie line power errors to zero in steady state; but the transient response is not satisfactory.

The linear decentralized load frequency control using pole-placement linear control theory has been investigated to improve the transient response. However the realization of such controllers is difficult and expensive because the feedback portion of the above controller is a function of complete state vector of the system. Generally, all the state variables are not achievable. Even if state estimation techniques are used to estimate the inescapable state variables the data needs to be transferred over long distance. This involves additional cost of telemeter.

The third type controller called "variable structure system" (VSS) controller has been used by some of the investigators. This controller algorithm requires only two measurable variables for example frequency deviation and tie-line power deviation. This controller is tactically as simple as that of conventional controller and can be implemented with very little additional cost. However, the VSS controller needs the switching strategy which may not be simple for large and complex system. To select the best controller algorithm the requirement must be known. These requirements are discussed below.

6.3.2 Requirements for Selecting Controller Strategy

The following requirements must be satisfied (Jaleey et al. 1990):

1. Control loop must be characterized by sufficient degree of stability.
2. Following a step load change, the frequency error should return to zero. This is referring to as isochronous control. The magnitude of transient frequency deviation must be minimized or in other words frequency should not exceed $\pm 0.02\,\mathrm{Hz}$.
3. The static change in the tie line power flow following a step load change either area must be zero.
4. The integral of the frequency error should be minimized so that accuracy of synchronous clocks is minimized. The time error should not exceed $\pm 3\,\mathrm{s}$.

6.3.3 Modelling of Thermal Power Plant (Single Area System):

Let us consider the problem of controlling the power output of the generator of a closely-knit area so as to maintain the schedule frequency. All the generators in such area constitute a coherent group so that the generators speed up and slow down together maintaining their relative power angles. Such an area is defined as *control area*.

The boundaries of control area will generally coincide with that of an individual electricity board.

To understand a load frequency control problem, let us consider a single turbo generator system supplying an isolated load. Figures 6.11 and 6.12 show schematically the speed governing system of a steam turbine. The system consists of the following components:

1. *Fly ball speed governor.* This is the heart of the system, which senses the change in speed (frequency). As the speed increases the fly ball moves outwards and point B on the linkage mechanism moves downwards. The reverse happens when the speed decreases.
2. *Hydraulic amplifier.* It comprises a pilot valve and main piston arrangement. Low power level pilot valve movement is converted into high power valve movement. This is necessary in order to open nor close the steam valve against high-pressure steam.

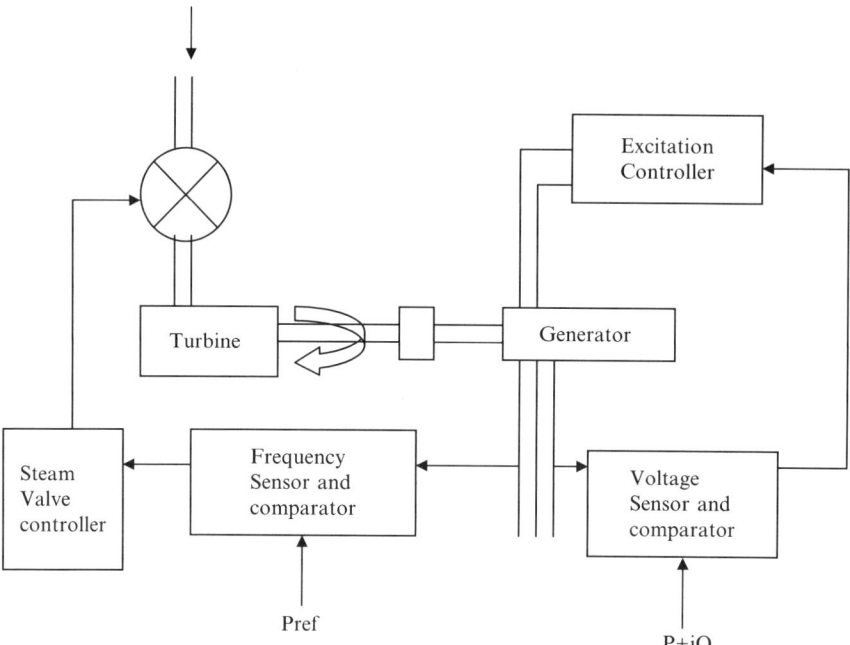

Fig. 6.11. Block diagram of load frequency and excitation voltage regulation

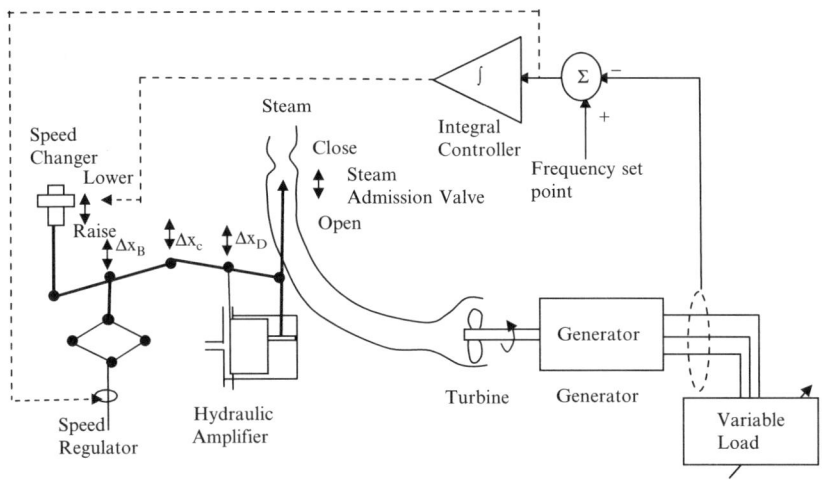

Fig. 6.12. Functional diagram of single area system

3. *Linkage mechanism.* ABC is a rigid link pivoted at B and CDE is another rigid link pivoted at D. This link mechanism provides a movement to control valve in proportion to change in speed. It also provides a feed back from the steam valve movement (link4).

4. *Speed changer.* It provides a steady state power output setting for the turbine. Its downward movement opens the upper pilot valve so that more steam is admitted to the turbine under steady conditions (hence more steady power output). The reverse is true for upward movement of speed changer.

6.3.3.1 Model of Speed Governing System

Assume the system is initially operating under steady conditions – the linkage mechanism stationary and pilot valve closed, steam valve opened by a definite magnitude, turbine running at constant speed with turbine output balancing the generator load. Let the operating conditions be characterized by

f^0 = system frequency
pg = generator output = turbine output (neglecting generator losses)
Ye = steam valve setting.

We shall obtain a linear incremental model around these operating conditions. Let a point A on the linkage mechanism be moved downwards by a small amount Δya. It is a command which causes the turbine power output to change and can therefore can be written as

$$\Delta ya = Kc\Delta Pc \tag{6.11}$$

where ΔPc is the commanded increase in power.

The command signal ΔPc (i.e. ΔYe) sets into a sequence of events – the pilot valve moves upwards, high pressure oil flows on to the top of the main piston moving it downwards; the steam valve opening consequently increases, the turbine generator speed increases, i.e. the frequency goes up. Let us model these events mathematically.

The two factors contribute to the movement of C:

1. Δya contributes $-(i2/i1)\Delta$ya or $-k1\Delta$ya (i.e. upwards) or $-k1kc\Delta$Pc.
2. Increase in frequency Δf causes the fly balls to move outwards so that B moves downwards by a proportional amount k2$'\Delta$f. The consequent movement if C with a remaining fixed at Δya is $+(i1 + i2/i1)K_2'DF = K_2'DF$ (i.e. downwards).

The net movement of C is therefore

$$\Delta yc = -k1kc\Delta Pc + k2\Delta f \tag{6.12}$$

The movement of D, Δyd is the amount by which the pilot valve opens. it is contributed by Δyc and Δye and can be written as

$$\Delta Yd = [I1/(I3 + I4)]\Delta Yc + [I3/(I3 + I4)]\Delta Ye$$
$$= K3\Delta yc + K4\Delta Ye. \tag{6.13}$$

The movement Δyd depending upon its sign opens one of the ports of the pilot valve admitting high-pressure oil into the cylinder there by moving the main piston and opening the steam valve by Δyd certain justifiable simplifying assumptions, which can be made at this, are:

1. Inertial reaction forces of main piston and steam valve are negligible compared to the forces exerted on the piston by high-pressure oil.
2. Because of (1) above the rate of oil admitted to the cylinder is proportional to port opening Δyd.

The volume of oil admitted to the cylinder is thus proportional to the time integral of Δyd. The movement Δye is obtained by dividing the oil volume by the area of cross-section of the piston. Thus

$$\Delta Ye = K5 \int_0^t -(\Delta Yd)dt. \tag{6.14}$$

It can be verified from the schematic diagram that a positive movement yd, causes negative (upward) movement Δye accounting for the negative sign used in (6.14).

Taking the Laplace transform of (6.12), (6.13) and (6.14), we get

$$\Delta Yc(s) = -K1Kc\Delta Pc(s) + K2\Delta F(s) \tag{6.15}$$
$$\Delta Yd(s) = K3\Delta Yc(s) + K4\Delta Ye(s) \tag{6.16}$$
$$\Delta Ye(s) = -K5\frac{1}{s}\Delta Yd(s) \tag{6.17}$$

Eliminating $\Delta Yc(s)$ & $\Delta Yd(s)$, we can write:

$$\Delta Ye(s) = \frac{K1K3\Delta Pc(s) - K2K3\Delta F(s)}{K4 + \frac{S}{K5}}$$

$$= \left[\Delta Pc(s) - \frac{1}{R}\Delta F(s)\right] * \left\{\frac{Ksg}{1 + Tsq}\right\} \qquad (6.18)$$

where

Ksg = gain of speed governor
R = speed regulation of the governor
Tsg = time constant of speed governor

The speed governing system of a hydro turbine is more involved. An additional feedback loop provides temporary droop compensation to prevent instability. This is necessitated by the large inertia of the penstock gate, which regulates the rate of water input to the turbine as given in Fig. 6.13.

6.3.3.2 Turbine Model

Let us now relate the dynamic response of a steam turbine in terms of changes in power output to change in steam valve opening ye. Figure 6.14 shows a two stage steam turbine with a reheat unit. The dynamic response is largely influenced by two factors:

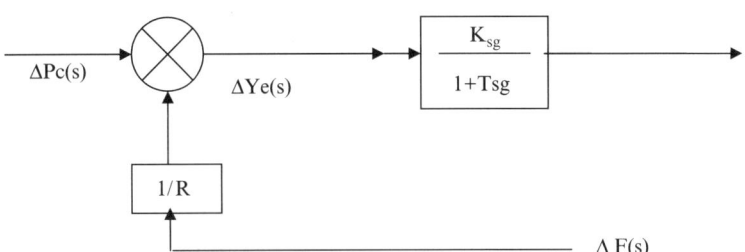

Fig. 6.13. The block diagram speed governing system

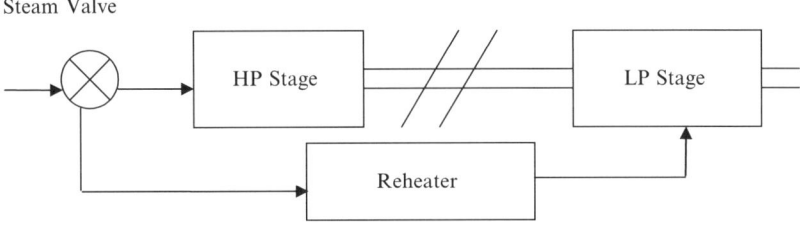

Fig. 6.14. Two stage steam turbine

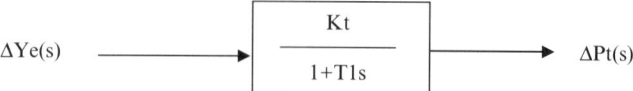

Fig. 6.15. Turbine transfer function model

(1) Entrained steam between the inlet steam valve and first stage of turbine
(2) The storage action in the reheater, which causes the output of low pressure stage to lag behind that of the high pressure stage.

Thus, the turbine transfer function is characterized by two time constants. For ease of analysis, it will be assumed here that turbine can be modelled to have a single equivalent time constant, Fig. 6.15 shows the transfer function model of a steam turbine. Typically the time constant T1 lies in the range 0.2–2.5 s.

Steam valve

Writing the power balance equation, we have

$$\Delta P_G - \Delta P_D = \frac{2HP_r}{f^0}\frac{d}{dt}(\Delta f) + B\Delta f.$$

Dividing throughout by Pr and rearranging, we get

$$\Delta P_G(pu) - \Delta P_D(pu) = \frac{2H}{f^0}\frac{d}{dt}(\Delta f) + B(pu)\Delta f$$

Taking the Laplace transform, we can write $\Delta F(s)$

$$\Delta F(s) = \frac{\Delta Pg(s) - \Delta Pd(s)}{B + \frac{2H}{f^0}(s)}$$

$$= [\Delta Pg(s) - \Delta Pd(s)] * \left(\frac{Kp(s)}{1 + Tp(s)}\right) \tag{6.19}$$

where

$\mathrm{Tps} = \frac{2H}{Bf^0} =$ power system time constant
$\mathrm{Kps} = \frac{1}{B} =$ power system gain.

Equation (6.13) can be represented in block diagram form as shown in Fig. 6.16.

6.3.3.3 Generator Load Model

The increment in power input to the generator load system is $\Delta Pg - \Delta Pd$ where $\Delta Pg = \Delta Pd$ incremental turbine power output (assuming generator incremental loss to be negligible) and ΔPd is the load increment.

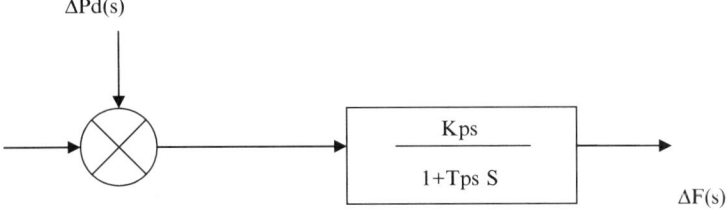

Fig. 6.16. Block diagram model of generator load model

This increment in power input to the system is accounted for in two ways:

1. Rate of stored kinetic energy in the generator rotor. At scheduled frequency (f), the stored energy is

$$Wke = H^*Pr \qquad kW - s \ (kj)$$

Where Pr is the kW rating of the turbo-generator and H is defined as its inertia is constant. The kinetic energy being proportional to square of speed (frequency), the kinetic energy at a frequency of $(f+\Delta f)$ is given by:

$$Wke = Wke^0(f^0 + \Delta f)^2/(f^0)$$
$$= H\,Pr(1 + (2\Delta f/f^0)) \qquad (6.20)$$

Rate of change of kinetic energy is therefore

$$\frac{d}{dt}(W_{ke}) = \frac{2H\,Pr}{f^0}\frac{d}{dt}(\Delta f) \qquad (6.21)$$

2. As the frequency changes, the motor load changes being sensitive to speed, the rate of change of load with respect to frequency, i.e. can be regarded as nearly constant for small changes in frequency Δf can be expressed as

$$\left(\frac{\partial P_D}{\partial f}\right)\Delta f = B\Delta f \qquad (6.22)$$

where the constant B can be determined empirically. B is positive for a predominantly motor load.

6.3.4 Response of Load Frequency Control of an Isolated (Single Area) Power System

Now the analysis of load frequency control of a simple power system is presented. The complete system to be considered for the design of controller is shown in Fig. 6.17 the system response has been obtained for uncontrolled and controlled cases in the following sections:

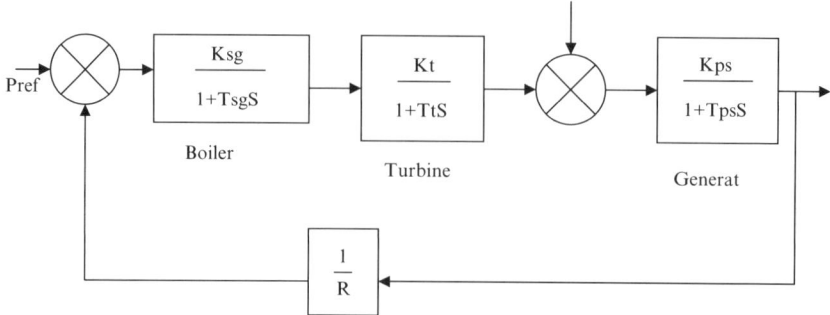

Fig. 6.17. Block diagram model of load frequency control (isolated power system)

6.3.4.1 Uncontrolled Case

Steady state response:

In uncontrolled case, speed changer has fixed settings, i.e. ΔPref = 0
For step load change ΔPD = M
Laplace transform of it is ΔPref = 0
Now from block diagram in Fig. 6.14. We obtain by inspection

$$\left[\left(\Delta P_{ref} - \frac{1}{R}\Delta f\right)G_H G_T - \Delta P_D\right] = \Delta f \tag{6.23}$$

In Laplace transform

$$\Delta f(s) - \frac{G_P}{1 + \left(\frac{1}{R}\right)G_P G_H G_T}\Delta P_D(s) \tag{6.24}$$

Using the final value theorem

$$\Delta f_{Ss} = \lim_{s-0}[S\Delta f(s)] = \frac{SG_P}{1 + \left(\frac{1}{R}\right)G_P G_P G_T} * \frac{M}{S}$$

$$= -\frac{K_P M}{1 + \left(\frac{K_P}{R}\right)} = -\frac{M}{D + \frac{1}{R}}Hz \tag{6.25}$$

If $\beta\Delta[D + 1/R]$p.u. MWHz
Then

$$\Delta fs = -(M/\beta) \tag{6.26}$$

where β is called area freq. response characteristic (AFRC). Thus, in uncontrolled case the steady state response has constant error.

Dynamic response:

Finding the dynamic response (for a step load) is quite straight forward. By taking inverse laplace transform of (6.27) gives an expression for $\Delta f(t)$. However, as GH, GT, GP contain at least one time constant each, the denomination will be of third order, resulting in unwieldy algebra.

We can simplify the analysis considerably by making the reasonable assumption that the action of the speed governor plus the turbine generation is instantaneous compared with the rest of power system. (Tg $\ll$ Tt $\ll$ Tp) where Tp is generally 20 s. Tg $\approx$ Tt $\leq$ 1 s, thus assume Tg = Tt = 0 and gain equal to

$$\Delta f(s) = \frac{\frac{K_P}{1+ST_P}}{1 + \frac{1}{R}\frac{K_P}{1+ST_P}} * \frac{M}{S} \tag{6.27}$$

Above equation can also be written as

$$\Delta F(s) = -M\frac{RK_P}{R+K_P}\left(\frac{1}{S} - \frac{1}{S + \frac{R+K_P}{RT_P}}\right) \tag{6.28}$$

Taking inverse Laplace transform of above equation we get

$$\Delta f(s) = -\Delta P_D\frac{RK_P}{R+K_P}\left[1 - e^{-t}\left[\frac{K_P+R}{RT_P}\right]\right] \tag{6.29}$$

Thus the error $= e^{-t}[\frac{K_P+R}{RT_P}]$. This persists in uncontrolled case.

A simulated response of uncontrolled single area non-reheat is shown in Fig. 6.19.

6.3.4.2 Controlled Case

Control area:

The power pools in which all the generators are assumed to be tightly coupled with change in load. Such as area, where all the generators are running coherently is termed as control area.

Integral area:

By using the control strategy, we can control the intolerable dynamic frequency changes with changes in load and also the synchronous clocks run on time but not without error during transient period. We have added to the uncontrolled system in Fig. 6.20 an integral controller which actuates the speed changer by real power command signal ΔPc.

$$\Delta P_C = -K_t \int \Delta f\, dt \tag{6.30}$$

The negative polarity must be chosen so as to cause a positive frequency error to give rise to a negative, or "decrease" command.

Here Kg and Kt are such that KgKt Δ 1.

In central load frequency control of a given area, the signal fed into the integrator is referred to as area control error (ACE), i.e.

$$\text{ACE} \underset{=}{\Delta} \Delta f \tag{6.31}$$

Taking Laplace transform of above (6.30). We get $\Delta P_C(s) = \frac{K_1}{s}\Delta f(s)$

And for step input load:

$$\Delta P_D(s) = \frac{M}{S}$$

$$\Delta f(s) = \frac{K_P}{(1 + ST_P) + \left(\frac{1}{R} + \frac{K_1}{S}\right) * \frac{K_P}{(1+ST_g)(1+ST_t)}} * \frac{M}{S} \qquad (6.32)$$

$$= \frac{RK_P S(1 + ST_g)(1 + ST_t)}{S(1 + ST_g)(1 + ST_t)(1 + ST_P)R + K_P(RK_1 + S)} * \frac{M}{S}$$

By using final value theorem, we readily obtain from the above equation the static frequency droops:

$$\Delta f_{steady} = \lim_{s-0}[S\Delta f(S)] = 0, \text{ i.e. no error.}$$

6.3.4.3 Single Area Thermal Plant

The block diagram representation of a single area Thermal plant is shown in Fig. 6.20. The system response has been obtained for uncontrolled and controlled systems.

(a) *Uncontrolled case*:

The block diagram of the plant shown in the Fig. 6.16 can be modeled in the state space form with the help of the following matrix equations

$$x = Ax + Bu + Fp$$

Where

$$A = \begin{bmatrix} \frac{1}{T_P} & \frac{K_P}{T_P} & 0 \\ 0 & \frac{-1}{T_T} & \frac{1}{T_T} \\ \frac{-1}{RT_H} & 0 & \frac{-1}{T_H} \end{bmatrix}, B = \begin{bmatrix} 0 \\ 0 \\ \frac{1}{T_H} \end{bmatrix}, F = \begin{bmatrix} -\frac{K_P}{T_P} \\ 0 \\ 0 \end{bmatrix}, X = \begin{bmatrix} \Delta f \\ \Delta P_T \\ \Delta P_H \end{bmatrix}, p = [\Delta P_E]$$

u = control vector, it is zero in uncontrolled case.
P = disturbance vector (disturbance in load).

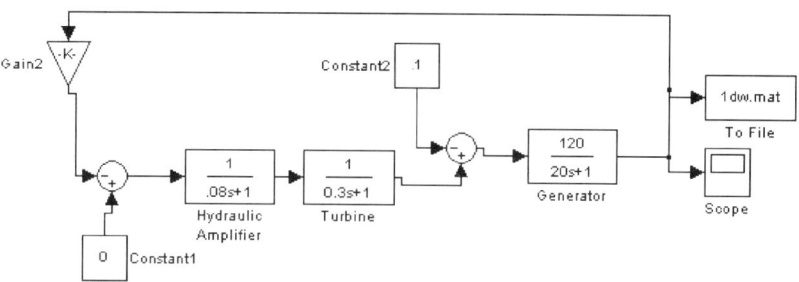

Fig. 6.18. Single area representation of thermal plant (uncontrolled)

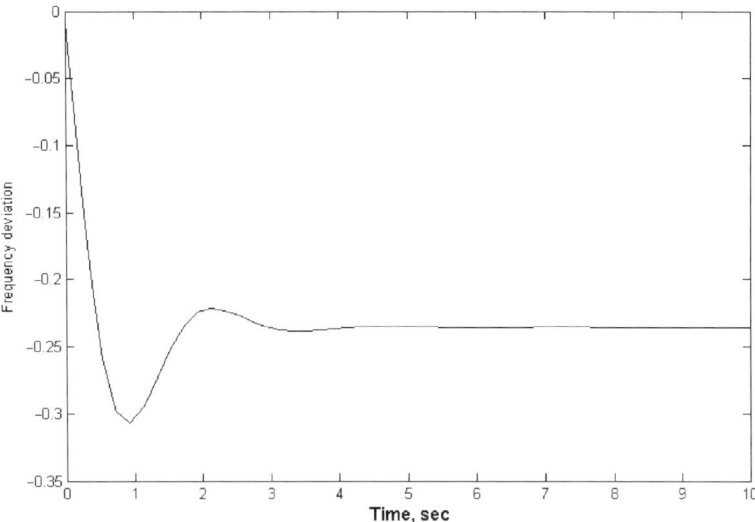

Fig. 6.19. Single area thermal plant (frequency deviation for 10% disturbance in load)

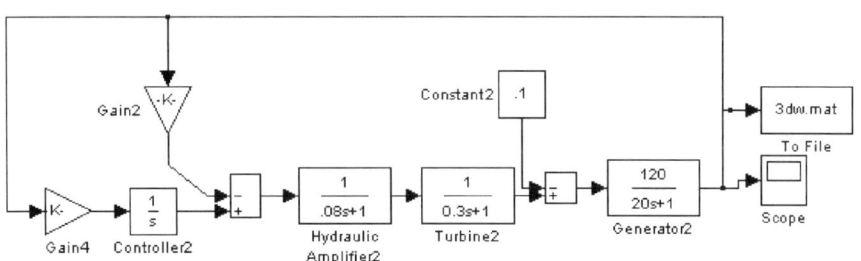

Fig. 6.20. Single area representation of thermal plant (controlled)

The response is taken with the help of MATLAB for 10% step change in load. Figure 6.19 shows that there is a constant state error in the response for the uncontrolled case.

(b) *Controlled case with integral controller:*

The steady state error of the uncontrolled system can be made zero by using an integral controller in the forward path as shown in Fig. 6.20. When an integral controller is used then the command signal, which actuates the speed changer, is

$$\Delta P_{ref} = \Delta P_c = -K_i \int \Delta f dt$$

The signal that is fed to the controller is called area control error (ACE).

$$\text{ACE} = \Delta f$$

The response is taken with the help of MATLAB 6.5 for 10% step change in load. Figure 6.21 shows the response of the single area thermal plant with integral controller.

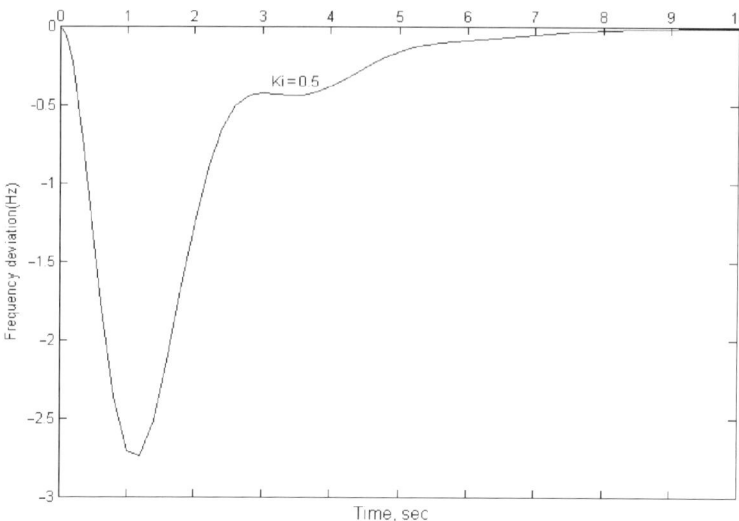

Fig. 6.21. Single area thermal plant (frequency deviation for 10% disturbance in load)

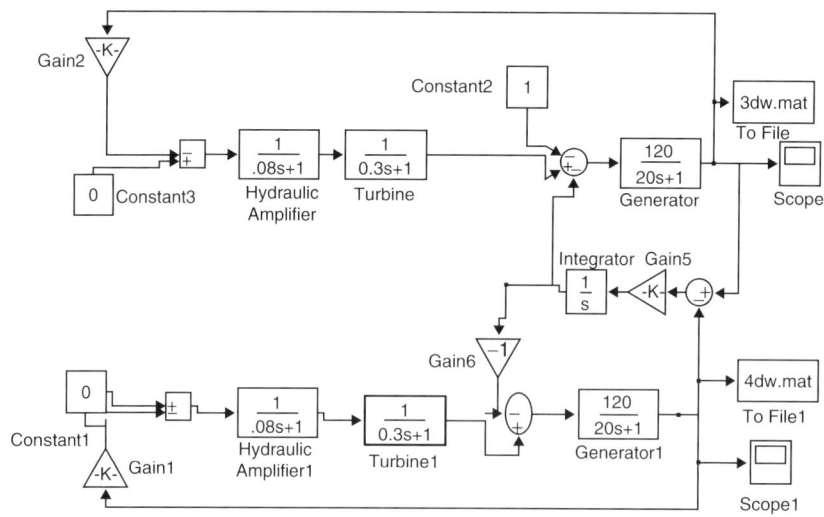

Fig. 6.22. Two-area representation of thermal plant (uncontrolled)

6.3.4.4 Two-Area Thermal System

The response of two-area system is taken for uncontrolled as well as controlled case with the help of MATLAB.

(a) *Uncontrolled case*:

The block diagram representation of two-area thermal plant without controller is shown in Fig. 6.22.

The block diagram of the plant can be modeled in the state space form with the help of following equations.

$$x = Ax + Bu + Fp,$$

Where

$$A = \begin{bmatrix} \frac{-1}{T_{P1}} & \frac{K_{P1}}{T_{P1}} & 0 & \frac{-K_{P1}}{T_{P1}} & 0 & 0 & 0 \\ 0 & \frac{-1}{T_{T1}} & \frac{K_{T1}}{T_{T1}} & 0 & 0 & 0 & 0 \\ \frac{-K_{H1}}{T_{H1}R_1} & 0 & \frac{-1}{T_{H1}} & 0 & 0 & 0 & 0 \\ 2\pi T_0 & 0 & 0 & 0 & -2\pi T_0 & 0 & 0 \\ 0 & 0 & 0 & \frac{K_{P2}}{T_{P2}} & \frac{-1}{T_{P2}} & \frac{K_{P2}}{T_{P2}} & 0 \\ 0 & 0 & 0 & 0 & 0 & \frac{-1}{T_{T2}} & K_{T2}T_{T2} \\ 0 & 0 & 0 & 0 & \frac{-K_{H2}}{T_{H2}R_2} & 0 & \frac{-1}{T_{H2}} \end{bmatrix}, \quad B = \begin{bmatrix} 0 & 0 \\ 0 & 0 \\ \frac{1}{T_{H1}} & 0 \\ 0 & 0 \\ 0 & \frac{1}{T_{H2}} \\ 0 & 0 \\ 0 & 0 \end{bmatrix}$$

$$F = \begin{bmatrix} \frac{-K_{P1}}{T_{P1}} & 0 \\ 0 & 0 \\ 0 & 0 \\ 0 & 0 \\ 0 & \frac{-K_{P2}}{T_{P2}} \\ 0 & 0 \\ 0 & 0 \end{bmatrix}, \quad X = \begin{bmatrix} \Delta f_1 \\ \Delta P_{T1} \\ \Delta P_{H1} \\ \Delta P_{tie} \\ \Delta f_2 \\ \Delta P_{T2} \\ \Delta P_{H2} \end{bmatrix}, \quad P = \begin{bmatrix} \Delta P_{E1} \\ \Delta P_{E2} \end{bmatrix}, \quad u = \begin{bmatrix} \Delta P_{ref1} \\ \Delta P_{ref2} \end{bmatrix}$$

state vectors
u = control vector, it is zero in uncontrolled case.
p = disturbance vector (load disturbance).

The frequency changes in area-1 and area-2 for the 10% step load perturbation in area-1 is shown in Fig. 6.23. There is a steady state error in the response.

(b) *Controlled case with integral controller:*

When an integral controller is added to each area of the uncontrolled plant in forward path the steady state error in the frequency becomes zero.

The state space equation for the block diagram, shown in the Fig. 6.24 is $X = Ax + Bu + Fp$.

Where

$$A = \begin{bmatrix} \frac{-1}{T_{P1}} & \frac{K_{P1}}{T_{P1}} & 0 & \frac{-K_{P1}}{T_{P1}} & 0 & 0 & 0 & 0 & 0 \\ 0 & \frac{-1}{T_{T1}} & \frac{K_{T1}}{T_{T1}} & 0 & 0 & 0 & 0 & 0 & 0 \\ \frac{-K_{H1}}{T_{H1}R_1} & 0 & \frac{-1}{T_{H1}} & 0 & 0 & 0 & 0 & 0 & 0 \\ 2\pi T_0 & 0 & 0 & 0 & -2\pi T_0 & 0 & 0 & 0 & 0 \\ 0 & 0 & 0 & \frac{K_{P2}}{T_{P2}} & \frac{-1}{T_{P2}} & \frac{K_{P2}}{T_{P2}} & 0 & 0 & 0 \\ 0 & 0 & 0 & 0 & 0 & \frac{-1}{T_{T2}} & \frac{K_{T2}}{T_{T2}} & 0 & 0 \\ 0 & 0 & 0 & 0 & \frac{-K_{H2}}{T_{H2}R_2} & 0 & \frac{-1}{T_{H2}} & 0 & 0 \\ -K_{i1}B_1 & 0 & 0 & -K_{i1} & 0 & 0 & 0 & 0 & 0 \\ 0 & 0 & 0 & K_{i2} & -K_{i2}B_2 & 0 & 0 & 0 & 0 \end{bmatrix}, \quad B = \begin{bmatrix} 0 & 0 \\ 0 & 0 \\ \frac{K_{H1}}{T_{H1}} & 0 \\ 0 & 0 \\ 0 & 0 \\ 0 & 0 \\ 0 & \frac{K_{H2}}{T_{H2}} \\ 0 & 0 \\ 0 & 0 \end{bmatrix}$$

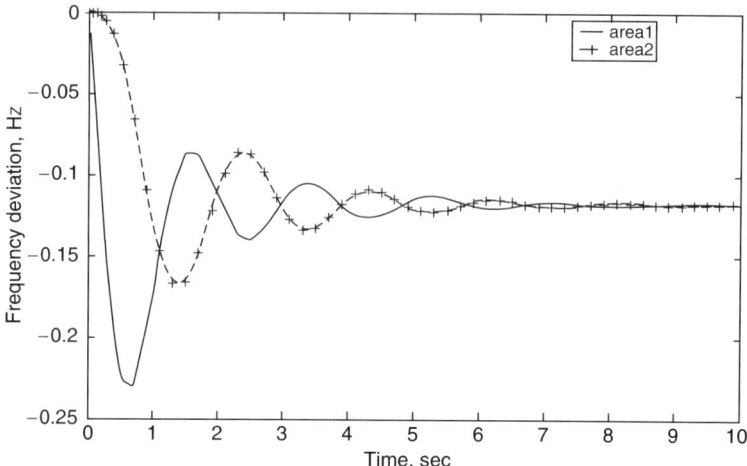

Fig. 6.23. Two area thermal plant (frequency deviation for 10% disturbance in area1)

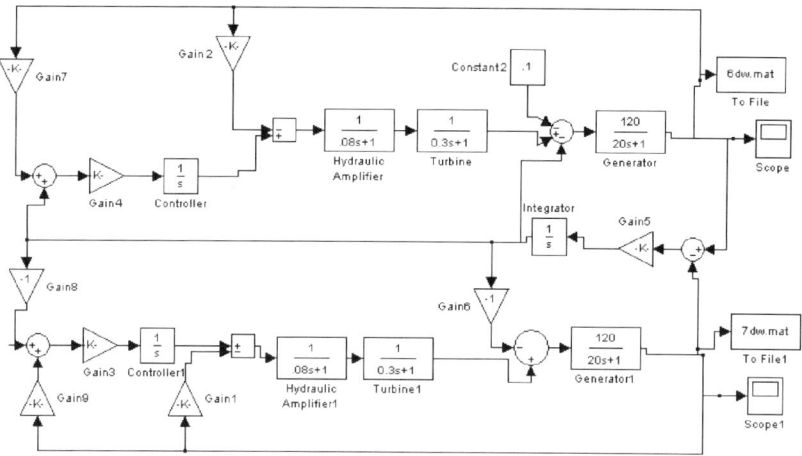

Fig. 6.24. Two-area representation of thermal plant (controlled)

$$F = \begin{bmatrix} \frac{-K_{P1}}{T_{P1}} & 0 \\ 0 & 0 \\ 0 & 0 \\ 0 & 0 \\ 0 & \frac{-K_{P2}}{T_{P2}} \\ 0 & 0 \\ 0 & 0 \\ 0 & 0 \\ 0 & 0 \end{bmatrix}, \ p = \begin{bmatrix} \Delta P_{E1} \\ \Delta P_{E2} \end{bmatrix}, \ u = \begin{bmatrix} \Delta P_{ref1} \\ \Delta P_{ref2} \end{bmatrix}$$

x^T = state vectors, $[\Delta f_1 \ \Delta P_{T1} \ \Delta P_{H1} \ \Delta P_{tie} \ \Delta f_2 \ \Delta P_{T2} \ \Delta P_{H2}]$
u = control vector, it is zero in uncontrolled case and p = disturbance vector (change in load).

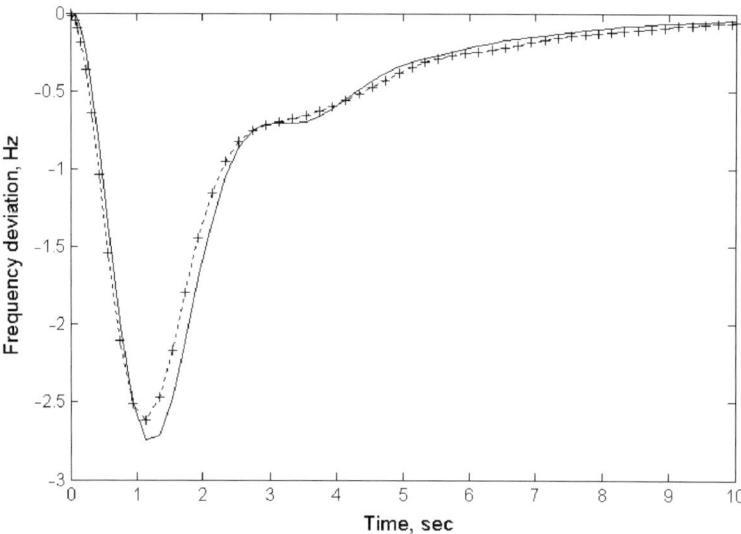

Fig. 6.25. Two area thermal plant (frequency deviation for 10% disturbance in area1)

The frequency changes in area-1 and area-2 for the value of integral gain as is shown in Fig. 6.25.

6.3.4.5 Hydro – Thermal Plant

The system is simulated in MATLAB and the response is obtained for controlled and uncontrolled case.

(a) *Uncontrolled case*

The hydro-thermal interconnected power system shown in Fig. 6.26 can be represented by the following state space equation.

The system state equation can be written as $X' = Ax + Bu + Fp$, Where

$$A = \begin{bmatrix} \frac{-1}{T_{P1}} & \frac{K_{P1}}{T_{P1}} & 0 & \frac{-K_{P1}}{T_{P1}} & 0 & 0 & 0 & 0 & 0 \\ 0 & \frac{-1}{T_{r1}} & \frac{1}{T_{r1}} - \frac{K_{T1}}{T_{T1}} & \frac{K_{T1}K_{r1}}{T_{T1}} & 0 & 0 & 0 & 0 & 0 \\ 0 & 0 & \frac{-1}{T_{T1}} & \frac{1}{T_{T1}} & 0 & 0 & 0 & 0 & 0 \\ \frac{-1}{T_{H1}R_1} & 0 & 0 & \frac{-1}{T_{H1}} & 0 & 0 & 0 & 0 & 0 \\ 2\pi T_0 & 0 & 0 & 0 & 0 & -2\pi T_0 & 0 & 0 & 0 \\ 0 & 0 & 0 & 0 & \frac{K_{P2}}{T_{P2}} & \frac{-1}{T_{P2}} & \frac{K_{P2}}{T_{P2}} & 0 & 0 \\ 0 & 0 & 0 & 0 & 0 & 0 & \frac{-1}{T_{r2}} & \frac{K_{r2}K_{T2}}{T_{T2}} & \frac{1}{T_{r2}} - \frac{K_{T2}}{T_{T2}} \\ 0 & 0 & 0 & 0 & 0 & 0 & 0 & \frac{K_{T2}}{T_{T2}} & \frac{-1}{T_{T2}} \\ 0 & 0 & 0 & 0 & 0 & \frac{-1}{R_2 T_{T2}} & 0 & 0 & \frac{-1}{T_{T2}} \end{bmatrix},$$

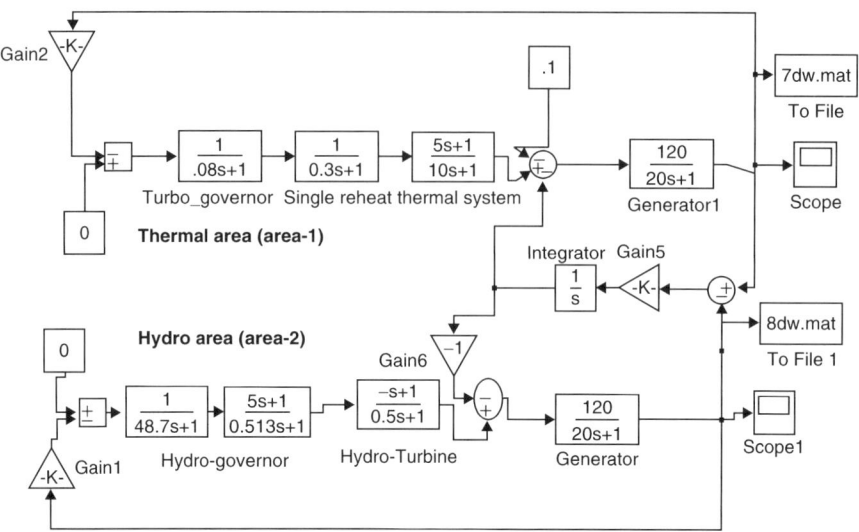

Fig. 6.26. Two area hydro thermal plant (uncontrolled)

$$
B = \begin{bmatrix} 0 & 0 \\ 0 & 0 \\ \frac{1}{T_{H1}} & 0 \\ 0 & 0 \\ 0 & 0 \\ 0 & \frac{-2T_{r2}}{T_1 T_2} \\ 0 & 0 \\ 0 & \frac{T_{r2}}{T_1 T_2} \\ 0 & T_{T1} \end{bmatrix}, \ F = \begin{bmatrix} \frac{-K_{P1}}{T_{P1}} & 0 \\ 0 & 0 \\ 0 & 0 \\ 0 & 0 \\ 0 & \frac{-K_{P2}}{T_{P2}} \\ 0 & 0 \\ 0 & 0 \\ 0 & 0 \\ 0 & 0 \end{bmatrix}, \ u = \begin{bmatrix} \Delta P_{ref1} \\ \Delta P_{ref2} \end{bmatrix}, \ P = \begin{bmatrix} \Delta P_{e1} \\ \Delta P_{e2} \end{bmatrix}
$$

$$x^T = [\Delta f_1 \ \Delta P_{G1} \ \Delta P_{R1} \ \Delta X_{e1} \ T_{1e} \ \Delta f_2 \ \Delta P_{G2} \ \Delta X_{e2} \ \Delta P_{R2}]$$

u = control vector, it is zero in uncontrolled case and p= disturbance vector.

Figure 6.27 shows the response of uncontrolled hydro-thermal system. There is a steady state error in the system. The disturbance of 10% step load change is given in area-1 (thermal area).

a) *Controlled case with integral controller:*

The block diagram with integral controller is shown in Fig. 6.28. It can also be modeled by following state space equation:

$$x = Ax + Bu + Fp.$$

The frequency response of two-area hydrothermal plant is shown in Fig. 6.29.

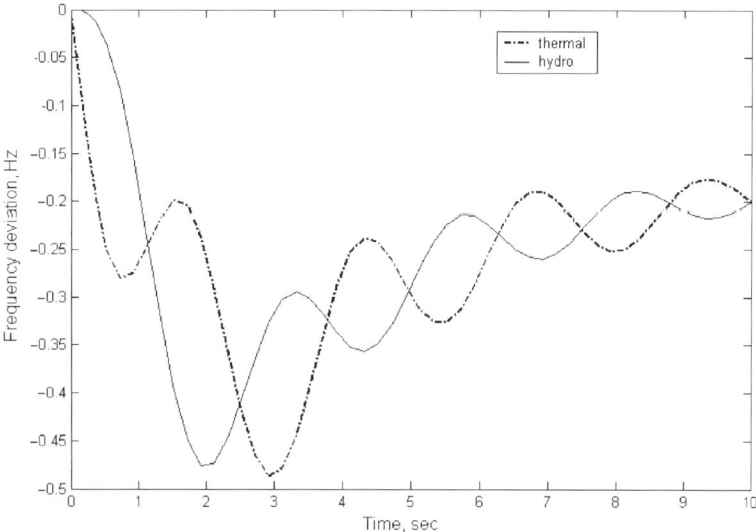

Fig. 6.27. Two area hydro-thermal plant (frequency deviation for 10% disturbance in thermal area)

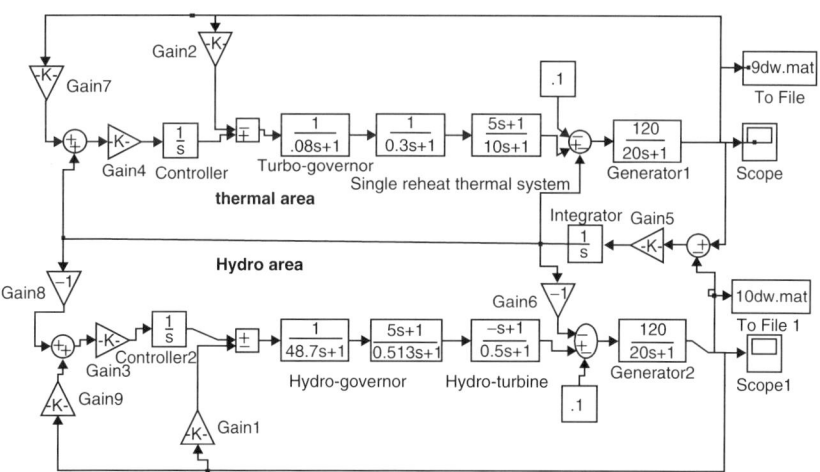

Fig. 6.28. Two areas hydrothermal plant (controlled)

6.3.4.6 Single Area Hydro System

(a) *Uncontrolled case*

The block diagram is shown in Fig. 6.30.

Frequency response of single area hydro system with 10% disturbance in load is shown in Fig. 6.31.

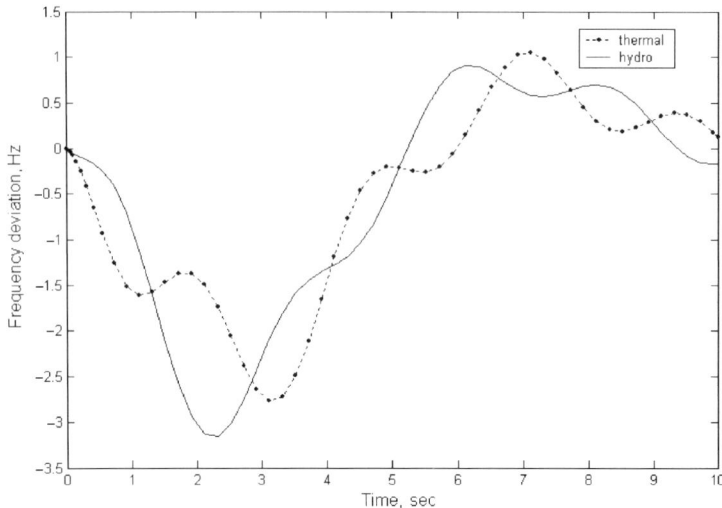

Fig. 6.29. Two area hydro-thermal plant (frequency deviation for 10% disturbance in thermal area)

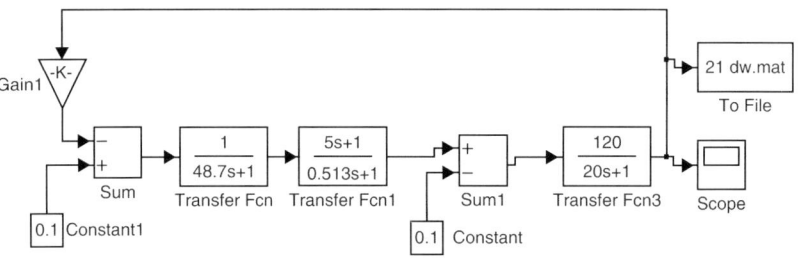

Fig. 6.30. Single area hydro plant (uncontrolled)

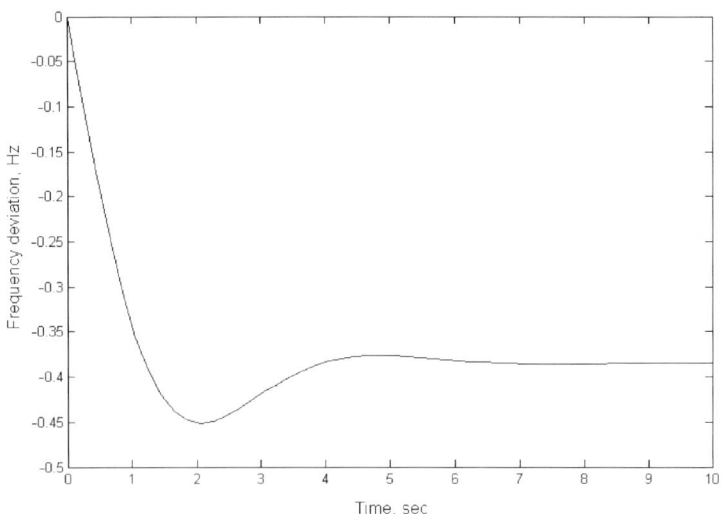

Fig. 6.31. Single area hydro plant (frequency deviation for 10% disturbance)

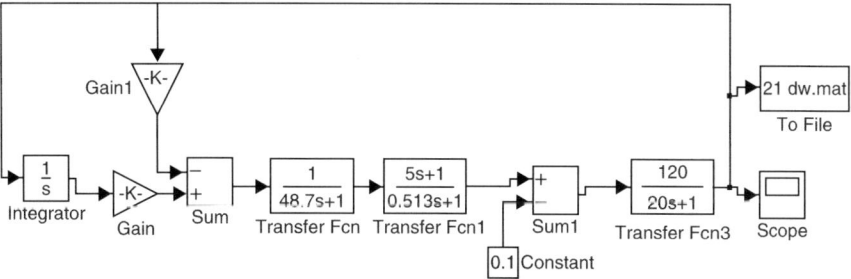

Fig. 6.32. Single area hydro plant (uncontrolled)

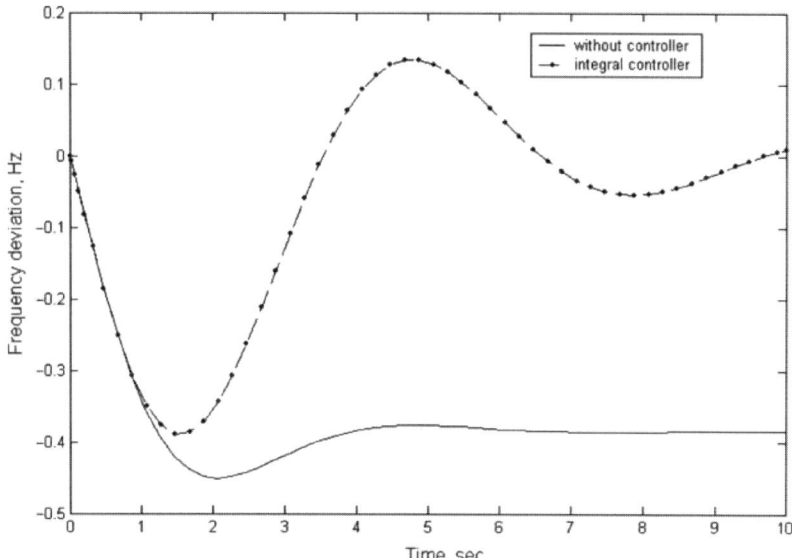

Fig. 6.33. Single area hydro plant (frequency deviation for 10% disturbance)

(b) *Controlled case*

The block diagram is shown in Fig. 6.32.

The frequency response of single area hydro system when controlled by integral controller is shown in Fig. 6.33.

6.3.4.7 Results and Discussion

It has been found that the use of ACE as control signal reduces the frequency and tie line power error to zero in steady state, but the transient response is not satisfactory. It can be seen in Fig. 6.33 that overshoot is more than 0.02 Hz and settling time is also more compared to VSS controller.

6.3.5 Development of GN Based Load Frequency Controller

Many investigations have been reported in the field of load frequency control using neural network (Beaufays et al. 1999; Djukanovic et al. 1995; Zeynelgil et al. 2002). A net interchange tie line bias control strategy has been widely accepted by utilities. The frequency and the interchanged power are kept at their desired values by means of feedback of the integral of area control error (ACE), containing the frequency deviation and the error of the tie-line power and controlling the prime mover input. The fixed gain controller based on classical control theories in literature are insufficient because of changes in operating point during a daily cycle (Demiroren et al. 2001). Load frequency controller primarily consists of integral controller. The integrator gain is set to a level that compromises between fast transient recovery and low overshoot in the dynamic response of the overall system. This type of controller is slow and does not allow the designer to take into account possible nonlinearities in the generator unit. The inherent nonlinearities in system components and synchronous machines have led researchers to consider neural network techniques to build a non-linear ANN controller with high efficiency of performance (Davison and Tripathi 1978; Djukanovic et al. 1995). Here ANN controller is used because the controller provides faster control than the others. Beaufays et al. (1999) and Zeynelgil et al. (2002) used neural network to act as the intelligent load frequency control scheme. The control scheme guarantees that steady state error of frequencies and inadvertent interchange of tie-lines are maintained in a given tolerance limitation.

There are some inherent problems of ANN as controller like long training time is required, huge network is needed for a good controller. To overcome these drawbacks Chaturvedi et al. (1999) has used a generalized neuron (GN) model for modeling the rotating electrical machines and load forecasting problem. The ANN training is mainly depending on the input–output mapping, range of normalization of training data, neuron structure and error function used in training algorithms. Chaturvedi et al. (1996) already studied the effect of mapping of input-output data, normalization range of training data and the neuron structure. In this paper, the effect of different error functions on GN based load frequency controller has been studied.

The power system considered here includes two different areas connected through a tie-line. First area of the power system consists of the steam turbines, which includes reheater. The second area consists of hydro turbine. First, a step loading increasing in the first area of power system is considered. The two area system is controlled using GN based load frequency controller using back-propagation learning algorithm and results are compared with conventional integral controller. The aim of this controller is to restore the frequency to its nominal value in the shortest time whenever there is a change in demand. The action of the controller should be coupled with minimum frequency transients and zero steady state error.

Conventional integral controller action is based on the change in the frequency which make delayed than that of a GN controller whose action is based on the rate of change of frequency. It also makes use of the rate of change of frequency to estimate the electric load perturbation. The load perturbation estimate could be obtained either by a linear estimator or by a non-linear GN estimator.

6.3.5.1 Single Area System

It has been shown that the level variable vector [df dpt dph] of a single area system controlled by an integral controller eventually converged to a steady state value equal to [0 dpe dpe/K_1] but this convergence was slow. The GNN controller that replaces the integral controller should make a plant converge to the same steady state vector, while limiting the duration and magnitude of the transients. Such an operation cannot be performed instantaneously. Besides, the value of the desired control action is not known beforehand.

The dynamic GN controller and plant structure is shown in Fig. 6.34. It is quite clear from the figure that the frequency can be sensed at every time instant. The GN controller output u can be determined from the past value of u and the frequency variation and the load perturbation. The load perturbation of large systems is not directly measurable. It must therefore be estimated by linear estimator or by non-linear network estimator. If the non-linearities in

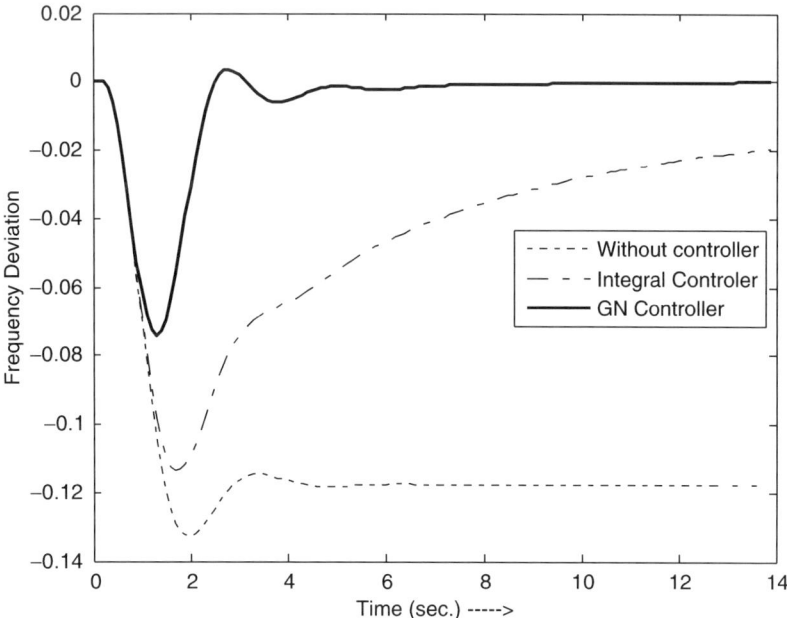

Fig. 6.34. Performance comparison of different controller for a single area system

the system justify, such an estimator takes as inputs a series of K-samples of frequency fluctuations at the output of generator (df (t−1) df(t−2) . . . df(t−n) and estimates the instantaneous value of load perturbation dpe based on this input vector. The estimate dpe is then used to drive the plant controller. This can also be implemented with the help of GN estimator. We assume that the electric perturbation is a step function of amplitude. When the step load perturbation hits the system, the plant state changes which is necessary to control to be achieved by GN controller. The one way to implement GN controller is to built a neural network emulator for the plant and back propagating error gradients through it is nothing other than approximating the true Jacobian matrix of the plant using neural network training. Whenever the equations of the plant are known a priori and they can be used to compute, analytically or numerically, the elements of the Jacobian matrix. Error gradient at the input of the plant is then obtained by multiplying the output error gradient by the Jacobian matrix. This approach avoids the introduction and training of neural network emulator, which brings a substantial saving in development time. This approach is used by Beaufays.

The GN controller uses the frequency variation sample for predicting the load, disturbance and that load disturbance is used by GN controller to control the plant dynamics.

Computer stimulations have been conducted to illustrate the behaviour of single area system to step load perturbation and the performance of GNN controller is compared with the integral controller as shown in Figs. 6.35 and 6.36. ANN and GN based load frequency controller also compared and GN based controller found better as shown in Fig. 6.37.

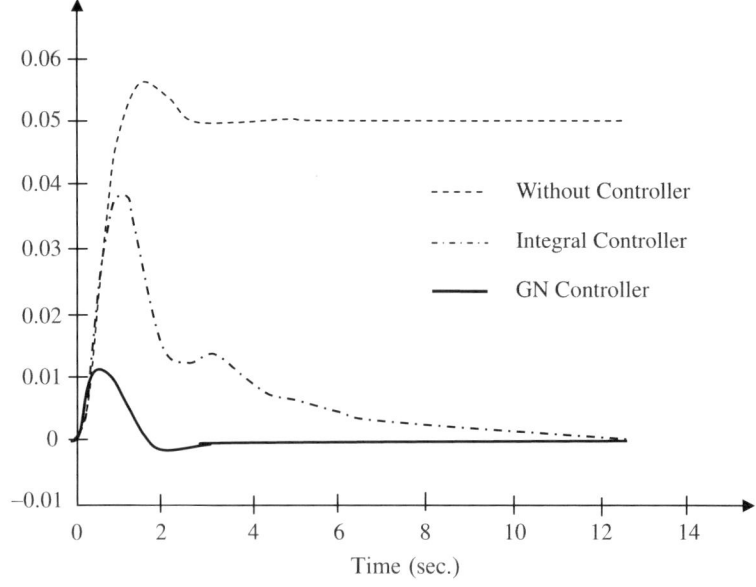

Fig. 6.35. Variation in tie line power of two area system

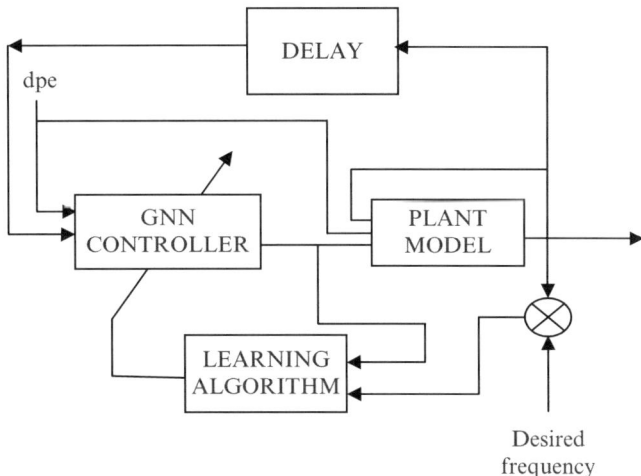

Fig. 6.36. Block diagram of load frequency control of power systems using GNN controller

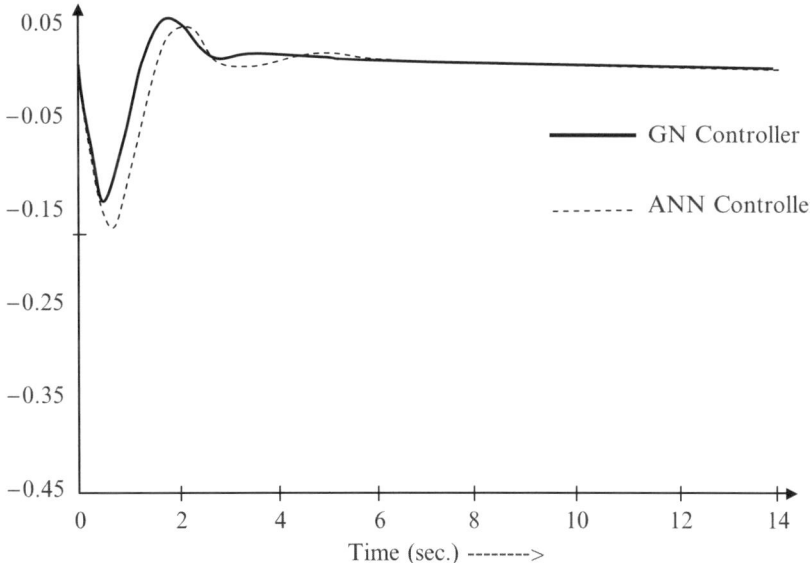

Fig. 6.37. Comparison of ANN and GN controller performance for a single area system for df

6.3.5.2 GN Controller of a Two Area System

The GN control scheme for a two-area system is basically the same as for one area system. The level variable vector of two area system was [df1 df2 dp12] after step load perturbation has occurred in one area or simultaneously in

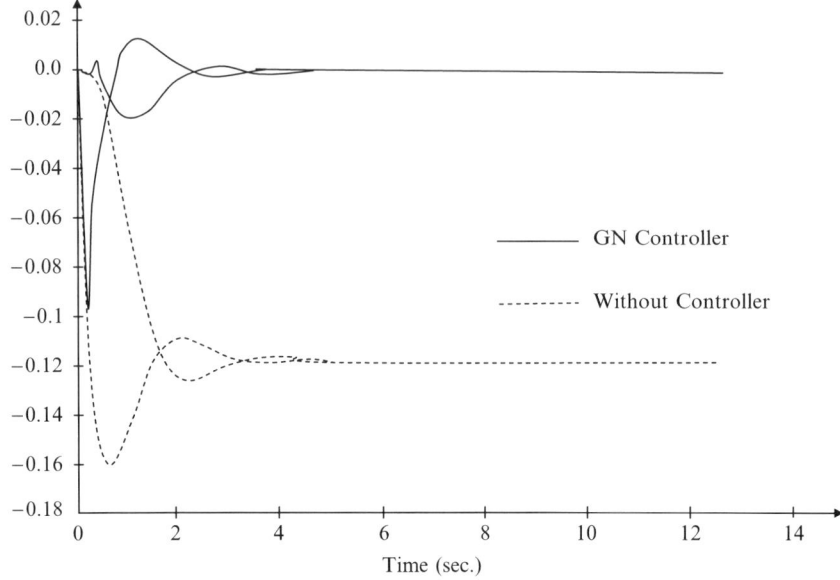

Fig. 6.38. Frequency variation in two area system with and without controller

both areas the level variable vector deviates from its steady state value and the controller will control these variables but the controller output is not instantaneous. The controller starts functioning after some delay time so that the GN estimator can estimate the load perturbation and that signal is available for the controller. The controllers of both areas start controlling when there is any variation in the frequency from its normal value. The performance of GN controller of two area system is simulated on computer and the results are compared without controller and with integral controller. The performances of GN controller is also tested under different values of regulation parameter R. it is seen that when the value of R increases the frequency oscillations also increases, but die out after certain time. Beaufays et al. (1999), in his work controls the same plant dynamics using conventional ANN consisting of 20 hidden neurons and one output neuron. In the present work the same plant dynamics can be controlled by one generalized neuron. The frequency deviations for both the areas are shown in Fig. 6.38.

6.3.5.3 Results and Discussion

GN model has been successfully applied to control the turbine reference power of a single area system. The same principle has been applied to a simulated to a two-area system. The GN controllers have been trained using back propagation through time. The GN controller is found very suitable for controlling the plant dynamics in relatively less time. Each GN controller receives only local information about the system (frequency in that specific area). Such architecture decentralizes the control of the overall system and reduces the amount of information to be exchanged between different modes of the power grid. The successful application of this generalized neural network for the load frequency control of power system motivates to use this technique for estimation of load disturbance on the basis of frequency deviation (df) and voltage variations (dv) at the different buses.

The GN controller with different error functions is also simulated. Table 6.28 shows the frequency deviation for single area system when controlled by conventional controller and GN based controller with different error functions. Figure 6.39 shows the frequency deviation and change in turbine

Table 6.28. Comparison of conventional controller and GN based controller for single area system

Settling time (Ts) s	Frequency Dip (Hz)	Controller	
8.00	0.270	Conventional controller	
2.20	0.240	Sum squared EF	
2.25	0.200	Cauchy EF	
2.05	0.210	Geman EF	GN
2.50	0.245	Mean fourth power EF	controller

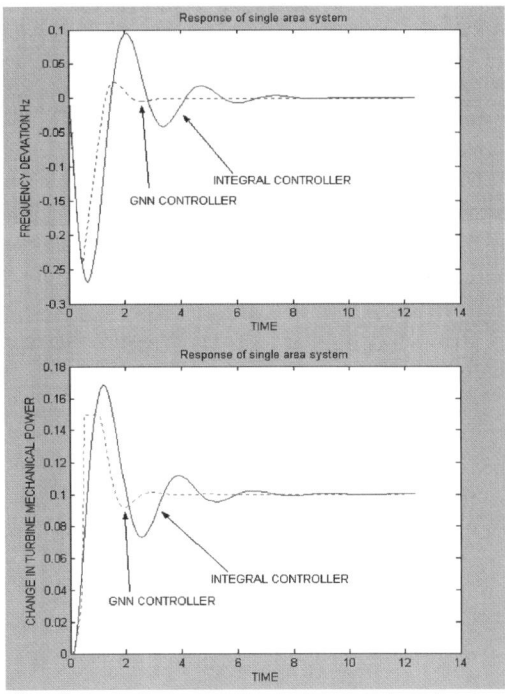

Fig. 6.39. Response of single area system using sum squared error function

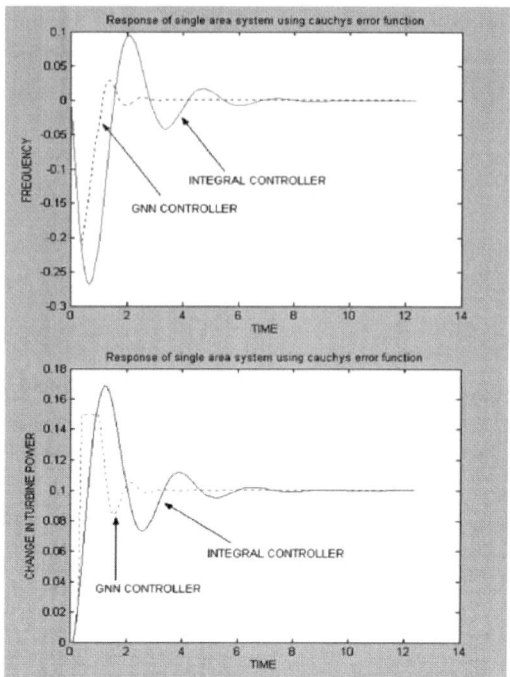

Fig. 6.40. Response of single area system using Cauchy's error function

power during disturbance. It is clear from the figure that GN controller with sum squared EF during training has less overshoot and less settling time as compared to integral controller.

Figure 6.40 represents the performance of GN controller with Cauchy's EF. In this case, there are slightly more oscillations in both frequency deviation and change in turbine power as compared GN controller with sum squared EF, although much less than integral controller. At the same time the dip in frequency deviation is less.

Figure 6.41 portrays the performance of GN controller with Geman's EF is much better than GN controller with sum squared EF and integral controller in terms of settling time, overshoot and oscillations. It is also comparable to the GN controller with Cauchy's EF.

The results show that frequency deviation and change in turbine power in the single area system when the disturbance is applied when GN controller with mean fourth power EF is used (refer Fig. 6.42). The performance of GN controller with this EF is better than GN controller with sum squared EF, but not as good as GN controller with Cauchy's and Geman's EFs. The performance of GN controller with different EF is also tried for two area system and the results are given in Table 6.29 and Figs. 6.43–6.47.

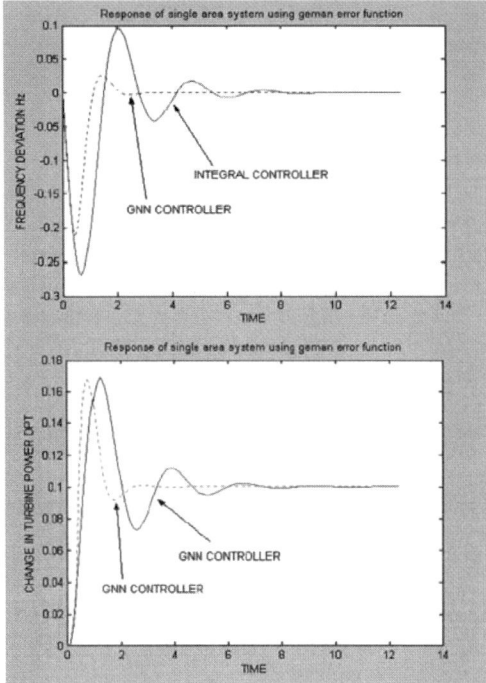

Fig. 6.41. Single area system using Geman EF

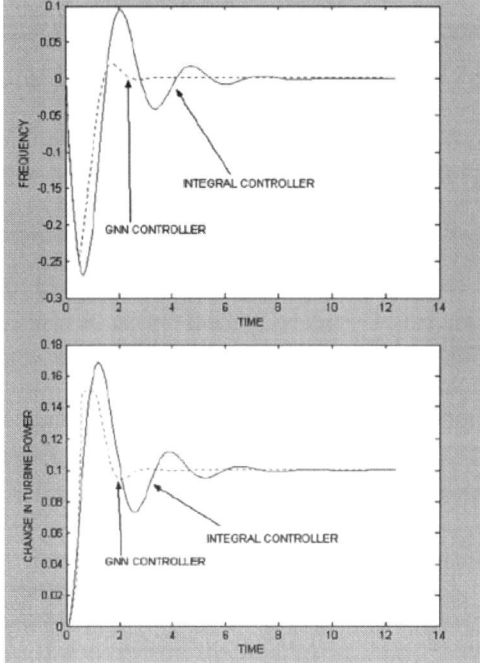

Fig. 6.42. Response of single area system using MPFE error function

Table 6.29. Comparison of conventional controller and GN based controller for two area system

Frequency of areas #1		Frequency of areas #2		Controller	
Ts (s)	Dip (Hz)	Ts (s)	Dip (Hz)		
				Conventional controller	
3.0	0.155	2.80	0.050	Sum squared EF	
2.8	0.180	2.20	0.045	Cauchy EF	GN based
2.9	0.180	2.40	0.040	Geman EF	Controller
3.1	0.175	2.35	0.050	Mean fourth power EF	

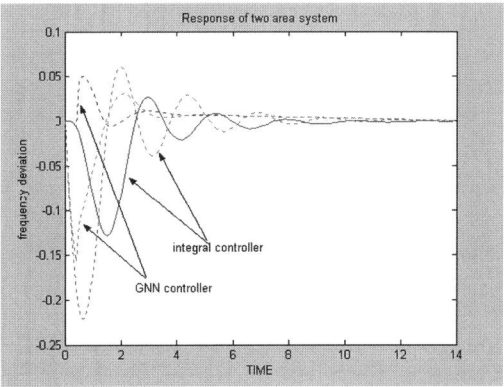

Fig. 6.43. Response of two area system using SSE error function

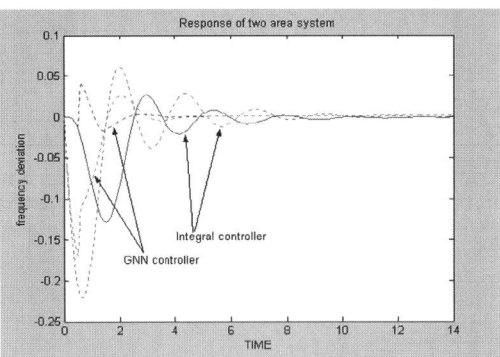

Fig. 6.44. Response of two area system using Cauchy's error function

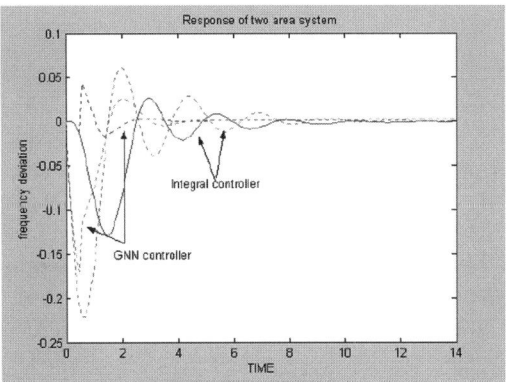

Fig. 6.45. Response of two area system using GEMAN error function

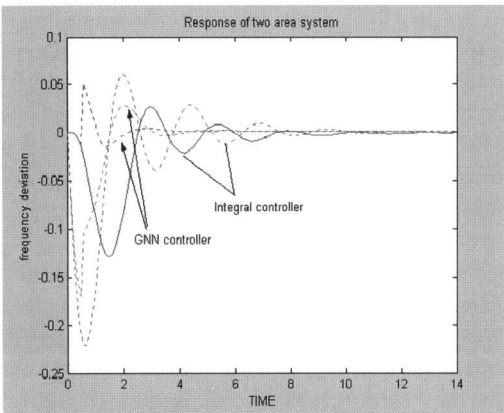

Fig. 6.46. Response of two area system using MPFE error function

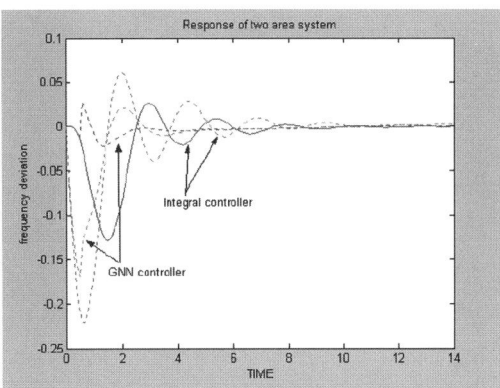

Fig. 6.47. Response of two area system using combined error function

6.3.5.4 Conclusions

Generalized neuron model has been successfully applied to control the turbine reference power of a computer – simulated generator unit. The GN based controller has been made adaptive using on-line back-propagation learning through time. GN based controller is found quite suitable for controlling plant dynamics in relatively less time. The same GN controller can also be used to control two area system. The GN controllers use local information about the system frequency in that area. Such architecture decentralises the control of the overall system and reduce the amount of information to be exchanged between different areas of the power grid.

6.4 Power System Stabilizer Problem

Earlier individual area requirements were met by a single generating station. An operator was quite capable of manually adjusting the generator outputs to suit the needs of customers in surrounding area. The evolution and revolution of electric supply technology from small, isolated, multiple – kilowatt generating plants of late 19th century to today's multiple gigawatt generating plants is one of the outstanding development of the last century. Due to increased power demands, small isolated system were unable to supply power with efficient load frequency and voltage control and a reasonable degree of reliability. In order to properly meet the large fluctuations in the load due to varied number of power users, to lower overall production cost and save through the utilization of a diversified system, individual generating stations and eventually, utility companies joined together over a transmission and distribution network to form larger power pools. The deregulated environment is changed the power sector completely. Now the power quality and reliability is one of the most important aspects along with the cost in the power market.

To improve the power quality, in the late 1950s most of the generating units added to electric utility system equipped with continuously acting automatic voltage regulators (AVRs). These generators have a detrimental impact upon the steady state stability of power system. Oscillations of small magnitude and low frequency often persisted for a long time and in some cases presented limitations on power transfer capability. These electro-mechanical oscillations between parallel connected generators were of concern because they:

1. Gave rise to periodic variations in electrical voltages and phase angles at the load buses in the system,
2. Caused excessive wear of mechanical control components,
3. Caused inadvertent operation of protection devices on the system or on connected equipment.
4. Excited sub harmonic torsional shaft oscillations on large multistage turbine units, or
5. Led to dynamic instability.

It is well known that the AVRs introduced negative damping to weakly damped interconnected systems. A supplementary control signal in the excitation system and/or the governor system of a generating unit can provide extra damping for the system and thus improve the unit's dynamic performance (DeMello and Laskowski 1979). Power system stabilizers (PSSs) aid in maintaining power system stability and improving dynamic performance by providing a supplementary signal to the excitation system. This is an easy, economical and flexible way to improve power system stability. Over the past few decades, PSSs have been extensively studied and successfully used in the industry.

The conventional PSS (CPSS) was first proposed in the 1950s based on a linear model of the power system at some operating point to damp the low frequency oscillations in the system. Linear control theory was employed as the design tool for the CPSS. After decades of theoretical studies and field experiments, this type of PSS has made a great contribution in enhancing the operating quality of the power system (DeMello et al., 1978; Larsen and Swann 1981).

With the development of power systems and increasing demand for quality electricity, it is worthwhile looking into the possibility of using modern control techniques. The linear optimal control strategy is one possibility that has been proposed for supplementary excitation controllers (Ohtsuka et al. 1986). Preciseness of the linear model to represent the actual system and the measurement of some variables are major obstacles to the application of the optimal controller in practice.

A more reasonable design of the PSS is based on the adaptive control theory as it takes into consideration the non-linear and stochastic characteristics of the power systems (Pierre 1987; Zhang et al. 1993). This type of stabilizer can adjust its parameters on-line according to the operating condition. Many years of intensive studies have shown that the adaptive stabilizer can not only provide good damping over a wide operating range but more importantly, also can solve the coordination problem among stabilizers.

Power systems being dynamic systems, the response time of the controller is the key to a good closed loop performance. Many adaptive control algorithms have been proposed in the recent years. Generally speaking, the better the closed loop system performance is, the more complicated the control algorithm becomes, thus needing more on-line computation time to calculate the control signal.

6.4.1 Conventional PSS

A practical CPSS [6] with the shaft speed input may take the form as shown in the following transfer function

$$u_{pss} = -k_g \frac{sT_w}{(1 + sT_w)} \frac{(1 + sT_1)(1 + sT_3)}{(1 + sT_2)(1 + sT_4)} \Delta\omega(s)$$

The commonly used PSS (CPSS) is a fixed parameter device designed using the classical linear control theory and a linear model of the power system at a specific operating point. It uses a lead/lag compensation network to compensate for the phase shift caused by the low frequency oscillation of the system during perturbation. By appropriately tuning the parameters of the lead/lag network, it is possible to make a system have the desired damping ability. Although this type of PSS has made great contribution in enhancing the operating quality of the power systems, it suffers from some problems.

Power systems are highly non-linear systems. They operate over a wide range of operating conditions and are subject to multi-modal oscillations (Larsen et al. 1981). The linearized system models used to design fixed parameter CPSS can provide optimal performance only at the operating point used to linearize the system. Therefore, the following problems are presented in the design of the CPSS:

1. Selection of a proper transfer function that covers the frequency range of interest.
2. Automatic tracking of the system operating conditions.
3. Maintaining properly tuned parameters as system changes.

Application of the adaptive control theory can take into consideration the non-linear and stochastic characteristics of the power system (Ohtsuka et al. 1986). Parameters of an adaptive stabilizer are adjusted on-line according to the operating conditions. Many years of intensive studies have shown that the adaptive stabilizer can provide good damping over a wide operating range (Ohtsuka et al. 1986; Zhang et al. 1993; Swidenbank et al. 1999; Hiyama and Lim 1989) and can also work in coordination with CPSSs (Changaroon et al. 2000; Segal et al. 2000).

More recently, artificial neural networks (ANNs) and fuzzy set theoretic approach have been proposed for power system stabilization problems (Swidenbank et al. 1999; Abido and Abdel-Magid 1999; Hosseinzadeh and Kalam 1999; Hsu and Chen 1991; Hornik et al. 1989). Both techniques have their own advantages and disadvantages. The integration of these approaches can give improved results.

The commonly used neuron model has been modified to obtain a generalized neuron (GN) model using fuzzy compensatory operators as aggregation operators to overcome the problems such as large number of neurons and layers required for complex function approximation, that not only affect the training time but also the fault tolerant capabilities of the ANN (Fausett Laurence 1994). Application of this GN as an adaptive PSS (GNAPSS) is described in the following section.

6.4.2 GN Based PSS and its Training

A block diagram of the GN controller and power system is shown in Fig. 6.48. The power system consists of a single machine connected to infinite bus

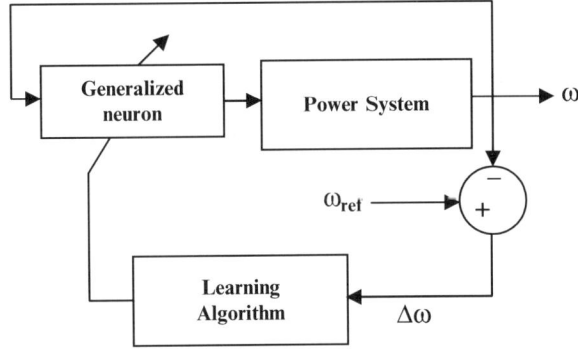

Fig. 6.48. Block diagram of GN based PSS

through a double circuit transmission line. The angular speed of synchronous machine, sensed at a fixed time intervals is used as input to the GN based PSS (GNPSS). The GNPSS calculates the output or control action. The dynamic model of the synchronous machine infinite bus system and its parameters are given in the Appendix.

Training of an ANN is a major exercise. Performance of GN based PSS depends upon the training of the GN. Data used for training must cover most of the working range and working conditions in order to get good performance. Of course it is impossible to train any GN under all working conditions that the controller is likely to meet. Still most of the working conditions must be included in the training. The current and past three generator speed signals (i.e. $\omega(t)$, $\omega(t\text{-}T)$, $\omega(t\text{-}2^*T)$, and $\omega(t\text{-}3^*T)$, where T is the sampling period), and past three values of the PSS output are used as inputs to the GN. Hence, the input vector for the GN can be written as:

$$X_i = [\omega(t), \omega(t\text{-}T), \omega(t\text{-}2^*T), \omega(t\text{-}3^*T),$$
$$u(t\text{-}T), u(t\text{-}2^*T), u(t\text{-}3^*T)] \tag{6.33}$$

where ω – angular speed in rad/s.

The output of the GN is the control signal u, which is a function of the angular speed and past control signals.

Training data for the GN is acquired from the system controlled by the CPSS, which is tuned for each operating condition. The GN is trained off-line over a wide working range of the generator operating conditions i.e. output ranging from 0.1 to 1.0 pu and the power factor ranging from 0.7 lag to 0.8 lead. Similarly, a variety of disturbances are also included in the training, like change in reference voltage, governor input torque variation, one transmission line outage and three phase fault on one circuit of the double circuit transmission line.

6.4.3 Comparison of GN and ANN PSS

The generalized neuron model is much less complex compared to a three-layered ANN proposed earlier for PSS. These ANNs were 30-10-1 (Zhang et al. 1995), 20-20-1 (Zhang et al. 1993) and 35-1 (Zhang et al. 1993a). Taking, for illustration purposes, an ANN with one hidden layer and much smaller number of neurons, a comparison of structural complexity associated with ANN and generalized neuron model is given in Table 6.14.

It is clear from Table 6.30 that the number of interconnections for a GNM is very small as compared to ANN. Hence the number of unknown weights is reduced drastically, which ultimately reduces the training time and training data required. A comparison of the performance of the GN and ANN based PSS is given in Fig. 6.49.

6.4.4 Simulation Results of GN Based PSS

A number of simulation studies were first performed to study the performance of the GNPSS.

A. *CPSS parameter tuning*

With the generator operating at $P = 0.9$ pu and $Q = 0.4$ pu lag, a 100 ms three phase to ground fault is applied at 0.5 s at the generator bus. The CPSS is carefully tuned under the above conditions to yield the best performance and its parameters are kept fixed for all studies.

B. *Performance under three-phase to ground fault*

The results have been compared for the GNPSS and conventional PSS for a 100 ms three-phase to ground fault at generator bus under the following operating conditions:

  1. $P = 0.9$ pu and $Q = 0.4$ pu lag,
  2. $P = 0.7$ pu and $Q = 0.3$ pu lead and
  3. $P = 0.2$ pu and $Q = 0.2$ pu lead

The results are shown in Fig. 6.46 for deviation in angular speed. Because CPSS has been tuned for $P = 0.9$ pu, $Q = 0.4$ pu lag, performance at

Table 6.30. Comparison of network complexity involved in ANN and GNM

Components	ANN	GNM
Structure of ANN/GN	(7-7-1)	(1)
Number of neurons used	08	01
Number of layers containing processing neurons	02	01
Number of interconnections	56	15
Training epochs required to reach error level of 0.001	64,7000	8,700

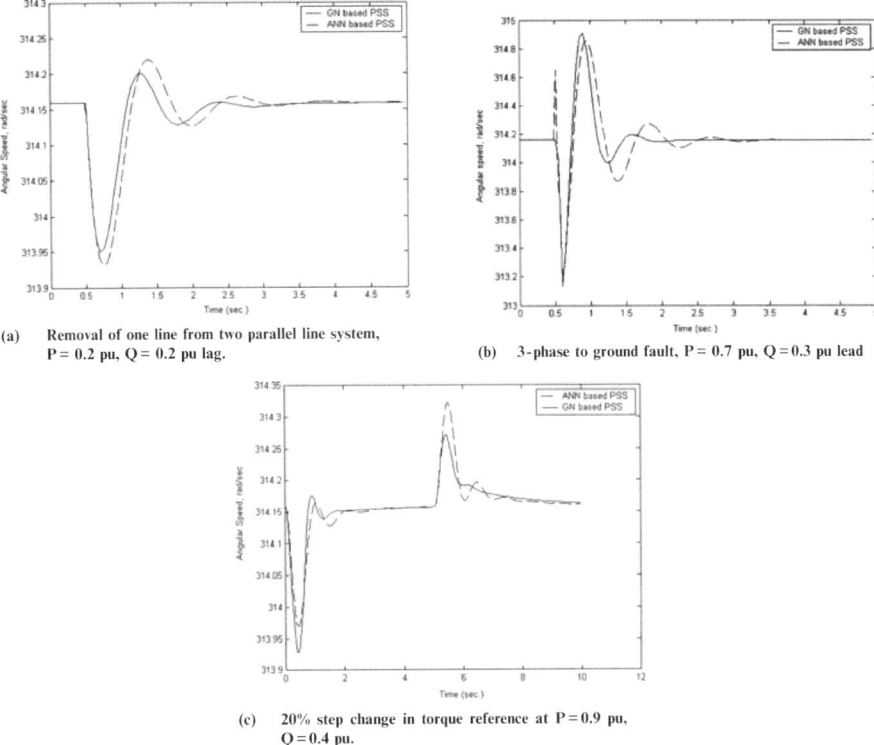

(a)   Removal of one line from two parallel line system,
       P = 0.2 pu, Q = 0.2 pu lag.

(b)   3-phase to ground fault, P = 0.7 pu, Q = 0.3 pu lead

(c)   20% step change in torque reference at P = 0.9 pu,
       Q = 0.4 pu.

Fig. 6.49. Comparison of the performance of ANN and GN based PSSs

this operating condition as shown in Fig. 6.50a is practically the same for
both the GNPSS and the CPSS. System performance at other operating
conditions, as given in Figs. 6.50b and c, is better with GNPSS than a
fixed parameter CPSS.

C. Performance with one line removed

The results have been compared under different operating conditions such
as P = 0.9 and Q = 0.4 pu lead, P = 0.7 pu and Q = 0.3 pu lead, P =
0.2 pu and Q = 0.2 pu lag when one line is removed from the system
with two parallel transmission lines operating initially. It can be seen
from Fig. 6.51 that the GNPSS damps out the oscillations very effectively.
Because the GNPSS is trained for a wide range of operating conditions,
it is able to adjust the control output to that suitable for the working
conditions.

D. Performance under reference operating point changes

Applying the change in several small steps instead of a large step can reduce the severity of reference changes. It is also possible to apply a ramp
with a small gradient in order to change the system reference settings.

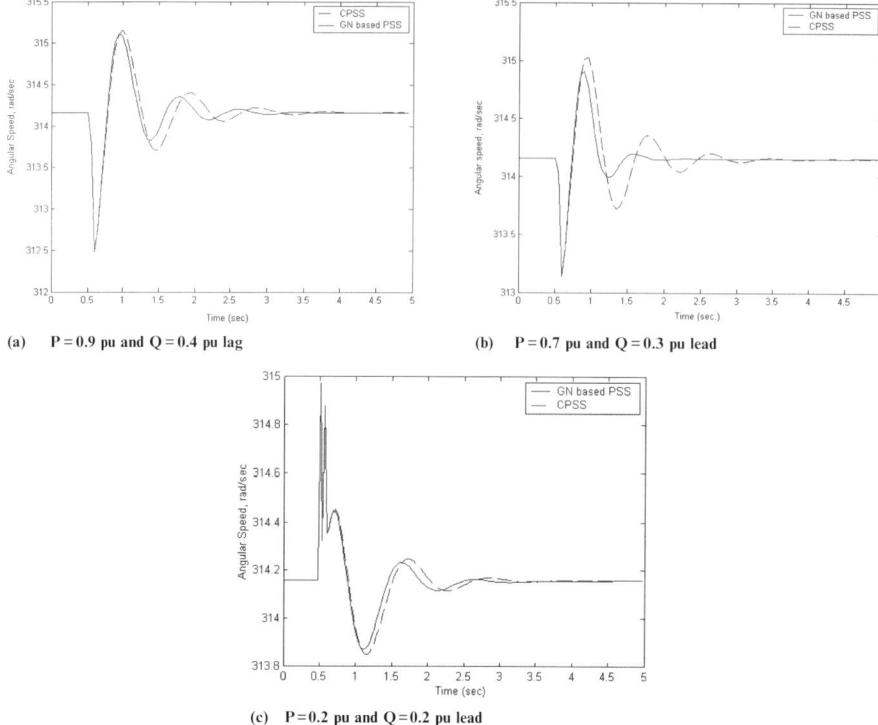

(a)   P = 0.9 pu and Q = 0.4 pu lag

(b)   P = 0.7 pu and Q = 0.3 pu lead

(c)   P = 0.2 pu and Q = 0.2 pu lead

Fig. 6.50. Performance of GN-PSS and CPSS for a 3-phase ground fault

Both these things are done by the GNPSS to reduce the severity of reference changes. The performance of GNPSS has been evaluated for step changes in reference setting in Vref and Pref.

1. *Step change in Governor reference (Pref)*
   The GNPSS performance is studied for a sudden change in the governor reference by 20% to its initial value. The results given in Fig. 6.52 show that the angular speed deviations are damped quickly with the GNPSS.

2. *Step change in voltage reference (Vref) of AVR*
   A step change of 5% to its initial value was applied to Vref under the same operating condition as in the case of Pref change. The variation in angular speed is shown in Fig. 6.53.

E. *Performance under different H values*
   Performance of the GNPSS under different H values varying from 5 to 25 for a 20% step change in Pref is shown in Fig. 6.54. The results are consistently good.

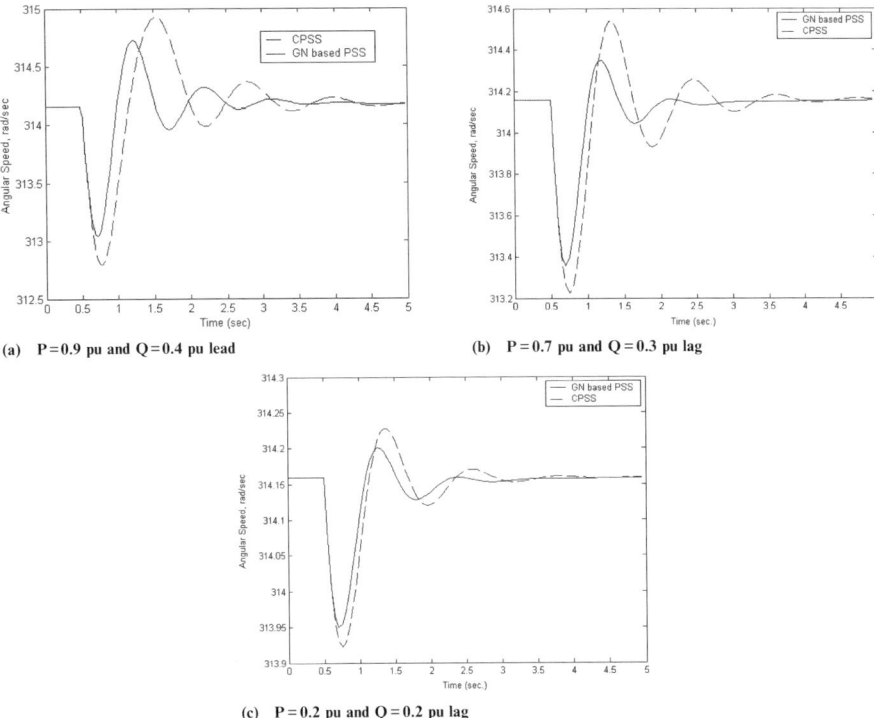

(a)   P = 0.9 pu and Q = 0.4 pu lead

(b)   P = 0.7 pu and Q = 0.3 pu lag

(c)   P = 0.2 pu and Q = 0.2 pu lag

Fig. 6.51. Performance of GN-PSS based PSS when one line is removed from the circuit

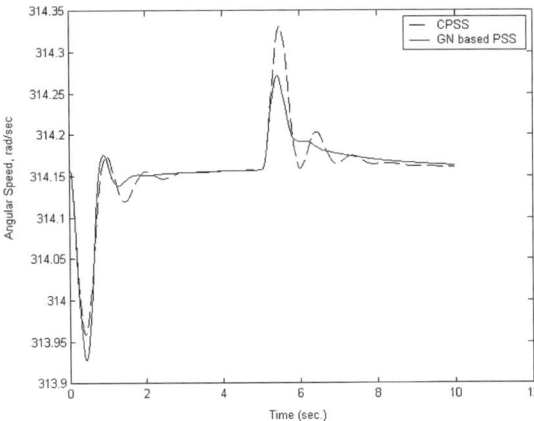

Fig. 6.52. Performance of GN based PSS when 20% step change in Pref at P = 0.9 pu and Q = 0.4 pu lag

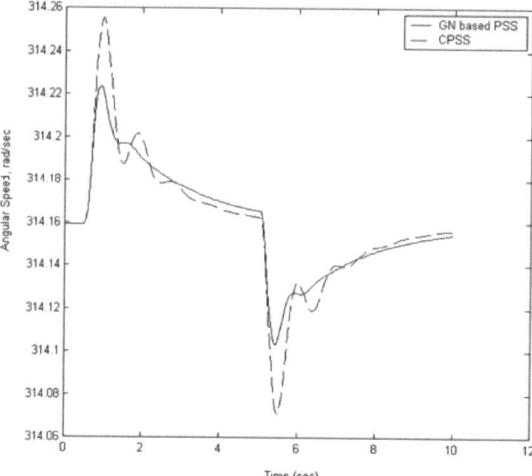

Fig. 6.53. Performance of GN based PSS when 5% step change in Vref at P = 0.9 pu and Q = 0.4 pu lag

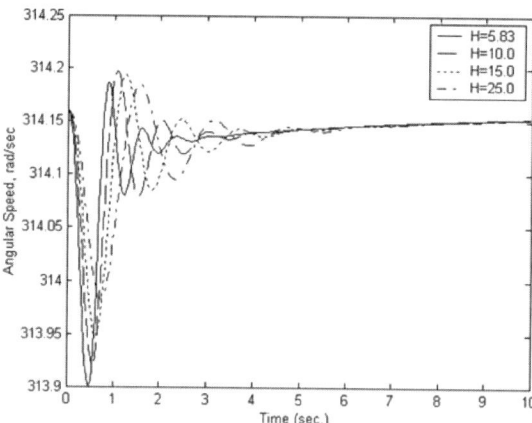

Fig. 6.54. Performance of GN based PSS for different values of H under step change in Pref

6.4.5 Experimental Test

The behavior of the proposed GNPSS has been further investigated on a physical model in the Power System Research Laboratory at the University of Calgary, Alberta, Canada. The physical model consists of a three-phase 3-kVA micro-synchronous generator connected to a constant voltage bus through a double circuit transmission line model. The transmission lines are modeled by six Π sections, each section is equivalent of 50 km length. The transmission line parameters are the equivalent of 1,000 MVA, 300 km and 500 kV. A field time constant regulator has been employed to adjust the transient field time constant (Tdo') to the desired value (Huber et al. 1972).

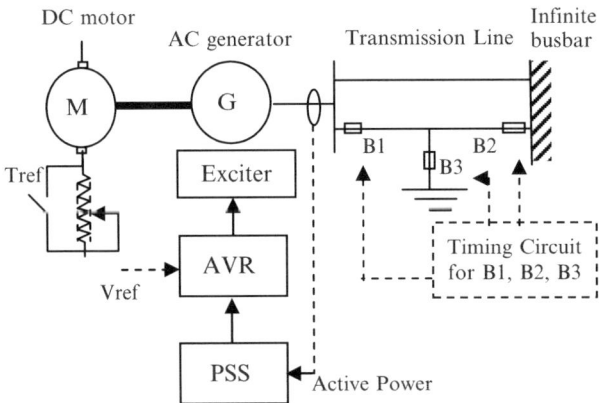

Fig. 6.55. Experimental setup for Laboratory Power System model

The governor turbine characteristics are simulated using the micro-machine prime mover. It can be achieved by dc motor which is controlled as a linear voltage to torque converter. An overall schematic diagram of this physical model is given in Fig. 6.55. The Laboratory model mainly consists of the turbine M, the generator G, the transmission line model, the AVR, DSP *board and Man-machine interface.*

The GNPSS control algorithm is implemented on a single board computer, which uses a Texas Instruments TMS320C31 digital signal processor (DSP) to provide the necessary computational power. The DSP board is installed in a personal computer with the corresponding development software and debugging application program. The analog to digital input channel of DSP board receives the input signal and control signal output is converted by the digital to analog converter. The IEEE type PSS1A CPSS is also implemented on the same DSP, with a 1ms sampling period. The following tests have been performed on the experimental set up to study the performance of the GNPSS and CPSS.

A. *Step change in power reference (pref)*

The experiment is performed on the micro-synchronous generator under the following operating conditions:

1. 0.67 pu active power and 0.9 lagging power factor, and
2. 0.25 pu active power and 0.8 leading power factor.

A disturbance of 30% step decrease in reference power was applied at 0.5 and again increased to the same initial value at 4.5. The change in generator electrical power with GNPSS and CPSS is shown in Fig. 6.56. The proposed controller exhibits fast and well-controlled damping.

B. *Transient faults*

To investigate the performance of the GNPSS under transient conditions caused by transmission line faults various tests on the experimental set up have been conducted.

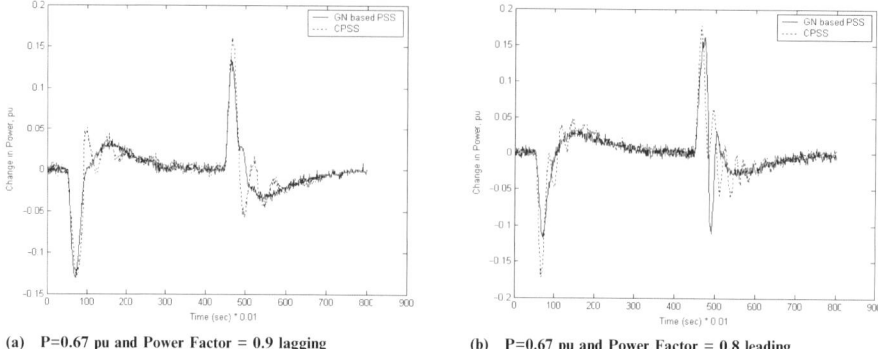

(a)   **P=0.67 pu and Power Factor = 0.9 lagging**   (b)   **P=0.67 pu and Power Factor = 0.8 leading**

Fig. 6.56. Experimental results under 30% step change in power reference (Pref)

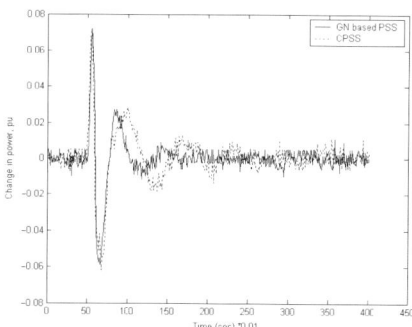

Fig. 6.57. Experimental results of single-phase fault at P = 0.25 pu and power factor = 0.8 lagging

1. *Single-phase to ground fault test*
   In this experiment, the generator was operated at P = 0.25 pu and 0.8 pf lag. At this operating condition and with both lines in operation, a single-phase to ground fault was applied in the middle of one transmission line for 100 ms. The system performance is shown in Fig. 6.57. It can be observed that the GNPSS provides faster settling.
2. *Two-phase to ground fault test*
   The two-phase to ground fault test has been performed for the following two operating conditions:
   1. P = 0.25 pu and 0.8 lagging power factor, and
   2. P = 0.67 pu and 0.8 leading power factor
   at the middle of one transmission line. The results of these experiments shown in Fig. 6.58 are consistently better with the GNPSS.
3. *Three-phase to ground fault test*
   A 100 ms three-phase to ground fault was applied at different operating conditions at the middle of one transmission line at 0.5.

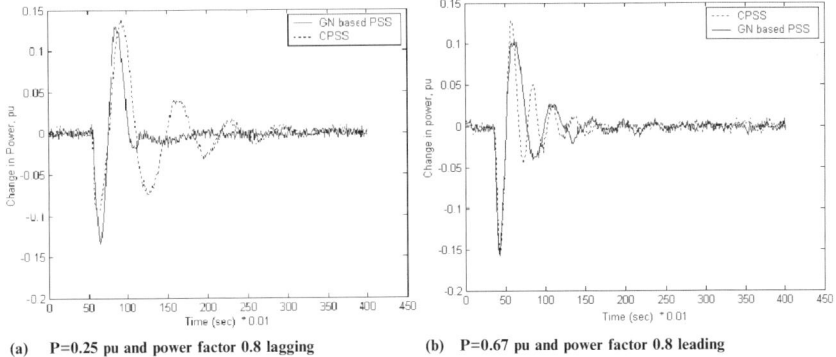

(a)   P=0.25 pu and power factor 0.8 lagging

(b)   P=0.67 pu and power factor 0.8 leading

Fig. 6.58. Experimental results of 2-phase ground fault

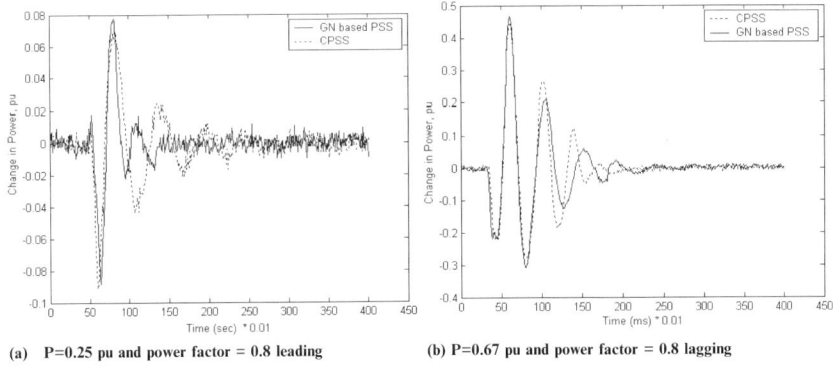

(a)   P=0.25 pu and power factor = 0.8 leading

(b) P=0.67 pu and power factor = 0.8 lagging

Fig. 6.59. Experimental results of 3-phase to ground fault

Illustratative results for two tests at P $=$ 0.25 pu, 0.8 pf lead, and P $=$ 0.67 pu, 0.8 pf lag are given in Fig. 6.59. The results show that the GNPSS provides consistently good performance.

4. *Successful re-closing*
A three-phase to ground fault is applied on one of the transmission lines and the faulty line is opened. After clearing the fault, transmission line is automatically re-closed. The results are shown in Fig. 6.60. Both the overshoot and settling time for GNPSS are smaller.

C. *Removal of one line*
One line is removed at 0.5 s and again connected at 5.8 s. Figure 6.61 shows that in this type of fault also the system performance with the GNPSS is very good in terms of damping the oscillations.

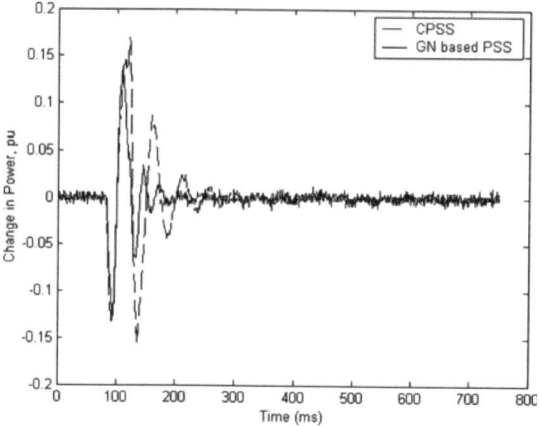

Fig. 6.60. Successful re-closing at P = 0.5 pu and pf = 0.8 leading

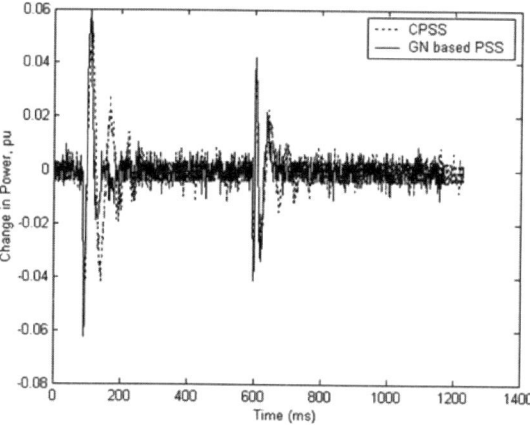

Fig. 6.61. Removal of one line and reconnected at P = 0.5 pu and pf = 0.8 lagging

6.4.6 Adaptive GN Based Power System Stabilizer

Most of the work done on adaptive PSS uses self-tuning adaptive control approach as it is a very effective adaptive control scheme. The structure of a self-tuning adaptive controller has two parts: an on-line plant model predictor and a controller.

The plant model is updated by the on-line predictor each sampling period to track the dynamic behavior of the plant. Then a suitable control strategy is used to calculate the control signal based on the updated plant model. Any one out of a number of control strategies, such as minimum variance, generalized minimum variance, pole assignment, pole shift (PS) control (Bollinger et al. 1975; Clarke 1981) can be used in the self-tuning adaptive control.

Studies have shown that an adaptive PSS can adjust its parameters on-line according to the changes in environment, and maintain desired control ability over a wide operating range of the power system. Taking advantage of the neural networks to easily accommodate non-linearities and time dependencies of non-linear dynamic systems, a GN is used to develop an adaptive PSS.

6.4.6.1 GN Predictor

Identification procedure includes setting up a suitably parameterized identification model and adjusting the parameters of the model to optimize a performance function based on the error between the plant and the identified model output.

A schematic diagram of the GN based plant predictor using forward modeling is shown in Fig. 6.62. A GN predictor is placed in parallel with the system and has the following inputs:

$$X_i(t) = [y\_vector, \ u\_vector] \tag{6.33}$$

where
  y_vector=[y(t), y(t-T), y(t-2T), y(t-3T)]
  u_vector=[u(t-T), u(t-2T), u(t-3T)]

T is the sampling period, y is the plant output and u is the controller output.

For application as a PSS, the plant output, y, may be the generator speed deviation or the deviation of the generator output power. The PSS output is the plant input, u, added at the input summing junction of the AVR.

The dynamics of the plant can be viewed as a non-linear mapping as below:

$$y \ (t+T) = f_i(X_i(t)) \tag{6.34}$$

Therefore, the GN-predictor for the plant can be represented by a non-linear function Fi.

$$y_i(t + T) = F_i(X_i(t), \ W_i(t)) \tag{6.35}$$

where, $W_i(t)$ is the matrix of GN predictor weights at time instant t.

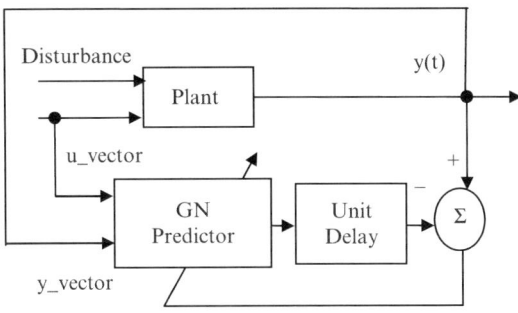

Fig. 6.62. Schematic diagram of the proposed GN predictor

6.4.6.2 Training of GN-Predictor

Training of the proposed GN Predictor has two steps, off-line training and on-line update.

Off-line training of the GN predictor for the PSS was performed with data acquired from simulation studies on a generating unit model, equipped with a governor and an AVR, and connected to a constant voltage bus through a double circuit transmission line. The seventh order model of the synchronous machine, the transfer functions of the AVR and governor and the parameters are given in the Appendix. In off-line training, the GN Predictor is trained for a wide range of operating conditions, i.e. output power ranging from 0.1 to 1.0 pu and the power factor ranging from 0.7 lag to 0.8 lead. Similarly, a variety of disturbances, such as change in reference voltage, input torque variation, one transmission line outage and three phase fault on one circuit of the double circuit transmission line, are also included in the training.

Error between the system output and the GN predictor output at a unit delay, called the performance index, $J_i(t)$, of the GN predictor is used as the GN predictor training signal:

$$J_i(t) = \frac{1}{2}[y_i(t) - y(t)], \tag{6.36}$$

where

$y_i(t)$ is the GN predictor output with unit delay.
$y(t)$ is the actual plant output.

The weights of the GN predictor are updated as

$$W_i(t) = W_i(t - T) + \Delta W_i(t),$$

where $\Delta W_i(t)$, change in weight depending on the instantaneous gradient, is calculated by

$$\Delta W_i(t) = -\eta_i J_i(t + T)\frac{\partial J_i(t)}{\partial W_i(t)} + \alpha \Delta W_i(t - T), \tag{6.37}$$

where

η_i – learning rate and
α_i – momentum factor for GN predictor.

The off-line training is performed with 0.1 learning rate and 0.4 momentum factor. After off-line training is finished, i.e. the average error between the plant and the GN-predictor outputs converges to a small value, the GN-predictor represents the plant characteristics reasonably well, i.e.

$$y(t + T) = f_i (X_i(t)) \approx y_i(t + T) = F_i (X_i(t), W_i(t)),$$

it is connected to the power system for on-line update of weights. The learning rate and momentum factor are very crucial factors in on-line updating and greatly affect the performance of the GN predictor. If the value of learning rate is high then the response of the GN may go unstable, and if it is too low then time required to modify its behavior is large. The momentum factor is used to overcome the problem of local minima of GN. Hence, for on-line training, the values of these factors were chosen carefully based on the previous experience.

On-line performance of the GN predictor on a physical model of the generating unit connected to a constant voltage bus for a transient three-phase to ground fault of 100 ms duration is shown in Fig. 6.63. Performance in response to the removal of one line from the double circuit transmission line and re-energized after 5 s is shown in Fig. 6.64. The error, difference between the speed predicted by the GN predictor and the system speed, can be seen to be very small. GN identifier is also experimentally test for 23% step change in reference torque change, voltage reference change and 3-phase to ground fault on one of the transmission line from double circuit transmission as shown in Fig. 6.65.

6.4.6.3 GN Controller

A schematic diagram of the GN controller is shown in Fig. 6.66. The plant consists of the single machine connected to the constant voltage bus as described above. The last four sampled values of the output are used as input to the GN controller. Besides the output, the past three control actions are

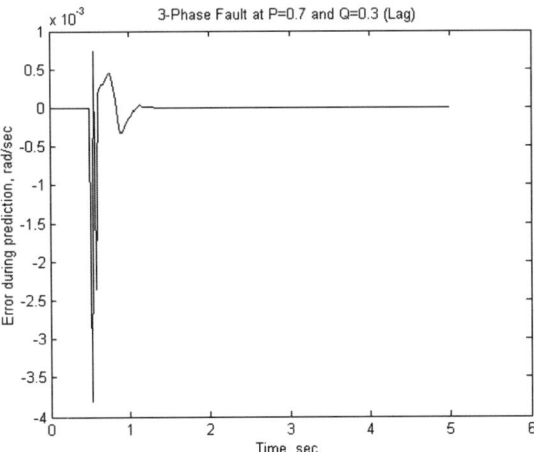

Fig. 6.63. Performance of GN predictor for a 100 ms 3-phase to ground transient fault at generator bus

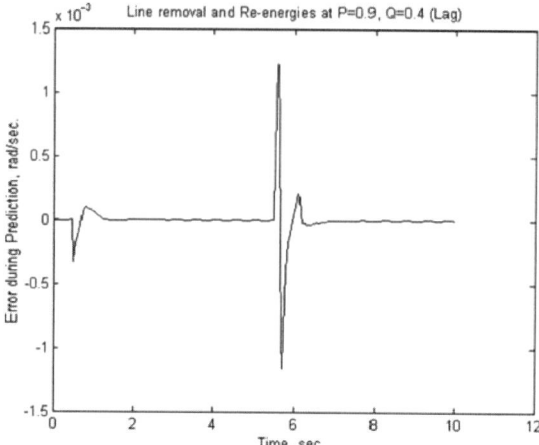

Fig. 6.64. Performance of GN predictor when one line is removed at 0.5 s and reenergized at 5.5 s

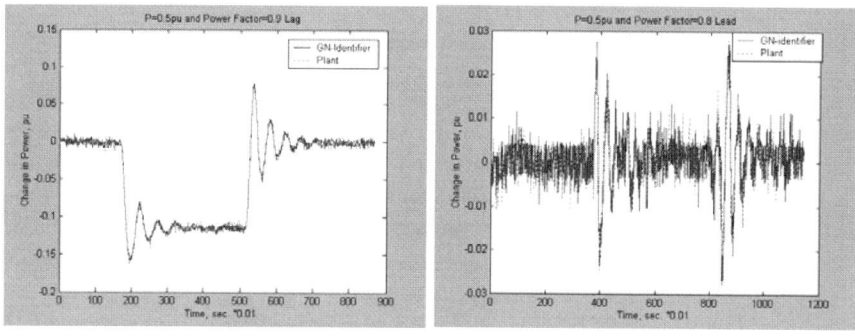

(a) 23% step change in Torque reference.   (b) 15% step change in voltage reference.

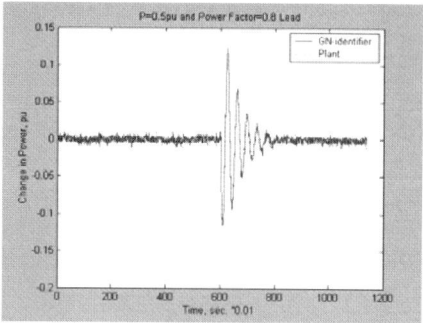

(c) Three-phase to ground fault for 100ms on one transmission line

Fig. 6.65. Experimental results of GN-predictor

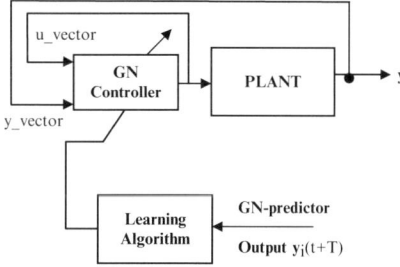

Fig. 6.66. Schematic diagram for GN controller

also given to the GN controller as inputs. These inputs are normalized in the range 0.1–0.9. The output of the GN-controller is the control signal u(t).

$$u(t) = F_c(X_i(t), W_c(t)), \tag{6.38}$$

where, Wc(t) is the matrix of neural controller weights at time instant t. The u(t) is de-normalized to get the actual control action and then sent to the plant and the GN-predictor simultaneously.

D. Training of GN controller
Training of the proposed GN controller is also done in two steps – off-line training and on-line update. In off-line training, the GN controller is trained for a wide range of operating conditions and a variety of disturbances similar to those used in training the predictor. Off-line training data for the GN controller has been acquired from the system controlled by the CPSS. For this purpose, the CPSS was tuned for each operating condition.

The performance index of the neural controller is

$$J_c(t) = \frac{1}{2}[y_i(t+T) - y_d(t+T)]^2, \tag{6.39}$$

where $y_d(t+T)$ is the desired plant output at time instant $(t+T)$. In this study it is set to be zero.

The weights of the GN controller are updated as

$$W_c(t) = W_c(t-T) + \Delta W_c(t) \tag{6.40}$$

ΔWc(t), change in weight depending on the instantaneous gradient, is calculated by

$$\Delta W_c(t) = -\eta_c \omega_i(t+T)\frac{\partial J_c(t)}{\partial u(t)}\frac{\partial u(t)}{\partial W_c(t)} + \alpha_c \Delta W_c(t-T), \tag{6.41}$$

where

η_c learning rate for GN controller.
α_c – momentum factor for GN controller.

Off-line training is started with small random weights (± 0.01) and then updated with relatively high learning rate and momentum factor ($\eta_c = 0.1$ and $\alpha_c = 0.4$).

After off-line training is finished, the proposed controller is connected to the power system for on-line update with learning rate ($\eta_c = 0.001$) and momentum factor ($\alpha_c = 0.01$). In on-line updating of GN-controller weights, expected error is calculated from the one step ahead predicted output, $y_i(t + T)$, of the GN-predictor. The expected error is then used to update the weights on-line. Parameters of the GN predictor and controller are adjusted every sampling period. This allows the controller to track the dynamic variations of the power system and provide the best control action.

GN adaptive PSS has been tested experimental for various operating conditions and shown good performance.

1. *Single-phase to ground fault test*
   In this experiment, with the generator operating at P = 0.8 pu, 0.9 pf lead, a transient 100 ms single-phase to ground fault was applied in the middle of one transmission line. The system performance is shown in Fig. 6.67. It can be observed that the GNAPSS is able to reduce the magnitude of system oscillations.
2. *Two-phase to ground fault test*
   Results of a transient two-phase to ground fault test at P = 0.8 pu, 0.9 pf lead at the middle of one transmission line are shown in Fig. 6.68.
3. *Three-phase to ground fault test*
   A transient three-phase to ground fault was applied for 100 ms at different operating conditions at the middle of one transmission line at 0.5 s.

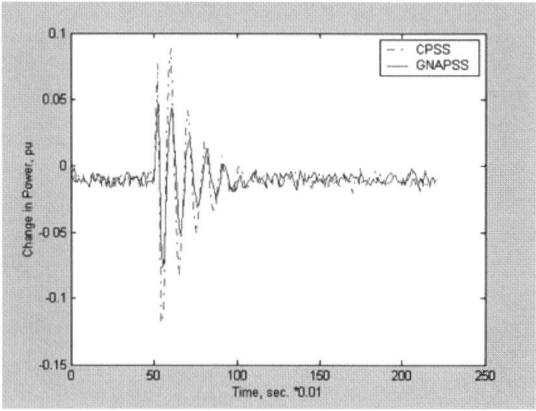

Fig. 6.67. Experimental results for a single-phase fault at P = 0.8 pu, 0.9 pf lead

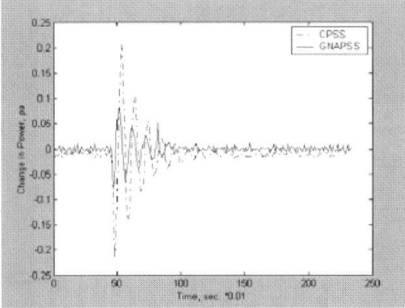

Fig. 6.68. Experimental results for a two-phase to ground fault at P = 0.8 pu, 0.9 pf lead

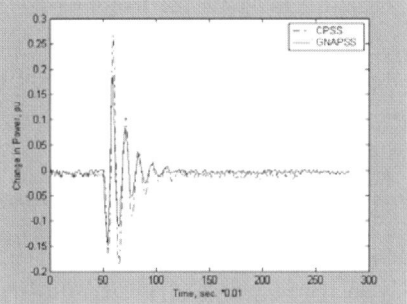

Fig. 6.69. Experimental results for a 3-phase to ground fault at P = 0.5 pu, 0.9 pf lag

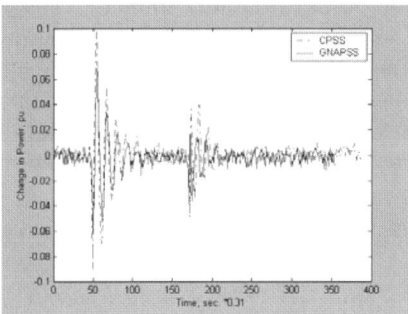

Fig. 6.70. Removal of one line and reconnection

An illustrative result at P = 0.5 pu, 0.9 pf lag is given in Fig. 6.69. Results show that the GNAPSS provides consistently good performance even though the disturbance changes significantly in severity.

4. *Removal of one line*

In this test, with the generator operating at P = 0.8 pu, 0.9 pf lag, one circuit of the double circuit line was disconnected at 0.5 s and re-connected at 1.75 s. The results in Fig. 6.70 show that in this type of fault also the

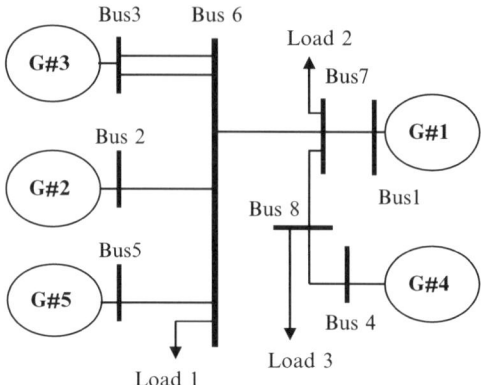

Fig. 6.71. Schematic diagram of a five-machine power system

system performance with the GNAPSS is very good in terms of damping the oscillations.

6.4.6.4 GN Based Adaptive PSS for Multi-Machine System

A five-machine power system without infinite bus, that exhibits multi-mode oscillations, Fig. 6.71, is used to study the performance of the previously trained GNAPSS. In this system, generators #1, #2 and #4 are much larger than generators #3 and #5. All five generators are equipped with governors, AVRs and exciters. This system can be viewed as a two area system connected through a tie line between buses #6 and #7. Generators #1 and #4 form one area and generators #2, #3 and #5 form another area. Parameters of all generators, Transmission line parameters, loads and operating conditions are given in the Appendix. Under normal operating condition, each area serves its local load and is almost fully loaded with a small load flow over the tie line.

6.4.6.5 Performance of GNAPSS with Torque Disturbance

1. *Simulation studies with GNAPSS installed on one generator:*
   The GNAPSS is trained for a single machine infinite bus system and the same parameters (weights) are used for GNAPSS with multi-machine system. The proposed GNAPSS is installed only on generator #3 and CPSSs with the following transfer function are installed on generator #1, and #2:

$$u_{pss} = K_s \frac{sT_5}{(1 - sT_5)} \frac{(1 + sT_1)}{(1 + sT_2)} \frac{(1 + sT_3)}{(1 + sT_4)} \Delta P_e(s) \qquad (6.42)$$

The GNPSS was trained by data obtained from the system controlled by a CPSS following the procedure explained in Sect. IIB. The following

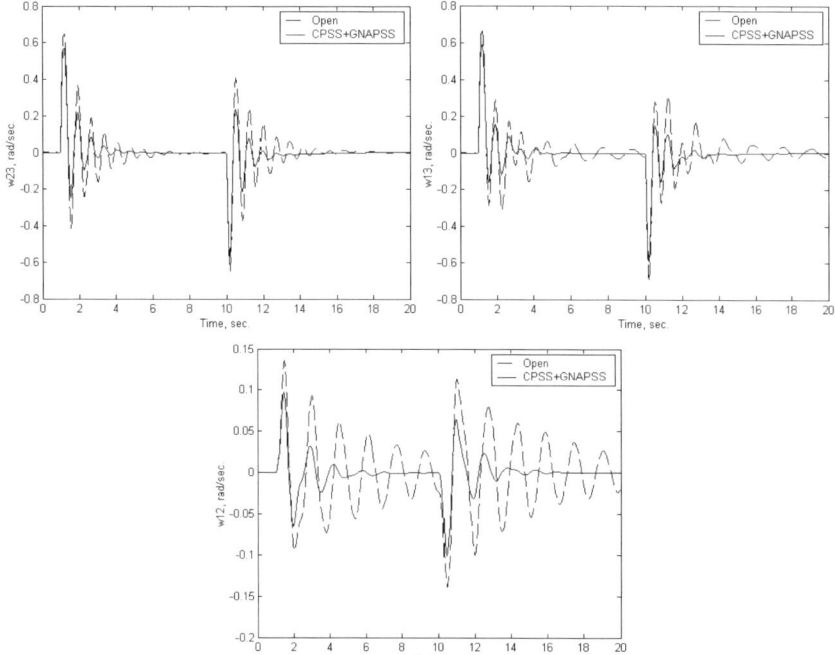

Fig. 6.72. System response with CPSS installed at generators #1, #2 and GNAPSS at #3 for 30% Step change in Pref

parameters are set for the fixed parameter CPSS for all studies in the multi-machine environment:

$$Ka = 0.2, \ T1 = T3 = 0.07, \ T2 = T4 = 0.03, \ T5 = 2.5.$$

Speed deviation of generator #3 is sampled at a fixed time interval of 30 ms. The system response is shown in Fig. 6.72 for the operating conditions given in the Appendix. Each part of the figure shows difference in speed between two generators.

2. *GNAPSS installed on three generators*:
   In this test, GNAPSSs are installed on generators #1, #2 and #3. A 30% step decrease in mechanical input torque reference of generator #3 was applied at 1 s and returns to its original level at 10 s. The simulation results of only GNAPSSs and only CPSSs applied at generators #1, #2, and #3 are shown in Fig. 6.73. It is clear from the results that both modes of oscillations are damped out very effectively.

6.4.6.6 Three Phase to Ground Fault

In this test, a three phase to ground fault is applied at the middle of one transmission line between buses #3 and #6 at 1 s and the faulty line is removed

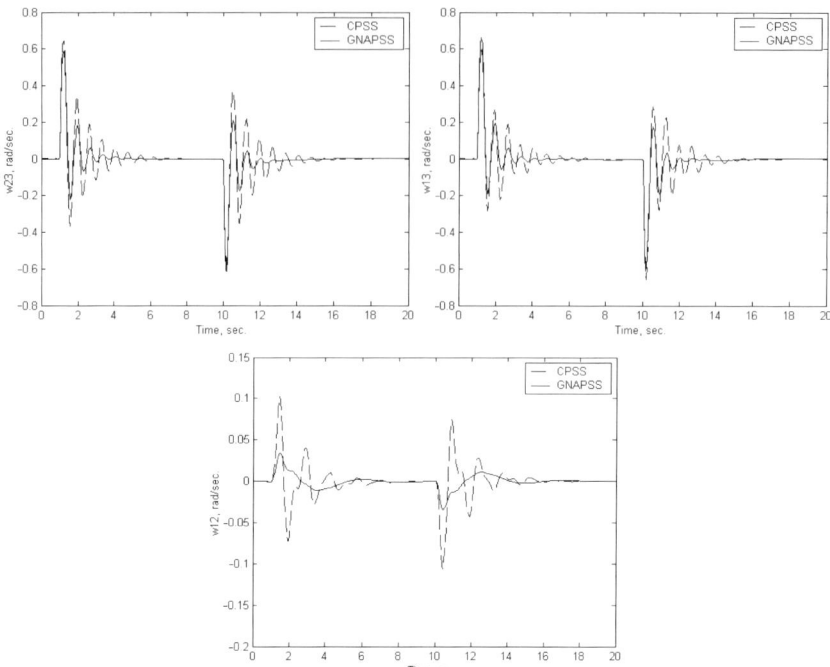

Fig. 6.73. System response under change in Tref with only GNAPSS and only CPSS installed on G1, G2, and G3

100 ms later. At 10 s, the faulty line is restored successfully. The GNAPSSs are installed on all five generators. The system responses are shown in Fig. 6.74. The results with CPSSs installed on the same generators are also shown in the same figures. From the system responses, it can be concluded that although the CPSS can damp the oscillations caused by such a large disturbance; the proposed GNAPSS has much better performance.

6.4.6.7 Coordination Between GNAPSS and CPSS

The advanced PSSs would not replace all CPSSs being operated in the system at the same time. Therefore, the effect of the GNAPSS and CPSSs working together needs to be investigated. In this test, the proposed GNAPSS is installed on generators #1 and #3 and CPSSs on generators #2, #4 and #5. The operating conditions are the same as given in the Appendix. A 0.2 pu step decrease in the mechanical input torque reference of generator #3 is applied at 1 s and returns to its original level at 10 s. The system responses are shown in Fig. 6.75. The results demonstrate that the two types of PSSs can work cooperatively to damp out the oscillations in the system. The proposed GNAPSS input signals are local signals. The GNAPSS coordinates itself with the other PSSs based on the system behaviour at the generator terminals.

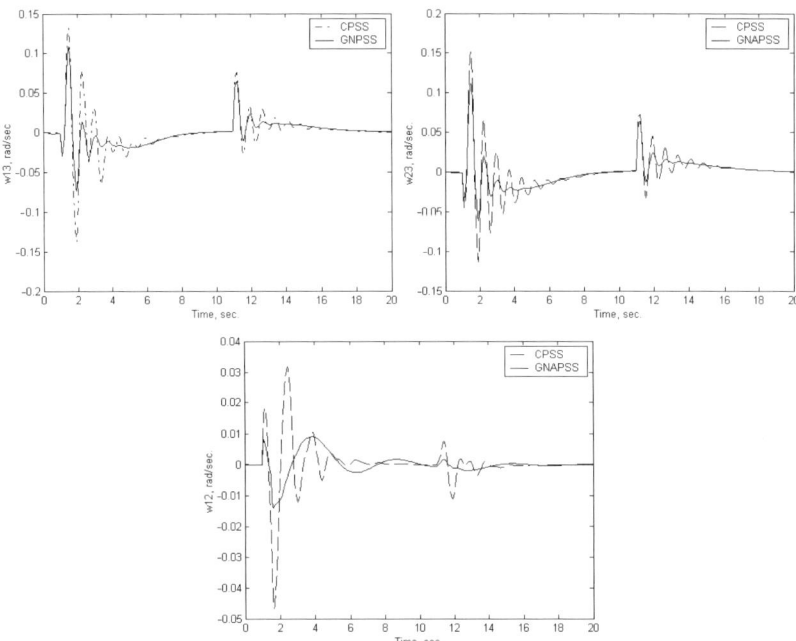

Fig. 6.74. System response with only GNAPSS and only CPSS installed on all five machines for 3-phase to ground fault

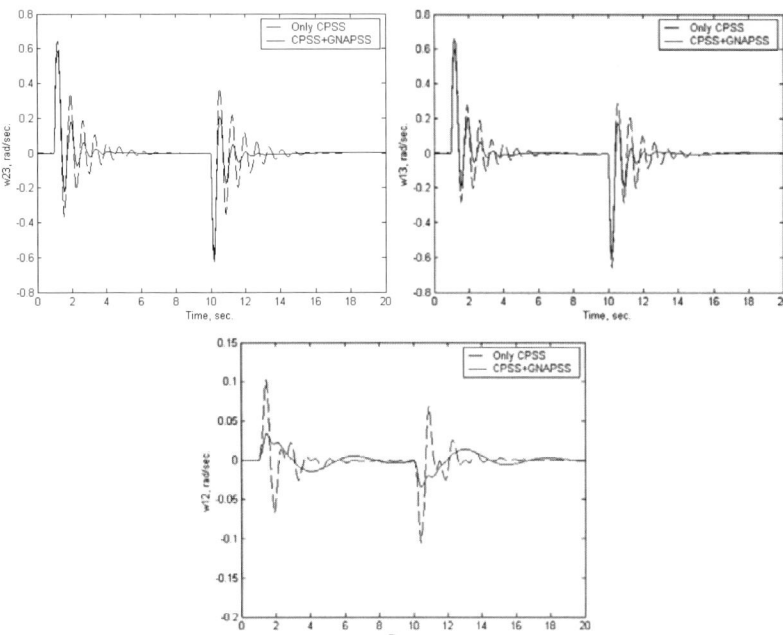

Fig. 6.75. System response with GNAPSS at G1, G3 and CPSS on G2, G4, G5. for ±0.3 pu step change in torque reference

6.4.7 Conclusions

A GN based PSS is adaptive in nature and having learning capabilities is described in this paper. It can incorporate the non-linearities involved in the system. It uses only one neuron and is trained using back-propagation learning algorithm. Because it has a much smaller number of weights than the common multi-layer feed-forward ANN, the training data required is drastically reduced. Training time is also significantly reduced, because the number of weights to be determined is much less than an ANN.

GN has been employed to perform the function of a PSS to improve the stability and dynamic performance of the power system. Computer simulation studies described in the paper show that the performance of the GN based PSS can provide very good performance over a wide range of operating conditions.

The proposed GN based PSS has been implemented on a DSP and its performance investigated on a physical model of a single machine infinite bus system under various operating conditions and disturbances such as transient faults, one line removal from double circuit transmission, change in reference point, etc. It is found that the system performance with the GN based PSS is consistently good indicating that it can adapt to changing operating conditions.

6.5 Aircraft Landing Control System Using GN Model

It is observed that landing performance is the most typical phase of an aircraft performance. During landing operation the stability and controllability are the major considerations. To achieve a safe landing, an aircraft has to be controlled in such a way that its wheels touch the ground comfortably and gently within the paved surface of the runway.

The conventional control theory found very successful in solving well defined problems which are described precisely with definite and clearly mentioned boundaries. In real life systems the boundaries can not be defined clearly and conventional controller does not give satisfactory results.

Whenever, an aircraft deviates from its glide path (gliding angle) during landing operation, it will affect the landing field, landing area as well as touch down point on the runway. To control correct gliding angle (glide path) of an aircraft while landing, various traditional controllers like PID controller or state space controller as well as manoeuvring of pilots are used, but due to the presence of non-linearities of actuators and pilots these controllers do not give satisfactory results.

Since artificial neural network can be used as an intelligent control technique and are able to control the correct gliding angle, i.e. correct gliding path of an aircraft while landing through learning which can easily accommodate the aforesaid non-linearities. The existing neural network has various drawbacks such as large training time, large number of neurons and hidden layers required to deal with complex problems. To overcome these drawbacks and develop a non-linear controller for aircraft landing system a generalized neural network has been developed.

6.5.1 Introduction

The basic limitation of conventional control theory is the need to know the precise mathematical model of the system to be controlled. This information is seldom available. The effect of disturbances and unmodeled dynamics on the performance of the system must also be taken into account for real-life systems.

Practically, one rarely has precise knowledge of the system model. Furthermore, the system model may vary with time, e.g., the dynamic equations of an aircraft near sea-level are very different from those of the same aircraft at high altitudes. In such cases, to maintain the controller performance, it is necessary to use an adaptive controller, so that it could adopt the variations in system parameters and operating conditions.

6.5.2 Aircraft Landing System

The aircraft performance characteristics such as maximum speed, rate of climb, time to climb, range and take-off are all predicted from estimates of the variation of the lift, drag and thrust forces as functions of angle of attack, altitude and throttle setting respectively. The forces acting on aircraft during flying are lift (L), drag (D), thrust (T) and weight (W) of the aircraft. The flight path of the aircraft can be controlled within the limitations of its aerodynamic characteristics and structural strength, through control over the equilibrium angle of attack (α), angle of side slip (β), angle of bank (ϕ) and the output of the power plant.

The APPROACH AND LANDING exercise deals with landing of the aircraft from the turn on to downwind position to the completion of the landing run. A good landing follows a steady approach, hence it is important that the circuit be standardised so that the final approach is such as to facilitate a good landing. The final landing approach begins when the flight path is aligned with the runway in preparation for straight ahead descent and landing and ends when aircraft contacts the landing surface. Thus final approach may be considered to have five distinct phases listed below and also shown in Fig. 6.76 (Codvin 1942; Kermode 1970, 1984).

1. Level approach
2. Approach descent
3. Roundout
4. Float
5. Touchdown

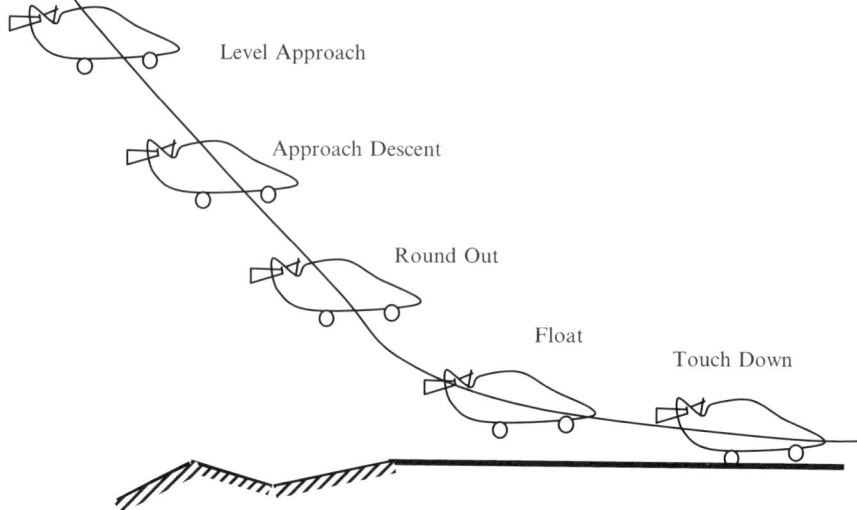

Fig. 6.76. Final Approach extents from line upto touch down in aircraft landing

The final approach involves a descent from the circuit height (1,000 feet) to the landing check height. The pilot should plan the approach so that the pilot gets sufficient time to judge the descent and position himself to carry out the straight final approach from at least 400 ft. This gives sufficient time to retrim the aircraft and concentrate on the final approach. The final turn-in should be adjusted so as to line up with the landing path (glide path) and s-turn should be avoided on the final approach. The adjustment of the rate of descent will depend on the pilot's judgment and type of approach. There are two types of landing approaches (Neil 1957; Nelson 1990; Pallet 1983; Robert 1978):

1. *Glide landing approach* in which the pilot must judge the point from where to start the approach, taking the prevailing wind into consideration. The steepness of the descent can be adjusted by intelligent use of flaps. Full flaps being lowered only when positive of making the intended touch-down point safely. On no account should be allow his speed to drop below that recommended, in an effort to stretch his glide. This approach is done under the influence of the force of gravity and without the use of the engine.
2. *Engine assisted landing approach*, in which the descent should be started slightly earlier than for the glide approach. The approach path is adjusted with the throttle, speed being maintained by the elevators. This approach is recommended for all modern aircraft due to following advantages:
   1. Permits a flatter approach as the glide path of some aircraft is uncomfortably steep.
   2. The approach path can be adjusted as required.
   3. The stalling speed is low, the approach can be carried at a speed, thus reducing the landing run.

4. Use of engine improves elevator and rudder control of propeller-driven aircraft.
5. Landing is safer, quicker and more accurate.

6.5.3 Drawbacks of Existing Landing Control System

1. The ILS is a pilot interpreted system, pilot is directly depending on the signals transmitted by ILS, he has to blindly follow the path shown. Even slight mistake can lead to disastrous landing.
2. The ILS serviceability checks and calibrated are to be carried out by the flying aircraft only and ground technicians cannot ensure its serviceability merely on ground.
3. GCA system requires a high resolution radar and high talented ground controller to guide pilot (Taylor and Parmar 1983; Houghton and Corpenter 1996; Webb 1971).
4. In GCA system radar picks up the aircraft position and displays it on the screen and ground controller in turn guides the pilot, but little parallax error in this may take away from the actual centre line of the runway (Barres and Cormic 1995; Courtland et al. 1949; Nagaraja 1975).
5. Atmospheric conditions are drastically changing and conventional controller are unable to cope up with these conditions.
6. The inherent non-linearities in system components and cognition level of human being have led researchers to consider neural network techniques to build a non-linear **ANN** controller a with high efficiency of performance.

The existing simple neural networks have numerous deficiencies as stated below:

1. The number of neurons required in hidden layers is large for complex function approximation.
2. The number of hidden layers required for complicated functions may be greater than three. Though it has been reported that a network with only three layers can approximate any functional relation (Mingye et al. 1998; Irie and Miyake 1988), it is found that the training time required is very large, which can be computationally very expensive.
3. The fault tolerant capabilities of the existing neural networks are very limited.
4. Existing neural networks require a large number of unknowns to be determined for complex function approximation. This increases the requirement of the minimum number of input–output pairs.

In the present work an automatic controller for aircraft landing system has been developed using Neural Networks to increase the accuracy and efficiency of the control system at the time of landing (Nguyen and Widrow 1990). The aim of this controller is to restore the glide path, if it is deviating from the desired path in the shortest time.

6.5.4 Mathematical Model Development of Aircraft During Landing

There are four forces lift (L), drag (D), thrust (T) and weight (w) are acting through centre of gravity of an aircraft along different axes under different level of flight as shown in Fig. 6.77. Summation of forces along X and Z axes and total moments acting about the Y axis, yields the equations of static equilibrium for the aircraft in straight symmetric flight (Taylor and Parmar 1983).

$$\Sigma F_X = 0, \qquad \Sigma F_Z = 0, \text{ and} \qquad \Sigma Mcg = 0.$$

Considering the above cases following equations can be derived

$$T * \cos\alpha - D + W * \sin\beta = 0 \tag{6.43}$$
$$- T * \sin\alpha - L + W * \cos\beta = 0 \tag{6.44}$$
$$\Sigma Mcg = 0 \tag{6.45}$$
$$\Sigma Mcg = Cmcg * q * s * c \tag{6.46}$$

From the above equations following conclusions have been drawn.

1. The pitching moment coefficient (Cmcg) is a function of Cl. The equilibrium can be stabilized if the components of aircraft are proportioned to allow Cmcg = 0. Mcg = 0 at some useful lift coefficient (Cl). This useful lift coefficient can be calculated from the required lift which is used for equilibrium.
2. Since thrust T is a function of aircraft speed and throttle control, the rate of climb (R/C) or rate of descent (R/S) is regulated through throttle control. Hence the rate of climb or rate of descent is a function of thrust.
3. Velocity of aircraft (V) for a given wing loading (W/s) and altitudes are purely a function of lift coefficient and the lift coefficient is a function of angle of attack (α).

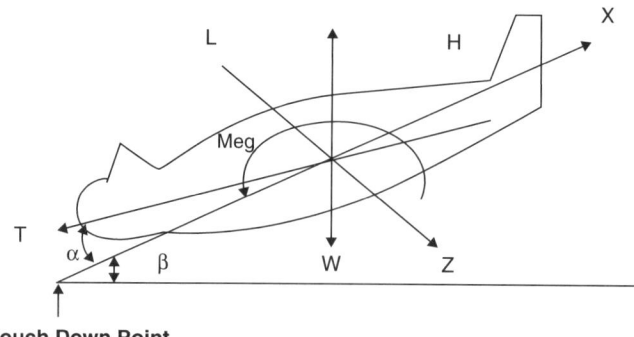

Touch Down Point

Fig. 6.77. Forces acting on an airplane during landing

4. Along shallow path of curve aircraft velocity (v) is a function of the angle of attack or lift coefficient.

In this work only gliding approach (engines are cut off) landing performances (i.e. round out, float and touch down) are considered. The aircraft will glide under the influence of the force of gravity and without the use of engine. The velocity of aircraft will be controlled by changing elevator angle with the help of actuator operation. By changing the elevator angle the angle of attack will be changed and hence, the gliding angle is controlled.

Since the angle of attack is relatively small angle, therefore $\cos \alpha = 1$ and $\sin \alpha = 0$.

Equations (6.43)–(6.46) can be rewritten as

$$D = W \sin \beta \text{ or } \sin \beta = D/W \tag{6.47}$$
$$L = W \cos \beta \text{ or } \cos \beta = L/W \tag{6.48}$$

The lift forces, drag and drag coefficient can be given as

$$L = 0.5 * Cl * v^2 * \sigma * s \tag{6.49}$$
$$D = 0.5 * Cd * v^2 * \sigma * s \tag{6.50}$$
$$Cd = Cdf + Cl^2/3.141 * A * e \tag{6.51}$$

Divide (6.47) to (6.48) and (6.50) to (6.49)

$$\tan \beta = D/L = Cd/Cl \tag{6.52}$$
$$\text{or} \quad \beta = \tan^{-1}(D/L) = \tan^{-1}(Cd/Cl) \tag{6.53}$$

From (6.48) and (6.49)

$$W * \cos \beta = 0.5 * Cl * v^2 * \sigma * s$$
$$v = \sqrt{[(2 * W^* \cos \beta p/(Cl * \sigma * s)]} \tag{6.54}$$

6.5.5 Develpoment of Landing Control System Using GN Model

The generalized neural network developed here is used to control the aircraft during landing using backpropagation through time learning algorithm (Werbos 1990; Widrow and Lehr 1990). The following assumptions have been taken for the development of the generalized neural network controller:

1. The gliding angle should be 5° to 7° and angle of attack should be 2° to 4°.
   a. The landing performance can be divided into two main phases:
   b. Transition from threshold to touch down including round out and float.
   c. Braked ground run.
2. Here only transition from threshold to touch down is considered.

a. The aircraft should start gliding after approaching the screen height with constant gliding angle.
b. Air density is constant at all altitude, since speed taken is indicated air speed.

The block diagram of generalized neural network controller and aircraft model structure is shown in Fig. 6.78. The controller should make the aircraft to follow the correct glide path. The aircraft variables and GNN controller parameters are given in Table 6.31. The gliding angle is sensed at every time instant and if there is any deviation in the gliding angle will be corrected by the GNN controller of aircraft. The inputs and output of GNN controller are given in Table 6.32 and received from the aircraft model block.

The vector M consisting of [β v L D] and fed to the GNN controller after a unit delay (i.e. at (t-1) time instant) as shown in Fig. 6.79.

The generalized neural network controller output u can be determined from the past value of vector M. The GNN controller modifies the lift coefficient depending on the control action u and the old value of Cl, and drag coefficient by the following equation:

$$Cl = Cl + u$$
$$Cd = Cdf + Cl^2/3.141 * A * e.$$

which in turn affect the lift and drag forces at every time instant (Δt). The ratio of D/L which also equals to Cd/Cl is compared with desired ratio of

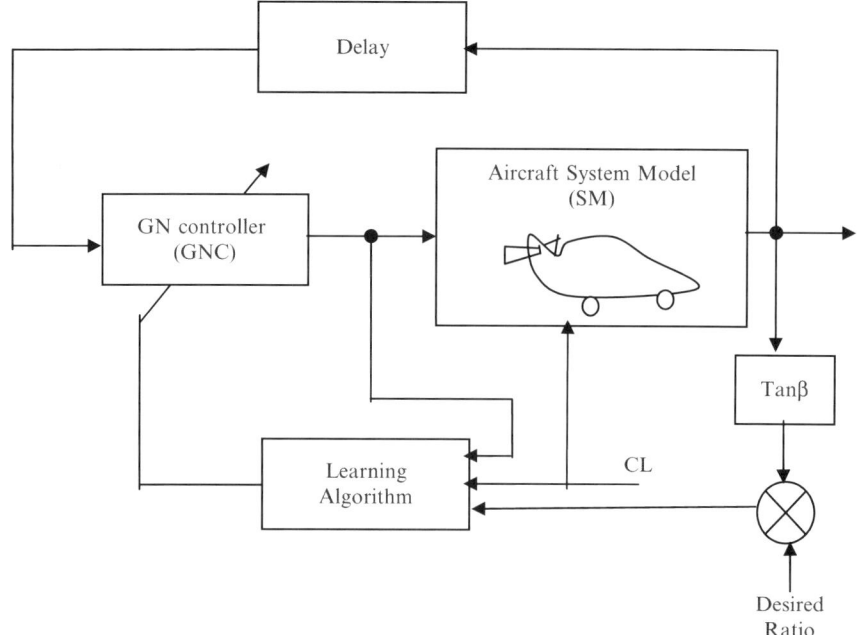

Fig. 6.78. Block Diagram for GN controller for aircraft landing

Table 6.31. Values of variables and parameters

Variable	Value
Lift coefficient	0.96
Desired L/D ratio	0.0875
Aspect ratio	9
Weight	3,000 lb
Cdf	0.02
Efficiency	0.8
Wing area	$100\,\text{ft}^2$
Air density	0.00248
Generalized neural network controller parameters	
Learning rate	0.0001
Lamda	0.10
Momentum	0.3

Table 6.32. GNN controller variables

Input variables	Output variable
Lift force (L)	Control action for lift coefficient (Cl)
Drag force (D)	
Velocity (v)	
Gliding angle (β)	

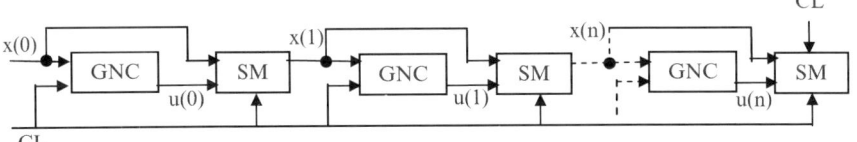

Fig. 6.79. GN controller and aircraft system model unfolded in time

Cd/Cl (i.e. 0.0875) and if there is any error then the controller will take a corrective action by changing the weights of the controller during learning.

One way to implement GNN controller is to build a neural network emulator of the aircraft model and backpropagation error gradient learning algorithm can be used. This is nothing but approximating the true Jacobean matrix of the plant using neural network.

In this work, a slightly different approach has been adopted by which the introduction and training of a huge neural network for copying the aircraft model can be avoided using aircraft equations. The basic idea is that instead of building a neural network copy of aircraft equations using backpropagation error gradients learning algorithm, which brings a substantial saving in development time. In addition, the true derivatives bring more precise than the one obtained by approximately with a neural network emulator of aircraft model. The controller training is faster and more precise.

6.5.6 Simulation Results

The above developed GNN controller for aircraft landing control system has been simulated on computer with initial lift coefficient = 0.96 and other variables given in Table 6.31. The GNN controller uses the four input variables (i.e. lift force, drag force, velocity and gliding angle) for calculating the control action **u**, which is finally modify the lift coefficient Cl in turn drag coefficient Cd. Then the ratio of Cd to Cl is compared with the desired ratio and the discrepancy is used for changing the weights of GNN controller as mentioned above. This process will be repeated till the discrepancy becomes zero. When the discrepancy is zero means the aircraft is following the correct glide path.

The computer simulation results obtained from the aircraft model with GNN controller are given in Table 6.33. Figures 6.80–6.83 show the variations of lift coefficient with respect to drag coefficient, Cl^2 and Cd, time profiles of gliding angle and velocity respectively.

Also the system is simulated using PID controller and the results obtained with it is given in Table 6.34.

Table 6.33. Simulation results with GN controller

Cl	Cd	β	v	Time
4.289745	0.833179	0.189972	74.813024	0.1
4.569534	0.925332	0.201850	72.327003	1.0
3.807501	0.660654	0.172148	79.033864	2.0
3.052516	0.431755	0.140769	88.494391	3.0
2.557884	0.309124	0.122011	96.089215	4.0
2.259584	0.245621	0.108369	103.118216	5.0
2.074728	0.210215	0.101035	107.672751	6.0
1.956167	0.189097	0.096404	110.925259	7.0
1.878235	0.175892	0.093399	113.227821	8.0
1.826201	0.167374	0.091412	114.846160	9.0
1.791118	0.161766	0.090082	115.970648	10.0
1.767323	0.158024	0.089185	116.762429	11.0
1.751122	0.155505	0.088576	117.306505	12.0
1.740066	0.153663	0.088128	117.712575	13.0
1.732510	0.152640	0.087879	117.940931	14.0
1.727340	0.151850	0.087686	118.118959	15.0
1.723801	0.151310	0.087554	118.241297	16.0
1.721377	0.150941	0.087463	118.325304	17.0
1.719716	0.150688	0.087401	118.382960	18.0
1.718578	0.150515	0.087359	118.422516	19.0
1.717799	0.150397	0.087330	118.449647	20.0
1.717264	0.150316	0.087310	118.468254	21.0
1.76898	0.150260	0.087296	118.481013	22.0
1.76646	0.150222	0.087287	118.489761	23.0
1.76474	0.150196	0.087280	118.495759	24.0

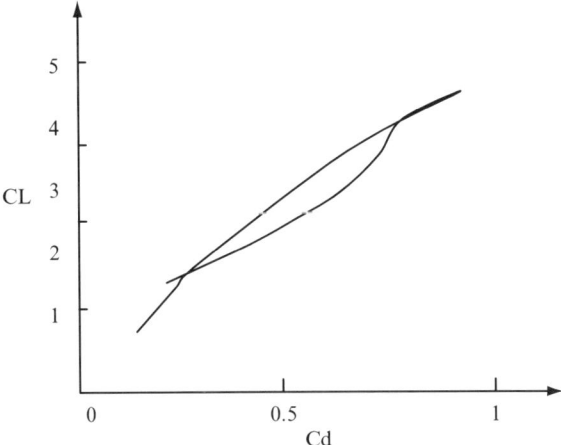

Fig. 6.80. Coefficient of lift vs. coefficient of drag

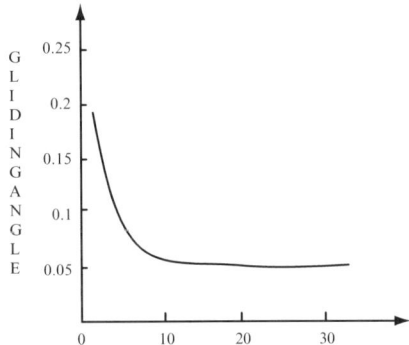

Fig. 6.81. Time profile of gliding angle

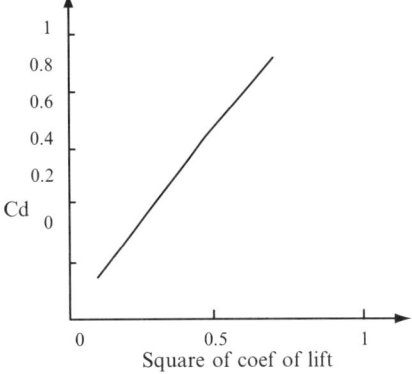

Fig. 6.82. Square of coef of lift vs coef of drag square of coef of lift

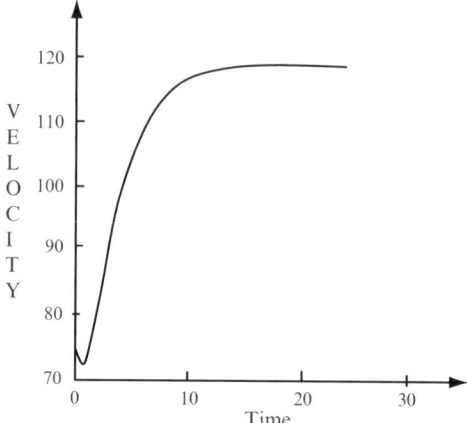

Fig. 6.83. Time profile of gliding angle

Table 6.34. Simulation results with PID controller

Cl	Cd	β	V	Time
4.289745	0.833179	0.189972	74.813024	0.1
5.467310	0.961323	0.210890	73.717020	1.0
4.791444	0.953321	0.201850	70.702300	2.0
4.175710	0.865065	0.182148	76.145864	3.0
3.552528	0.531345	0.151669	85.494391	4.0
3.057884	0.409124	0.132069	94.089215	5.0
2.859584	0.345621	0.118249	100.282160	6.0
2.294728	0.280215	0.103535	104.812751	7.0
2.156166	0.2181097	0.098704	109.192525	8.0
1.907823	0.1958911	0.095699	112.22782	9.0
1.842630	0.187374	0.093412	112.946160	10.0
1.790208	0.171766	0.092182	116.970648	11.0
1.777303	0.168024	0.091850	117.262429	12.0
1.762122	0.165505	0.090136	117.516405	13.0
1.751155	0.163663	0.089128	117.671250	14.0
1.735676	0.162640	0.088679	117.931561	15.0
1.730091	0.162350	0.087586	118.121233	16.0
1.728831	0.162110	0.087554	118.223197	17.0
1.722367	0.161989	0.087550	118.332304	18.0
1.721971	0.161985	0.087491	118.378960	19.0
1.719978	0.161965	0.087399	118.432456	20.0
1.718796	0.161917	0.087350	118.450767	21.0
1.718572	0.161816	0.087320	118.461254	22.0
1.716898	0.161760	0.087300	118.479803	23.0
1.716646	0.160211	0.087298	118.488895	24.0

6.5.7 Conclusions

The generalized neural network control has been developed for aircraft landing control system, which controls the lift coefficient Cl and drag coefficient Cd, which in turns controls the lift and drag forces and ultimately correct the angle of attack by changing elevator angle as per the velocity and deviation in the gliding angle of aircraft. The following conclusions have been drawn:

1. The GNN controller is adaptive in nature and consisting of learning capabilities make the controller superior than conventional controllers.
2. The GNN controller can incorporate the non-linearities involved in the system and cognition level of human beings.
3. The GNN controller is relatively efficient and accurate than conventional controllers.
4. The GNN technique can be used for design and parameter calculation of aircraft after the successful application in landing control system.

6.6 Bibliography and Historical Notes

Readers interested in gaining deep knowledge of other archetectures of ANN may go through the book written by Sundararajan and Saratchandran (1998). **Hippert et al. (2001) gave an interesting review on ANN applications in load forecasting**. Mandal et al. (2006) used ANN for electricity price and short term load forecasting. **Benaouda et al. (2006) used wavelet- based model for electricity load forecasting**. Benaouda and Murtagh (2007) explained neuro-wavelet approach to time-series load forecasting.

Enrique and Pasi (2006) used soft approaches to information retrieval. Pal (2004) used them for data mining.

7

Introduction to Fuzzy Set Theoretic Approach

*It is the mark of an instructed mind to rest satisfied with that degree
of precision which the nature of the subject admits, and not to seek
exactness where only an approximation of the truth is possible.*

Aristotle, 384–322 BC
Ancient Greek philosopher

*So far as the laws of mathematics refer to reality, they are not certain
and so far as the laws they are certain, they do not refer to reality.*

Albert Einstein
Geometrie and Erfahrung

We must exploit our tolerance for imprecision.

Lotfi Zadeh
Professor, Systems Engineering, UC Berkeley, 1973

7.1 Introduction

The scientists have been trying to develop an intelligent machine similar to
human beings since last many years. There are many points of similarity and
differences between computers and human processing. A comparison between
computer (machine system) and human system is given in Table 7.1. The
computer is a logical machine, works on the basis of precise logic. Today
we have very fast computers with large memory to store data. Then also
these computers could not help us in answering simple questions as given
below.

D.K. Chaturvedi: *Soft Computing Techniques and its Applications in Electrical Engineering*,
Studies in Computational Intelligence (SCI) **103**, 223–293 (2008)
www.springerlink.com   © Springer-Verlag Berlin Heidelberg 2008

Table 7.1. Comparison between computer and human working

S. No.	Items	Computer	Human
1.	Input	From physical sensors	From six senses (see, touch, hear, taste, smell and intuition)
2.	Type of input/output	Precise	Imprecise
3.	Logic	Binary logic	Fuzzy logic
4.	Processing	Sequential	Parallel
5.	Information required	Quantitative information	Qualitative information
6.	Repeatability	Good for repetitive work	Not good
7.	Emotions	No emotions	Good in emotions
8.	Intuitions	No	Good
9.	Learning or adaptability	No	Good
10.	Knowledge stored	In the form of instructions	Thumb rules
11.	Wisdom	Absent	Yes
12.	Handling ill defined and complex problems	Can not handle	It can handle
13.	Knowledge	Structured	Unstructured
14.	Processing speed	Very high	slow
15.	Reasoning	With 0 or 1	From 0 to 1

Why is this boy running?

Can computer recognize the state of a person from facial expression?

How to indicate sex or age by drawing hair style?
Or
Can computer recognize the ethnic origin of a person?

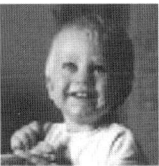

but Human Being may answer these questions. Hence, there is some thing missing in our logical computing. Human logic is not the same as binary logic. We work with imprecise and inaccurate information. Human beings always work with fuzzy logic. Consider the following examples:

Example 1. Consider a heap of wheat as shown below.

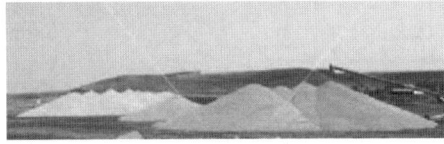

Is it still a heap if we remove one grain of wheat?
How about two grains?
Three?
. . . .

If it is argued bivalent by induction, we eventually remove all grains of wheat and still conclude that the heap remains or that it has suddenly vanished. No single grain takes us from heap to non-heap.

Example 2. Similarly, if we pluck out hairs from a non-bald headed person. Here, the transition is gradual and not abrupt, from a thing (non-bald) to its opposite (bald). Physically it is experienced that there is a degree of occurrence or degree of truth, rather than simply true or false.

Example 3. There is a glass full with water, if remove one drop of water, glass will remain full. If we remove two drops of water the glass will again remain full. If we continue to does this then after removing whole water from the glass, as per this logic glass remains full, which is not true.

From the above examples it is quite clear that the conventional (Binary) logic does not work well to handle the real life situations. Fuzzy logic is the best option to handle these situations. The term "fuzzy" was first coined by Prof. Lotfi A. Zade, University of California Barkley in the Engineering Journal, "Proceedings of the IRE," in 1962. This is the time for paradigm shift from crisp set to fuzzy set. Fuzzy set can deal with Uncertainty in terms of imprecision, nonspecific, vagueness, inconsistency, etc. Earlier uncertainty is undesirable in science and should be avoided by all possible means and science strives for certainty in all manifestation (precision, specific, sharpness, consistency).

Alternative view, which is tolerant of uncertainty and insists that science, can not avoid it. Warren Weaver (1948) mentioned that problems of organized simplicity and disorganized complexity (randomness). Very few problems laying in these categories and most of the problems laying in between, he called them organized complexity (nonlinear systems with large no. of components and rich interactions), which may non-deterministic but not as a results of randomness.

7.2 Uncertainty and Information

Only a small portion of the knowledge (information) for a typical problem might be regarded as certain, or deterministic as shown in Fig. 7.1. Unfortunately, the vast majority of the material taught in engineering classes is based on the presumption that the knowledge involved is deterministic. Most processes are neatly and surreptitiously reduced to closed-form algorithms – equations and formulas. When students graduate, it seems that their biggest fear upon entering the real world is "forgetting the correct formula." These formulas typically describe a deterministic process, one where there is no uncertainty in the physics of the process (i.e. the right formula) and there is no

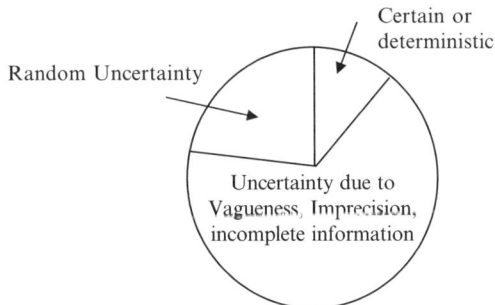

Fig. 7.1. World of information

(a) Unbaked Cookies   (b) Baked Cookies   (c) Over Baked Cookies

Fig. 7.2. Baking of cookies

uncertainty in the parameters of the. It is only after we leave the university, it seems, that we realize we were duped in academe, and that the information we have for a particular problem virtually always contains uncertainty. For how many of our problems can we say that the information content is known absolutely, i.e. with no ignorance, no vagueness, no imprecision, no element of chance? Uncertain information can take on many different forms. There is uncertainty that arises because of complexity; for example, the complexity in the reliability network of a nuclear reactor. There is uncertainty that arises from ignorance, from various classes of randomness, from the inability to perform adequate measurements, from lack of knowledge, or from vagueness, like the fuzziness inherent in our natural language.

The nature of uncertainty in a problem is a very important point that engineers should ponder prior to their selection of an appropriate method to express the uncertainty. Fuzzy sets provide a mathematical way to represent vagueness and fuzziness in humanistic systems. For example, to bake cookies as shown in Fig. 7.2, how you give instructions. You could say that when the temperature inside the cookie dough reaches 375°F, or you could advise to take out when the tops of the cookies turn *light brown*. Which instruction would you prefer to give? Most likely, the second instruction. The first instruction is too precise to implement practically; in this case precision is not useful. The vague term *light brown* is useful in this context and can be acted upon even by a child. We all use vague terms, imprecise information, and other fuzzy data just as easily as we deal with situations governed by chance, where probability techniques are warranted and very useful. Hence, our sophisticated computational methods should be able to represent and manipulate a variety of uncertainties.

7.3 Types of Uncertainty

1. *Stochastic uncertainty*
   It is the uncertainty towards the occurrence of a certain event, e.g. the probability of hitting the target is 0.8. (well defined).
2. *Lexical uncertainty*
   It is the uncertainty lies in human languages like hot days, stable occurrence, a successful financial year and so on.

Following is the base on which fuzzy logic is built:

As the complexity of a system increases, it becomes more difficult and eventually impossible to make a precise statement about its behavior, eventually arriving at a point of complexity where the fuzzy logic method born in humans is the only way to get at the problem.

(Originally identified and set forth by Prof. Lotfi A. Zadeh, University of California, Berkeley)

What do you mean by the term Fuzzy?
   English meaning of the word fuzzy is indistinct, blurred or not properly focused and the technical meaning is imprecise, uncertain or unreliable knowledge, uncertain/noisy/incomplete Information, ambiguity (vague/fuzzy concepts) or partial truth (Fig. 7.3).

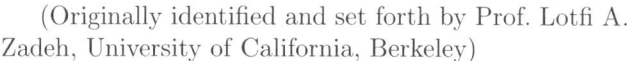

Fig. 7.3. Difference between binary logic and fuzzy logic

Real-world vagueness

In our day-today conversation we often say such as

"Maria is tall," or

"It is very hot today."

Such statements are difficult to translate into precise language without losing some of their semantic value. If we mention the statement

"Maria's height is 1.60 m." or

"Maria's height is 1.25 standard deviations about the mean height for women of her age in her country"

It does not convey directly that she is tall, because here the concept and context both required like Maria belongs to which country and what is the average height there?

Imprecision

The imprecision is nonetheless a form of information that can be quite useful to humans. The ability to embed such reasoning in hitherto intractable and complex problems is the criterion by which the efficacy of fuzzy logic is judged. Undoubtedly this ability cannot solve problems that require precision – problems such as shooting precision laser beams over tens of kilometers in space; milling machine components to accuracies of parts per billion; or focusing a microscopic electron beam on a specimen the size of a nanometer. The impact of fuzzy logic in these areas might be years away, if ever. But not many human problems require such precision – problems such as parking a car, backing up a trailer, navigating a car among others on a freeway, washing clothes, controlling traffic at intersections, judging beauty contestants, and a preliminary understanding of a complex system.

Requiring precision in engineering models and products translates to requiring high cost and long lead times in production and development. For other than simple systems, expense is proportional to precision: more precision entails higher cost. When considering the use of fuzzy logic for a given problem, an engineer or scientist should ponder the need for *exploiting the tolerance for imprecision*. Not only does high precision dictate high costs but also it entails low tractability in a problem.

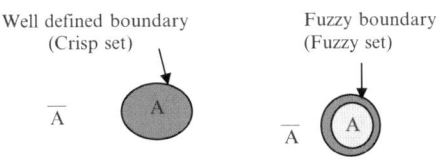

$A \cap \overline{A} = \phi$ and $A \cup \overline{A} = X$ for crisp sets

$A \cap \overline{A} \neq \phi$ and $A \cup \overline{A} \neq X$ for fuzzy sets

7.4 Introduction of Fuzzy Logic

Fuzzy logic is a superset of conventional (Boolean) logic that has been extended to handle the concept of partial truth–truth values between "completely true" and "completely false".

The value zero is used to represent non-membership, and the value one is used to represent membership. The truth or falsity of the statement "x is in U" is determined by finding the ordered pair whose first element is x. The statement is true if the second element of the ordered pair is 1, and the statement is false if it is 0.

Similarly, a fuzzy subset F of a set S can be defined as a set of ordered pairs, each with a first element that is an element of the set S, and a second element that is a value in the interval [0, 1], with exactly one ordered pair present for each element of S. This defines a mapping between elements of the set S and values in the interval [0, 1]. The value zero is used to represent complete non-membership, the value one is used to represent complete membership, and values in between are used to represent intermediate degrees of membership. The set S is referred to as the universe of discourse for the fuzzy subset F.

Let's define a fuzzy subset TALL, which will answer the question "to what degree is person x tall?" To each person in the universe of discourse, we have to assign a degree of membership in the fuzzy subset TALL. The easiest way to do this is with a continuous membership function based on the person's height as illustrated in Fig. 7.4.

$$\text{TALL(x)} = \begin{cases} 0, & \text{if height(x)} < 5\,\text{ft.,} \\ (\text{height(x)} - 5\text{ft})/2\text{ft.,} & \text{if}\,5\,\text{ft.} <= \text{height (x)} <= 7\,\text{ft.,} \\ 1, & \text{if height(x)} > 7\,\text{ft.} \end{cases}$$

Fuzzy set also sometimes represented by discrete fuzzy membership functions like good friends as shown in Fig. 7.5.

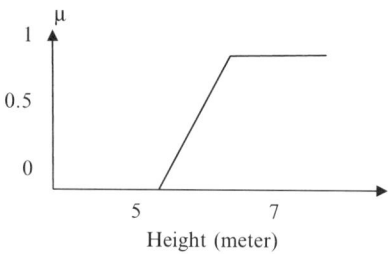

Fig. 7.4. Continuous membership function for tall

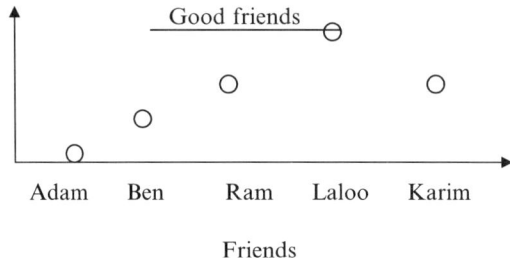

Fig. 7.5. Discrete fuzzy membership function

7.5 Historical Development of Fuzzy Logic

Fuzzy set was specifically designed to mathematically represent uncertainty and vagueness and to provide formalized tools for dealing with the imprecision intrinsic to many problems. However, the story of fuzzy logic started much earlier.

- To devise a concise theory of logic, and later mathematics, *Aristotle* posited the so-called "Laws of Thought". One of these, the "Law of the Excluded Middle," states that every proposition must either be *True* (**T**) or *False* (**F**).
- It was *Plato* who laid the foundation for what would become fuzzy logic, indicating that there was a third region (beyond **T** and **F**) where these opposites "tumbled about."
- A systematic alternative to the bi-valued logic of Aristotle was first proposed by *Lukasiewicz* around 1920 (Klir and Yuan 1995), when he described a three-valued logic, along with the mathematics to accompany it.
- The third value, he proposed, can best be translated as the term "possible," and he assigned it a numeric value between **T** and **F**. Eventually, he proposed an entire notation and axiomatic system from which he hoped to derive modern mathematics.
- Later, he felt that three- and infinite-valued logics were the most intriguing, but he ultimately settled on a four-valued logic because it seemed to be the most easily adaptable to Aristotelian logic (Klir and Folger 2000).
- The notion of an infinite-valued logic was introduced in Zadeh's seminal work "Fuzzy Sets" (Zadeh 1965) where he described the mathematics of fuzzy set theory, and by extension fuzzy logic.
- This theory proposed making the membership function (or the values **F** and **T**) operate over the range of real numbers [0, 1]. New operations for the calculus of logic were proposed, and showed to be in principle in 1968.
- Fuzzy logic provides an inference morphology that enables approximate human reasoning capabilities to be applied to knowledge-based systems. The theory of fuzzy logic provides a mathematical strength to capture the

uncertainties associated with human cognitive processes, such as thinking and reasoning.

- 1972 An Association was formed called "Japan Fuzzy Systems Research" which is now called International Fuzzy Systems Association (IFSA) to promote the activities in this direction.
- 1973 Prof. L.A. Zadeh developed systematic treatment for Fuzzy Logic.

- Prof. Ebrahim H. Mamdani, U.K. (1974a,b) and Mamdani and Assilian (1975) used fuzzy sets with an adaptive feedback control strategy to control a small toy steam engine. This was the first practical applications of fuzzy logic. Human Reasoning is based on Fuzzy reasoning. The steam generating system in a power plant is proved to be a most complicated non-linear system and conventional controllers could not give very good results. Hence, fuzzy logic controller was used.

- In 1976 it was applied in the automatic control system of a rotary furnace for cement production by Mamdani (1976).
- In 1980 Smith & Co. (Denmark) had used fuzzy Controller commercially for Cement Kiln. Larsen (1980) used fuzzy logic for various industrial applications.
- Fuji Electric Co. Ltd., Japan designed a fuzzy controller for water purification in 1983.
- First Fuzzy International Conference was held in 1985.
- Prof. Yamakawa (1987) designed a super high speed fuzzy controller for the Sendai underground railways, which was utilized by Hitachi company in Japan. This system automatically decreased the speed of a train on entering a station, ensuring that the train stopped at a predetermined place. It also had the benefit of being a highly comfortable ride through mild acceleration and braking.
- 1989 Foundation of the Laboratory for Industrial Fuzzy Engineering (LIFE) Research.

Today, there are number of products in the market which are controlled with fuzzy (Yager and Filev 2002) and neuro-fuzzy (Nie and Linkens 1995; Jang et al. 1997) techniques. Recently, Sony uses fuzzy logic to recognize Kanji characters in the Palm Top Computers (Abraham et al. 2002). In Tokoyo motor show, Mitsubishi had a computer in an WSR-IV prototype that uses fuzzy logic to imitate the information processing in a driver's brain. It studies the driving habits and responses of driver under different situation. Then assist the driver while driving, if the driver does not respond as the computer predicted, then computer can automatically take control of the brakes to avoid a collision. In these applications precision is not so important.

Precision   Significance

Fig. 7.6. Difference between precision and significance

7.6 Difference Between Precision and Significance

The two valued logic has proved very effective and successful in solving well defined problems, which are characterized by precise description of the system being dealt with in quantitative form. But, in everyday life there are many situations, which can not be dealt satisfactorily with simple "yes" or "no" basis, but some shade of gray is required. Fuzzy sets allow the description of the situation in the linguistic terms rather than in terms of precise and numeric values. Consider an example of an industry, a crane carrying a load and a person standing in the way as shown in the diagram (Fig. 7.6). There are two statements to express the situation, one is crisp and other is fuzzy, as mentioned in diagram. Now, which one is better and effective?

7.7 Fuzzy Set

A Fuzzy set is a any set that allows its member to have a different grades of membership function in the interval of [0,1]. Fuzzy set theory is a methodology which shows how to tackle uncertainty, and to handle imprecise information in a complex situation. Let X be a collection of objects or a universe of discourse then a fuzzy set A in X is a set of ordered pairs $A = \{\mu_A(x)/x\}$ where $\mu_A(x)$ is the characteristics function (or membership function) of x in A (Fig. 7.7). If the membership function of A is discrete then A is written as

$$A = [\mu_1(x)/x_1 + \mu_2(x)/x_2 + \mu_3(x)/x_3 + \cdots\cdots + \mu_n(x)/x_n]$$
$$= \Sigma[\mu_i(x)/x_i]$$

where "+" sign or Σ denotes union.
$\mu_i(x)$ – Membership value/grade
x_i – Variable value

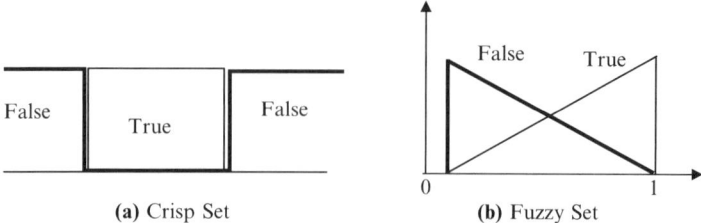

Fig. 7.7. Characteristic (membership) functions

When the membership function is continuous then the fuzzy set A is written as

$$A = \int \mu_A(x)/x$$

where the integral denotes the fuzzy singletons.

Examples:

1. High temperature.
2. A set of all tall persons.
3. A set of all educated persons.

There are three basic methods by which a set can be defined:

1. *List of elements*
   A set may be defined by all the elements along with their membership value or membership grade associated with it.

   $$A = [\mu_1(x)/x_1 + \mu_2(x)/x_2 + \mu_3(x)/x_3 + \cdots\cdots + \mu_n(x)/x_n] \text{ or}$$
   $$= [\mu_1(x)/x_1, \mu_2(x)/x_2, \mu_3(x)/x_3, \cdots\cdots, \mu_n(x)/x_n]$$

2. *Rule method*
   A set may be represented by some rule which all the elements follow. A is defined by the following notation as the set of all elements of x for which the proposition P(x) is true.

   $$A = \{x \mid p(x)\}$$

   where | denote the phrase "such that"
   p(x) − x has the property p.

3. *Characteristics (membership) function*
   A set may also be defined by a characteristic (membership) function which will give the membership value of each element.

   $$\mu_A(x) = \begin{array}{ll} f(x) & \text{for } x \in A \\ 0 & \text{for } x \notin A \end{array}$$

   where f(x) could be any function like $f(x) = 0.5x + 1$ or $f(x) = 1.5{*}e^{-1.2x}$.

Some main useful features of the fuzzy sets:

1. Fuzzy logic provides a systematic basis for quantifying uncertainty due to vagueness and incompleteness of the information.
2. Classes with no sharp boundaries can be easily modeled using fuzzy sets.
3. Fuzzy reasoning is an informalism that allows the use of expert knowledge and is able to process this expertise in a structured and consistent way.
4. There is no broad assumption of complete independence of the evidence to be combined using fuzzy logic, as required for the other subjective probabilistic approach.
5. When the information is inadequate to support a random definition, the use of probabilistic methods may be difficult. In such cases the use of fuzzy set is promising.

Subset

If every member of set A is also a member of set B (i.e. $x \in A$ implies $x \in B$), then A is called subset of B and written as $A \subseteq B$. Subset has the following properties:

1. Every set is subset of itself and every set is a subset of the universal set.
2. If $A \subseteq B$ and $B \subseteq A$ then $A = B$ (equal set).
3. If $A \subseteq B$ and $B \neq A$ then $A \subset B$ (A is proper subset of B).

Power set

The family of all subsets of a given set A is called the power set of A and denoted by $P(A)$. The family of all subsets of $P(A)$ is called second order power set of A, denoted by $P^2(A)$.

$$P^2(A) = P(P(A)).$$

7.8 Operations on Fuzzy Sets

7.8.1 Fuzzy Intersection

Fuzzy intersection of two sets A and B is interpreted as "A **AND** B" which takes the minimum value of two membership functions.

$$A \cap B(x) = \sum \{\mu_A(x) \wedge \mu_B(x)\}$$
$$= \min(\mu_A(x), \mu_B(x))$$

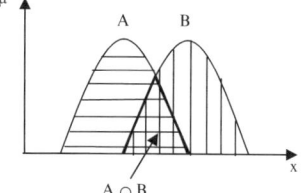

7.8.2 Fuzzy Union

The fuzzy union is interpreted as "A OR B" which takes maximum value of two membership functions i.e.

$$A \vee B(x) = \sum \mu_A(x) \vee \mu_B(x)$$
$$= \max(\mu_A(x), \mu_B(x))$$

Table 7.2. Different averaging operators

S. No.	Name of averaging operator	Formula
1.	Harmonic mean	$2\mu_A \cdot \mu_B/(\mu a_1 + \mu_B)$
2.	Geometric mean	$\sqrt{(\mu_A \cdot \mu_B)}$
3.	Arithmetic mean	$(\mu_A + \mu_B)/2$
4.	Dual of geometric mean	$1 - \sqrt{((1 - \mu_A) + (1 - \mu_B))}$
5.	Dual of harmonic mean	$(\mu_A + \mu_B - 2\mu_A.\mu_B)/(2 - \mu_A - \mu_B)$
6.	Median	$\text{med}(\mu_A, \mu_B, \alpha), \ \alpha E(0, 1)$
7.	Generalized mean	$((\mu_A{}^\alpha + \mu_B{}^\alpha)/2)^{1/\alpha}, \ \alpha \geq 1$
8.	Weighted generalized mean	$((w_1.\mu_A{}^\alpha + w_2.\mu_B{}^\alpha)/2)^{1/\alpha}, \ \alpha \geq 1$

Union and intersection operators qualify as aggregation operations on fuzzy sets. The results of aggregation will be maximum or minimum value in all sets. It do not produce value between $\min(a_1, a_2, a_3, \ldots a_n)$ and $\max(a_1, a_2, a_3, \ldots a_n)$. Hence the operator whose output value lies between these limits are called compensating/averaging/mean operators. Mathematically, it can be written as:

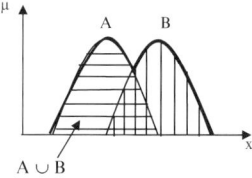

$$\min(\mu a_1, \mu a_2, \mu a_3, \ldots \mu a_n) \leq h(\mu a_1, \mu a_2, \mu a_3, \ldots \mu a_n) \leq \max(\mu a_1, \mu a_2, \mu a_3, \ldots \mu a_n).$$

In decision process the idea of trade-offs corresponds to viewing the global evaluation of an action as lying between worst and the best local ratings. This occurs in the presence of conflicting goals, when a compensation between corresponding capabilities is allowed. Averaging operators realize trade offs between objectives, by allowing a positive compensation between ratings. Different types of averaging operators are given in Table 7.2.

7.8.3 Fuzzy Complement

The complement of a fuzzy set A, which is understood as "NOT(A)", is defined by

$$\bar{A} = \sum(1 - \mu_A(x))$$

where $\bar{A}$ stands for the complement of A. The relative compliment of a set A with respect to set B is the set containing all members of B that are not members of A denoted by B-A.

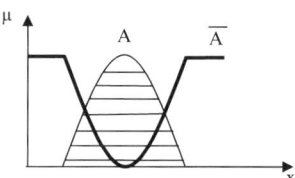

$$(B - A) = \{x \mid x \in B \text{ and } x \notin A\}$$

If set B is universal set then compliment is absolute. The absolute compliment is always involutive. Absolute compliment of empty set is always equal to universal set and vice versa.

$$\overline{\phi} = X \text{ and } \overline{X} = \phi$$

7.8.4 Combination

The convex combination is an operator which combines different fuzzy sets into a single fuzzy set using the weights assigned to each fuzzy set. The total membership function $\mu_T(x)$ as a result of convex combination of the membership functions μ_{A1}, μ_{A2}, ..., μ_{An}, is defined by

$$\mu_T(x) = w_1(x)\mu_{A1}(x) + w_2(x)\mu_{A2}(x) + \cdots + w_n(x)\mu_{An}(x)$$

where $w_1, w_2, \ldots, w_n$ are the weights of the fuzzy sets $A_1, A_2, \ldots, A_n$, respectively such that:

$$w_1(x) + w_2(x) + \cdots + w_n(x) = 1$$

where the "+" sign in the above equations denotes an arithmetic addition.

7.8.5 Fuzzy Concentration

The concentration of the fuzzy sets produces a reduction in membership value $\mu_i(x)$ by taking power more than 1 to the membership value of that fuzzy set. If a fuzzy set A is written as:

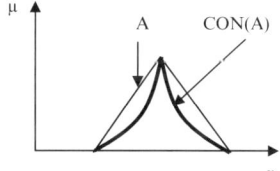

$$A = \{\mu_1/x_1 + \mu_2/x_2 + \cdots + \mu_n/x_{n,}\}$$

then fuzzy concentrator applied to a fuzzy set A is defined:

$$\text{CON}(A) = A^m = \{\mu_1{}^m/x_1 + \mu_2{}^m/x_2 + \cdots + \mu_n{}^m/x_{n,}\}$$

where CON(A) represents the concentrator applied to A.
m is any number greater than 1.

7.8.6 Fuzzy Dilation

The Fuzzy dilation is an operator which increases the degree of belief in each object of fuzzy set by taking the power less than 1 of the membership value. The dilation has an opposite effect to that of concentration.

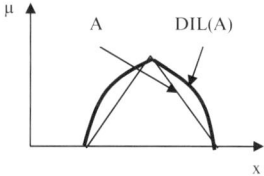

$$\text{DIL}(A) = A^m = \{\mu_1{}^m/x_1 + \mu_2{}^m/x_2 + \cdots\cdots\cdots\cdots + \mu_n{}^m/x_n\}$$

where m is any number lesser than 1.

The fuzzy operations of plus and minus applied to a fuzzy set give an intermediate effect of CON(A) and DIL(A) where the value of m is 1.25 and 0.75, respectively. If a set A is fuzzy set then the plus and minus may be defined as:

$$\text{Plus}(A) = A^{1.25} \text{ and Minus}(A) = A^{0.75}$$

7.8.7 Fuzzy Intensification

Intensification is an operator which increases the membership function of a set above the cross point (at $\alpha = 0.5$) and decreases below the cross point.

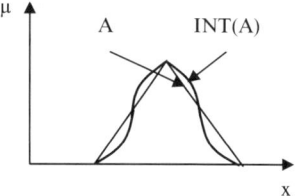

If A is a fuzzy set, $x \in A$ then the intensification applied to A is defined as-

$$\mu_{\text{INT}(A)}(x) \geq \mu_A(x), \ \mu_A(x) \geq 0.5$$
$$\mu_{\text{INT}(A)}(x) \leq \mu_A(x), \ \mu_A(x) \leq 0.5$$

Bounded Sum

The bounded sum of two fuzzy sets A and B in the universes X and Y with the membership functions $\mu_A(x)$ and $\mu_B(y)$ respectively is defined by

$$A \oplus B = \mu_{A \oplus B}(x) = 1 \wedge (\mu_A(x) + \mu_B(y))$$
$$= \min(1, \ (\mu_A(x) + \mu_B(y))$$

where the "+" sign is an arithmetic operator.

Bounded product

The bounded product of two fuzzy sets A and B in the universes X and Y with membership functions $\mu_A(x)$ and $\mu_B(y)$ respectively is defined as

$$A \otimes B = \mu_{A \otimes B}(x) = 0 \vee (\mu_A(x) + \mu_B(x) - 1)$$
$$= \max(0, \ (\mu_A(x) + \mu_B(x) - 1)).$$

7.8.8 α-Cuts

α-cuts are the slices through a fuzzy set producing regular (non-fuzzy) sets. If $\bar{A}$ is a fuzzy subset of some set Ω, then an α-cut of $\bar{A}$, written $\bar{A}[\alpha]$ is defined as (Fig. 7.8)

Let

$$\text{Young} = \{1/5, 1/10, 0.8/20, 0.5/30, 0.2/40, 0.1/50\}$$
$$\text{Young}^{0.8} = \{1/5, \ 1/10, \ 0.8/20\}$$

The distinct α-cut of a fuzzy set A is called a level set (A_α) of A. Hence, the fuzzy set is the union of all possible level set, i.e. $A = U \ A_\alpha$. This is also called first decomposition theorem.

$$\bar{A}\,[\alpha] = \{x \in \Omega \,|\bar{A}(x) \geq \alpha\} \qquad \text{for all } \alpha,\ 0 < \alpha \leq 1.$$

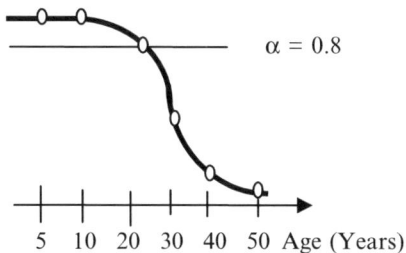

Fig. 7.8. α-cuts for a given fuzzy set A

$$\alpha = 0.8$$

5   10   20   30   40   50   Age (Years)

Strong α-cut

Strong α-cut of a fuzzy set A is also a crisp set $A^{\alpha+}$ that contains all the elements of the universal set X that have a membership grade in A greater to the specified values of α between 0 and 1.

$$A^{\alpha+} = \{x | A(x)A > \alpha\}$$

For example $\text{Young}^{0.8+} = \{1/5,\ 1/10\}$.

The α-cut of a given fuzzy set A for two distinct values of α say α_1 and α_2 such that $\alpha_1 < \alpha_2$ are $A^{\alpha 1}$ and $A^{\alpha 2}$ such that $A^{\alpha 1} \supseteq A^{\alpha 2}$. Similarly for strong α-cut $A^{\alpha 1+} \supseteq A^{\alpha 2+}$.

Properties of α-cut

a. $A^{\alpha+} \subseteq A^{\alpha}$
b. $(A \cap B)^{\alpha} = (A^{\alpha} \cap B^{\alpha})$ and $(A \cup B)^{\alpha} = (A^{\alpha} \cup B^{\alpha})$
c. $(A \cap B)^{\alpha+} = (A^{\alpha+} \cap B^{\alpha+})$ and $(A \cup B)^{\alpha} = (A^{\alpha} \cup B^{+\alpha})$
d. $\bar{A}^{\alpha} = \bar{A}^{(1-\alpha)+}$

7.8.9 Fuzzy Quantifier/Modifier/Hedges

These are special terms by which the fuzzy sets are modified. Mathematically it can be written as:

$$h(A) = \mu_A{}^n$$

where μ_A membership value of fuzzy set A.

n - is value of modifier.

Every modifier h satisfy the the following conditions:

a. Boundary conditions, i.e. modifier operator (h) does not change the membership values at extreme points. $h(0) = 0$ and $h(1) = 1$
b. h is a continuous function.
c. If h is strong modifier then h^{-1} is weak modifier and vice versa.
d. Composition of two modifier operators is also a modifier.

Basically there are two types of fuzzy modifier depending on relativity:

1. Absolute modifier
2. Relative modifier

1. *Absolute modifier*
   These are the modifiers with respect to the absolute quantity like about 10, much more than 100, at last about 5, etc.
2. *Relative modifier*
   These are the modifiers which modify the fuzzy set with respect to some other fuzzy set/quantity like almost all, about half, most of it, etc.

Fuzzy modifier may also be classified based on the strength of modifier:

1. Strong modifier
2. Weak modifier
3. Identity modifier

1. *Strong modifier*
   If the value of modifier $n \geq 1$ then fuzzy modifier is called a strong modifier (concentrator), because the area under fuzzy membership function get reduced (ref. Fig. 7.9).

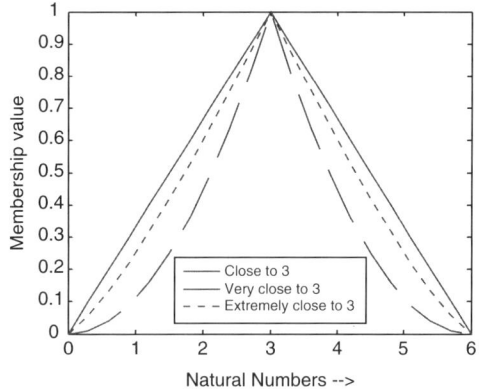

Fig. 7.9. Strong fuzzy modifier

Let

close to '3' = {0/1, 0.5/2, 1/3, 0.5/4, 0/5}

Very close to '3' = $\mu_A{}^2$ = {0/1, 0.25/2, 1/3, 0.25/4, 0/5}

Extremely close to '3' = $\mu_A{}^{1.25}$ = {0/1, 0.25/2, 1/3, 0.25/4, 0/5}

% Matlab Program for Strong Fuzzy Modifiers

```
% Close_to_3=A
A=[0:.1:1, .9:-0.1:0];
x=[0:6/20:6];
Very_Close_to_3=A.^2;
Extermey_Close_to_3=A.^1.25;
plot(x,A)
hold
plot(x,Very_Close_to_3, '--')
plot(x,Extermey_Close_to_3, ':')
legend('Close to 3', 'Very close to 3',
'Extremely close to 3')
xlabel('Natural Numbers -->')
ylabel('Membership value')
```

2. *Weak fuzzy modifier*

   If the value of modifying parameter n ≤ 1 then, modifier is called weak modifier (also called dilution, because the area under fuzzy membership function is increased as shown in Fig. 7.10). For example:

   More or less close to '3'= $\mu_A^{0.75}$ = {0/1, 0.5946/2, 1/3, 0.5946/4, 0/5}

   Extremely close to '3'= $\mu_A^{0.5}$ = {0/1, 0.7071/2, 1/3, 0.7071/4, 0/5}

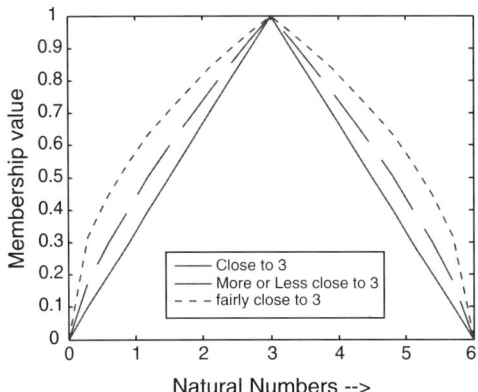

Fig. 7.10. Weak fuzzy modifier

% Matlab Program for Weak Fuzzy Modifiers

```
clear all
% Close_to_3=A
A=[0:.1:1, .9:-0.1:0];
x=[0:6/20:6];
More_or_less_Close_to_3=A.^0.75;
fairly_Close_to_3=A.^0.5;
plot(x,A)
hold
plot(x,More_or_less_Close_to_3, '--')
plot(x,fairly_Close_to_3, ':')
legend('Close to 3', 'More or Less close to 3',
'fairly close to 3')
xlabel('Natural Numbers -->')
ylabel('Membership value')
```

3. *Identity modifier*

   If the value of modifier n = 1, then it is called identity modifier. Under this condition there is no change in the area under fuzzy membership function.

Intensification

It is the combination of strong and weak modifiers. If the membership value of fuzzy set is less than 0.5, the strong fuzzy modifier is used and when the membership value of fuzzy set is greater than 0.5, the weak fuzzy modifier is used as shown in Fig. 7.11. This is an operator which make a fuzzy set nearly a crisp set (Table 7.3).

$$h(A) = \begin{matrix} \mu_A{}^{n_1} & n_1 \geq 1 \text{If } 0 \leq \mu_A \leq 0.5 \\ \mu_A{}^{n_2} & n_2 \leq 1 \text{ If } 0.5 \leq \mu_A \leq 1.0 \end{matrix}$$

For example intensified close to 3 $= \begin{matrix} 2\mu_A{}^2 & \text{If } 0 \leq \mu_A \leq 0.5 \\ 1 - 2(1 - \mu_A)^2 & \text{If } 0.5 \leq \mu_A \leq 1.0 \end{matrix}$

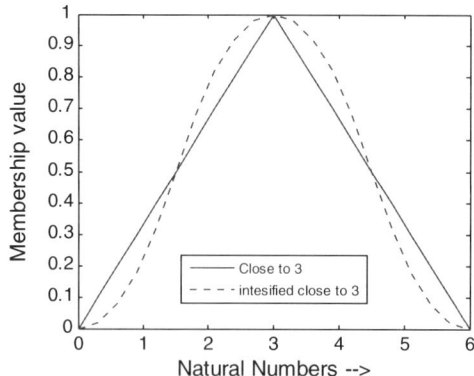

Fig. 7.11. Intensified modifier

Table 7.3. Summary of operations on fuzzy sets

S. No.	Operations	Symbol	Formula
1.	Intersection	$A \cap B$	$\min(\mu_A(x), \mu_B(x))$
2.	Union	$A \cup B$	$\max(\mu_A(x), \mu_B(x))$
3.	Compliment	$\bar{A}$	$1 - \mu_A(x)$
4.	Concentration	$CON(A)$	μ_i^m if $m \geq 1$
5.	Dilution	$DIL(A)$	μ_i^m if $m \leq 1$
6.	Intensification	$INT(A)$	μ_i^m if $m \geq 1$ for $\mu_i < 0.5$
		$\bigcirc$	μ_i^m if $m \leq 1$ for $\mu_i > 0.5$
7.	Bounded sum	$A \oplus B$	$\min(1, (\mu_A(x) + \mu_B(y))$
8.	Bonded difference	$A - B$	$\max(0, (\mu_A(x) - \mu_B(x)))$
9.	Bounded product	$A \otimes B$	$\max(0, (\mu_A(x) + \mu_B(x) - 1))$
10.	Equality	$A = B$	$\mu_A(x) = \mu_B(x)$

% Matlab Program for intensified fuzzy modifiers

```
clear all
% Close_to_3=A
A=[0:.1:1, .9:-0.1:0];
x=[0:6/20:6];
[l n]=size(A);
  for i=1:n
      if A(i)<=0.5
        A1(i)=2*A(i).^2
      else
        A1(i)=1-2*(1-A(i)).^2
      end
end
plot(x,A); hold
plot(x,A1, ':')
legend('Close to 3', 'intensified close to 3')
xlabel('Natural Numbers -->')
ylabel('Membership value')
```

Linguistic hedges

Linguistic hedges such as very, much, more or less, etc. modify the meaning of atomic as well as composite terms and thus serve to increase the range of linguistic variable from a small collection of primary term.

A hedge h may be regarded as an operator which transforms the fuzzy set M (u), representing the meaing of u, into the fuzzy set M (hu). For example, by using the hedge very in conjunction with not, and the primary term tall, we can generate the fuzzy sets very tall, very very tall, not very tall, tall and not very tall, etc.

$$very\ x = x^2$$
$$very\ very\ x = (very\ x)^2 = x^4$$
$$not\ very\ x = \overline{(very\ x)}$$
$$plus\ x = x^{1.25}$$
$$minus\ x = x^{0.75}$$

From last two, we have the approximate identity

$$plus\ plus\ x = min\ us\ very\ x$$

Example

If $U = 1 + 2 + 3 + 4 + 5$ and $small = \frac{1}{1} + \frac{0.8}{2} + \frac{0.6}{3} + \frac{0.4}{4} + \frac{0.2}{5}$

Then $very\ small = \frac{1}{1} + \frac{0.64}{2} + \frac{0.36}{3} + \frac{0.16}{4} + \frac{0.04}{5} = \mu_A^2/A$

$very\ very\ small = \frac{1}{1} + \frac{0.4}{2} + \frac{0.1}{3}$ (Neglecting small terms)

$not\ very\ small = \frac{0}{1} + \frac{0.36}{2} + \frac{0.64}{3} + \frac{0.84}{4} + \frac{0.96}{5}$

$plus\ small = \frac{1}{1} + \frac{0.76}{2} + \frac{0.53}{3} + \frac{0.32}{4} + \frac{0.13}{5}$

$min\ us\ small = \frac{1}{1} + \frac{0.85}{2} + \frac{0.68}{3} + \frac{0.5}{4} + \frac{0.3}{5}$

7.9 Characteristics of Fuzzy Sets

7.9.1 Normality

A fuzzy set is said to be normal if the greatest value of its membership function is unity. Vx $\mu(x) = 1$, where Vx stands for the supremum of the $\mu(x)$ (least upper bound) otherwise the set is subnormal as shown in Fig. 7.12.

7.9.2 Convexity

A convex fuzzy set is described by a membership function whose membership values are strictly monotonically increasing, or strictly monotonically decreasing, or strictly monotonically increasing then strictly monotonically decreasing with the increasing value for elements in the universe of discourse X, i.e. if x, y, z ∈ A and x < y < z then $\mu_A(y) \geq min\ [\mu_A(x),\ \mu_A(z)]$ (ref. Fig. 7.13).

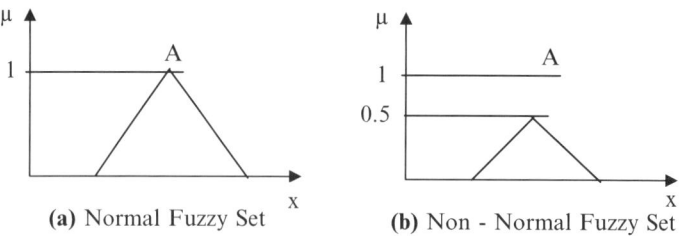

(a) Normal Fuzzy Set   **(b) Non - Normal Fuzzy Set**

Fig. 7.12. Normality of fuzzy set

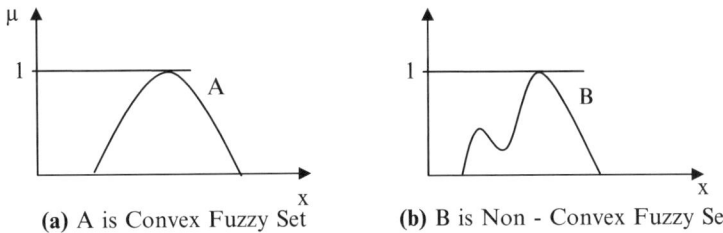

(a) A is Convex Fuzzy Set   **(b)** B is Non - Convex Fuzzy Set

Fig. 7.13. Convexity of fuzzy set

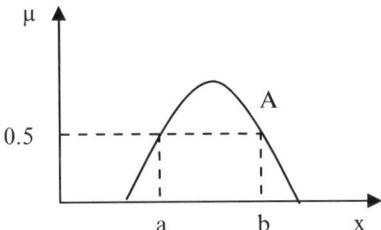

Fig. 7.14. Cross over points

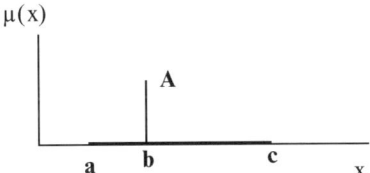

Fig. 7.15. Fuzzy singletone

7.9.3 Cross Over Point

The cross over points of a membership function are defined as the elements in the universe for which a particular fuzzy set $\tilde{A}$ has values equal to 0.5, i.e. for which $\mu(x) = 0.5$. For example a, b are the cross over points as given in Fig. 7.14.

7.9.4 Fuzzy Singletone

A fuzzy single tone is a fuzzy set which only has a membership grade for a single value. Let A be a fuzzy single tone of a universe of discourse X, $x \varepsilon X$, then A is written as $A = \mu/x$. With this definition, a fuzzy set can be considered as the union of the fuzzy single tone. In the given Fig. 7.15 range from a to c is the support and the element b only has the membership value greater than zero. Fuzzy sets of this type are called Fuzzy Singletone.

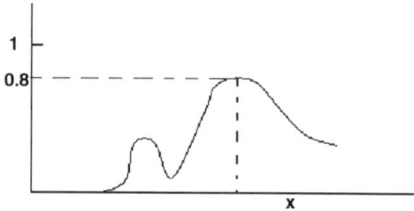

Fig. 7.16. Height of membership function

7.9.5 Height

The height of a fuzzy set $\tilde{A}$ is the maximum value of the membership function, i.e. max $\{\mu(x)\}$. The height of fuzzy set in the given example is 0.8 (ref. Fig. 7.16).

7.9.6 Cardinality

The number of members of a finite discrete fuzzy set A is called cardinality of A and denoted by $|A|$. The number of possible subsets of a set A is equal to $2^{|A|}$ which is also called the power set.

7.10 Properties of Fuzzy Sets

7.10.1 Commutative Property

The fuzzy sets follow commutative property, i.e.

$A \cup B = B \cup A$ or max $[\mu_A(x), \ \mu_B(y)] =$ max $[\mu_B(x), \ \mu_A(y)]$
and $A \cap B = B \cap A$ or min $[\mu_A(x), \ \mu_B(y)] =$ min $[\mu_B(x), \ \mu_A(y)]$.

7.10.2 Associative Property

The fuzzy sets also follow associative property, i.e.

$A \cup (B \cup C) = (A \cup B) \cup C$ or
max $[\mu_A(x), \ \text{max}(\mu_B(x), \ \mu_c(z)] =$ max$[\mu_A(x), \ \text{max} \ \{\mu_B(y), \ \mu_c(z)\}]$
and $A \cap (B \cap C) = (A \cap B) \cap C$ or
min$[\mu_A(x), \ \text{min} \ \{\mu_B(y), \ \mu_C(z)\}] =$ min$[\text{min} \ \{\mu_A(x), \ \mu_B(y)\}, \ \mu_C(z)]$

7.10.3 Distributive Property

$A \cup (B \cap C) = (A \cup B) \cap (A \cup C)$ or
max $[\mu_A(x), \ \text{min}(\mu_B(x), \ \mu_c(z)] =$ min$[\text{max} \ \{\mu_A(x), \ \mu_B(y)\},$
$\text{max} \ \{\mu_A(x), \ \mu_c(z)\}]$.

7.10.4 Idem Potency

$A \cup A = A$ means $\qquad \max \{\mu_A(x), \mu_A(x)\} = \mu_A(x)$

and $A \cap A = A$ means $\qquad \min \{\mu_A(x), \mu_A(x)\} = \mu_A(x).$

7.10.5 Identity

$A \cup \phi = A$ means $\qquad \max \{\mu_A(x), 0\} = \mu_A(x)$

$A \cap X = A$ means $\qquad \min \{\mu_A(x), 1\} = \mu_A(x)$

$A \cap \phi = A$ means $\qquad \min \{\mu_A(x), 0\} = 0$

and $A \cup X = X$ means $\qquad \max \{\mu_A(x), 1\} = 1.$

7.10.6 Involution

$A = A$ i.e. $1\text{-}(1\text{-}\mu_A(x)) = \mu_A(x).$

7.10.7 Excluded Middle Law

This law is only valid for classical set theory and not true for fuzzy set theory.
$\qquad A \cup \bar{A} \neq X$ means $\max \{\mu_A(x), \mu_{\bar{A}}(x)\} \neq 1.$

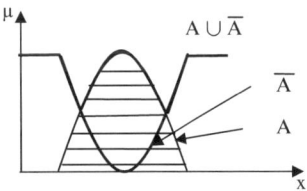 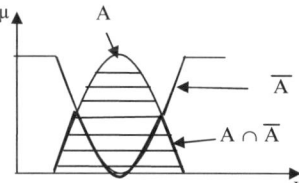

7.10.8 Law of Contradiction

This is also true for classical sets and not for fuzzy sets.
$\qquad A \cap \bar{A} \neq \varphi$ means $\min \{\mu_A(x), \mu_{\bar{A}}(x)\} \neq 0.$

7.10.9 Demorgan's Law

1. $\overline{(A \cup B)} = \bar{A} \cap \bar{B}$

   A circuit of diodes as shown in the following figure can physically represent this.

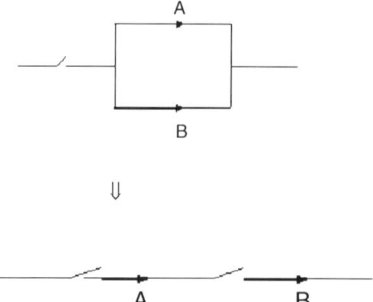

2. $\overline{(A \cap B)} = \overline{A} \cup \overline{B}$ A circuit of diodes as shown in the following figure can physically represent this.

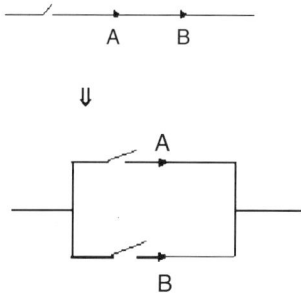

7.10.10 Transitive

If $A \subseteq B$ and $B \subseteq C$ Then $A \subseteq C$ (Table 7.4).

Table 7.4. Summary of laws of fuzzy set

S. No.	Property	Formula
1.	Commutative	$A \cup B = B \cup A$
2.	Associative	$A \cup (B \cup C) = (A \cup B) \cup C$
3.	Distributive	$A \cup (B \cap C) = (A \cup B) \cap (A \cup C)$
4.	Idem potency	$A \cup A = A$ and $A \cap A = A$
5.	Identity	$A \cup \varphi = A,\ A \cap X = A$
		$A \cap \varphi = A$, and $A \cup X = X$
6.	Involution	$\overline{\overline{A}} = A$
7.	Demorgan's Law	$\overline{(A \cup B)} = \overline{A} \cap \overline{B}$ and $\overline{(A \cap B)} = \overline{A} \cup \overline{B}$
8.	Convexity	Strictly monotonic
9.	Transitive	If $A \subseteq B$ and $B \subseteq C$ Then $A \subseteq C$
10.	Reflexive	
11.	Equivalence	If set is convex, transitive and reflexive
12.	Absorption	$(A \cap B) = A$ and $A \cap (A \cup B) = A$
		$A \cup X = X$ and $A \cap \varphi = \varphi$

7.11 Fuzzy Cartesian Product

Let A be a fuzzy set on universe X and B be a fuzzy set on universe Y; then the Cartesian product between fuzzy sets A and B will result in a fuzzy relation R, which is contained within the full Cartesian product space, or

$$A \times B = R \subset X \times Y$$

where the fuzzy relation R has membership function

$$\mu_R(x, y) = \mu_{A \times B}(x, y) = \min(\mu_A(x), \mu_B(y))$$

The Cartesian product defined by $A \times B = R$ above is implemented in the same fashion as it is the cross product of two vectors. The Cartesian product is not the same operation as the arithmetic product. Each of the fuzzy sets could be thought of as a vector of membership values; each value is associated with a particular element in each set.

For example, a fuzzy set (vector) A that has four elements, hence column vector of size 4×1, and for a fuzzy set (vector) B that has five elements, hence a row vector size of 1×5, the resulting fuzzy relation, R will be represented by a matrix of size 4×5; i.e. R will have four rows and five columns.

Example 1. Suppose we have two fuzzy sets, A defined on a universe of four discrete temperatures, $X = \{x_1, x_2, x_3, x_4\}$, and B defined on a universe of three discrete pressures, $Y = (y_1, y_2, y_3)$, and we want to find the fuzzy Cartesian product between them. Fuzzy set A could represent the "ambient" temperature and fuzzy set B the "near optimum" pressure for a certain heat exchanger, and the Cartesian product might represent the conditions(temperature-pressure pairs) of the exchanger that are associated with "efficient" operations.

Let

$$A = 0.2/x_1 + 0.5/x_2 + 0.8/x_3 + 1/x_4$$
$$\text{and } B = 0.3/y_1 + 0.5/y_2 + 0.9/y_3$$

Here A can be represented as a column vector of size 4×1 and B can be represented as a row vector of size 1×3. Then the fuzzy Cartesian product, results in a fuzzy relation R (of size 4×3) representing "efficient" conditions,

$$
A \times B = R =
\begin{array}{c}
x_1 \\ x_2 \\ x_3 \\ x_4
\end{array}
\begin{bmatrix}
0.2 & 0.2 & 0.2 \\
0.3 & 0.5 & 0.5 \\
0.3 & 0.5 & 0.8 \\
0.3 & 0.5 & 0.9
\end{bmatrix}
\begin{array}{ccc}
y_1 & y_2 & y_3
\end{array}
$$

7.12 Various Shapes of Fuzzy Membership Functions

It can be any real valued function, but in general normalized membership functions with values between 0 and 1.

a. *Triangular membership function*

Fuzzy number is defined as a convex single-point normal fuzzy set defined on the real line as shown in Fig. 7.17. Let us consider a fuzzy one. Its maximum membership will be at one and reduces on both sides as number increases or decreases as shown below. It is also called triangular fuzzy number.

b. *Trapezoidal fuzzy membership function*

A fuzzy set A is called trapezoidal fuzzy number with tolerance interval [a, b], left width α and right width β if its membership function has the following form

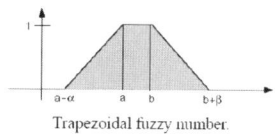

Trapezoidal fuzzy number.

$$A(t) = \begin{cases} 1 - \frac{a-t}{\alpha} & if \ a - \alpha \le t \le a \\ 1 & if \ a \le t \le b \\ 1 - \frac{t-b}{\beta} & if \ a \le t \le b + \beta \\ 0 & otherwise \end{cases}$$

and we use the notation A = (a, b, α, β). It can easily be shown that

$$[A]_\gamma = [a - (1 - \gamma)\alpha, b + (1 - \gamma)\beta], \forall \gamma \in [0, 1]$$

The support of A is (a − α, b + β).

A trapezoidal fuzzy number may be seen as a fuzzy quantity "x is approximately in the interval [a, b]".

Features of the membership function:
The following are the some of the features that are defined for a membership function.

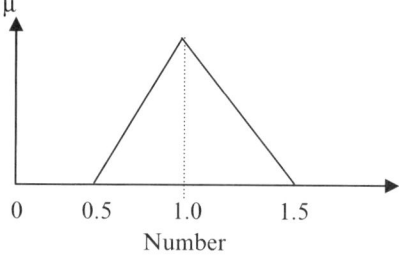

Fig. 7.17. Fuzzy one

Core
The core of a membership function for some fuzzy set is defined as that region of the universe that is characterized by complete or full membership value in the set.

Support
Support of a membership function for some fuzzy set is defined as that region of the universe that is characterized by nonzero membership value in the set.

Boundaries
Boundaries of a membership function for some fuzzy set is defined as that region of the universe containing elements that have a nonzero membership but not complete membership value that is this value lies between 0 and 1 (Fig. 7.18).

c. *Gaussian or bell shaped membership function*
The Gaussian or bell shaped membership function is also used in fuzzy set theory. The general expression for bell shaped membership function is given by following equation:

$$f(x, \alpha, \beta, \gamma) = \gamma \, \exp(-(x - \alpha)^2 / \beta),$$

where α, β, γ – are the parameters of bell shaped functions.

d. *Generalized membership function*
It is a membership function by which any shape can be generated. It is consisting of minimum four segments A, B, C, and D as shown in Fig. 7.19. The length and angle of these segments can be altered depending on the need to develop a membership function.

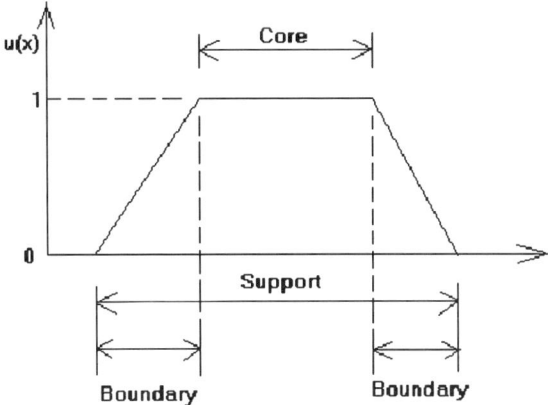

Fig. 7.18. Core, support and boundaries of a membership function

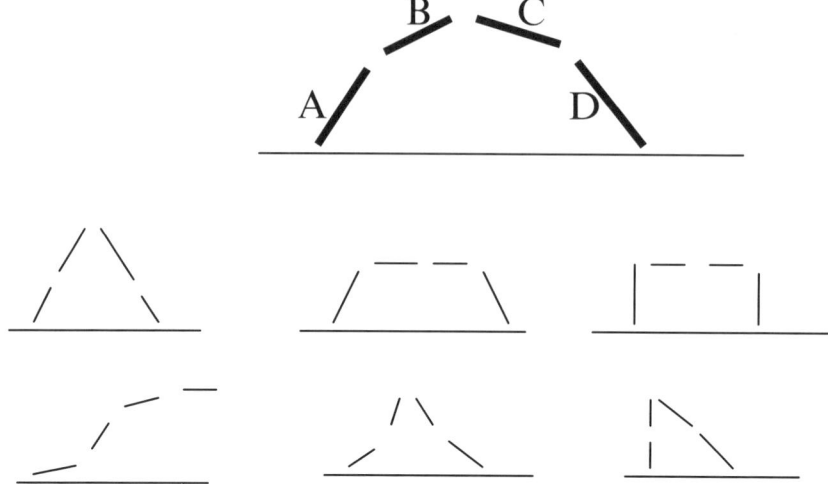

Fig. 7.19. Generalized membership function

%Matlab Program for plotting bell shaped membership function.

```
clear all,
clc;
a=1;
b=0.16;
gamma=1.1;
x=0:.1:3;
f=gamma*exp(-(x-a).^2/b);
plot(x,f);
```

Gaussian membership function

Properties of membership function:

1. The membership function should be strictly monotonically increasing, or strictly monotonically decreasing, or strictly monotonically increasing then strictly monotonically decreasing with the increasing value for elements in the universe of discourse X, i.e.

$$\text{If } x, y, z \in A \text{ and } x < y < z \text{ then}$$
$$\mu_{A(y)} \geq \min[\mu_A(x), \mu_A(z)].$$

2. The membership function should be continuous/piecewise continuous.
3. The membership function should be differentiable to provide smooth results.
4. The membership function should be of simple straight segments to make the process of fuzzy models easy and to high accuracy.

5. The membership function should satisfy the condition of a partition of unity i.e. the sum of the memberships of each element x from the universe of discourse X is equal to 1, i.e. $\sum \mu_A(x) \equiv 1, \ \forall x \in X$.

7.13 Methods of Defining of Membership Functions

There are numerous methods of constructing membership functions. They are constructed either intuitively or based on some algorithmic or logical operations. These methods can be classified as:

- Direct methods
- Indirect methods

Both direct and indirect methods are further classified to methods that involve one expert and methods that require multiple experts.

Direct methods:

In direct methods, each expert is expected to assign to each given element x, a membership grade A (x) that according to his opinion best captures the meaning of the linguistic term represented by the fuzzy set A.

In case of multiple experts, the most common method to aggregate the opinions of individual expert is based on probabilistic interpretation of membership functions.

Let,

n = Number of experts
a_i (x) = degree of belongingness of x to A in the opinion of expert-i
c_i = degree of competence of the expert-i

$$A(x) = \sum_{i=1}^{n} c_i a_i(x) \quad \text{where,} \quad \sum_{i=1}^{n} c_i = 1.$$

Indirect methods:

In indirect methods, experts are required to answer simple questions, easier to answer and less sensitive to the various biases of subjective judgments, which pertain to the constructed membership functions only implicitly. The answers are subject to further processing based on various assumptions to get the membership function.

The direct method with single expert gives very inconsistent result as it is based on a single person's opinion in a detailed manner. It is simply derived from the capacity of humans to develop membership functions through their own innate intelligence and understanding. On the other hand, the indirect method with multiple expert results in a very accurate and consistent dataset but the procedure is complicated.

Construction of membership from sample data:

It is assumed that n sample data $(x1, a1), (x2, a2), (x3, a3) \ldots \ldots (x_n, a_n)$ are given where a_i is a given grade of membership of x_i in a fuzzy set A; i.e. $a_i = A(x_i)$. The problem is to determine the whole membership function A.

- Artificial neural network
- Interpolation
- Curve fitting

Artificial neural network:

The multi-layer artificial neural networks is used for constructing membership functions based on learning patterns from sample data. The backpropagation algorithm is used to train the above neural network.

Interpolation:

Interpolation means to find the values of a function $f(x)$ for an x between different x values $x_0, x_1, \ldots \ldots x_n$ at which the values of $f(x)$ are given. With the interpolation we can approximate continuous function with desired accuracy.

There are many interpolation methods such as:

- Lagrange interpolation
- Newton's interpolation
- Newton–Gregory forward interpolation
- Spline interpolation, etc.

Curve fitting:

In case of least square curve fitting for bell shaped function with $\Upsilon = 1$, it gives a normal membership function. For minimization of sum square error with the function parameters, steepest descent method, or some optimization method may be used. Some more advanced technique, e.g. Genetic algorithm for minimizing the error can also be used.

7.14 Fuzzy Compositional Operators

Let R be the fuzzy relation from X to Y and s be the fuzzy relation from Y to Z then the composition of R and S is a fuzzy relation from X to Z and is represented by RoS. The fuzzy compostion RoS is given by

(i) *MAX-MIN composition*:

$$RoS \Rightarrow \int \vee (\mu_R(x, y) \wedge \mu_S(y, z))/(x, z)$$
$$XxZ \quad \text{where } \vee = \max \text{ and } \wedge = \min$$

Example: Let fuzzy relations R and S are given as:

$$R = \begin{bmatrix} 0.3 & 0.5 & 0.4 \\ 0.5 & 0.6 & 0.3 \\ 0.9 & 0.7 & 1.0 \end{bmatrix} \quad S = \begin{bmatrix} 0.8 & 0.6 & 0.7 \\ 0.1 & 0.4 & 0.6 \\ 0.5 & 0.9 & 1.0 \end{bmatrix}$$

Then fuzzy composition RoS is given by:

$$\text{RoS} = \begin{bmatrix} 0.3 & 0.5 & 0.4 \\ 0.5 & 0.6 & 0.3 \\ 0.9 & 0.7 & 1.0 \end{bmatrix} \text{o} \begin{bmatrix} 0.8 & 0.6 & 0.7 \\ 0.1 & 0.4 & 0.6 \\ 0.5 & 0.9 & 1.0 \end{bmatrix}$$

$$= \begin{bmatrix} 0.3 \cup 0.1 \cup 0.4 & 0.3 \cup 0.4 \cup 0.4 & 0.3 \cup 0.5 \cup 0.4 \\ 0.5 \cup 0.1 \cup 0.3 & 0.5 \cup 0.4 \cup 0.3 & 0.5 \cup 0.6 \cup 0.3 \\ 0.8 \cup 0.1 \cup 0.5 & 0.6 \cup 0.7 \cup 0.9 & 0.7 \cup 0.6 \cup 1 \end{bmatrix}$$

$$= \begin{bmatrix} 0.4 & 0.4 & 0.5 \\ 0.5 & 0.5 & 0.6 \\ 0.8 & 0.9 & 1 \end{bmatrix}$$

$$\text{SoR} = [\text{S}]\text{o}[\text{R}] = \begin{bmatrix} 0.8 & 0.6 & 0.7 \\ 0.1 & 0.4 & 0.6 \\ 0.5 & 0.9 & 1.0 \end{bmatrix} \text{o} \begin{bmatrix} 0.3 & 0.5 & 0.4 \\ 0.5 & 0.6 & 0.3 \\ 0.9 & 0.7 & 1.0 \end{bmatrix}$$

$$= \begin{bmatrix} 0.3 \cup 0.5 \cup 0.7 & 0.5 \cup 0.6 \cup 0.7 & 0.4 \cup 0.3 \cup 0.7 \\ 0.1 \cup 0.4 \cup 0.6 & 0.1 \cup 0.4 \cup 0.6 & 0.1 \cup 0.3 \cup 0.6 \\ 0.3 \cup 0.5 \cup 0.9 & 0.5 \cup 0.6 \cup 0.7 & 0.4 \cup 0.3 \cup 1 \end{bmatrix}$$

$$= \begin{bmatrix} 0.7 & 0.7 & 0.7 \\ 0.6 & 0.6 & 0.6 \\ 0.9 & 0.7 & 1 \end{bmatrix}$$

Now it is clear that RoS $\neq$ SoR.

(ii) *MIN-MAX composition*: This composition is defined as:

$$\text{RoS} \Rightarrow \int \wedge (\mu_R(x, y) \vee \mu_S(y, z))/(x, z)$$
$$\text{XxZ} \quad \text{where } \vee = \text{max and } \wedge = \text{min}$$

Example: Let fuzzy relations R and S are given as:

$$R = \begin{bmatrix} 0.3 & 0.5 & 0.4 \\ 0.5 & 0.6 & 0.3 \\ 0.9 & 0.7 & 1.0 \end{bmatrix} \quad S = \begin{bmatrix} 0.8 & 0.6 & 0.7 \\ 0.1 & 0.4 & 0.6 \\ 0.5 & 0.9 & 1.0 \end{bmatrix}$$

Then fuzzy composition RoS is given by:

$$RoS = \begin{bmatrix} 0.3 & 0.5 & 0.4 \\ 0.5 & 0.6 & 0.3 \\ 0.9 & 0.7 & 1.0 \end{bmatrix} \circ \begin{bmatrix} 0.8 & 0.6 & 0.7 \\ 0.1 & 0.4 & 0.6 \\ 0.5 & 0.9 & 1.0 \end{bmatrix}$$

$$= \begin{bmatrix} 0.8 \cap 0.5 \cap 0.5 & 0.6 \cap 0.5 \cap 0.9 & 0.7 \cap 0.6 \cap 1 \\ 0.8 \cap 0.6 \cap 0.3 & 0.6 \cap 0.6 \cap 0.9 & 0.7 \cap 0.6 \cap 1 \\ 0.9 \cap 0.7 \cap 1 & 0.9 \cap 0.7 \cap 1 & 0.9 \cap 0.7 \cap 1 \end{bmatrix}$$

$$= \begin{bmatrix} 0.5 & 0.5 & 0.6 \\ 0.3 & 0.6 & 0.6 \\ 0.7 & 0.7 & 0.7 \end{bmatrix}$$

(iii) *MAX-MAX composition*: This composition is defined as:

$$RoS \Rightarrow \int \vee(\mu_R(x,y) \vee \mu_S(y,z))/(x,z)$$

$$\text{where } \vee = \max \text{ and } \wedge = \min$$

Example: Let fuzzy relations R and S are given as:

$$R = \begin{bmatrix} 0.3 & 0.5 & 0.4 \\ 0.5 & 0.6 & 0.3 \\ 0.9 & 0.7 & 1.0 \end{bmatrix} \quad S = \begin{bmatrix} 0.8 & 0.6 & 0.7 \\ 0.1 & 0.4 & 0.6 \\ 0.5 & 0.9 & 1.0 \end{bmatrix}$$

Then fuzzy composition RoS is given by:

$$RoS = \begin{bmatrix} 0.3 & 0.5 & 0.4 \\ 0.5 & 0.6 & 0.3 \\ 0.9 & 0.7 & 1.0 \end{bmatrix} \circ \begin{bmatrix} 0.8 & 0.6 & 0.7 \\ 0.1 & 0.4 & 0.6 \\ 0.5 & 0.9 & 1.0 \end{bmatrix}$$

$$= \begin{bmatrix} 0.8 \cup 0.5 \cup 0.5 & 0.6 \cup 0.5 \cup 0.9 & 0.7 \cup 0.6 \cup 1 \\ 0.8 \cup 0.6 \cup 0.3 & 0.6 \cup 0.6 \cup 0.9 & 0.7 \cup 0.6 \cup 1 \\ 0.9 \cup 0.7 \cup 1 & 0.9 \cup 0.7 \cup 1 & 0.9 \cup 0.7 \cup 1 \end{bmatrix}$$

$$= \begin{bmatrix} 0.8 & 0.9 & 1 \\ 0.8 & 0.9 & 1 \\ 1 & 1 & 1 \end{bmatrix}$$

(iv) *MIN-MIN composition*: This type of fuzzy composition is defined as following:

$$RoS \Rightarrow \int \wedge(\mu_R(x,y) \wedge \mu_S(y,z))/(x,z)$$

$$\text{where } \vee = \max \text{ and } \wedge = \min$$

Example: Let fuzzy relations R and S are given as:

$$R = \begin{bmatrix} 0.3 & 0.5 & 0.4 \\ 0.5 & 0.6 & 0.3 \\ 0.9 & 0.7 & 1.0 \end{bmatrix} \quad S = \begin{bmatrix} 0.8 & 0.6 & 0.7 \\ 0.1 & 0.4 & 0.6 \\ 0.5 & 0.9 & 1.0 \end{bmatrix}$$

Then fuzzy composition RoS is given by:

$$RoS = \begin{bmatrix} 0.3 & 0.5 & 0.4 \\ 0.5 & 0.6 & 0.3 \\ 0.9 & 0.7 & 1.0 \end{bmatrix} o \begin{bmatrix} 0.8 & 0.6 & 0.7 \\ 0.1 & 0.4 & 0.6 \\ 0.5 & 0.9 & 1.0 \end{bmatrix}$$

$$= \begin{bmatrix} 0.3 \cap 0.1 \cap 0.4 & 0.3 \cap 0.4 \cap 0.4 & 0.3 \cap 0.5 \cap 0.4 \\ 0.5 \cap 0.1 \cap 0.3 & 0.5 \cap 0.4 \cap 0.3 & 0.6 \cap 0.6 \cap 0.3 \\ 0.8 \cap 0.1 \cap 0.5 & 0.6 \cap 0.7 \cap 0.9 & 0.7 \cap 0.6 \cap 1 \end{bmatrix}$$

$$= \begin{bmatrix} 0.1 & 0.3 & 0.3 \\ 0.1 & 0.3 & 0.3 \\ 0.1 & 0.6 & 0.6 \end{bmatrix}$$

(v) *MAX-product*: This class of composition is defined as:

$$RoS \Rightarrow \int \vee(\mu_R(x,y) \cdot \mu_S(y,z))/(x,z)$$

where $\vee = \max$ and $\wedge = \min$.

Example: Let fuzzy relations R and S are given as:

$$R = \begin{bmatrix} 0.3 & 0.5 & 0.4 \\ 0.5 & 0.6 & 0.3 \\ 0.9 & 0.7 & 1.0 \end{bmatrix} \quad S = \begin{bmatrix} 0.8 & 0.6 & 0.7 \\ 0.1 & 0.4 & 0.6 \\ 0.5 & 0.9 & 1.0 \end{bmatrix}$$

Then fuzzy composition RoS is given by:

$$RoS = \begin{bmatrix} 0.3 & 0.5 & 0.4 \\ 0.5 & 0.6 & 0.3 \\ 0.9 & 0.7 & 1.0 \end{bmatrix} o \begin{bmatrix} 0.8 & 0.6 & 0.7 \\ 0.1 & 0.4 & 0.6 \\ 0.5 & 0.9 & 1.0 \end{bmatrix}$$

$$= \begin{bmatrix} 0.24 \cup 0.05 \cup 0.20 & 0.18 \cup 0.20 \cup 0.36 & 0.21 \cup 0.30 \cup 0.4 \\ 0.40 \cup 0.06 \cup 0.15 & 0.30 \cup 0.24 \cup 0.27 & 0.35 \cup 0.36 \cup 0.30 \\ 0.72 \cup 0.07 \cup 0.5 & 0.54 \cup 0.28 \cup 0.9 & 0.63 \cup 0.42 \cup 1 \end{bmatrix}$$

$$= \begin{bmatrix} 0.24 & 0.36 & 0.40 \\ 0.40 & 0.30 & 0.36 \\ 0.72 & 0.36 & 1 \end{bmatrix}$$

(vi) *MAX-average*: This is defined as following:

$$RoS \Rightarrow \tfrac{1}{2}[\int \vee(\mu_R(x,y) + \mu_S(y,z))/(x,z)]$$
$$XxZ \quad \text{where } \vee = \max \text{ and } \wedge = \min.$$

Example: Let fuzzy relations R and S are given as:

$$R = \begin{bmatrix} 0.3 & 0.5 & 0.4 \\ 0.5 & 0.6 & 0.3 \\ 0.9 & 0.7 & 1.0 \end{bmatrix} \quad S = \begin{bmatrix} 0.8 & 0.6 & 0.7 \\ 0.1 & 0.4 & 0.6 \\ 0.5 & 0.9 & 1.0 \end{bmatrix}$$

Then fuzzy composition RoS is given by:

$$
\text{RoS} = \begin{bmatrix} 0.3 & 0.5 & 0.4 \\ 0.5 & 0.6 & 0.3 \\ 0.9 & 0.7 & 1.0 \end{bmatrix} \text{o} \begin{bmatrix} 0.8 & 0.6 & 0.7 \\ 0.1 & 0.4 & 0.6 \\ 0.5 & 0.9 & 1.0 \end{bmatrix}
$$

$$
= 1/2 \begin{bmatrix} 1.1 \cup 0.6 \cup 0.9 & 0.9 \cup 0.9 \cup 1.3 & 1.0 \cup 1.1 \cup 1.0 \\ 1.3 \cup 0.7 \cup 0.8 & 1.1 \cup 1.0 \cup 1.2 & 1.2 \cup 1.2 \cup 1.3 \\ 1.7 \cup 0.8 \cup 1.5 & 1.5 \cup 1.1 \cup 1.9 & 1.6 \cup 1.4 \cup 2 \end{bmatrix}
$$

$$
= 1/2 \begin{bmatrix} 1.1 & 1.3 & 1.1 \\ 1.3 & 1.2 & 1.3 \\ 1.7 & 1.9 & 2 \end{bmatrix} = \begin{bmatrix} 0.55 & 0.65 & 0.55 \\ 0.65 & 0.60 & 0.65 \\ 0.85 & 0.95 & 1 \end{bmatrix}
$$

7.15 Relation

If A and B are two sets and there is a specific property between elements x of A and y of B, this property can be described using the ordered pair (x, y). A set of such (x, y) pairs, $x \in A$ and $y \in B$, is called a relation $\boldsymbol{R}$.

$$\boldsymbol{R} = \{(x, y) | x \in A, \ y \in B\}$$

$\boldsymbol{R}$ is a binary relation and a subset of $A \times B$.

The term "x is in relation $\boldsymbol{R}$ with y" is denoted as

$(x, y) \in R$ or $x\boldsymbol{R}y$ with $R \subseteq A \times B$.
If $(x, y) \notin \boldsymbol{R}$, x is not in relation $\boldsymbol{R}$ with y.
If $A = B$ or $\boldsymbol{R}$ is a relation from A to A, it is written $(x, x) \in \boldsymbol{R}$ or $x\boldsymbol{R}x$ for $R \subseteq A \times A$.

n-ary relation

For sets $A_1, A_2, \ldots, A_n$ the relation among elements $x_1 \in A_1, x_2 \in A_2 \ldots, x_n \in A_n$ can be described by n-tuple $(x_1, x_2, \ldots, x_n)$. A collection of such n-tuples $(x_1, x_2, \ldots, x_n)$ is a relation $\boldsymbol{R}$ among $A_1, A_2, \ldots, A_n$. That is

$$(x_1, x_2, \ldots, x_n) \in \boldsymbol{R}$$
$$\boldsymbol{R} \subseteq A_1 \times A_2 \times \ldots \times A_n$$

Domain and range

Let $\boldsymbol{R}$ stand for a relation between A and B. The domain and range of this relation are defined as follows (Fig. A.2):

$$dom(\boldsymbol{R}) = \{x | x \in A, \ (x, y) \in \boldsymbol{R} \text{ for some } y \in B\}$$
$$ran(\boldsymbol{R}) = \{y | y \in B, \ (x, y) \in \boldsymbol{R} \text{ for some } x \in A\}$$

Here we call set A as support of $dom(\boldsymbol{R})$ and B as support of $ran(\boldsymbol{R})$.

$dom(\boldsymbol{R}) = A$ results in completely specified and $dom(\boldsymbol{R}) \subseteq A$ incompletely specified. The relation $\boldsymbol{R} \subseteq A \times B$ is a set of ordered pairs $(x,\ y)$. Thus, if we have a certain element x in A, we can find y of B, i.e. the mapped image of A. We say "y is the mapping of x" (Fig. 7.20).

7.15.1 Representation Methods of Relations

There are four methods of expressing the relation between sets A and B.

(1) *Bipartigraph*

The first is by illustrating A and B in a Fig. 7.21 and representing the relation by drawing arcs or edges.

(2) *Coordinate diagram*

The second is to use a coordinate diagram by plotting members of A on x axis and that of B on y axis, and then the members of $A \times B$ lie on the space. Figure 7.22 shows this type of representation for the relation $\boldsymbol{R}$, namely $x^2 + y^2 = 9$ where $x \in A$ and $y \in B$.

(3) *Matrix*

The third method is by manipulating relation matrix. Let A and B be finite sets having m and n elements respectively. Assuming $\boldsymbol{R}$ is a relation

Fig. 7.20. Domain and range

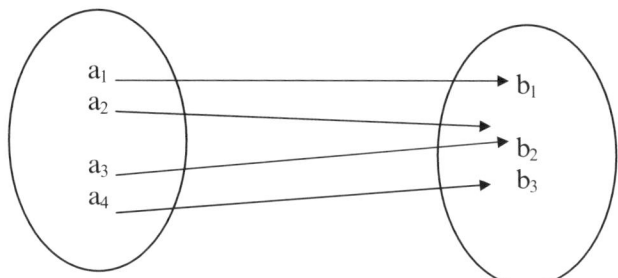

Fig. 7.21. Binary relation from A to B

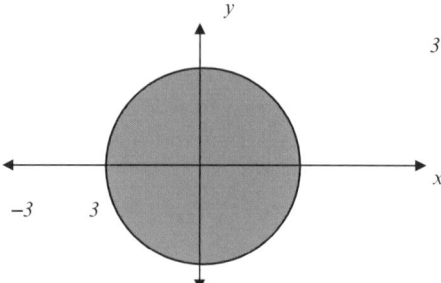

Fig. 7.22. Relation of $x^2 + y^2 = 9$

between A and B, we may represent the relation by matrix $M_R = (m_{ij})$ which is defined as follows:

$$M_R = (m_{ij})$$
$$m_{ij} = \begin{cases} 1, & (a_i, b_j) \in R \\ 0, & (a_i, b_j) \notin R \end{cases}$$
$$i = 1, 2, 3, \ldots\ldots, m$$
$$j = 1, 2, 3, \ldots\ldots, n$$

Such matrix is called a relation matrix, and that of the relation in Fig. 7.21 is given in the following:

R	b_1	b_2	b_3
a_1	1	0	0
a_2	0	1	0
a_3	0	1	0
a_4	0	0	1

(4) *Digraph*

The fourth method is the directed graph or digraph method. Elements are represented as nodes, and relations between elements as directed edges. $A = \{1, 2, 3, 4\}$ and $R = \{(1, 1), (1, 2), (2, 1), (2, 2), (1, 3), (2, 4), (4, 1)\}$ for instance. Figure 7.23 shows the directed graph corresponding to this relation. When a relation is symmetric, an undirected graph can be used instead of the directed graph.

7.15.2 Fundamental Properties of a Relation

Now we shall see the fundamental properties of relation defined on a set, that is, $R \subseteq A \times A$. We will review the properties such as reflexive relation, symmetric relation, transitive relation, closure, equivalence relation, compatibility relation, pre-order relation and order relation in detail.

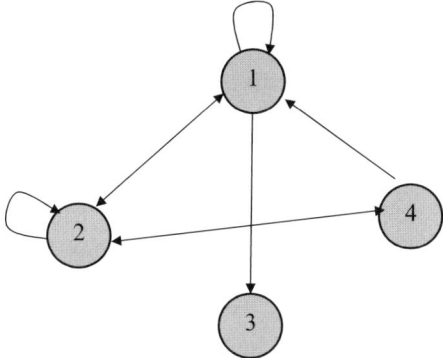

Fig. 7.23. Directed graph

(1) *Reflexive relation*
   If for all $x \in A$, the relation xRx or $(x,\ x) \in R$ is established, we call it reflexive relation. the reflexive relation might be denoted as

$$x \in A \rightarrow (x,x) \in or\ \mu_R(x,x) = 1, \forall x \in A,$$

where the symbol "$\rightarrow$" means "implication",
If it is not satisfied for some $x \in A$, the relation is called "irreflexive". If it is not satisfied for all $x \in A$, the relation is "antireflexive".
When you convert a reflexive relation into the corresponding relation matrix, you will easily notice that every diagonal member is set to 1. A reflexive relation is often denoted by D.

(2) *Symmetric relation*
   For all $x,\ y \in A$, if $xRy = yRx$, R is said to be a symmetric relation and expressed as

$$(x,y) \in R \rightarrow (y,x) \in R\ or$$
$$\mu_R(x,y) = \mu_R(y,x), \forall x, y \in A$$

The relation is "asymmetric" or "nonsymmetric" when for some $x,\ y \in A$, $(x,\ y) \in R$ and $(y,\ x) \notin R$.
It is an "antisymmetric" relation if for all $x,\ y \in A$, $(x,\ y) \in R$ and $(y,\ x) \notin R$.

(3) *Transitive relation*
   This concept is achieved when a relation defined on A verifies the following property. For all $x,\ y,\ z \in A$

$$(x,y) \in R, (y,z) \in R \rightarrow (x,z) \in R$$

(4) *Closure*

When relation R is defined in A, the requisites for closure are:
(1) Set A should satisfy a certain specific property.
(2) Intersection between A's subsets should satisfy the relation R.
The smallest relation R' containing the specific property is called closure of R.

Example 1. If R is defined on A, assuming R is not a reflexive relation, then $R' = D \cup R$ contains R and reflexive relation. At this time, R' is said to be the reflexive closure of R.

Example 2. If R is defined on A, transitive closure of R is as follows (Fig. 7.24), which is the same as R^∞ (reachability relation).

$$R^\infty = R \cup R^2 \cup R^3 \cup \ldots.$$

The transitive closure R^∞ of R for $A = \{1,\ 2,\ 3,\ 4\}$ and $R = \{(1,\ 2),\ (2,\ 3),\ (3,\ 4),\ (2,\ 1)\}$ is, $R^\infty = \{(1,\ 1),\ (1,\ 2),\ (1,\ 3),\ (1,4),\ (2,1),\ (2,2),\ (2,\ 3),\ (2,\ 4),\ (3,\ 4)\}$.

Equivalence relation

Relation $R \subseteq A \times A$ is an equivalence relation if the following conditions are satisfied.

(1) *Reflexive relation*
$$x \in A \rightarrow (x, x) \in R$$

(2) *Symmetric relation*
$$(x, y) \in R \rightarrow (y, x) \in R$$

(3) *Transitive relation*

$$(x, y) \in R, (y, z) \in R \rightarrow (x, z) \in R$$

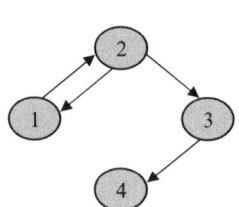
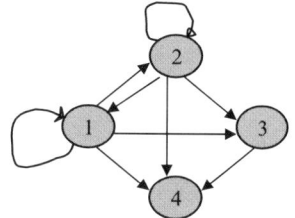

Fig. 7.24. Transitive closure

7.15.3 Fuzzy Relation

If a crisp relation R represents that of from sets A to B, for $x \in A$ and $y \in B$, its membership function $\mu_R(x, y)$ is,

$$\mu_R(x, y) = \begin{cases} 1 \ \textit{iff} \ (x, y) \in R \\ 0 \ \textit{iff} \ (x, y) \notin R \end{cases}$$

This membership function maps $A \times B$ to set $\{0, 1\}$.

$$\mu_R : A \times B \rightarrow \{0, 1\}$$

We know that the relation R is considered as a set. Recalling the previous fuzzy concept, we can define ambiguous relation.

Fuzzy relation has degree of membership whose value lies in $[0, 1]$.

$$\mu_R : A \times B \rightarrow [0, 1]$$
$$R = \{((x, y), \mu_R(x, y)) | \mu_R(x, y) \geq 0, x \in A, y \in B\}$$

Here $\mu_R(x, y)$ is interpreted as strength of relation between x and y. When $\mu_R(x, y) \geq \mu_R(x', y')$, (x, y) is more strongly related than (x', y').

When a fuzzy relation $R \subseteq A \times B$ is given, this relation R can be thought as a fuzzy set in the space $A \times B$.

Let's assume a Cartesian product space $X_1 \times X_2$ composed of two sets X_1 and X_2. This space makes a set of pairs (x_1, x_2) for all $x_1 \in X_1$, $x_2 \in X_2$. Given a fuzzy relation R between two sets X_1 and X_2, this relation is a set of pairs $(x_1, x_2) \in R$. Consequently, this fuzzy relation can be presumed to be a fuzzy restriction to the set $X_1 \times X_2$. Therefore, $R \subseteq X_1 \times X_2$.

Fuzzy binary relation can be extended to n-ary relation. If we assume $X_1, X_2, \ldots, X_n$ to be fuzzy sets, fuzzy relation $R \subseteq X_1 \times X_2 \times \cdots \times X_n$ can be said to be a fuzzy set of tuple elements $(x_1, x_2, \ldots, x_n)$, where $x_1 \in X_1$, $x_2 \in X_2$, $\ldots$, $x_n \in X_n$.

Example 1. Figure 7.25 for instance, crisp relation R in the figure (a) reflects a relation in $A \times A$. Expressing this by membership function, $\mu_R(a, c) = 1$, $\mu_R(b, a) = 1$, $\mu_R(c, b) = 1$ and $\mu_R(c, d) = 1$.

If this relation is given as the value between 0 and 1 as in Fig. 7.25(b), this relation becomes a fuzzy relation. Expressing this fuzzy relation by membership function yields,

$$\mu_R(a, c) = 0.8, \mu_R(b, a) = 1, \mu_R(c, b) = 0.9 \ and \ \mu_R(c, d) = 1.$$

It's corresponding fuzzy matrix is as follows.

A \ A	a	b	c	d
a	0.0	0.0	0.8	0.0
b	1.0	0.0	0.0	0.0
c	0.0	0.9	0.0	1.0
d	0.0	0.0	0.0	0.0

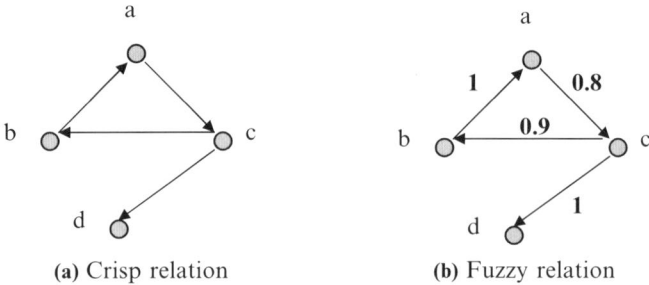

(a) Crisp relation   **(b)** Fuzzy relation

Fig. 7.25. Crisp and fuzzy relation

Fuzzy relation is mainly useful when expressing knowledge. Generally, the knowledge is composed of rules and facts. A rule can contain the concept of possibility of event b after event a has occurred. For instance, let us assume that set A is a set of events and R is a rule. Then by the rule R, the possibility for the occurrence of event c after event a occurred is 0.8 in the previous fuzzy relation. When crisp relation R represents the relation from crisp sets A to B, its domain and range is defined as:

$$dom(R) = \{x | x \in A, y \in A, \mu_R(x, y) = 1\}$$
$$ran(R) = \{y | x \in A, y \in A, \mu_R(x, y) = 1\}$$

Domain and range of fuzzy relation
     When fuzzy relation R is defined in crisp sets A and B, the domain and range of this relation are defined as:

$$\mu_{dom(R)}(x) = \min_{y \in B} \mu_R(x, y)$$
$$\mu_{ran(R)}(y) = \max_{x \in A} \mu_R(x, y)$$

The *"height of the relation"* is defined as follows:

$$h(R) = \max_{y \in B} \max_{y \in A} \mu_R(x, y)$$

Set A becomes the support of $dom(R)$ and $dom(R) \subseteq A$. Set B is the support of $ran(R)$ and $ran(R) \subseteq B$.

Example 2. Let

$$X = \{1, 2, \ldots\ldots, 100\}, \ Y = \{50, 51, \ldots\ldots, 100\}.$$

And Binary fuzzy relation R is defined as "x is much smaller than y"
Its membership function defined as:

$$\mu_R(x, y) = \begin{cases} 1 - \frac{x}{y}, & for \ x \leq y \\ 0, & otherwise \end{cases} \quad where \ x \in X \ and \ y \in Y$$

Domain of R : $\mu_{dom(R)}(x) = \max\limits_{y \in Y} \mu_R(x,\ y) = \max\limits_{y \in Y}\left(1 - \dfrac{x}{y}\right) = 1 - \dfrac{x}{100}$

Range of R : $\mu_{ran(R)}(y) = \max\limits_{x \in X} \mu_R(x,\ y) = \max\limits_{x \in X}\left(1 - \dfrac{x}{y}\right) = 1 - \dfrac{1}{y}$

Height of R : $h(R) = \max\limits_{y \in B} \max\limits_{x \in A} \mu_R(x,\ y) = 1 - \dfrac{1}{100} = \dfrac{99}{100} = 0.99.$

Fuzzy matrix

Given a certain vector, if an element of this vector has its value between 0 and 1, we call this vector a fuzzy vector. Fuzzy matrix is a gathering of such vectors. Given a fuzzy matrix $A = (a_{ij})$ and $B = (b_{ij})$, we can perform operations on these fuzzy matrices.

(1) *Sum*

$$A + B = \mathrm{Max}[a_{ij}, b_{ij}].$$

(2) *Max product*

$$A \bullet B = AB = \underset{k}{Max}[\mathrm{Min}(a_{ik}, b_{kj})]$$

Example 3. The followings are examples of sum and max product on fuzzy sets A and B.

$$A = \begin{array}{c|ccc} & a & b & c \\ \hline a & 0.2 & 0.5 & 0.0 \\ b & 0.4 & 1.0 & 0.1 \\ c & 0.0 & 1.0 & 0.0 \end{array} \qquad B = \begin{array}{c|ccc} & a & b & c \\ \hline a & 1.0 & 0.1 & 0.0 \\ b & 0.0 & 0.0 & 0.5 \\ c & 0.0 & 1.0 & 0.1 \end{array}$$

$$A + B = \begin{array}{c|ccc} & a & b & c \\ \hline a & 1.0 & 0.5 & 0.0 \\ b & 0.4 & 1.0 & 0.5 \\ c & 0.0 & 1.0 & 0.1 \end{array} \qquad A \bullet B = \begin{array}{c|ccc} & a & b & c \\ \hline a & 0.2 & 0.1 & 0.5 \\ b & 0.4 & 0.1 & 0.5 \\ c & 0.0 & 0.0 & 0.5 \end{array}$$

Here let's have a closer look at the product $A \bullet B$ of A and B. For instance, in the first row and second column of the matrix $C = A \bullet B$, the value 0.1 $(C_{12} = 0.1)$ is calculated by applying the max–min operation to the values of the first row (0.2, 0.5 and 0.0) of A, and those of the second column (0.1, 0.0 and 1.0) of B.

$$\begin{array}{ccc} 0.2 & 0.5 & 0.0 \\ Min \Downarrow \quad 0.0 & 0.5 & 0.1 \\ \hline 0.0 & 0.5 & 0.0 \\ \Rightarrow & 0.5 \\ Max \end{array}$$

And for all i and j, if $a_{ij} \le b_{ij}$ holds, matrix B is bigger than A.

$$a_{ij} \le b_{ij} \Leftrightarrow A \le B.$$

Also when $A \le B$ for arbitrary fuzzy matrices S and T, the following relation holds from the max-product operation.

$$A \le B \Leftrightarrow SA \le SB, \quad AT \le BT$$

Fuzzy relation matrix

If a fuzzy relation R is given in the form of fuzzy matrix, its elements represent the membership values of this relation. That is, if the matrix is denoted by M_R, and membership values by $\mu_R(i,j)$, then $M_R = (\mu_R(i,j))$.

7.15.4 Operation of Fuzzy Relation

We know now a relation is one kind of sets. Therefore we can apply operations of fuzzy set to the relation. We assume $R \subseteq A \times B$ and $S \subseteq A \times B$.

(1) *Union relation*

Union of two relations R and S is defined as follows:

$$\forall(x,y) \in A \times B$$
$$\mu_{R \cup S}(x,y) = Max[\mu_R(x,y), \mu_S(x,y)]$$
$$= \mu_R(x,y) \vee \mu_S(x,y).$$

We generally use the sign $\vee$ for Max operation. For n relations, we extend it to the following.

$$\mu R_1 \cup R_2 \cup \ldots \ldots \cup R_n(x,y) = \vee_{R_i} \mu_{Ri}(x,y).$$

If expressing the fuzzy relation by fuzzy matrices, i.e. M_R and M_S, matrix $M_{R \cup S}$ concerning the union is obtained from the sum of two matrices.

$$M_{R \cup S} = M_R + M_S.$$

(2) *Intersection relation*

The intersection relation $R \cap S$ of set A and B is defined by the following membership function.

$$\forall(x,y) \in A \times B$$
$$\mu_{R \cap S}(x,y) = Min[\mu_R(x,y), \mu_S(x,y)]$$
$$= \mu_R(x,y) \wedge \mu_S(x,y)$$

The symbol $\wedge$ is for the min operation. In the same manner, the intersection relation for n relations is defined by

$$\mu R_1 \cap R_2 \cap \ldots \ldots \cap R_n(x,y) = \wedge_{R_i} \mu_{Ri}(x,y)$$

(3) *Complement relation*

Complement relation $\overline{R}$ for fuzzy relation R shall be defined by the following membership function.

$$\forall(x,\ y) \in A \times B$$
$$\mu\overline{R}(x,y) = 1 - \mu_R(x,y)$$

Example 4. Consider two binary relations R and S on $X \times Y$ defined as

R : x is considerably smaller than y
S : x is very close to y

The fuzzy relation matrices M_R and M_S are given as

M_R	y_1	y_2	y_3
x_1	0.3	0.2	1.0
x_2	0.8	1.0	1.0
x_3	0.0	1.0	0.0

M_S	y_1	y_2	y_3
x_1	0.3	0.0	0.1
x_2	0.1	0.8	1.0
x_3	0.6	0.9	0.3

Fuzzy relation matrices $M_{R \cup S}$ and $M_{R \cap S}$ corresponding to R∪S and R∩S yield the followings.

$M_{R \cup S}$	y_1	y_2	y_3
x_1	0.3	0.2	1.0
x_2	0.8	1.0	1.0
x_3	0.6	1.0	0.3

$M_{R \cap S}$	y_1	y_2	y_3
x_1	0.3	0.0	0.1
x_2	0.1	0.8	1.0
x_3	0.0	0.9	0.0

Where the union of R and S means that "x is considerably smaller than y" *OR* "x is very close to y".
And the intersection of R and S means that "x is considerably smaller than y" *AND* "x is very close to y".
Also complement relation of fuzzy relation R shall be

$M_{\overline{R}}$	a	b	c
1	0.7	0.8	0.0
2	0.2	0.0	0.0
3	1.0	0.0	1.0

(4) *Inverse relation*
When a fuzzy relation $R \subseteq A \times B$ is given, the inverse relation of R^{-1} is defined by the following membership function.
For all $(x, y) \subseteq A \times B$, $\mu_R^{-1}(y, x) = \mu_R(x, y)$

Composition of fuzzy relation
Two fuzzy relations R and S are defined on sets A, B and C. That is, $R \subseteq A \times B$, $S \subseteq B \times C$. The composition $S \bullet R = SR$ of two relations R and S is expressed by the relation from A to C, and this composition is defined by the following.

$$\text{For } (x, y) \in A \times B, (y, z) \in B \times C,$$
$$\mu_S \bullet_R (x, z) = \underset{y}{Max}[Min(\mu_R(x, y), \mu_S(y, z))]$$
$$= \underset{y}{\vee}[\mu_R(x, y) \wedge \mu_S(y, z)]$$

$S \bullet R$ from this elaboration is a subset of $A \times C$. That is, $S \bullet R \subseteq A \times C$.

If the relations R and S are represented by matrices M_R and M_S, the matrix $M_{S \bullet R}$ corresponding to $S \bullet R$ is obtained from the product of M_R and M_S.

$$M_{S \bullet R} = M_R \bullet M_S$$

Example 5. Consider fuzzy relations $R \subseteq A \times B$, $S \subseteq B \times C$. The sets A, B and C shall be the sets of events. By the relation R, we can see the possibility of occurrence of B after A, and by S, that of C after B. For example, by M_R, the possibility of $a \in B$ after $1 \in A$ is 0.1. By M_S, the possibility of occurrence of 7 after a is 0.9.

R	a	b	c	d
1	0.1	0.2	0.0	1.0
2	0.3	0.3	0.0	0.2
3	0.8	0.9	0.1	0.4

S	α	β	γ
a	0.9	0.0	0.3
b	0.2	1.0	0.8
c	0.8	0.0	0.7
d	0.4	0.2	0.3

Here, we can not guess the possibility of C when A is occurred. So our main job now will be the obtaining the composition $S \bullet R \subseteq A \times C$. The following matrix $M_{S \bullet R}$ represents this composition and it is also given in Fig. 7.26.

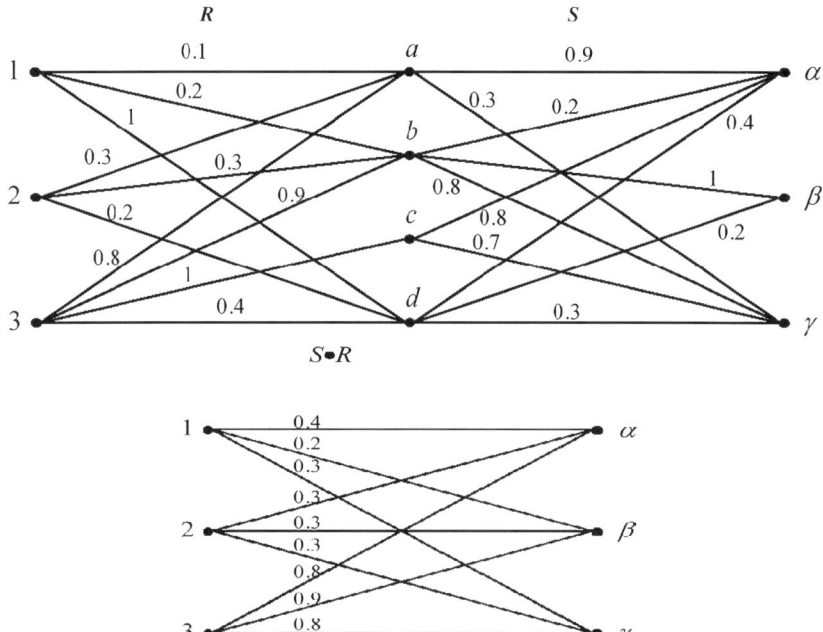

Fig. 7.26. Composition of fuzzy relation

$S \bullet R$	α	β	γ
1	0.4	0.2	0.3
2	0.3	0.3	0.3
3	0.9	0.9	0.8

Now we see the possibility of occurrence of $\alpha \in C$ after event $1 \in A$ is 0.4, and that for $\beta \in C$ after event $2 \in A$ is 0.3, etc.

Presuming that the relations R and S are the expressions of rules that guide the occurrence of event or fact. Then the possibility of occurrence of event B when event A is happened is guided by the rule R. And rule S indicates the possibility of C when B is existing. For further cases, the possibility of C when A has occurred can be induced from the composition rule $S \bullet R$. This manner is named as an "inference" which is a process producing new information.

7.15.5 Projection and Cylindrical Extension

We can project a fuzzy relation $R \subseteq A \times B$ with respect to A or B as in the following manner.

For all $x \in A$, $y \in B$,
$$\mu R_A(x) = \underset{y}{Max}\ \mu_R(x, y) : \text{projection to A}$$
$$\mu R_B(y) = \underset{x}{Max}\ \mu_R(x, y) : \text{projection to B}.$$

Here the projected relation of R to A is denoted by R_A, and to B is by R_B.

Example 6. There is a relation $R \subseteq A \times B$. The projection with respect to A or B shall be

A \ B	b_1	b_2	b_3
a_1	0.1	0.2	1.0
a_2	0.6	0.8	0.0
a_3	0.0	1.0	0.3

$$M_{R_A} = \begin{array}{c|c} a_1 & 1.0 \\ a_2 & 0.8 \\ a_3 & 1.0 \end{array} \qquad M_{R_B} = \begin{array}{c|ccc} & b_1 & b_2 & b_3 \\ \hline & 0.6 & 1.0 & 1.0 \end{array}$$

In the projection to A, the strongest degree of relation concerning a_1 is 1.0, that for a_2 is 0.8 and that for a_3 is 1.0.

Projection in n dimension

So far has been the projection in two-dimensions relation. Extending it to n-dimensional fuzzy set, assume relation R is defined in the space of $X_1 \times X_2 \times \ldots \ldots \times X_n$. Projecting this relation to subspace of $X_{i1} \times X_{i2} \times \ldots \ldots \times X_{ik}$, gives a projected relation : $R_{X_{i1} \times X_{i2} \times \ldots \ldots \times X_{ik}}$

$$\mu_{R X i 1_{i1} \times X_{i2} \times \ldots \times X_{ik}}(x_{i1}, x_{i2}, \ldots, x_{ik})$$
$$= \underset{X_{j1}, X_{j2}, \ldots, X_{jm}}{Max} \mu_R(x_1, x_2, \ldots, x_n).$$

Here X_{j1}, X_{j2}, ..., X_{jm} represent the omitted dimensions, and X_{i1}, X_{i2}, ..., X_{ik} gives the remained dimensions, and thus

$$\{X_1, X_2, \ldots, X_n\} = \{X_{i1}, X_{i2}, \ldots, X_{ik}\} \cup \{X_{j1}, X_{j2}, \ldots, X_{jm}\}.$$

Cylindrical extension

As the opposite concept of projection, cylindrical extension is possible. If a fuzzy set or fuzzy relation R is defined in space $A \times B$, this relation can be extended to $A \times B \times C$ and we can obtain a new fuzzy set. This fuzzy set is written as $C(R)$.

$$\mu_{C(R)}(a, b, c) = \mu_R(a, b)$$
$$a \in A, \ b \in B, \ c \in C$$

Example 7. In the previous example, relation R_A is the projection of R to direction A. If we extend it again to direction B, we can have an extended relation $C(R_A)$. For example

$$\mu_{C(R_A)}(a_1, b_1) = \mu_{R_A}(a_1) = 1.0$$
$$\mu_{C(R_A)}(a_1, b_2) = \mu_{R_A}(a_1) = 1.0$$
$$\mu_{C(R_A)}(a_2, b_1) = \mu_{R_A}(a_2) = 0.8$$

$$M_{C(R_A)} = \begin{array}{c|ccc} & b_1 & b_2 & b_3 \\ \hline a_1 & 1.0 & 1.0 & 1.0 \\ a_2 & 0.8 & 0.8 & 0.8 \\ a_3 & 1.0 & 1.0 & 1.0 \end{array}$$

The new relation $C(R_A)$ is now in $A \times B$.

Let two fuzzy relations be defined as follows:

$$R \subseteq X_1 \times X_2, S \subseteq X_2 \times X_3$$

Even though we want to apply the intersection operation between R and S, it is not possible because the domains of R and S are different each other. If we obtain cylindrical extensions $C(R)$ and $C(S)$ to space of $X_1 \times X_2 \times X_3$, and then $C(R)$ and $C(S)$ have the same domain. We can now apply operations on the two extended sets $C(R)$ and $C(S)$. Therefore join (or intersection) of R and S can be calculated by the intersection of $C(R)$ and $C(S)$.

$$\text{join } (R, S) = C(R) \cap C(S)$$

The projection and cylindrical extension are often used to make domains same for more than one fuzzy sets.

7.16 Approximate Reasoning

In 1979 *Zadeh* introduced the theory of approximate reasoning. This theory provides a powerful framework for reasoning in the face of imprecise and uncertain information.

Central to this theory is the representation of propositions as statements assigning fuzzy sets as values to variables.

Suppose we have two interactive variables $x \in X$ and $y \in Y$ and the causal relationship between x and y is completely known. Namely, we know that y is a function of x.

$y = f(x)$ Then we can make inferences easily

$$\text{premise } y = f(x)$$
$$\text{fact } x = x'$$
$$\text{consequence } y = f(x')$$

This inference rule says that if we have $y = f(x), \forall x \in X$ and we observe that $x = x'$ then y takes the value $f(x')$

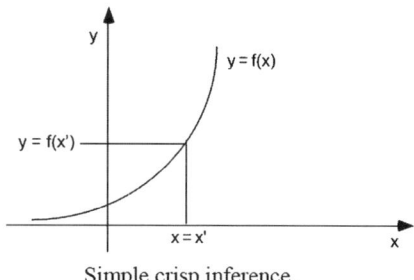

Simple crisp inference.

Suppose that we are given an $x' \in X$ and want to find $y' \in Y$ which correponds to x' under the rule-base.

> **Rule 1**: If $x = x_1$ Then $y = y_1$
> and
> **Rule 2**: If $x = x_2$ Then $y = y_2$
> and
> **Rule 3**: If $x = x_3$ Then $y = y_3$
> and
>
> . . . .
> **Rule n**: If $x = x_n$ Then $y = y_n$
> **Fact**: $x = x'$
>
> **Consequence**: $y = y'$

Let x and y be linguistic variables, e.g. "x is Big" and "y is Medium". The basic problem of approximate reasoning is to find the membership function of the consequence C from the rule-base and the fact A.

Rule 1: If $x = A_1$ Then $y = C_1$
and
Rule 2: If $x = A_2$ Then $y = C_2$
and

. . . .

Rule n: If $x = A_n$ Then $y = C_n$
Fact: $x = A$

Consequence: $y = C$

Zadeh introduced a number of translation rules which allow us to represent some common linguistic statements in terms of propositions in our language.

Entailment rule:

x is A	*Anjali is very beautiful*
$A \subset B$	*very beautiful $\subset$ beautiful*
x is B	*Anjali is beautiful*

Conjuction rule:

x is A	*Anjali is beautiful*
and	
x is B	*Anjali is intelligent*
x is A $\cap$ B	*Anjali is beautiful and intelligent*

Disjunction rule:

x is A	*Anjali is married*
or	
x is B	*Anjali is bachelor*
x is $A \cup B$	*Anjali is married or bachelor*

Projection rule:

(x, y) have relation R: x is $\Pi_X(R)$
(x, y) have relation R: y is $\Pi_Y(R)$,

e.g.

(x, y) is close to (3, 2): x is close to 3
(x, y) is close to (3, 2): y is close to 2.

Negation rule:

not (x is A): x is $\overline{A}$,
e.g. not (x is high): x is not high.

Table 7.5. Truth table for classical implication

P	Q	P → Q
T	T	T
T	F	F
F	T	T
F	F	T

The classical implication

Let P = "x is in A" and Q = "y is in B" are crisp propositions, where A and B are crisp sets for the moment. The implication P → Q is interpreted as $\neg(p \cap \neg q)$.

"*P entails Q*" *means that it can never happen that P is true and Q is not true.*

It is easy to see that $P \rightarrow Q = \neg P \cup Q$

In 1930s Lukasiewicz, polish mathematician explored for the first time logics other than Aristotelian (classical or binary logic) (Rescher 1969; Ross 1995). In this implication the proposition P is the hypothesis or the antecedent, and the proposition Q is also referred to as the conclusion or the consequent. The compound proposition P → Q is true in all cases except where a true antecedent P appears with a false consequent Q, i.e. a true hypothesis cannot imply a false conclusion as given in the Truth Table 7.5.

Generalized Modus Ponens (GMP)

Classical logic elaborated many reasoning methods called tautologies. One of the best known is Modus Ponens. The reasoning process in Modus Ponens is known as follows:

Rule	IF X is A THEN Y is B
Fact	X is A
Conclusion	Y is B

In classical Modus Ponens tautology the truth value of the premise "X is A" and conclusion "Y is B" is allowed to assume only two discrete values 0 or 1 and fact considered "X is ...", must fully agree with the implication premise: *IF X is A THEN Y is B*.

Only then may the implication be used in the reasoning process. Both the premise and the rule conclusion must be formulated in a deterministic way. Statements with non-precise formulations as given below are not accepted.

$$X \text{ is about } A,$$
$$X \text{ is more than } A,$$
$$Y \text{ is more or less } B,$$

In fuzzy logic an approximate reasoning has been applied. It enables the use of fuzzy formulations in premises and conclusions. The approximate reasoning based on he *Generalized Modus Ponens* tautology is:

Rule	IF X is A THEN Y is B
Rule	
Fact	X is A′
Conclusion	Y is B′

Where **A′**, **B′** can mean e.g.: **A′** = more than A, **B′** = more or less B, etc. A reasoning example according to the *Generalized Modus Ponens* (GMP) tautology can be like:

Rule	IF (route of the trip is long) THEN (traveling time is long)
Fact	Route of the trip is very long
Conclusion	Travelling time is very long

Here, fuzzy implication is expressed in the following way:

$$\forall u \in U, \forall v \in V$$

$$\mu_R(u,v) = \begin{cases} 1 \ if \ \mu_A(u) \le \mu_B(v) \\ \mu_B(v) \ otherwise \end{cases}$$

$$\mu_{B'}(v) = \sup_{u \in U} \min(\mu_{A'}(u), \mu_R(u,v))$$

In fuzzy logic and approximate reasoning, the most important fuzzy implication inference rule is the *Generalized Modus Ponens* (GMP). The classical *Modus Ponens* inference rule says:

premise   if p then q
fact p

Consequence q

This inference rule can be interpreted as: If p is true and $p \to q$ is true then q is true.

The fuzzy implication inference is based on the compositional rule of inference for approximate reasoning suggested by Zadeh.

Compositional rule of inference

premise if x is A then y is B
fact   x is A′

consequence y is B′

Table 7.6. Truth table for modus ponens

P	Q	$P \to Q$	$[P \cap (P \to Q)]$	$[P \cap (P \to Q)] \to Q$
0	0	1	0	1
0	0.5	1	0	1
0	1	1	0	1
0.5	0	0	0	1
0.5	0.5	1	0.5	1
0.5	1	1	0.5	1
1	0	0	0	1
1	0.5	0.5	0.5	1
1	1	1	1	1

where the consequence B' is determined as a composition of the fact and the fuzzy implication operator

$$B' = A' \circ (A \to B) \text{ that is}$$
$$B'(v) = \sup_{u \in U} \min\{A'(u), (A \to B)(u, v)\}, v \in V$$

The consequence B' is nothing else but the shadow of $A \to B$ on A'.

The Generalized Modus Ponens (GMP), which reduces to classical modus ponens when $A' = A$ and $B' = B$, is closely related to the forward data-driven inference which is particularly useful in the fuzzy logic control. The truth table for GMP is given in Table 7.6.

The classical Modus Tollens inference rule says: If $p \to q$ is true and q is false then p is false. The Generalized Modus Tollens.

>**premise** if x is A then y is B
>**fact** y is B'
>
>**consequence** x is A'

which reduces to "Modus Tollens" when $B = \overline{B}$ and $A' = \overline{A}$, is closely related to the backward goal-driven inference. The consequence A' is determined as a composition of the fact and the fuzzy implication operator

$$\overline{A'} = \overline{B'} \circ (A \to B)$$

Fuzzy relation schemes

>**Rule 1**: If $x = A_1$ and $y = B_1$ Then $z = C_1$
>**Rule 2**: If $x = A_2$ and $y = B_2$ Then $z = C_2$
>
> . . . .
>
>**Rule n**: If $x = A_n$ and $y = B_n$ Then $z = C_n$
>**Fact**: x is x_0 and y is y_0
>
>$z = C$

The ith fuzzy rule from this rule-base

Rule i: If $x = A_i$ and $y = B_i$ Then $z = C_i$

is implemented by a *fuzzy relation Ri* and is defined as

$$R_i(u, v, w) = (A_i \times B_i \rightarrow C_i)(u, w)$$
$$= [A_i(u) \cap B_i(v)] \rightarrow C_i(w)$$
$$\text{For } i = 1, 2, \ldots, n$$

Find C from the input x_0 and from the rule base: $R = \{R_1, R_2, \ldots\ldots, R_n\}$
Interpretation of

- Logical connective "and"
- Sentence connective "also"
- Implication operator "then"
- Compositional operator "o"

We first compose x_0 X y_0 with each Ri producing intermediate result

$$C_i' = \bar{x}_0 \cap \bar{y}_0 \circ R_i$$

for $i = 1, \ldots, n$. Here $C'i$ is called the output of the ith rule

$$C_i'(w) = A_i(x_0) \cap B_i(y_0) \rightarrow C_i(w) \qquad \text{for each } w.$$

Then combine the C_i' component wise into C' by some aggregation operator:

$$C = \cup C_i' = \bar{x}_0 \, \bar{x} \, \bar{y}_0 \circ \bar{R}_1 \cup \ldots. x_0 \, x \, y_0 \circ R_n$$
$$C_i(w) = A_i(x_0) \cap B_i(y_0) \rightarrow C_i(w) \cup \ldots\ldots \cup A_n(x_0) \cap B_n(y_0) \rightarrow C_n(w)$$

Steps involved in approximate reasoning

- Input to the system is (x_0, y_0)
- Fuzzify the input (x_0, y_0)
- Find the firing strength of the ith rule after aggregating the inputs is

$$A_i(x_0) \cap B_i(y_0)$$

- Calculate the ith individual rule output is

$$C_i'(w) = A_i(x_0) \cap B_i(y_0) \rightarrow C_i(w)$$

- Determine the overall system output (fuzzy) is

$$C = C_1' \cup C_2' \ldots \cup C_n'$$

- Finally, calculate the defuzzified output.

We present five well-known inference mechanisms in fuzzy rule-based systems.

Mamdani: The fuzzy implication is modeled by Mamdani's minimum operator and the sentence connective *also* is interpreted as oring the propositions and defined by *max* operator.

The firing levels of the rules, denoted by α_i, $i = 1,\ 2$, are computed by

$$\alpha_1 = A_1(x_0) \cap B_1(y_0) \text{ and } \alpha_2 = A_2(x_0) \cap B_2(y_0).$$

The individual rule outputs are obtained by

$$C_1'(w) = (\alpha_1 \cap C_1(w)) \text{ and } C_2'(w) = (\alpha_2 \cap C_2(w)).$$

Then the overall system output is computed by calculating the individual rule outputs as shown in Fig. 7.27

$$C(w) = C_1'(w) \cup C_2'(w)$$
$$= (\alpha_1 \cap C_1(w)) \cup (\alpha_2 \cap C_2(w))$$

Finally, to obtain a deterministic control action, we employ any defuzzification strategy.

Tsukamoto: All linguistic terms are supposed to have monotonic membership functions.

The firing levels of the rules, denoted by α_i, $i = 1,\ 2$, are computed by

$$\alpha_1 = A_1(x_0) \cap B_1(y_0) \text{ and } \alpha_2 = A_2(x_0) \cap B_2(y_0).$$

In this mode of reasoning the individual crisp control actions z_1 and z_2 are computed from the equations

$$\alpha_1 = C_1(z_1), \quad \alpha_2 = C_2(z_2),$$

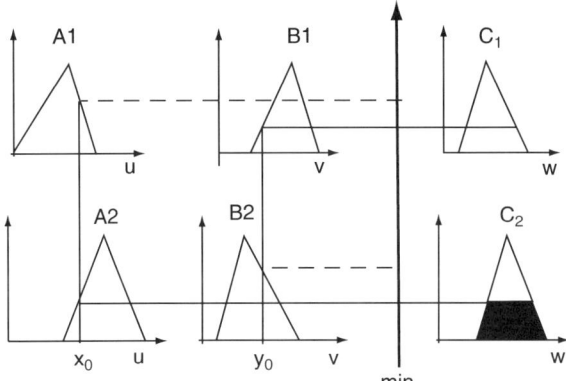

Fig. 7.27. Fuzzy inference

and the overall crisp control action is expressed as

$$z_0 = \frac{\alpha_1 z_1^* + \alpha_2 z_2^*}{\alpha_1 + \alpha_2}$$

i.e. z_0 is computed by the discrete center-of-gravity method.

If we have n rules in our rule-base then the crisp control action is computed as

$$z_0 = \frac{\displaystyle\sum_{i=1}^{n} \alpha_i z_i}{\displaystyle\sum_{i=1}^{n} \alpha_i}$$

where α_i is the firing level and z_i is the (crisp) output of the ith rule, $i = 1, \ldots, n$.

Sugeno: Sugeno and Takagi use the following architecture

Rule 1: If $x = A_1$ and $y = B_1$ Then $z_1 = a_1 x + b_1 y$
Rule 2: If $x = A_2$ and $y = B_2$ Then $z_2 = a_2 x + b_2 y$
. . . .
Rule n: If $x = A_n$ and $y = B_n$ Then $z_n = a_n x + b_n y$
Fact: x is x_0 and y is y_0

Consequence: Z_0

The firing levels of the rules are computed as shown in Fig. 7.28

$$\alpha_1 = A_1(x_0) \cap B_1(y_0) \text{ and } \alpha_2 = A_2(x_0) \cap B_2(y_0),$$

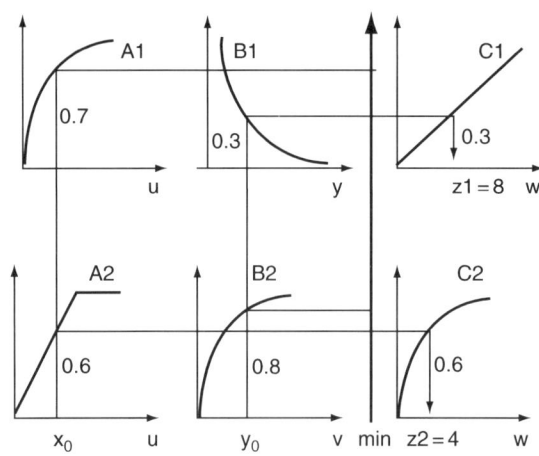

Fig. 7.28. Tsukamoto inference

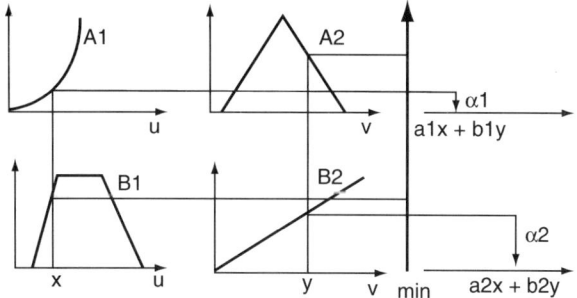

Fig. 7.29. Sugeno inference mechanism

then the individual rule outputs are derived from the relationships (Fig. 7.29)

$$z_1{}^* = a_1x_0 + b_1y_0, z_2{}^* = a_2x_0 + b_2y_0,$$

and the crisp control action is expressed as $z_0 = \frac{\alpha_1 z_1^* + \alpha_2 z_2^*}{\alpha_1 + \alpha_2}$.

If we have n rules in our rule-base then the crisp control action is computed as

$$z_0 = \frac{\displaystyle\sum_{i=1}^{n} \alpha_i z_i^*}{\displaystyle\sum_{i=1}^{n} \alpha_i}$$

where α_i denotes the firing level of the ith rule, $i = 1, \ldots, n$.

Larsen: The fuzzy implication is modeled by Larsen's product operator and the sentence connective *also* is interpreted as oring the propositions and defined by max operator.

Let us denote α_i the firing level of the i-th rule, $i = 1, 2$ as shown in Fig. 7.30

$$\alpha_1 = A_1(x_0) \cap B_1(y_0) \text{ and } \alpha_2 = A_2(x_0) \cap B_2(y_0)$$

Then membership function of the inferred consequence C is point-wise given by

$$C(w) = (\alpha_1 C_1(w)) \cup (\alpha_2 C_2(w)).$$

To obtain a deterministic control action, we employ any defuzzification strategy.

If we have n rules in our rule-base then the consequence C is computed as

$$C(w) = \bigvee_{i=1}^{n} (\alpha_i C_i(w)),$$

where α_i denotes the firing level of the ith rule, $i = 1, \ldots, n$.

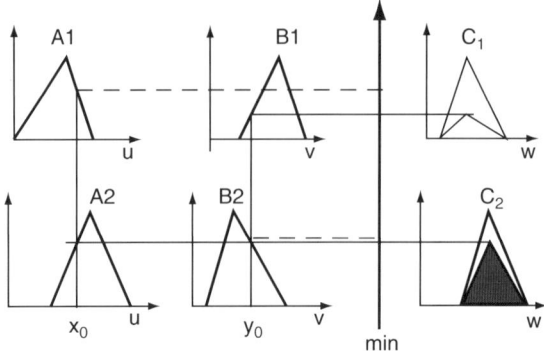

Fig. 7.30. Inference with Larsen's product operation rule

7.17 Defuzzification Methods

The output of the inference process so far is a fuzzy set, specifying a possibility distribution of control action. In the on-line control, a nonfuzzy (crisp) control action is usually required. Consequently, one must defuzzify the fuzzy control action (output) inferred from the fuzzy control algorithm, namely:

$$z_0 = defuzzifier(C),$$

where z_0 is the nonfuzzy control output and *defuzzifier* is the defuzzification operator.

Defuzzification is a process to select a representative element from the fuzzy output C inferred from the fuzzy control algorithm. The most often used defuzzification operators are:

1. *Center-of-area/gravity*
   The defuzzified value of a fuzzy set C is defined as its fuzzy centroid:

   $$z_0 = \frac{\int_w zC(z)dz}{\int_w C(z)dz}.$$

   The calculation of the center-of-area defuzzified value is simplified if we consider finite universe of discourse W and thus discrete membership function $C(w)$

   $$z_0 = \frac{\int_w z_j C(z_j)dz}{\int_w C(z_j)dz}.$$

2. *First-of-maxima*
   The defuzzified value of a fuzzy set C is its smallest maximizing element, i.e.

   $$z_0 = \min\{z \,\big|\, C(z) = \max_w C(w)\}$$

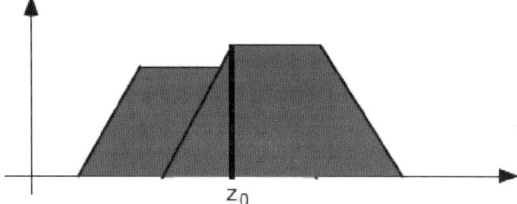

3. *Middle-of-maxima*

   The defuzzified value of a discrete fuzzy set C is defined as a mean of all values of the universe of discourse, having maximal membership grades

   $$z_0 = \frac{1}{N} \sum_{j=1}^{N} z_j,$$

   where $\{z_1, \ldots, z_N\}$ is the set of elements of the universe W which attain the maximum value of C.

   If C is not discrete then defuzzified value of a fuzzy set C is defined as

   $$z_0 = \frac{\int_G z\,dz}{\int_G dz},$$

   where G denotes the set of maximizing element of C.

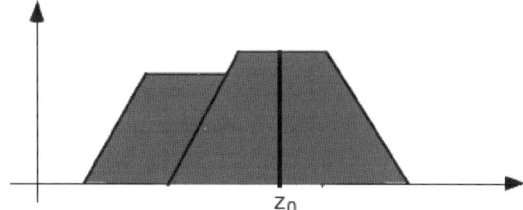

4. *Max-criterion*

   This method chooses an arbitrary value, from the set of maximizing elements of C, i.e.

   $$z_0 \in \{z \,\big|\, C(z) = \max_w C(w)\}.$$

5. *Height defuzzification*

   The elements of the universe of discourse W that have membership grades lower than a certain level α are completely discounted and the defuzzified value z_0 is calculated by the application of the center-of-area method on those elements of W that have membership grades not less than α:

   $$z_0 = \frac{\int_{[C]^\alpha} z C(z)\,dz}{\int_{[C]^\alpha} C(z)\,dz},$$

   where $[C]^\alpha$ denotes the α-level set of C as usually.

7.18 Fuzzy Rule Based System

The inputs of fuzzy rule-based systems should be given in fuzzy form, and therefore, we have to fuzzify the crisp inputs, i.e pre-processed sensor's output. Furthermore, the output of a fuzzy system is always a fuzzy output, and therefore to get appropriate crisp value we have to defuzzify and post process it. Fuzzy logic control systems usually consist of three major parts: *Fuzzification, Approximate reasoning* and *Defuzzification* (Fig. 7.31).

A fuzzification operator has the effect of transforming crisp data into fuzzy sets. In most of the cases we use fuzzy singletons as fuzzifiers

$$fuzzifier\,(x_0) = \overline{x_0},$$

where x_0 is a crisp input value from a process.

1. *Preprocessing module* does input signal conditioning and also performs a scale transformation (i.e. an input normalization) which maps the physical values of the current system state variables into a normalized universe of discourse (normalized domain).
2. *Fuzzification* block performs a conversion from crisp (point wise) input values to fuzzy input values, in order to make it compatible with the fuzzy set representation of the system state variables in the rule antecedent.
3. *Fuzzy knowledge Base* of a FKBS comprises knowledge of the application domain and the attendant control goals. It includes the following:
   1. Fuzzy sets (membership functions) representing the meaning of the linguistic values of the system state and control output variables.
   2. Fuzzy rules in the form of ***If* x is A *THEN* y is B**. x is state variable and y output variable. A and B are linguistic variables. The basic function of the fuzzy rule base is to represent in a structured way the control policy of an experienced process operator and/or control engineer in the form of set of production rules. Rules have two parts: antecedent and consequent.

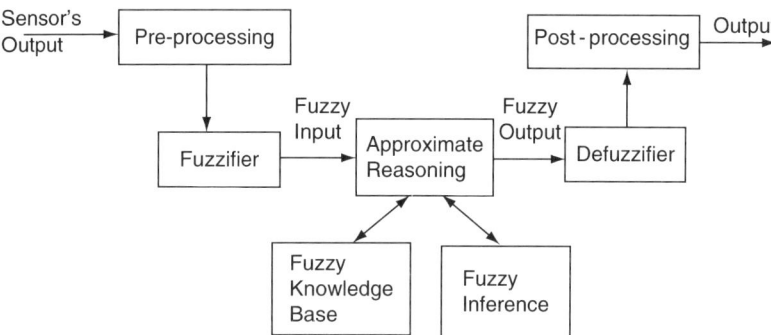

Fig. 7.31. Fuzzy knowledge Based System (FKBS)

4. *Fuzzy inference.* There are two basic types of approaches employed in the design of the inference engine of a FKBC: (1) Composition based inference (firing) and (2) individual rule based inference (firing). Mostly we use second type of inference. The basic function of inference engine of the second type is to compute the overall values of the control output variable based on the individual contribution of each rule in the rule base.

5. *Defuzzification* converts the set of modified control output values into a single point wise value. Six most often used defuzzification methods: centre of area (COA), centre of sum (COS), centre of largest area (CL), first maxima (FM), mean of maxima (MOM) and height (weighted average) defuzzification.

6. *Post processing module* involves de-normalization of fuzzy controller crisp output onto its physical domain.

7.19 Summary

- This chapter introduces the basics of fuzzy logic and different types of uncertaininty.
- The different types of operation on fuzzy sets are discussed, including different types of membership function and their representations.
- The chapter also discusses the fuzzy quantifiers/modifiers. Fuzzy modifier is the strength of fuzzy logic systems.
- Difference between precision and significance is also explained.
- Difference between fuzzy relation and crisp relation is described and representation of fuzzy relationship is mentioned in the chapter.
- The approximate inference techniques like generalized modus ponens and generalized modus tolence are explained.
- Various defuzzification schemes used in fuzzy systems are also explained in brief.
- In the last portion of the chapter, the Fuzzy system blocks are explained like fuzzification, approximate reasoning and defuzzification blocks. Sometimes the pre-processing and post-processing is also required, which is also discussed.

7.20 Bibliography and Historical Remarks

Prof. Lofti A. Zadeh is the father of fuzzy logic. He first time coined the word fuzzy by publishing a seminar paper on fuzzy set in 1965. He included fuzzy relation, projection. He is awarded by Prestigious IEEE award and IEEE published an article the describes Zadeh's personal journey along with the concept of fuzzy logic.

Research progress in fuzzy logic is nicely summarized in a collection edited by Gupta, Saridis and Gaines. Mamdani gives a survey of fuzzy logic control and discussed several important issues design, development and analysis of fuzzy logic controller. Bart Kasko introduced the geometric interpretation of fuzzy sets and investigated the related issues in late 1980s. Basu (2004) used a fuzzy-based simulated annealing technique for economic emission load dispatch. Thukaram and Yesuratnam (2007) explained fuzzy expert system technique for load curtailed to improved reactive power dispatch.

7.21 Exercises

1. Differentiate between fuzzy and crisp logic.
2. Discuss the situation where binary logic is not helpful?
3. What are the motivations for developing fuzzy logic?
4. What are the historical perspectives of fuzzy logic?
5. Explain the terms:
   a. Vagueness
   b. Imprecision
6. Mention the advantages and disadvantages of standard logic and fuzzy logic.
7. Explain the uncertainty in real life situations and how does it help in modeling complex situations.
8. Differentiate between lexical and stochastic uncertainty.
9. Find two examples of fuzzy sets in the news paper article and highlight the fuzzy words in it.
10. Draw the fuzzy membership function for "Tall Building"?
11. What are the different methods be used to assign the membership values for "Tall Building"?
12. Explain the difference between precision and significance.
13. The quality of students in a class could be divided in the range 0–10, which is the universe of discourse for this variable. How do the membership values be assigned for excellent, good, fair and bad students. Also draw their shapes.
14. Explain different operations which could be performed on fuzzy sets.
15. What do you mean by a fuzzy modifier or Hedges. Also explain the strong and weak modifiers.
16. Describe in brief about the concentration, dilution and intensification of fuzzy sets.
17. Construct the membership function for
    a. Not bad
    b. Very good
    c. not good or not bad
    d. Extremely fair and
    e. more or less excellent.

Use the membership function for excellent, good, fair and bad defined in question 7.13.

18. "Fuzzy sets are context and concept dependent". Justify the above mentioned statement.
19.   a. Let A and B be two fuzzy sets in the universe of U and V, respectively. Let R be the Cartesian product of A and B. Is the projection of R on U identical to A?
      b. Let R and S be two fuzzy relations defined as:

$$R = \begin{array}{c} x1 \\ x2 \end{array} \begin{bmatrix} \begin{array}{ccc} y1 & y2 & y3 \end{array} \\ 0.1 & 0.5 & 0.2 \\ 0.7 & 0.8 & 1.0 \end{bmatrix}$$

$$S = \begin{array}{c} y1 \\ y2 \\ y3 \end{array} \begin{bmatrix} \begin{array}{ccc} z1 & z2 & z3 \end{array} \\ 0.1 & 0.8 & 0.2 \\ 0.4 & 0.7 & 0.9 \\ 0.3 & 0.1 & 0.8 \end{bmatrix}$$

   (a) Find the result of R o S max-min composition.
   (b) Also find the result of R o S max-product and sum product method of composition.
20. Let the membership function of A is defined by $f(x) = 1/(1 + x^2)$, and $B = \exp(-x^2)$
    Draw these membership functions and find

   i. $A \cup B$
   ii. $A \cap B$
   iii. $\neg A$
   iv. $\neg B$.

21. Prove that $P \rightarrow Q$ is logically equivalent to $P \cup (P \cap Q)$.
22. Define convexity for fuzzy set.
23. Explain term coordinality based and fuzzy set based uncertainty.
24. Define various basic operation of fuzzy set. Also list out the properties satisfied by fuzzy set.
25. Draw typical membership function of approximately '3'.
26. Define "x plus" and "x minus".
27. Suppose that

$$u = 1 + 2, \quad v = 1 + 2 + 3,$$
$$A = 1/1 + 0.8/2 - B = 0.6/1 + 0.9/3 + 1/3$$

   Determine relation between A and B.
28. Explain projection and cylindrical extension principles.
29. Develop a reasonable membership function for the following fuzzy sets based on height measured in centimeter:

   (a) "Tall"
   (b) "Short"
   (c) "Not short"

30. Develop a reasonable membership function for the fuzzy color set "red" based on the wavelength of the color spectrum.

31. Develop a reasonable membership function for a square, based on the geometric properties of a rectangle. For this problem use L as the length of the longer side and l as the length of the smaller side.

32. For the cylindrical shapes shown in figure, develop a membership function for each of the following shapes using the ratio d/h, and discuss the reason for any overlapping among the three membership functions:

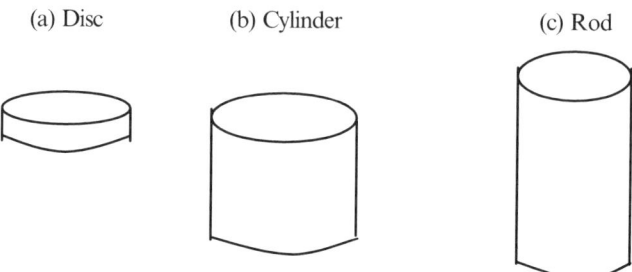

(a) Disc   (b) Cylinder   (c) Rod

33. The question of whether a glass of water is half - full or half - empty is an age old philosophical issue. Such descriptions of the volume of liquid in a glass depend on the state of mind of the person asked the question. Develop membership functions for the fuzzy sets "half-full", "empty", and "half-empty" using percentage volume as the element.

34. If u = 1 + 2 + 3 + 4 . . . 10
    A = 0.8/3 + 1/5 + 0.6/6
    B = 0.7/3 + 1/4 + 0.5/6
    Determine - (i) A'

    (b) 0.4 A
    (c) DIL (A)

35. Suppose "approximately 2" and "approximately 6" are defined as follows. Determine "approximately 12".
    approximately 2 = 1/2 + 0.6/1 + 0.8/3
    approximately 6 = 1/6 + 0.8/5 + 0.7/7.

36. The output can be affected by quality of camera, as well as the quality of film. A possible Universe of camera rating is X = {1, 2, 3, 4, 5}, where 1 represents the highest camera rating. A possible universe of picture ratings is Y = {1, 2, 3, 4, 5}, where once again, 1 is the highest rating for pictures. We now define two fuzzy sets

    A = "above average camera" = {.7/1 + 0.9/2 + 0.2/3 + 0/4 + 0/5}
    B = "above average picture quality" = {0.6/1+0.8/2+0.5/3+0.1/4+0/5}

    (a) From the proposition, IF A THEN B, find the relation using Mamdani implication,

(b) Suppose the camera manufacturer wants to improve camera and film sales by improving the quality of the camera. A new camera is rated as follows:

A' = "new and improved camera" = {0.8/1+0.8/2+0.1/3+0/4+0/5}

What might be resulting picture rating from this new camera?

37. Using your own intuition, develop fuzzy membership functions on the real line for the "approximately 2 and approximately 8", using the following function shapes:

(a) Symmetric triangle
(b) Trapezoids
(c) Gaussian functions.

38. Using your own intuition and your own definitions of the universe of discourse, plot fuzzy membership functions for the following variables:

(a) Weight of people
   (a) Very light
   (b) Light
   (c) Average
   (d) Heavy
   (e) Very heavy
(b) Age of people
   (a) Very young
   (b) Young
   (c) Middle-aged
   (d) Old
   (e) Very old
(c) Education of people
   (a) Fairly educated
   (b) Educated
   (c) Highly educated
   (d) Not highly educated
   (e) More or less educated

39. Given the continuous, noninteractive fuzzy sets A and B on universes X and Y, using Zadeh's notation for continuous fuzzy variables.

$$A = \int (1 - 0.1|x|)/x \quad \text{for } x \in [0, +10]$$
$$B = \int 0.2|y|/y \quad \text{for } y \in [0, +5]$$

(a) Construct a fuzzy relation R for the Cartesian product of A and B.
(b) Use max-min composition to find B', given the fuzzy sigleton A' = 1/3.

40. In the above question, calculate the relation between behaviour and reward using max-product composition show that a raise is easier to get than a compliment?

41. Given that

$$D1 = \{1/1.0 + 0.75/1.5 + 0.3/2.0 + 0.15/2.5 + 0/3\}$$
$$D2 = \{1/1 + 0.6/1.5 + 0.2/2 + 0.1/2.50/3\}$$

For these two fuzzy sets, find the following:

(a) $D1 \cup D2$
(b) $D1 \cap D2$
(c) $D1'$
(d) $D2'$
(e) $D1|D2$
(f) $(D1 \cup d2)'$
(g) $(D1 \cap D2)'$
(h) $D1 \cap D2'$
(i) $D2 \cup D1'$
(j) $D2 \cup D2'$

42. In the field of photography, the exposure time and development time of negative are the two factors that determine how the negative will come out after processing. Define two fuzzy sets, A = {exposure time of holographic plates}, and B = {Development times for the exposed plates}. The relative density or relative darkness, of the processed plate, which varies between 0 and 1, is shown in figure. As shown by the fuzzy sets A and B in the figure, the two variables exposure time and development time complement each other in determine the density of the negative.

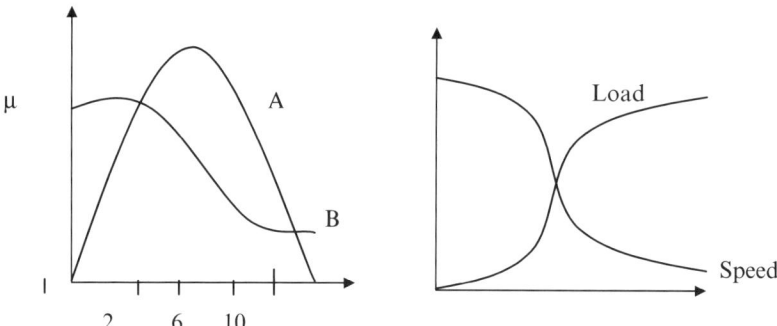

That is, if a negative is exposed for a shorter period of time, the plate can be developed for a longer time. Find the following graphically:

(a) $A \cup B$
(b) $A \cap B$
(c) $A' \cup B$
(d) $A \cap B'$

43. The speed of a hydraulic motor is a highly critical parameter, as it ultimately controls volume of fluid to be displaced. A typical problem in

the control of the hydraulic motor is that the load is placed on the motor can vary due to different circumstances. Define the load and speed as two fuzzy sets in the control of the hydraulic motor, with the membership functions shown in Fig. 7.6. Also assume that the load on the motor has an influence on the speed of the motor (finite torque output of the motor); e.g. when the motor load (L) increases, the motor speed (S) decreases, and vice versa. Graphically determine the following:

(a) $L \cup S$
(b) $L \cap S$
(c) $(L' \cup S)'$
(d) $(L \cap S')'$

44. A company sells a product called a video multiplexer, which multiplexes the vedio from 16 video cameras into a single video cassette recorder (VCR). The product has a motion detection feature that can increase the frequency with which a given camera's video is recorded to tape depending on the amount of motion that is present. It does this by recording more information from that camera at the expense of the amount of video that is recorded from the other 15 cameras. Define a universe X to be the speed of the objects that are present in the video of camera 1 (there are 16 cameras). For example, let X = {Low Speed, Medium Speed, High Speed} = {LS, MS HS}. Now, define a universe Y to represent the frequency with which the video from camera 1 is recorded to a VCR tape, i.e. the record rate of camera 1. Suppose, Y = {Slow Record Rate, Medium Record Rate, Fast Record Rate} = {SRR, MRR, FRR}.

Let us now define a fuzzy set A on X and fuzzy set B on Y, where A represents a fuzzy slow moving object present in video camera 1 and B represents a fuzzy slow record rate, biased to the slow side. For example,

$$A = \{1/LS+0.4/MS+0.2/HS\} \quad \text{and} \quad B = \{1/SRR+0.5/MRR+0.25/FRR\}$$

(a) Find the fuzzy relation for the Cartesian product A and B
(b) Suppose we introduce another fuzzy set C which represent a fuzzy fast moving object present in video camera 1, say for example, the following:

$$C = \{0.1/LS + 0.3/MS + 1/HS\}$$

Find the relation between C and B using Cartesian product
(c) Find C o R using max–min composition.
(d) Find C o S using max–min composition.
(e) Comment on the meaning of ports (c) and (d) and on the differences between the results.

45. All new jet aircraft are subjected to intensive flight simulation studies before they are ever tested under actual flight conditions. In these studies an important relationship is that between the mach number (% of the speed of sound) and the altitude of the aircraft. This relationship is

important to the performance of the aircraft and has a definite impact in making flight plans over populated areas. If certain mach levels are reached, breaking the sound barrier (sonic booms) can results in human discomfort and light damage to glass enclosures on the earth's surface. Current rules of thumb establish crisp breakpoints for the conditions that cause performance changes (and sonic booms) in aircraft, but in reality these breakpoints are fuzzy, because other atmospheric conditions such as the humidity and temperature also affect breakpoints in performance. For this problem, suppose the flight test data can be characterised as "near" or "approximately" or "in the region of" the crisp database breakpoints. Define a universe of aircraft speeds near the speed of sound as X = {0.73, 0.735, 0.74, 0.745, 0.75} and a fuzzy set on this universe for the speed "near mach 0.74" = M, where

$$M = \{0/0.73 + 0.8/0.735 + 1/0.74 + 0.8/0.745 + 0/0.75\}$$

and define a universe of altitudes as Y = {22.5, 23, 23.5, 24, 24.5, 25, 25.5} in k-feet, and a fuzzy set on this universe for the altitude fuzzy set "approximately 240,000 feet" = A, where

$$A = \{0/22.5k+0.2/23k+0.7/23.5k+1/24k+0.7/24.5k+0.2/25k+0/25.5k\}$$

(a) Construct the relation R = M x A
(b) for another aircraft speed, say M1 = "in the region of mach 0.74", where

$$M1 = \{0/0.73 + 0.8/0.735 + 1/0.74 + 0.6/0.745 + 0.2/0.75\}$$

find the relation S = M1 o R using max-min composition.

46. Three variables of interest in power transistors are the amount of current that can be switched, the voltage that can be switched, and the cost. The following membership functions for power transistors were developed from a hypothetical components catalog:

Average current (in amps) = I = {0.4/0.8 + 0.7/0.9 + 1/1 + 0.8/1.1 + 0.6/1.2}

Average voltage (in volts) = V = {0.2/30+0.8/45+1/60+0.9/75+0.7/90}

Note how the membership values in each set taper off faster toward the lower voltage and currents. These two fuzzy sets are related to the "power" of the transistor. Power in electronics is defined by an algebraic operation, P = VI, but let us deal with a general Cartesian relationship between voltage and current i.e. simply with P = V x I.

(a) Find the fuzzy Cartesian product P = V x I.
Now let us define a fuzzy set for the cost C in dollars, of a transistor, for example,

$$C = \{0.4/0.5 + 1/0.6 + 0.5/0.7\}$$

(b) Using a fuzzy Cartesian product, find $T = I \times C$. What would this relation T, represent physically.

(c) Using max–min composition, find $E = P \circ T$. What would this relation E, represent physically.

(d) Using max-product composition, find $E = P \circ T$. What would this relation E, represent physically.

47. The relation between temperature and maximum operating frequency R depends on various factors for a given electronic circuit. Let T be a temperature fuzzy set (in degrees Fahrenheit) and F represent a fuzzy set (in MHz), on the following universe of discourse:

$$T = \{-100, -50, 0, 50, 100\} \quad \text{and} \quad F = \{8, 16, 25, 33\}$$

Suppose a Cartesian product between T and F is formed that results in the following relation:

$$R = \begin{array}{c} \\ 8 \\ 16 \\ 25 \\ 33 \end{array} \begin{array}{ccccc} -100 & -50 & 0 & 50 & 100 \\ \begin{bmatrix} 0.2 & 0.5 & 0.7 & 1 & 0.9 \\ 0.3 & 0.5 & 0.7 & 1 & 0.8 \\ 0.4 & 0.6 & 0.8 & 0.9 & 0.4 \\ 0.9 & 1 & 0.8 & 0.7 & 0.4 \end{bmatrix} \end{array}$$

The reliability of the electronic circuit is related to the maximum operating temperature. Such a relation S can be expressed as Cartesian product between the reliability index, $M = \{1, 2, 4, 8, 16\}$ (in dimensionless units) and the temperature, as in the following example:

$$S = \begin{array}{c} \\ -100 \\ -50 \\ 0 \\ 50 \\ 100 \end{array} \begin{array}{ccccc} 1 & 2 & 4 & 8 & 16 \\ \begin{bmatrix} 1 & 0.8 & 0.6 & 0.3 & 0.1 \\ 0.7 & 1 & 0.7 & 0.5 & 0.4 \\ 0.5 & 0.6 & 1 & 0.8 & 0.8 \\ 0.3 & 0.4 & 0.6 & 1 & 0.9 \\ 0.9 & 0.3 & 0.5 & 0.7 & 1 \end{bmatrix} \end{array}$$

Composition can be performed on any two or more relations with compatible row-column consistency. To find relationship between frequency and the reliability index, use

(a) max-min composition

(b) max-roduct composition

48. Two fuzzy sets A and B both defined on X, are as follows:

$\mu(x_{ii})$	x_1	x_2	x_3	x_4	x_5	x_6
A	0.1	0.6	0.8	0.9	0.7	0.1
B	0.9	0.7	0.5	0.2	0.1	0

Express the following λ-cut sets using Zadeh's notation:
(a) $(A)_{0.7}$
(b) $(B)_{0.4}$
(c) $(A \cup B)_{0.7}$
(d) $(A \cap B)_{0.6}$
(e) $(A \cup A')_{0.7}$
(f) $(B \cap B')_{0.5}$
(g) $(A \cap B)'_{0.7}$
(h) $(A' \cap B')_{0.7}$

49. Which of the following are equivalence relations?

Set	Relation in the set
(a) People	Is the brother of
(b) People	Has the same parents
(c) Points on a map	Is connected by a road to
(d) Lines in plane geometry	Is perpendicular to
(e) Positive integers	For some integer k, equals 10^K times

Draw graphs of the equivalence relations with appropriate labels on the vertices.

50. The accompanying Sagittal diagram show two relations on the universe, $X = \{1, 2, 3\}$. Are these relations equivalence relations:

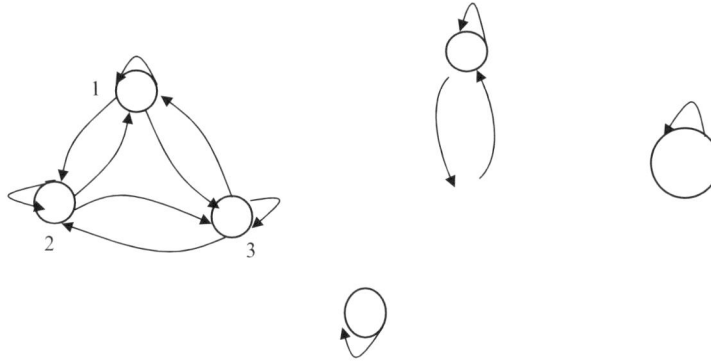

51. Two companies bid for a contract. A committee has to review the estimates of those companies and give reports to its chairperson. The reviewed reports are evaluated on a non-dimensional scale and assigned a weighted score that is represented by a fuzzy membership function, as illustrated by the two fuzzy sets B_1 and B_2 in figure. The chairperson is interested in the lowest bid, as well as metric to measure the combined "best" score. For the logical union of the membership functions shown here, find the defuzzified value z^*, using each of the seven methods. Comment on the differences of the methods.

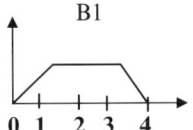

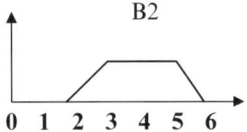

52. Under what conditions of P and Q is the implication P → Q a tautology?

53. The exclusive -or is given by the expression, $P \cup Q - (P' \cap Q) \cup (P \cap Q')$. Show that the logical - or, given by $P \cup Q$, gives a different results from the exclusive-or and comment on this difference using an example in your own field.

54. For a proposition R of the form P → Q, show the following:

  (a) R and its counter positive are equivalent, i.e. prove that $(P \rightarrow Q) \leftrightarrow (Q' \rightarrow P')$.

  (b) The converse of R and the inverse of R are equivalent, i.e. prove that $(Q \rightarrow P) \leftrightarrow (P \rightarrow Q')$.

55. (a) Show that the dual of the equivalence $(P \cup Q) \cup (P' \cap Q') \leftrightarrow X$ is also true.

  (b) Show that De Morgan's laws are duals.

56. You are faced with the problem of controlling a motor that is subjected to a variable load. The motor must maintain a constant speed, regardless of the load placed on it; therefore, the voltage applied to the motor must change to compensate for changes in load. Define fuzzy sets for motor speed (rpm) and motor voltage (volts) as follows:

  A = "motor speed OK" = $\{0.3/20 + 0.6/30 + 0.8/40 + 1/50 + 0.7/60 + 0.4/70\}$

  B = "motor voltage nominal" = $\{0.1/1 + 0.3/2 + 0.8/3 + 1/4 + 0.7/5 + 0.4/6 + 0.2/7\}$

  (a) Using classical implication, find a relation for the following compound proposition:

   "IF motor speed OK THEN motor voltage is nominal".

  (b) Now specify a new antecedent A' = "motor speed a little slow", where

   $$A' = \{0.4/20 + 0.7/30 + 1/40 + 0.6/50 + 0.3/60 + 0.1/70\}$$

   Using max-min composition (i.e. $B' = A' oR$), find the new consequent.

57. Fill in the following table using Zadah implication. Also show that how the results affected by the various other implications used.

A	B	A → B
0	0	
0	1	
1	0	
1	1	
0.2	0.3	
0.2	0.7	
0.8	0.3	
0.8	0.7	

8

Applications of Fuzzy Rule Based System

In almost every case you can build the same product without Fuzzy logic, but fuzzy is faster and cheaper.

Prof. LA Zadeh
University of California, Barkley

If you do not have a good plant model or if the system is changing, then fuzzy will produce a better solution than conventional control technique.

Bob Varley
Aerospace Company, Palmbay, Florida

8.1 Introduction

The basic concepts of fuzzy system were described in the earlier chapter. There are numerous applications of fuzzy systems in various fields such as operations research, modeling, evaluation, pattern recognition, control and diagnosis, etc. Fuzzy systems theory is the starting point for developing models of ambiguous thinking and judgment processes, the following fields of application are conceivable:

a. Human models for management and societal problems;
b. Use of high level human abilities for use in automation and information systems;
c. Reducing the difficulties of man-machine interface;
d. Other AI applications like risk analysis and prediction, development of functional device.

Fuzzy systems are quite popular in control applications. In standard control theory, a mathematical model is assumed for the controlled system, and control laws that minimize the evaluation functions are determined; but when the object is complicated, mathematical models cannot be determined and

D.K. Chaturvedi: *Soft Computing Techniques and its Applications in Electrical Engineering*, Studies in Computational Intelligence (SCI) **103**, 295–362 (2008)
www.springerlink.com © Springer-Verlag Berlin Heidelberg 2008

one cannot figure out how to decide on the evaluation functions. In these cases, skilled individuals perform control functions by using their experience and intuition to judge situations on the basis of what they think.

8.2 System's Modeling and Simulation Using Fuzzy Logic Approach

Modeling and simulation is playing important role in every field of engineering. There are various techniques available for modeling the systems. The conventional linear mathematical models of electrical machines are quite accurate in the operating range for which they were developed, but for slightly different operating conditions mathematical model output degrades, because mathematical model is unable to cope up the non-linearity of the system. If a non-linear model is developed for these systems, then model complexity is increases. It is difficult to simulate the complex system model and the computation time is unbearably long. To deal with system non-linearity non-conventional methods like ANN, Fuzzy systems, genetic algorithms etc. are used. The accuracy of ANN models depends on the availability of sufficient and accurate historical (numerical) data for training the network. Also there is no exact method to find the size of ANN and training method. The standard gradient descent training algorithm performance depends on initial weights, learning parameters, and quality of data. Most of the time in real life situations, it is difficult to get sufficient and accurate data for training ANN, but the operator experience may be helpful in these situations to develop fuzzy logic based models. Also, neither mathematical nor ANN models are transparent for illustration purpose, because in these models, information buried in model structure (differential or difference equations or ANN connections). Hence it is not easy to illustrate the model to others. Fuzzy model is quite simple and easy to explain due to its simple structure and complete information in the form of rules.

Fuzzy logic is applied to a great extent in controlling the processes, plants and various complex and ill defined systems, due to its inherent advantages like simplicity, ease in design, robustness and adaptability. Also it is established that this approach works very well especially when the systems are not clearly understood and transparent. In this section, this approach is used for modeling and simulation of D.C. machine to predict its characteristics. Also the effect of different connectives (aggregation operators) like intersection, union, averaging and compensatory operators, etc. different implications, different compositional rules, different membership functions of fuzzy set and their percentage overlapping, and different defuzzification methods have been studied.

The fuzzy logic model is developed on the basis of causal relationships between the variables. The causal relationship as depicted by the causal loop diagram of causal models is often fuzzy in nature. For example, there is no simple calculation to find the answer of 'how much load increase in a interconnected power system would results a given change in power angle? If one would like to answer this question it is necessary to simulate a complex differential or partial differential equations for the interconnected power system. This exercise is quite time expensive and cumbersome. It can be represented by a simple causal relationship as shown in Fig. 8.1 with appropriate fuzzy membership functions for load and power angle linguistic values. The system dynamics technique is used for modeling electrical machines (Chaturvedi 1992). The knowledge gathered in causal model is used to develop the fuzzy rule base. The flow chart of development of fuzzy simulator is shown in Fig. 8.2.

In fuzzy logic system development needs following vital decisions:

a. Selection of input and output variables
b. Normalization range of these variables
c. The number of fuzzy sets for each variable.

Load ⟶ Power angle δ

$+$

Fig. 8.1. Positive causal diagram

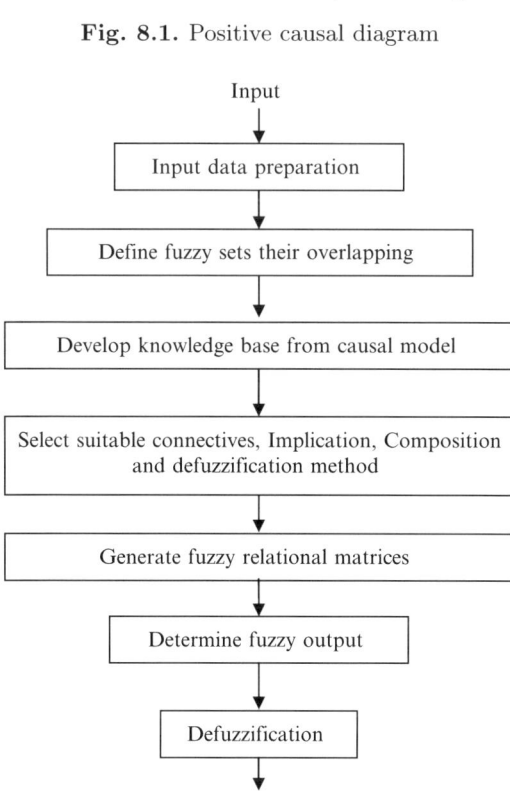

Fig. 8.2. Flow chart for development of fuzzy simulator

d. Selection of membership functions for each fuzzy set.
e. Determination of overlapping of fuzzy sets.
f. Acquire knowledge and define appropriate number of rules
g. Selection of intersection operators.
h. Selection of union operators.
i. Selection of implication methods.
j. Selection of compositional rules and
k. Selection of defuzzification methods.

8.2.1 Selection of Variables, their Normalization Range and the Number of Linguistic Values

The selection of variables depends on the problem situation and the type of analysis one would like to perform. For example, if one needs to conduct steady state analysis, the variables will be different from those for transient analysis. The reason, behind this is that in transient analysis it is necessary to take the flow rate variables into account, while in steady state analysis, theses variables can be neglected. In steady state analysis the analyzer is generally not interested in the manner the variable is attaining its steady state value. But in the study of transient behaviour the purpose is to determine how the variable achieves its final value. Hence, for transient analysis more variables have to be handled. After identifying the variables for a particular system one can draw the casual loop diagram for formulating the fuzzy rules as mentioned above.

Once the variables are identified, it is also necessary to decide about the normalization range for input and output variables. Then specify the number of linguistic values for each variable. The number of linguistic values also greatly affects the results. If there are more number of linguistic values, smoother output could be obtained, but the computational effort increases. For the basic commutating machine, the fuzzy model is simulated using three fuzzy sets, five fuzzy sets and seven fuzzy sets. The general experience is that as the number of fuzzy sets increases the computational time increases exponentially. Complex problems having numerous variables require enormous simulation time.

8.2.2 Selection of Shape of Membership Functions for Each Linguistic Value

The fuzzy membership function can be a triangular function, trapezoidal function, S-shaped function, π-function, or exponential function. For different fuzzy sets one can use different fuzzy membership functions. The most popular membership function is the triangular function due to the simplicity and ease in calculations. In the present research work, modeling and simulation of the basic commutating machine has been done using the triangular fuzzy membership function.

8.2.3 Determination of Overlapping of Fuzzy Sets

As one fuzzy set reaches its end, it is not necessary that the next fuzzy set should continue from where the previous one ended. In other words, a particular value of the variable may belong to more than one fuzzy set. As the domain of one fuzzy set finishes, another one may have already started. Thus there is an overlap of two or more fuzzy sets. It is observed that overlapping of neighboring fuzzy sets affects the results to a great extent.

8.2.4 Selection of Fuzzy Intersection Operators

The intersection of two fuzzy sets A and B is specified in general by a binary operation on the unit that is, a function of the form: $\mathbf{i} : [\mathbf{0},\ \mathbf{1}] \times [\mathbf{0},\ \mathbf{1}] \rightarrow [\mathbf{0},\ \mathbf{1}]$

For each element x of the universal set, this function takes as its argument the pair consisting of the element's membership grades in set A and in set B, and yields the membership grade of the element in the set constituting the intersection of A and B. Thus,

$$(\mathbf{A} \cap \mathbf{B})(\mathbf{x}) = \mathbf{i}[\mathbf{A}(\mathbf{x}), \mathbf{B}(\mathbf{x})] \qquad \forall \mathbf{x} \in \mathbf{X}$$

The various t-norms of the Dombi, Frank, Hamcher, Schweizer & Sklar 1,2,3,4, Dubious & Prade, Weber and Yu classes, which are defined by different choices of the parameters as given in Table 8.1, may be interpreted as performing fuzzy intersections of various strengths and shapes as shown in Figs. 8.3–8.8.

8.2.5 Selection of Fuzzy Union Operators

Fuzzy unions are close parallels of the fuzzy intersections. Like fuzzy intersection, the general fuzzy union of two fuzzy sets A and B is specified by a function $\mathbf{u} : [\mathbf{0},\ \mathbf{1}] \times [\mathbf{0},\ \mathbf{1}] \rightarrow [\mathbf{0},\ \mathbf{1}]$.

Table 8.1. Fuzzy intersection operators (T-norms)

Year	Name	Intersection opertor i	Parameter
-	Schweizer & Sklar 2	$1 - [(1-a)^w + (1-b)^w - (1-a)^w (1-b)^w]^{1/w}$	$w > 0$
-	Schweizer & Sklar 3	$\exp(-(|\ln a|^w + |\ln b|^w)^{1/w})$	$w > 0$
-	Schweizer & Sklar 4	$ab/[a^w + b^w - a^w b^w]^{1/w}$	$w > 0$
1963	Schweizer & Sklar 1	$\{\max(0, a^{-w} + b^{-w} - 1)\}^{-1/w},$	$-\infty < w < \infty$
1978	Hamcher	$ab/[w + (1-w)(a + b - ab)]$	$w > 0$
1979	Frank	$\log_s[1 + \{(w^a - 1)(w^b - 1)/(w - 1)\}]$	$w > 0,\ w \neq 1$
1980	Yager	$1 - \min\{1, [(1-a)^w + (1-b)^w]^{1/w}\}$	$w > 0$
1980	Dubois & Prade	$ab/\max(a,b,w)$	$w \notin [0,1]$
1982	Dombi	$[1 + \{(1/a) - 1)^w + ((1/b) - 1)^w\}^{1/w}]^{-1}$	$w > 0$
1983	Weber	$\max(0, \{(a + b + w \cdot ab - 1)/(1 + w)\})$	$w > -1$
1985	Yu	$\max[0, (1 + w)(a + b - 1) - w\, ab]$	$w > -1$

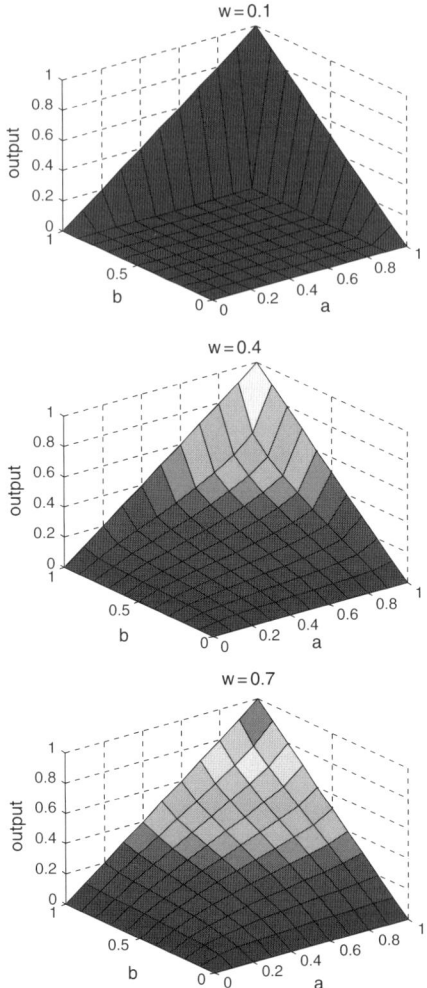

Fig. 8.3. Shapes and strength of Dombi t-norm operator changes as parameter value changes

The argument of this function is the pair consisting of the membership grade of some element x in fuzzy set A and the membership grade of that same element in fuzzy set B. The function returns the membership grade of the element in the set A ∪ B. Thus,

$$(\mathbf{A} \cup \mathbf{B})(\mathbf{x}) = \mathbf{u}[\mathbf{A}(\mathbf{x}), \mathbf{B}(\mathbf{x})] \qquad \mathbf{x} \in \mathbf{X}.$$

The various t-conorms of Dombi, Frank, Hamcher, Schweizer & Sklar 1,2,3,4, Dubois & Prade, Weber and Yu, defined by different choices of the parameters as mention in Table 8.2, can be interpreted as performing union operations of various strengths and shapes as shown in Figs. 8.9–8.14.

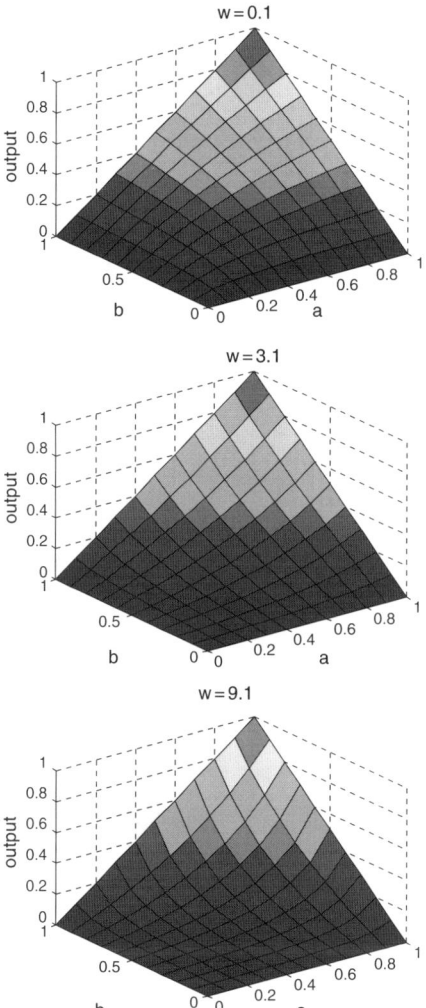

Fig. 8.4. Shapes and strength of Hamcher t-norm operator changes as parameter value changes

It is possible to define alternative compensatory operators (Mizumoto 1989) by taking the convex combination of min ($\cap$) and max ($\cup$) as mentioned in Table 5.3.

$$(\mathbf{x} \cap \mathbf{y})^{(1-w)} \cdot (\mathbf{x} \cup \mathbf{y})^{w}, \qquad \mathbf{0 \leq w \leq 1}$$

In general, we can obtain many kinds of compensatory operators by using t-norms T(x,y) and t-conorms S(x,y) dual to T(x,y).

Hence, in addition to the intersection (t-norm) and union (t-conorm) operators, the fuzzy simulator that has been developed also accommodates fuzzy

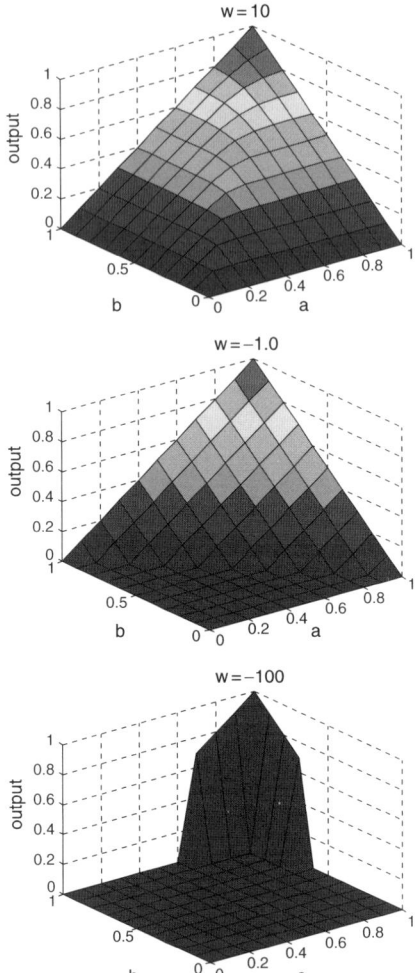

Fig. 8.5. Schweizer & Sklar 1 t-norm operator for different parameter values

compensatory operators, some of which can be obtained using t-norms, t-co norms, averaging operators and compensatory operators. The 3-D surfaces for compensatory operator for different parameters are shown in Figs. 8.15 and 8.16.

8.2.6 Selection of Implication Methods

To select an appropriate fuzzy implication for approximate reasoning under each particular situation is a difficult problem. The logical connective implication (Yaochu 2003; Ying 2002), i.e. P → Q (P implies Q) in classical theory $\mathbf{T}(\mathbf{P} \to \mathbf{Q}) = \mathbf{T}(\mathbf{P}' \cup \mathbf{Q})$.

Various types of implications operators summarized in Table 8.3.

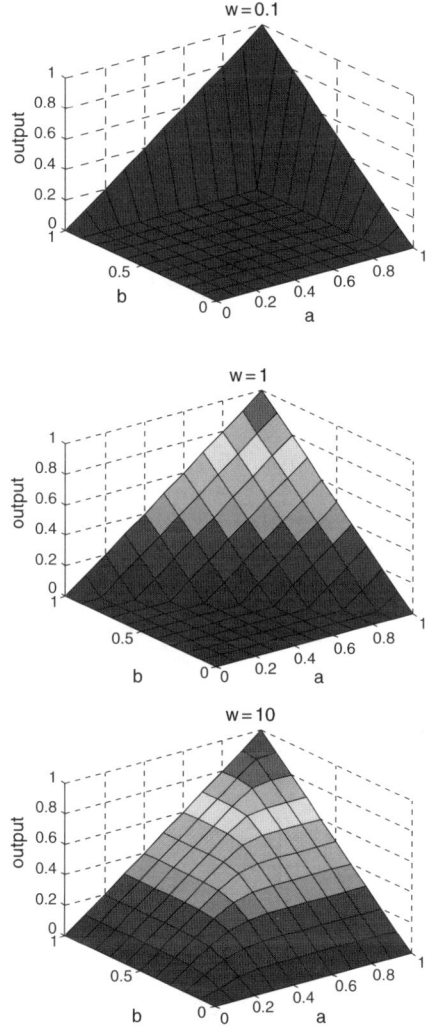

Fig. 8.6. Yager t-norm operator for different parameter values

8.2.7 Selection of Compositional Rule

The modus ponens deduction is used as a tool for making inferences in rule based systems. A typical if – then rule is used to determine whether an antecedent (cause or action) infers a consequent (effect or reaction). Suppose we have a rule of the form IF A THEN B, where A is a set defined on universe X and B is set defined by the universe Y. This can be translated into relation

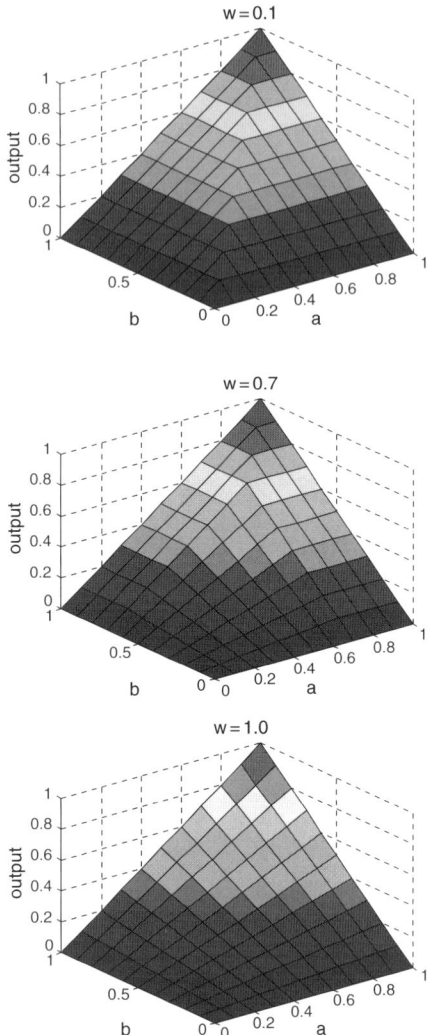

Fig. 8.7. Dubois & Prade t-norm operator for different parameter values

between sets A and B, that is, $R = (A \times B) \cup (A' \times Y)$. Now suppose a new antecedent say Aa is known and we want to find its consequence Bb.

$$Bb = Aa \text{ o } R = Aa \text{ o}((A \times B) \cup (A' \times Y)).$$

where symbol o denotes the composition operation. Modus ponens deduction can also be used for the compound rule, IF A, THEN B, ELSE C, where this compound rule is equivalent to the relation

$$R = (A \times B) \cup (A' \times C). \text{ or } R = \max(\min(A, B), \min(A', C)).$$

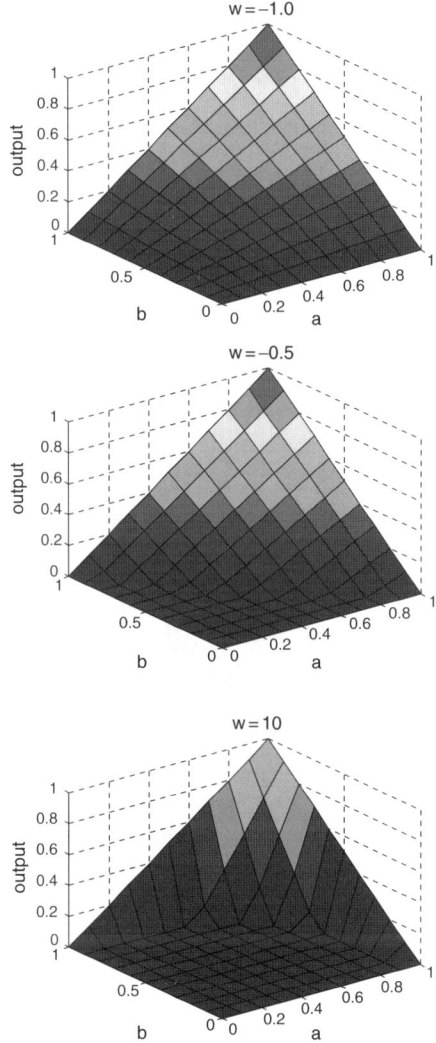

Fig. 8.8. Yu t-norm operator for different parameter values

This is also called Max – Min compositional rule. There are various other compositional rules also, which is given in Table 8.4 (Zimmermann 1991; Yager and Zadeh 1992).

8.2.8 Selection of Defuzzification Method

In modeling and simulation of systems, the output of a fuzzy model needs to be a single scalar quantity, as opposed to a fuzzy set. The conversion of a fuzzy quantity to a precise quantity is called defuzzification, just as fuzzification is

Table 8.2. Fuzzy unions (T-conorms)

Year	Name	Union operator u	Parameter
-	Schweizer & Sklar 2	$[a^w + b^w - a^w b^w]^{1/w}$	$w > 0$
-	Schweizer & Sklar 3	$1 - \exp(-(|\mathrm{In}(1-a)|^w + |\mathrm{In}(1-b)|^w)^{1/w})$	$w > 0$
-	Schweizer & Sklar 4	$1 - [(1-a)(1-b)/[(1-a)^w + (1-b)^w$	
		$-(1-a)^w(1-b)^w]^{1/w}$	$w > 0$
1963	Schweizer & Sklar 1	$1 - \{\max(0, (1-a)^w + (1-b)^w - 1)\}^{1/w}$	$w > 0$
1978	Hamcher	$[a + b + (w-2)ab]/[w + (1-w)ab]$	$w > 0$
1979	Frank	$1 - \log_s[1 + \{(w^a - 1)(w^b - 1)/(w - 1)\}]$	$w > 0, \ w \neq 1$
1980	Yager	$\min\{1, [a^w + b^w]^{1/w}\}$	$w > 0$
1980	Dubois & Prade	$1 - [(1-a)(1-b)/\max((1-a), (1-b), w)]$	$w \in [0, 1]$
1982	Dombi	$1/[1 + \{((1/a) - 1)^w + ((1/b) - 1)^w\}^{-1/w}]$	$w > 0$
1983	Weber	$\min 1, \{a + b + w.ab/(1-w)\}$	$w > -1$
1985	Yu	$\min[1, a + b + w.ab]$	$w > -1$

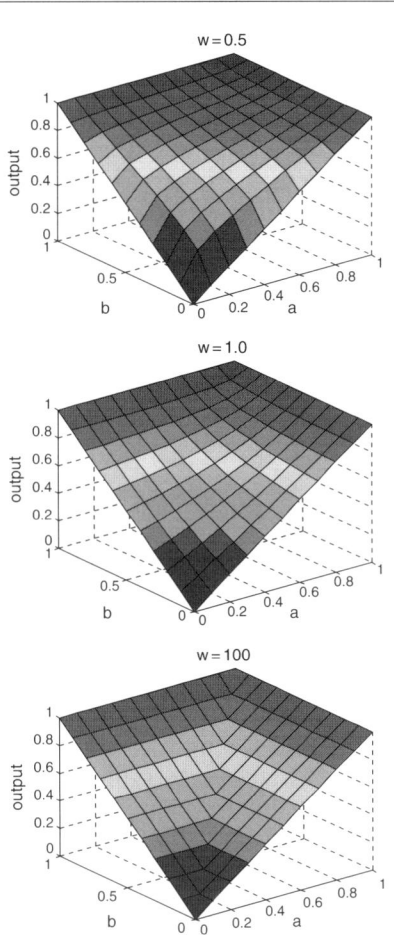

Fig. 8.9. Dombi t-conorm operator with different parameter value

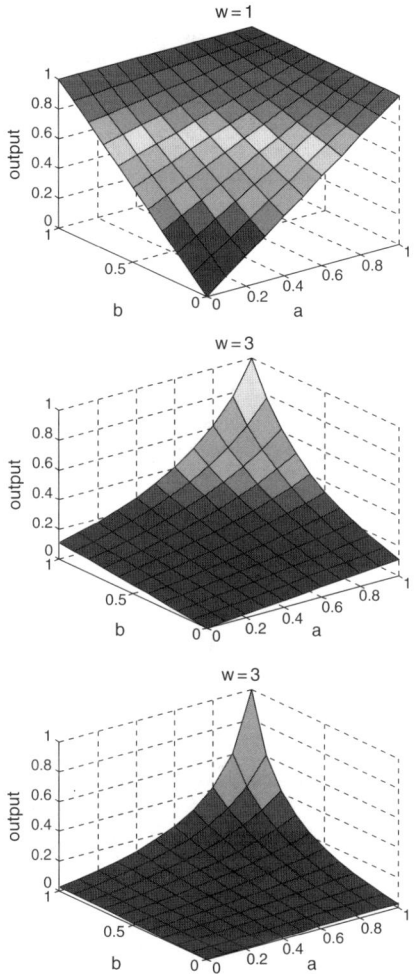

Fig. 8.10. Hamcher t-conorm operator with different parameter value

the conversion of a precise quantity to a fuzzy quantity. There are a number of methods in the literature, among them many that have been proposed by investigators in recent years, are popular for defuzzifying fuzzy output (Yager and Filev 1993].

The fuzzy logic system development software works as shown in Fig. 8.17.

8.2.9 Steady State D.C. Machine Model

The schematic diagram and the terminal graph of D.C. Machine is shown in Fig. 8.18. It is consisting of three ports, i.e. armature port, field port and shaft port. The measurements are done at each port for the pair of complimentary

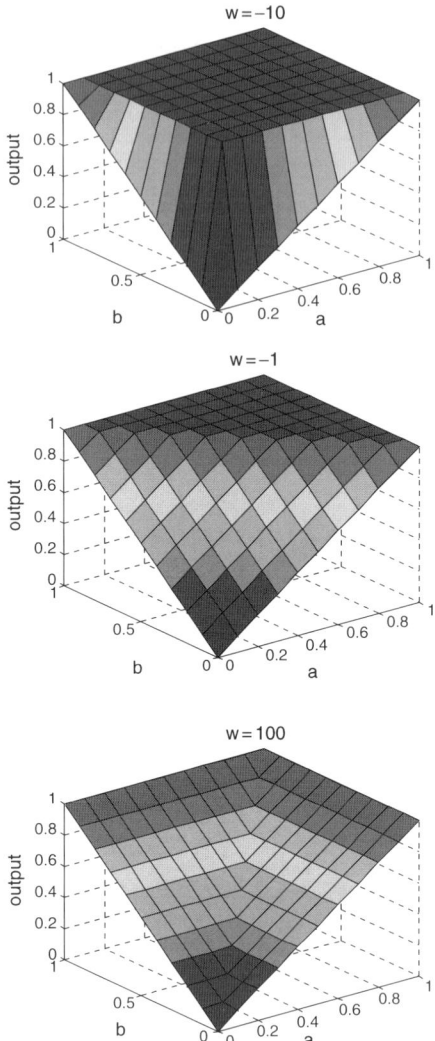

Fig. 8.11. Schweizer & Sklar 1 t-conorm operator for different parameter values

terminal variables with associated positive reference directions. In motoring
mode of operation of D.C. machine at field port field current is kept constant.

The static model of D.C. machine contains a fewer variables as compared to
the dynamic model. In case of D.C. machine static model, suppose we want
to find the effect of load torque on the motor speed. Then speed and load
torque are the variables of interest. Hence, the causal relationship is identified
between these variables. As load torque increases, machine speed reduces (ref.
Fig. 8.19), assuming that the other variables are constant.

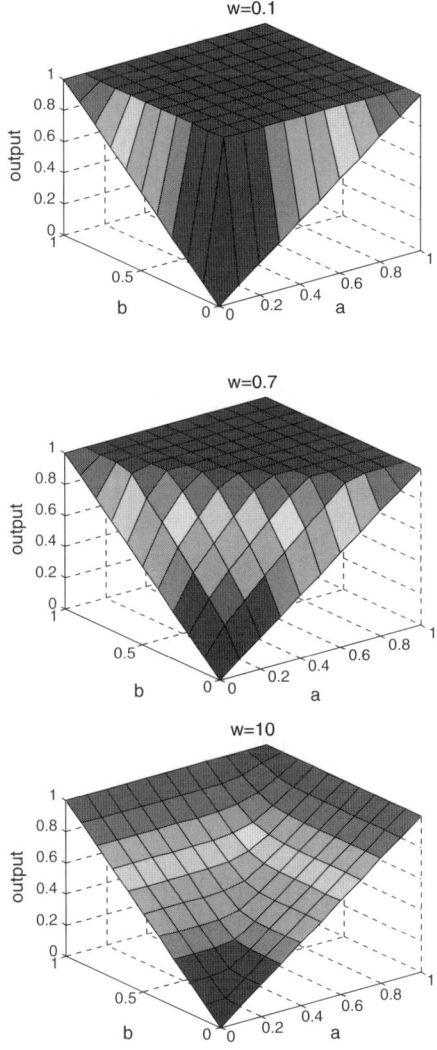

Fig. 8.12. Yager t-conorm operator for different parameter values

The causal relationships (links) are fuzzy in nature, hence fuzzy logic system could help in incorporating the beliefs and perception of the modeler in a scientific way. It also provides a methodology for qualitative analysis of system dynamics models. Since, most of the concepts in natural language are fuzzy, a fuzzy set theoretic approach provides the best solution for such problems. Normally the experts or operators in a particular domain can tell the relationship between variables, but it is difficult to tell the exact degree of relationship between variables. Most of the time, the degree of relationship is fuzzy like

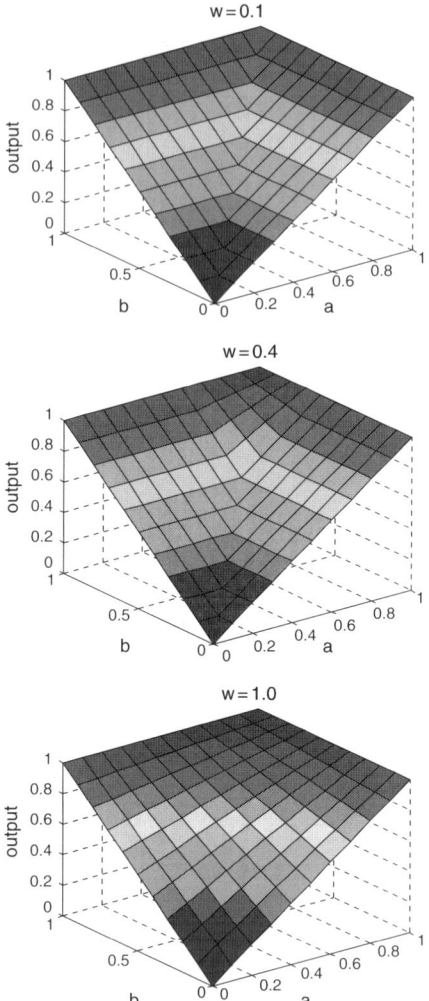

Fig. 8.13. Dubois & Prade t-conorm operator for different parameter values

change the relay setting *slightly*. It is also quite effective way of dealing the complex situations. Hence, integrated fuzzy logic approach and system dynamics technique is a natural choice for dealing with these type of circumstances. The system dynamics technique helps in identifying the relationship between variables and fuzzy logic helps in modeling these fuzzy related variables.

Step-1 Identifying linguistic variable

The very first step in modeling and simulation of this integrated approach is to identify variables for the given situation and then determine the causal

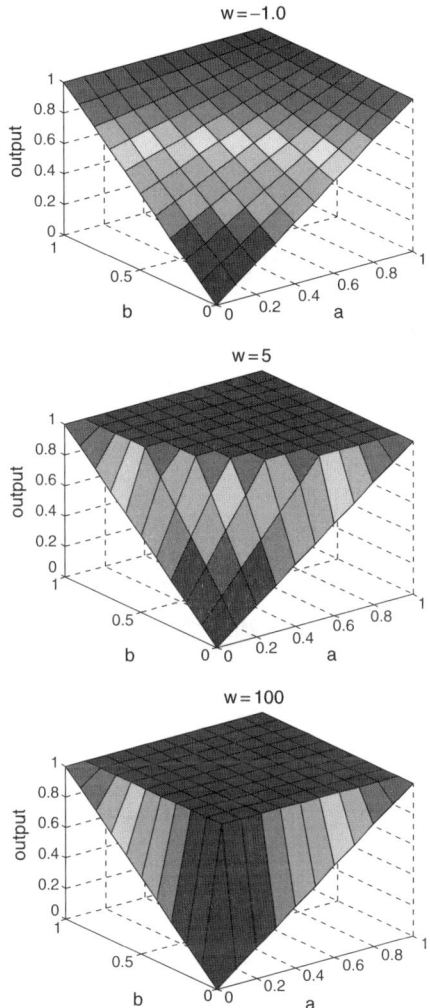

Fig. 8.14. Yu t-conorm operator for different parameter values

links (relationships) between them keeping the remaining system variables unchanged.

For D.C. machine modeling under steady state conditions the variables will be load torque, speed and armature current.

Step-2 Defining range of linguistic variables

Once the variables are identified, the next step is to find the possible meaningful range of variation of these variables. This could be found out from the experimental results or operator experience. The operator can easily tell

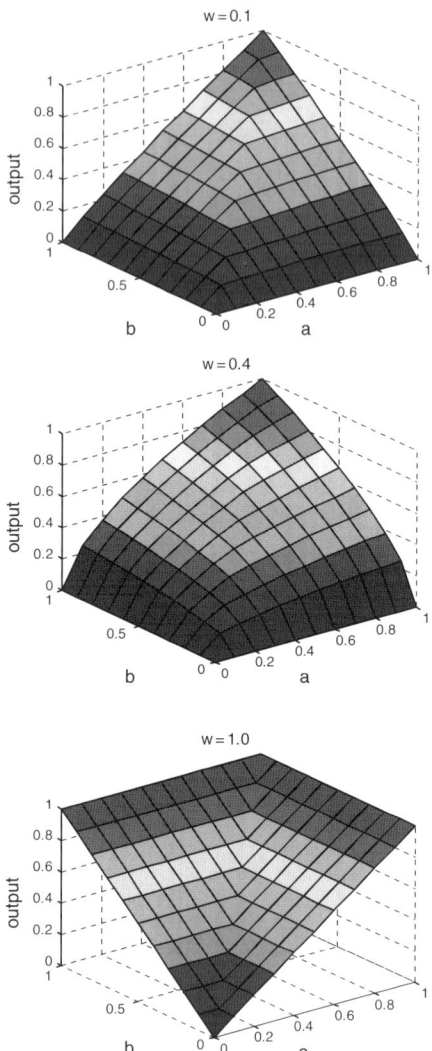

Fig. 8.15. Effect of change in parameter on compensatory (product) operator-1

about the variable range for a particular application and specific context. For
D.C. Machine modeling the range of the variables mentioned above could be
as shown in Table 8.5.

Step-3 Defining Linguistic values for variables

Define linguistic values for linguistic variables, i.e. load torque (T_L), speed (ω)
and armature current (I_a). Each of these variables subdivided into the optimal
number of linguistic values. In case of steady state model of D.C. machine five

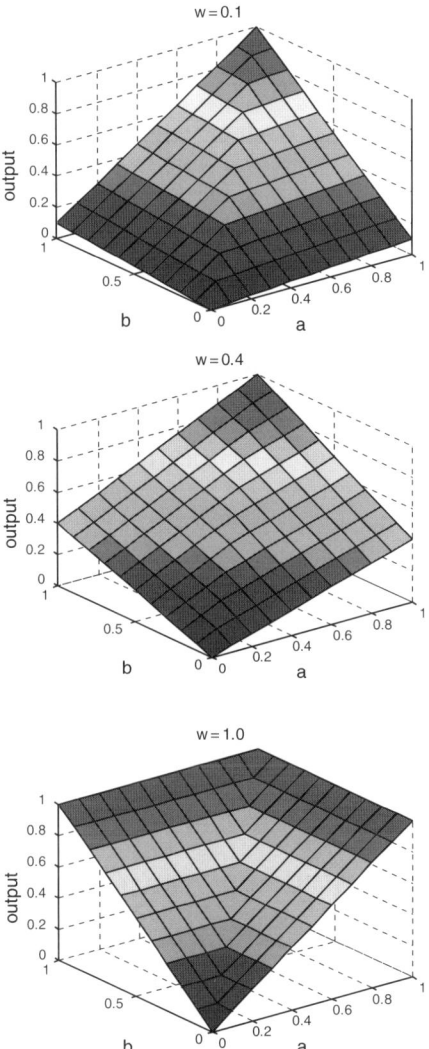

Fig. 8.16. Effect of change in parameter on compensatory (summation) operator-1

linguistic values are considered for each of these variables such as very low (VL), low (L), medium (M), high (H) and very high (VH). The linear shape (membership function) of these linguistic values can be represented in the form of Table 8.6. Graphically, it can be shown in Fig. 8.20.

Step-4 Defining of rules

As mentioned earlier, for defining rules, we have to draw causal links between identified variables. In case of D.C. Machine the causal link between variables

Table 8.3. List of fuzzy implications

Year	Name of implication	Implication function
1973	Zadeh	$\max\,[1-a,\,\min(a,b)]$
1969	Gaines – Rescher	1 if $a \leq b$
		0 if $a > b$
1976	Godel	1 if $a \leq b$
		0 if $a > b$
1969	Goguen	1 if $a \leq b$
		b/a if $a > b$
1938, 1949	Kleene – Dienes	$\max\,(1-a,b)$
1920	Lukasiewicz	$\min(1,\,1-a+b)$
1987	Pseudo – Lukasiewicz – 1	$\min[1,\,\{1-a+(1+w)b\}/(1+wa)]\;w > -1$
1987	Pseudo – Lukasiewicz – 2	$\min[1,(1-a^w+b^w)^{1/w}]\;w > 0$
1935	Reichenbach	$1-a+ab$
1980	Willmott	$\min[\max(1-a),\,\max(a,1-a),\,\max(b,1-b)]$
1986	Wu	1 if $a \leq b$
		$\min(1-a,b)$ if $a > b$
1980	Yager	1 if $a = b = 0$
		b^a others
1994	Klir and Yuan – 1	$1-a+a^2b$
1994	Klir and Yuan – 2	b if $a = 1$
		$1-a$ if $a \neq 1,\;b \neq 1$
		1 if $a \neq 1,\;b = 1$
	Mimdani	$\min(a,b)$
	Stochastic	$\min(1,\,1-a+ab)$
	Correlation product	a^*b
1993	Vadiee	$\max(ab, 1-a)$

may be drawn as shown in Fig. 8.21. The arrow of causal link represents direction of influence and $+$ or $-$ sign shows the effect. Depending of the sign causal links can be categorized as positive or negative causal links and fuzzy rules may be written from these causal links as shown in Fig. 8.22.

Let us derive the fuzzy relational matrix for rule – 1 after max–min operations.

Rule 1 = R (T_{L},ω)

$\quad = VL_{TL} \times VH_{\omega} + L_{TL} \times H_{\omega} + M_{TL} \times M_{\omega} + H_{TL} \times L_{\omega} + VH_{TL} \times VL_{\omega}$

Rule 1

$[1/0 + 0.75/1 + 0.5/2 + 0.25/3 + 0/4] \times [0/156.12 + 0.5/156.75 + 1/157.35 + 0.5/158.04$
$+ 0/158.85] + [0/2 + 0.50/3 + 1.0/4 + 0.50/5 + 0/6] \times [0/154.96 + 0.5/155.56 + 1/156.12$
$+ 0.5/156.75 + 0/157.35] + [0/4 + 0.50/5 + 1.0/6 + 0.50/7 + 0/7] \times [0/153.71 + 0.5/154.33$
$+ 1/154.96 + 0.5/155.56 + 0/156.12] + [0/6 + 0.50/7 + 1.0/8 + 0.5/9 + 0/10] \times [0/152.22$
$+ 0.5/153.00 + 1/153.71 + 0.5/154.33 + 0/154.96] + [0/8+0.25/9+0.5/10+0.75/11+1/12]$
$\times [1/150.88 + 0.75/151.46 + 0.5/152.22 + 0.25/153.00 + 0/153.71];$

Rule 1 =

	150.88	151.46	152.22	153.00	153.71	154.33	154.96	155.56	156.12	156.75	157.35	158.04	158.85
0										0.25	0.5	0.75	1.00
1										0.25	0.5	0.75	0.75
2										0.25	0.5	0.5	0.5
3							0.5000	0.5000	0.5000	0.2500	0.2500	0.2500	
4						0.5	1	1					
5					0.5000	0.5	0.5	0.5					
6				0.5	1	0.5							
7			0.5	0.5	0.5	0.5	0.5						
8			0.5	1	0.5								
9	0.2500	0.2500	0.5000	0.5000	0.5000	0.5000							
10	0.5	0.5	0.25	0.25									
11	0.75	0.75	0.5	0.25									
12	1	0.75	0.5	0.25									

Similarly, other relational matrices can also be determined as given below.

Rule – 2 $R(T_L, I_a)$

	3.0484	4.3210	5.5905	6.8601	8.1271	9.3944	10.664	11.933	13.202	14.471	15.747	17.0151	18.286
0	1	0.75	0.5	0.25									
1	0.75	0.75	0.5	0.25									
2	0.5	0.5	0.5	0.25									
3	0.2500	0.2500	0.2500	0.5000	0.5000	0.5000							
4				0.5	0.5	0.5							
5				0.5	0.5	0.5000	0.5000	0.5					
6						0.5	1	0.5					
7						0.5	0.5	0.5	0.5	0.5			
8							0.5	1	0.5				
9								0.2500	0.2500	0.5000	0.5000	0.5000	0.5000
10										0.25	0.25	0.5	0.5
11										0.25	0.5'	0.75	0.75
12										0.25	0.5	0.75	1

Table 8.4. Compositional rule

1.	Max-min	$b = \max\{\min(a, r)\}$
2.	Min- max	$b = \min\{\max(a, r)\}$
3.	Min – min	$b = \min\{\min(a, r)\}$
4.	Max-max	$b = \max\{\max(a, r)\}$
5.	Godel	$r = 1$ if $a \leq r$
		$r = r$ otherwise
		$b = \max(1, r)$
6.	Max- product (max – dot)	$b = \max(a, r)$
7.	Max – average	$b = \max((a + r)/2)$
8.	Sum - product	$b = f(\Sigma(a, r))$

Where a, b and r are the membership values $\mu_a(x)$, $\mu_b(y)$ and $\mu_r(x, y)$ and f – is a function.

Analysis

Let us take an example where the modeler interested in study the effect of load torque (say 4.55 Nm) on D.C. motor back EMF and armature current.

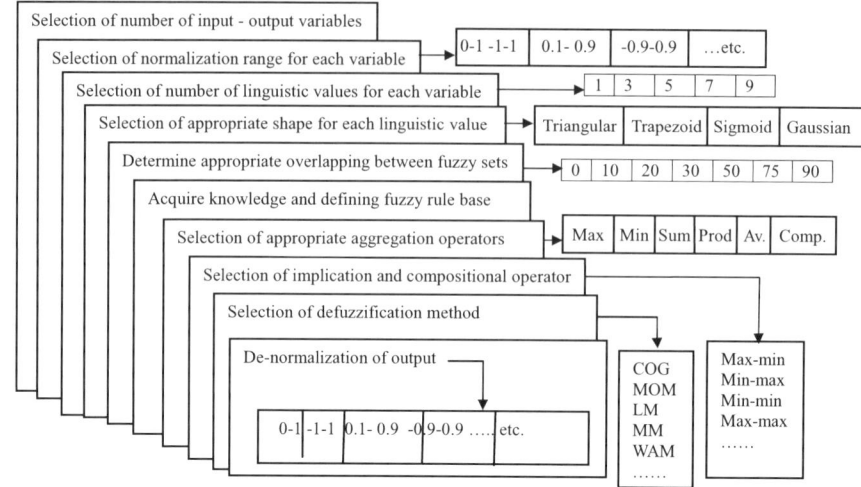

Fig. 8.17. Different windows of fuzzy system

Fig. 8.18. D.C. machine

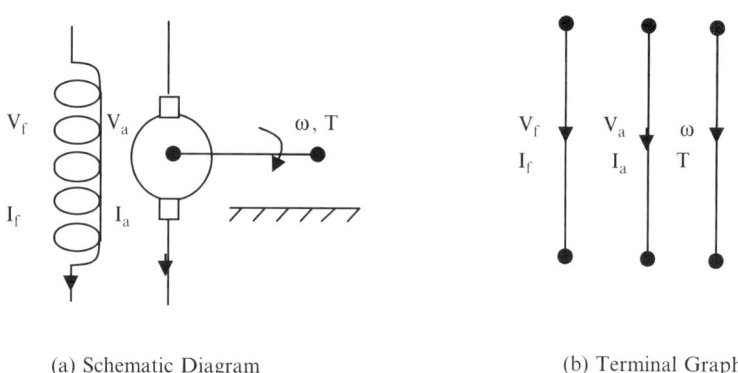

(a) Schematic Diagram   (b) Terminal Graph

Fig. 8.19. Negative causal diagram

First of all fuzzify the given crisp torque value into fuzzy linguistic vales (fuzzification). Project $T_L = 4.55\,\text{Nm}$ to find the percentage of fuzzy linguistic terms.

Table 8.5. Range of variables for steady state D.C. machine model

	Load torque	Speed	Armature current
Mini value	0	150.88	3.0484
Max value	12	156.85	18.286

Table 8.6. Membership function values of linguistic variables

Linguistic variables			Linguistic values				
T_L	ω	I_a	very low	Low	Medium	High	Very high
0	150.88	3.0484	1	0	0	0	0
1	151.46	4.3210	0.75	0	0	0	0
2	152.22	5.5905	0.5	0	0	0	0
3	153.00	6.8601	0.25	0.5	0	0	0
4	153.71	8.1271	0	1.0	0	0	0
5	154.33	9.3944	0	0.5	0.5	0	0
6	154.96	10.664	0	0	1	0	0
7	155.56	11.933	0	0	0.5	0.5	0
8	156.12	13.202	0	0	0	1.0	0
9	156.75	14.471	0	0	0	0.5	0.25
10	157.35	15.747	0	0	0	0	0.5
11	158.04	17.0151	0	0	0	0	0.75
12	158.85	18.286	0	0	0	0	1.0

Fuzzy load torque belongs to low and medium categories. It is 0.775 low and 0.275 medium torque as shown in Fig. 8.23.

$$\text{Fuzzy } (T_L) = [0/0 + 0/1 + 0/2 + 0.5/3 + 0.75/4 + 0.5/5 + 0.25/6$$
$$+0.25/7 + 0/8 + 0/9 + 0/10 + 0/11 + 0/12].$$

From these torque values speed can be inferred from Rule – 1.

$$\text{Fuzzy}(\omega) = \text{Fuzzy}(T_L) \text{ o Rule 1}$$
$$= [0.25/153.00 + 0.25/153.71 + 0.5/154.33 + 0.5/154.96$$
$$+ 0.5/155.56 + 0.75/156.12 + 0.75/156.75$$
$$+ 0.25/157.35 + 0.25/158.04 + 0.25/158.85];$$

This fuzzified speed output is defuzzified using defuzzification methods mentioned in earlier chapter to determine the crisp speed. One can use weighted average method as shown below:

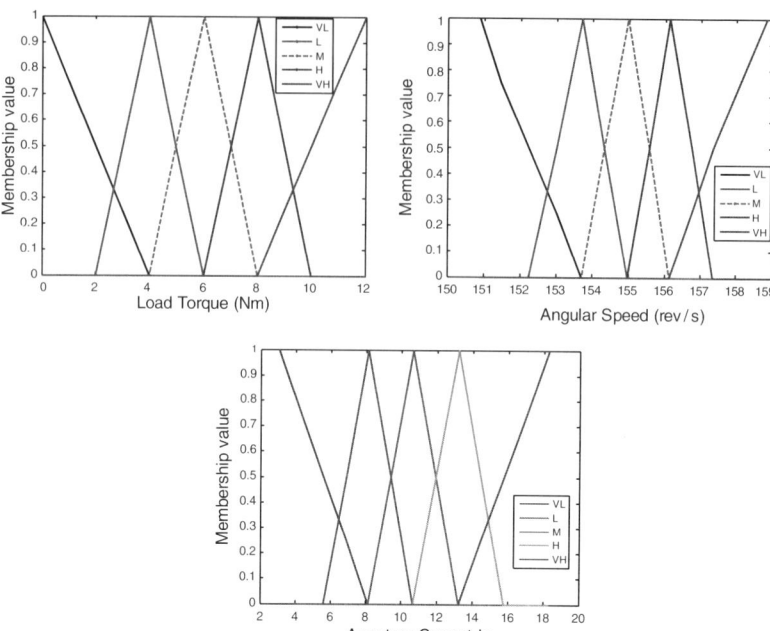

Fig. 8.20. Membership functions for different linguistic variables

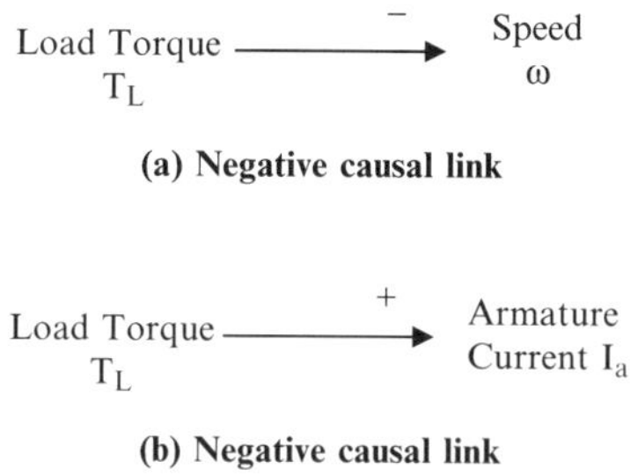

Load Torque ———— $^{-}$ ————▶ Speed
T_L ω

(a) Negative causal link

Load Torque ———— $^{+}$ ————▶ Armature
T_L Current I_a

(b) Negative causal link

Fig. 8.21. Causal links for D.C. machine model

Rule 1 *if* T_L is VL *then* ω is VH elseif T_L is L *then* ω is H elseif T_L is M *then* ω is M elseif T_L is H *then* ω is L *elseif* T_L is VH *then* ω is VL.

Rule 2 *if* T_L is VL *then* I_a is VL elseif T_L is L *then* I_a is L elseif T_L is M *then* I_a is M elseif T_L is H *then* I_a is H *elseif* T_L is VH *then* I_a is VH.

Fig. 8.22. Fuzzy rule derived from causal links

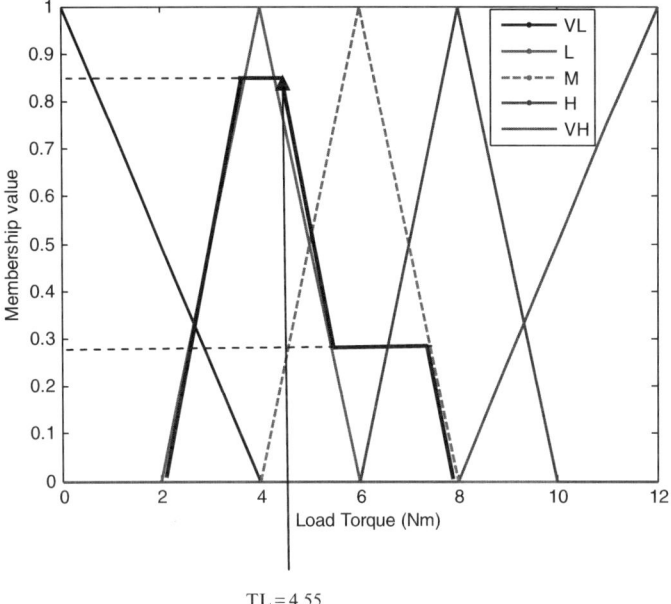

TL = 4.55

Fig. 8.23. Fuzzification process

$$\text{Crisp speed} = \frac{\sum\limits_{i=1}^{n} W_i X_i}{\sum\limits_{i=1}^{n} W_i}$$

$$= (0.25 * 153.00 + 0.25 * 153.71 + 0.5 * 154.33 + 0.5 * 154.96$$
$$+ 0.5 * 155.56 + 0.75 * 156.12 + 0.75 * 156.75 + 0.25 * 157.35$$
$$+ 0.256 * 158.04 + 0.25 * 158.85)/(0.25 + 0.25 + 0.5 + 0.5 + 0.5$$
$$+ 0.75 + 0.75 + 0.25 + 0.25 + 0.25) = 155.8388 \, \text{rev/s}.$$

Similarly, armature current can also be determined for the given load torque using rule 2.

$$\text{Rule} - 2R(T_{L}, I_{a}) => I_{a} = 8.98 \, \text{Amp}.$$

These results can be further improved by taking more points in discrete fuzzy sets. In the above calculation only 13 points are considered to specify the fuzzy membership value for each variable. If the number of points are more the accuracy of the results will be better, but at the same time the computational labor and time both increase.

8.2.10 Transient Model of D.C. Machine

The transient model provides the information about the variation of different variables with respect to time in the system (transient behavior). Hence, the transient model needs to capture the dynamics of the system. To do that one needs to develop all possible causal links in the system and combine them to get causal loop diagram. The causal loop diagram is the first step in modeling the system dynamics using system dynamics technique (Forrester 1961). The causal loop diagram for D.C. Machine is given in (Chaturvedi 1992, 1997) shown in Fig. 8.24.

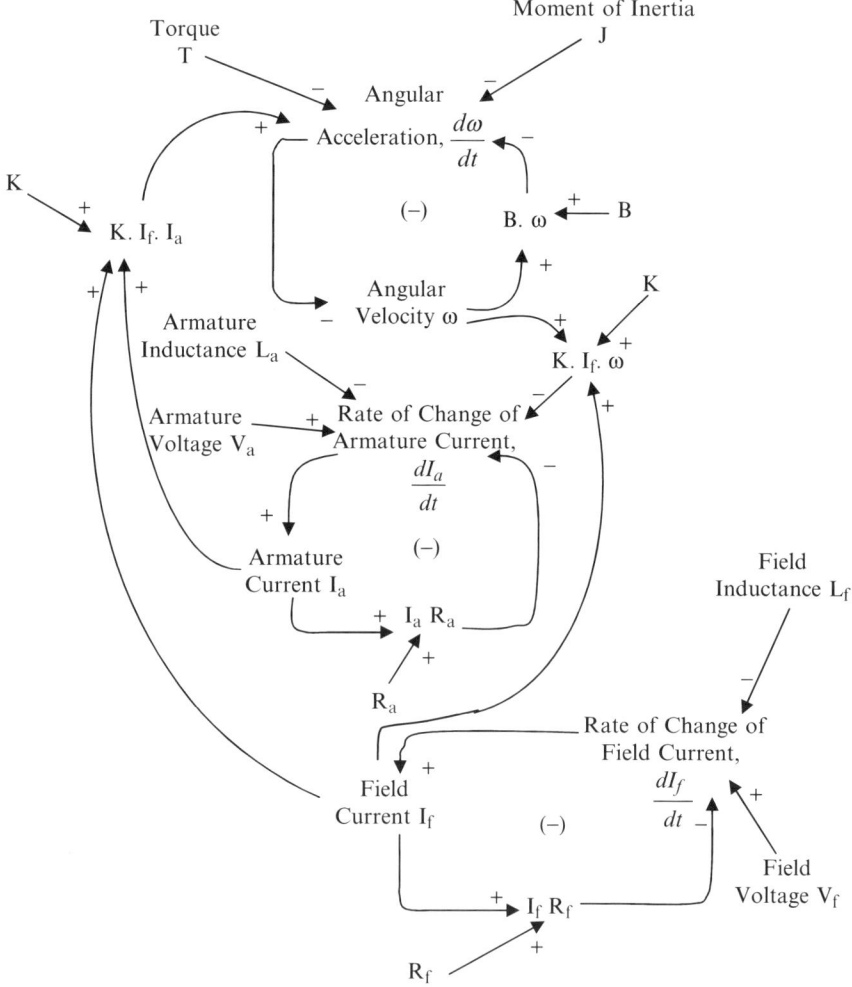

Fig. 8.24. Causal loop diagram for D.C. machine with four negative causal loops

The D.C. Machine (basic commutating machine) in electro-mechanical transducer mode of operation in which field port variables are constant. Hence, the causal loop diagram reduces to have only three causal loops after removing the causal loop of field port (bottom portion of Fig. 8.24). The reduced causal loop diagram consists of two level variables, i.e. armature current and angular speed and two rate variable (state variable) namely rate of change of armature current and angular acceleration, exogenous variables such as applied voltage and load torque, machine parameters like armature inductance, armature resistance, moment of inertia and damping coefficient of machine. From this causal loop diagram the fuzzy knowledge rule can be developed.

Rule 1 *if* **AC** is VL *then* **RAC** is VH elseif **AC** is L *then* **RAC** is H elseif **AC** is M *then* **RAC** is M elseif **AC** is H *then* **RAC** is L *elseif* **AC** is VH *then* **RAC** is VL.

Rule 2 *if* ω is VL *then* ώ is VH elseif ω is L *then* ώ is H elseif ω is M *then* ώ is M elseif ω is H *then* ώ is L *elseif* ω is VH *then* ώ is VL.

Rule 3 *if* **VA** is VL *then* **RAC** is VL elseif **VA** is L *then* **RAC** is L elseif **VA** is M *then* **RAC** is M elseif **VA** is H *then* **RAC** is H *elseif* **VA** is VH *then* **RAC** is VH.

Rule 4 *if* **AC** is VL *then* ώ is VL elseif **AC** is L *then* ώ is L elseif **AC** is M *then* ώ is M elseif **AC** is H *then* ώ is H *elseif* **AC** is VH *then* ώ is VH.

Rule 5 *if* **T**$_L$ is VL *then* ώ is VH elseif **T**$_L$ is L *then* ώ is H elseif **T**$_L$ is M *then* ώ is M elseif **T**$_L$ is H *then* ώ is L *elseif* **T**$_L$ is VH *then* ώ is VL.

Rule 6 *if* ω is VL *then* **RAC** is VH elseif ω is L *then* **RAC** is H elseif ω is M *then* **RAC** is M elseif ω is H *then* **RAC** is L *elseif* ω is VH *then* **RAC** is VL.

Rule 7 *if* **RAC** is VL *then* **AC** is VL elseif **RAC** is L *then* **AC** is L elseif **RAC** is M *then* **AC** is M elseif **RAC** is H *then* **AC** is H *elseif* **RAC** is VH *then* **AC** is VH.

Rule 8 *if* ώ is VL *then* ω is VL elseif ώ is L *then* ω is L elseif ώ is M *then* ω is M elseif ώ is H *then* ω is H *elseif* ώ is VH *then* ω is VH.

The above developed fuzzy logic model for D.C. Machine has been simulated for the machine parameters given in Table 8.7.

i. Study of different connectives

The fuzzy logic model is simulated for different t-norms and t-conorms as given in the litreture (Dubois and Prade 1980, 1985, 1992; Klir 1989, 1995; Zimmermann 1991) like Yager, Weber, Yu, Dobois and Parade, Hamcher, etc. with Mamdani Implication. Results as shown in Fig. 8.25–8.29. All, except Weber connectives are found to give satisfactory results in the range of load torque from 4 to 8 N m. Dubois and Prade connectives perform better in the range from 1 to 11 N m of load torque, but not in the complete range. It is considered expedient to try compensatory operators as connective. Hence the simulation has been performed for compensatory operators with Gains Rascher implication, max-min compositions and the results are shown in Fig. 8.30–8.32.

Table 8.7. Machine parameters of D.C. machine

Armature resistance	Ra = 0.40 ohm
Armature inductance	La = 0.1 H
Damping coefficient	B = 0.01 N ms
Moment of inertia	J = 0.0003 N ms^2
Torque constant	K = 13
Operating conditions	
Applied voltage	Va = 125 V
Load torque	TL = 1 N m

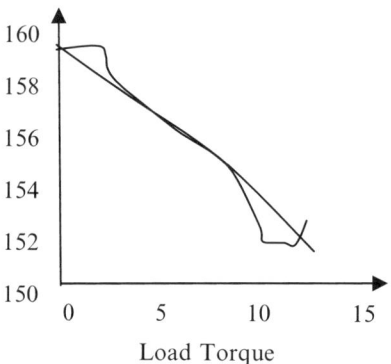

Fig. 8.25. Results with Mamdani implication and Yager connective (w = 0.9)

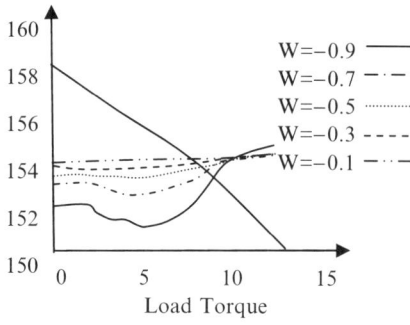

Fig. 8.26. Results with Mamdani implication and Weber connective

ii. Study of different implications methods

The D.C. Machine fuzzy logic model is simulated for different implication methods such as Goguen, Kleene-Dienes, Godel, Wu, etc. as shown in Fig. 8.34–8.36.

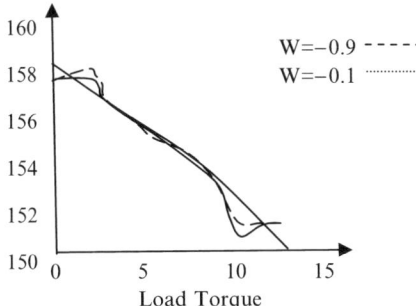

Fig. 8.27. Results with Mamdani implication and Yu connective

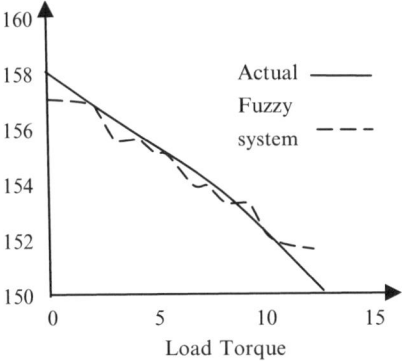

Fig. 8.28. Results with Mamdani impaction and max-min connective

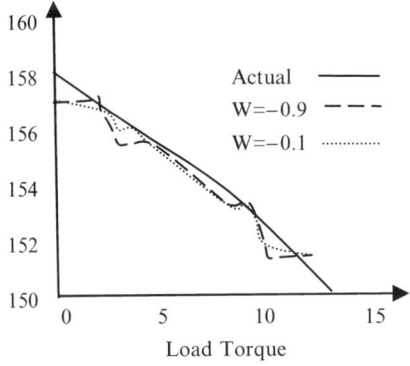

Fig. 8.29. Results with Mamdani implication and Dubois and Prade connective

iii. Different compositional rules

The effect of different compositional rules on the fuzzy model simulation of D.C. Machine have been studied. The results obtained for various compositional rules are shown in Fig. 8.37 and Table 8.8–8.10 with Mamdani implication. It is found that max-product composition gave good results.

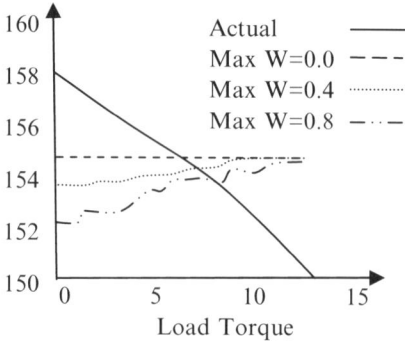

Fig. 8.30. Effect of Compensatory operator (summation type −5) with min = 0.1

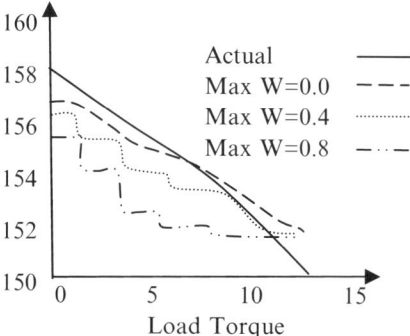

Fig. 8.31. Effect of compensatory operator (summation type −5) with min = 1.0

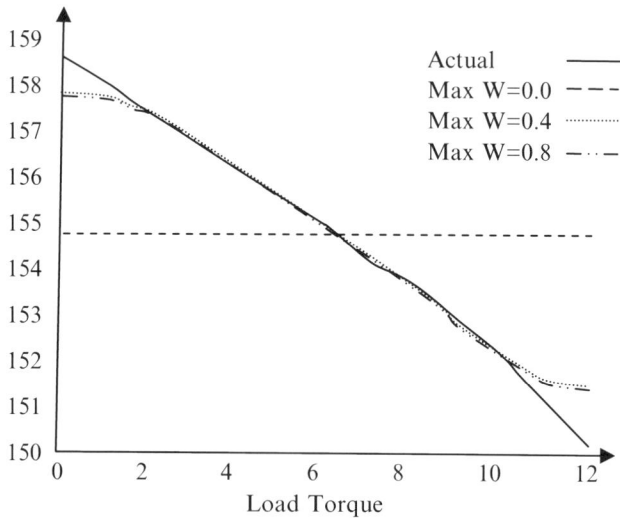

Fig. 8.32. Effect of Compensatory operator (product type) with min = 1.0

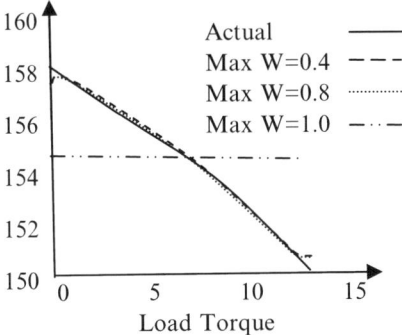

Fig. 8.33. Simulation results with Gaines-Rescher implication

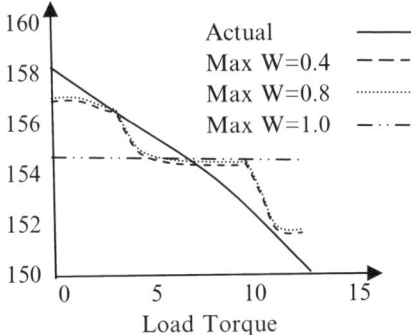

Fig. 8.34. Simulation results with Wu implication

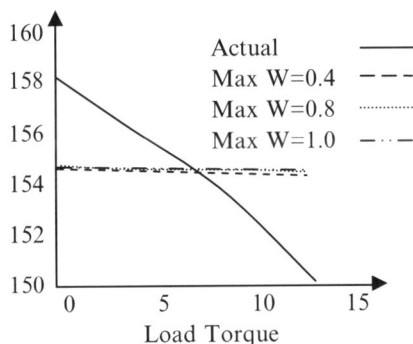

Fig. 8.35. Simulation results with Pseudo-Likasiewicz implication with w = 0.1 and min = 1.0

iv. Different defuzzification methods

It gives the crisp output from the fuzzy output. The effects of different defuzzification have been studied for D.C. Machine model with compensatory operator as connective, Gaines Rascher implication and max-min composition as shown in Fig. 8.38 and Fig. 8.40.

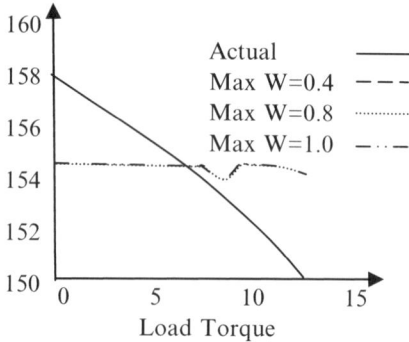

Fig. 8.36. Simulation results with Godel implication with w = 0.1 and min = 1.0

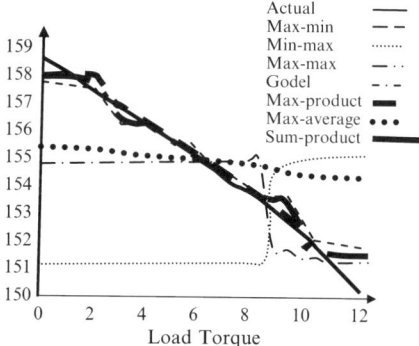

Fig. 8.37. Effect of compositional rules

Table 8.8. Effect of compositional rules for load torque $= 1\,\mathrm{N\,m}$

Compositions	Speed (actual value = 158.04)	Armature current (actual value = 4.321)
$\mathrm{Max}(a, b) - \mathrm{min}(a\,b)$	157.5641	10.67
$\mathrm{Max}(a\,b) - \mathrm{product}(a\,b)$	156.8617	7.3117
$\mathrm{Max}(a, b) - \mathrm{max}(0, a + b - 1)$	158.23083	3.9428
$\mathrm{Max}(a + b - ab) - \mathrm{min}(a, b)$	157.0845	7.7276
$\mathrm{Min}(1, a + b) - \mathrm{min}(a, b)$	157.4973	7.8753
$\mathrm{Min}(1, a + b) - \mathrm{product}(a, b)$	158.339	5.985

v. Effect of overlapping between membership functions

The effects of different degree of overlap between fuzzy sets have been studied and the simulation results are summarized in Tables 8.11. It is found that the slope of fuzzy set 0.5 and the membership variation of 0.1 gives relative better results (Fig. 8.30).

Table 8.9. Effect of compositional rules for load torque $= 4.55\,\mathrm{N}\,\mathrm{m}$

Compositions	Speed (actual value $= 155.76$)	Armature current (actual Value $= 8.7647$)
Max(a, b)$-$min(a, b)	155.805	8.98
Max(a, b)$-$product(a, b)	155.9298	8.7244
Max(a, b) $-$ max(0, a + b $-$ 1)	155.13403	8.12858
Max(a + b $-$ ab) $-$ min(a, b)	155.7086	10.2293
Min(1, a + b)$-$min(a, b)	155.635	10.6725
Min(1, a + b)$-$product(a, b)	155.7437	8.84

Table 8.10. Effect of compositional rules for load torque $= 6.55\,\mathrm{N}\,\mathrm{m}$

Compositions	Speed (actual value $= 154.46$)	Armature current (actual value $= 11.304$)
Max(a, b)$-$min(a,b)	154.2701	10.67
Max(a, b)$-$product(a,b)	154.6919	13.095
Max(a, b) $-$ max(0, a + b $-$ 1)	154.7079	10.6642
Max(a + b $-$ ab) $-$ min(a, b)	154.3165	11.09594
Min(1, a + b) $-$ min(a, b)	154.5325	9.345
Min(1, a + b) $-$ product(a, b)	154.610	11.334

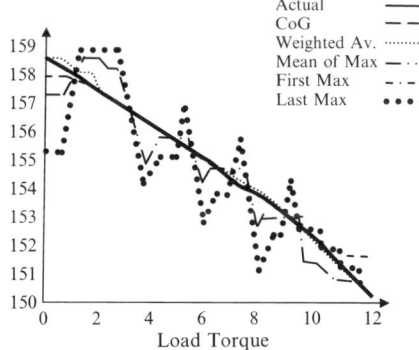

Fig. 8.38. Effect of defuzzification methods

8.2.11 Conclusions

In this integrated method for qualitative analysis:

a. Subjective beliefs and perceptions can be incorporated easily and scientifically in the model.
b. The system dynamics methodologies can give better insight about the system, specially when the system is ill defined.

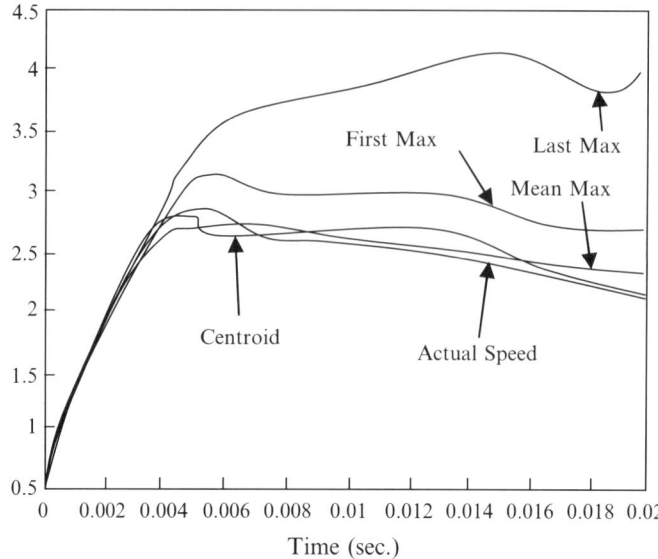

Fig. 8.39. Effect of different defuzzification method on Ia

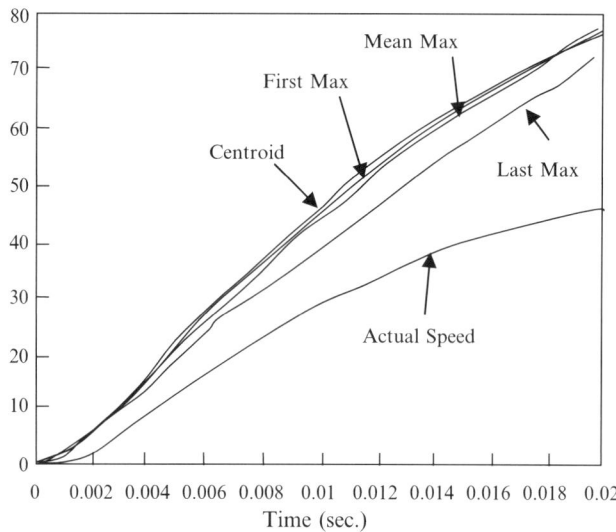

Fig. 8.40. Effect of different defuzzification method on Speed

c. This method provides a systematic approach for qualitative analysis system dynamics.

d. System dynamics which are not clear from the causal loop diagram can be understood with this methodology.

e. It is easy to write down the fuzzy rules once the causal relationship is established between variables.

Table 8.11. Effect of different overlapping and supports of fuzzy sets on fuzzy models

(a) *Load torque = 1 N m*			
Slope of fuzzy sets	Membership variations	Speed	Armature current
0.25	0–1	158.0175	4.3447
0.5	0–1	158.002	4.3452
0.375	0.25–1	157.8051	3.9294
0.25	0.5–1	157.653	3.7149
0.1875	0.25–1	157.4557	5.4967
0.125	0.5–1	157.3017	5.7984
(b) *Load torque = 4.55 N m*			
Slope of fuzzy sets	Membership variations	Speed	Armature current
0.25	0–1	155.48	9.4932
0.5	0–1	155.825	8.7613
0.375	0.25–1	155.707	9.015
0.25	0.5–1	155.6383	9.1841
0.1875	0.25–1	155.2684	9.93936
0.125	0.5–1	155.1192	10.2529
(c) *Load torque = 6.55 N m*			
Slope of fuzzy sets	Membership variations	Speed	Armature current
0.25	0–1	154.766	11.002
0.5	0–1	154.6325	11.2985
0.375	0.25–1	154.5292	11.5117
0.25	0.5–1	154.4275	11.7215
0.1875	0.25–1	154.8207	10.8849
0.125	0.5–1	154.8794	10.7594

f. The following inferences can be deduced from the results obtained from various simulations of fuzzy system models:

    a. 50% overlapping produces satisfactory results from mapping DC machines characteristics, whereas for induction machine different overlapping could map out different portions of the characteristics. Therefore, sliding mode overlapping model can be developed in which there is a facility that the overlapping could change as the speed changes.

    b. The connectives like Schweizer and Sklar – 1,2,3,4, Yu, Dubois and Prade, Frank connectives with Mamdani implication and correlation – product implication are found suitable for DC machine and induction machine models.

    c. Compensatory operators have also been used for modeling and simulation of machine characteristics and it is found that most of the time the compensatory operators (product-type) with Gaines Rascher implication and max-min composition gave good results.

d. Various compositions have also been tried with Mamdani implication. It is found that sum-product, max-min, and max product composition gave better results. With correlation-product implication max-product and max-min composition also gave optimal results.

e. Various defuzzification methods have been adapted and it is found that with Gaines Rascher implication and compensatory (product type)-1 as connective centre of gravity and weighted average defuzzification methods give satisfactory results.

8.3 Control Applications

Goal of control system is to enhance automation within a system while providing improved performance and robustness, for example cruise control for automobile to reduce drivers from tedious task of speed regulation while they are going on long trips.

Conventional controllers are derived from control theory techniques based on mathematical models of the open-loop process, called *system*, to be controlled. The purpose of the feedback controller is to guarantee a desired response of the output y. The process of keeping the output y close to the set point (reference input) $y*$, despite the presence of disturbances of the system parameters, and noise measurements, is called regulation. The output of the controller (which is the input of the system) is the control action u.

The general form of the discrete-time control law is

$$u(k) = f(e(k), e(k-1), \ldots, e(k-\tau), u(k-1), \ldots, u(k-\tau))$$

providing a control action that describes the relationship between the input and the output of the controller.
Where

e error between the desired set point $y*$ and the output of the system y,
τ defines the order of the controller, and
f is in general a nonlinear function.

The block diagram representing simple feedback control system is shown in Fig. 8.41.

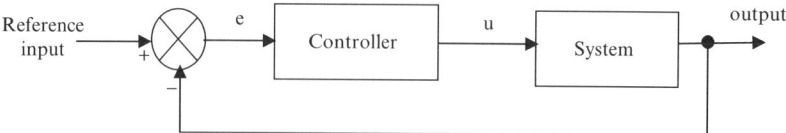

Fig. 8.41. Simple feedback control system

In a fuzzy logic controller (FLC), the dynamic behavior of a fuzzy system is characterized by a set of linguistic rules defined by expert knowledge. The expert knowledge is usually of the form **IF** (atencedent i.e. a set of conditions are satisfied) **THEN** (a set of consequences can be inferred). Since the antecedents and the consequents of these IF-THEN rules are associated with fuzzy concepts (linguistic terms), they are often called *fuzzy conditional statements*. Basically, fuzzy control rules provide a convenient way for expressing control policy and domain knowledge. Furthermore, several linguistic variables might be involved in the antecedents and the conclusions of these rules. When this is the case, the system will be referred to as a multi-input-multi-output (MIMO) fuzzy system. For example, in the case of two-input-single-output (MISO) fuzzy systems, fuzzy control rules have the form

Rule 1: If $x = A_1$ and $y = B_1$ Then $z = C_1$

Also

Rule 2: If $x = A_2$ and $y = B_2$ Then $z = C_2$

. . ..

Rule n: If $x = A_n$ and $y = B_n$ Then $z = C_n$

where x and y are the process state variables, z is the control variable, Ai, Bi, and Ci are linguistic values of the linguistic variables x, y and z in the universes of discourse U, V, and W, respectively, and an implicit sentence connective *also* links the rules into a rule set or, equivalently, a rule base.

We can represent the FLC in a form similar to the conventional control law

$$u(k) = F(e(k), e(k-1), \ldots, e(k-\tau), u(k-1), \ldots, u(k-\tau))$$

where the function F is described by a fuzzy rule base.

A typical FLC describes the relationship between the changes of the control action on the one hand,

$$\Delta u(k) = u(k) - u(k-1)$$

and the error $e(k)$ and its change $\Delta e(k)$ on the other hand.

$$\Delta e(k) = e(k) - e(k-1).$$

Such control law can be formalized as $\Delta u(k) = F(e(k), \ \Delta(e(k)))$

The actual output of the controller $u(k)$ is obtained from the previous value of control $u(k-1)$ that is updated by $\Delta u(k)$ as given in the expression $u(k) = u(k-1) + \Delta u(k)$.

Suppose now that we have two input variables x and y. A fuzzy control rule

Rule i: if (x is A_i and y is B_i) then (z is C_i)

is implemented by a *fuzzy implication* R_i and is defined as

$$R_i(u, v, w) = [A_i(u) \text{ and } B_i(v)] \rightarrow C_i(w) \text{ where } i = 1, 2, 3, \ldots n$$

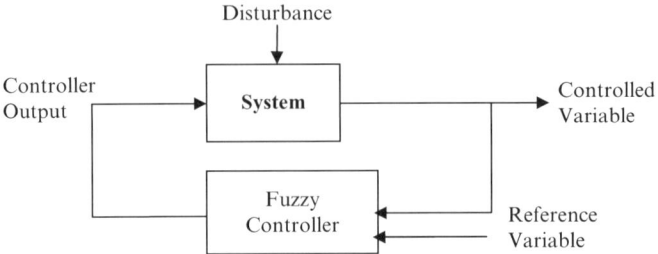

Fig. 8.42. Closed loop control system using fuzzy controller

where the logical connective *and* is implemented by the minimum operator, i.e.

$$[A_i(u) \text{ and } B_i(v)] \rightarrow C_i(w) = [A_i(u) \cap B_i(v)] \rightarrow C_i(w)$$
$$= \min\{A_i(u), B_i(v)\} \rightarrow C_i(w)$$

When fuzzy logic controller is appropriate?

Fuzzy logic controller is appropriate for systems with the following characteristics:

1. One or more continuous variables are there, which need control.
2. An exact mathematical model of the system does not exist. If it exists then it is too complex to use on line.
3. Serious non-linearities and delays are present in the system, which makes system model computationally expensive.
4. Time variant properties of system exists.
5. System requires operators intervention.
6. Human intuitions are required to control the system.
7. Written data base of system information is not sufficiently available, but system expertise are available (i.e. mental data base is there). Hence quantitative modeling is not possible, but qualitative model could be developed.
8. Where sufficient human experts knowledge available.

The computational structure of a fuzzy knowledge base controller (FKBC) in the closed loop is shown in Fig. 8.42.

8.3.1 Adaptive Control

Adaptation is fundamental characteristics of living organism (Human being, animal), since they attempt to maintain physiological equilibrium in the midst of changing environmental conditions.

An Adaptive controller can be designed so that it estimates some uncertainty within the system, then automatically designs a controller for the estimated plant uncertainty. In this way the control system uses information gathered on-line to reduce the model uncertainty, that is, to figure out exactly what the plant is at the current time so that good control can be achieved.

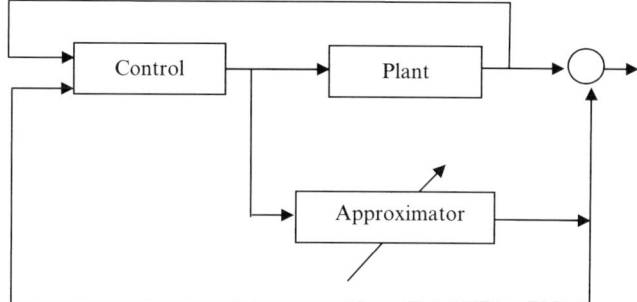

Fig. 8.43. Indirect adaptive controller

In conventional controller the gain of a controller are normally fixed, but in adaptive controllers it changes with the situation. Hence, adaptive controller modifies its behavior with changes. The basic difference between conventional adaptive control and adaptive fuzzy control systems stems from the use of the tuning of general non-linear structures (i.e. Fuzzy Systems) to match some unknown non-linearity (even if only the parameters that enter linearly are tuned). This shows that this type of controller has self organizing features in it.

There are two types of adaptive controllers: indirect adaptive and direct adaptive control systems.

A. *Indirect adaptive control scheme*

An indirect approach to adaptive control is made of an approximator (often referred to as an "identifier" in the adaptive control literature) that is used to estimate unknown plant parameters and a certainty equivalence control scheme in which the plant controller is defined (designed) assuming that the parameter estimates are their true values. The indirect adaptive approach is shown in Fig. 8.43. Here the adjustable approximator is used to model some of component of the system. If the approximation is good (i.e. we know how the plant should behave), then it is easy to meet our control objective. If, on the other hand, the plant output moves in the wrong direction, then we may assume that our estimate is incorrect and should be adjusted accordingly.

An example of indirect adaptive controller, consider the cruise control problem where we have an approximator that is used to estimate the vehicle mass and aerodynamic drag.

B. *Direct adaptive control scheme*

The Direct adaptive controller is shown in Fig. 8.44. Here the adjustable approximator acts as a controller. The adaptation mechanism is then designed to adjust the approximator causing it to match some unknown nonlinear controller that will stabilize the plant and make the closed loop system to achieve its performance objective. Note that we call this scheme "direct" since there is a direct adjustment of the parameters of the controller without identifying a model of the plant. This type of control is quite popular.

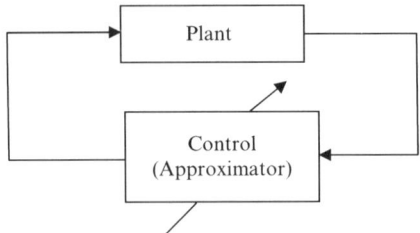

Fig. 8.44. Direct adaptive control

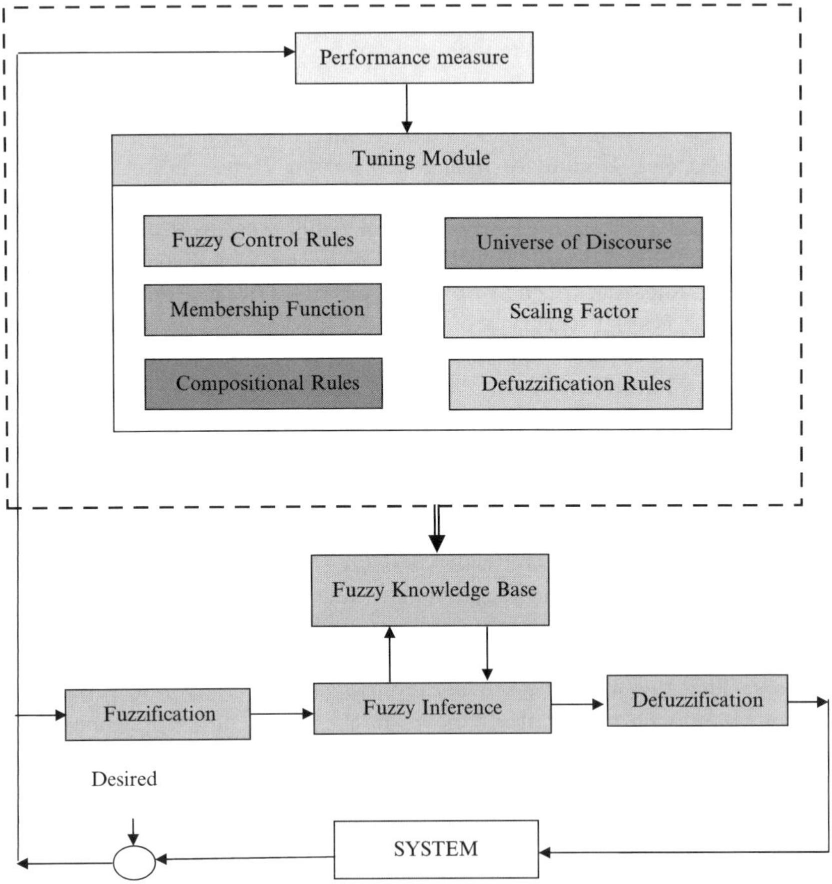

Fig. 8.45. Functional module of an adaptive fuzzy control

Neural networks and fuzzy systems can be used as the "approximator" in the adaptive scheme. Neural networks are parameterized nonlinear functions. Their parameters are, for instance, the weights and baises of the network and adjustment of these parameters results in different shaped nonlinearities.

Figure 8.45 shows the adaptive fuzzy logic based controller. The system output is compared with the desired output and error is calculated. This error and its rate of change is passed to fuzzy controller which controls the system behaviour. To make the controller adaptive the error and error rate is also given to the tunning module which tunes the the fuzzy rule base, shape and overlapping of membership functions, modifies the parameters of aggregation operators, compositional operator and implication operator, etc.

Some Examples on Control System:
    Let a linear second order system whose transfer function is given by G(s) and input to the system is unit step u(t) for t ≥ 0.

8.3.2 PID Control System

The block diagram of a system controlled by feedback error signal based PID controller is given in Fig. 8.46 and its simulink (Matlab) model is given in Fig. 8.47.

8.3.3 Fuzzy Control System

Controlling the second order system using fuzzy logic based controller, two inputs are taken (error and error rate) and based on these inputs controller action is determined as shown in Fig. 8.48.

    Input variables = {error, error rate} and output = {control action}

The linguistic values of these variables are as follows:

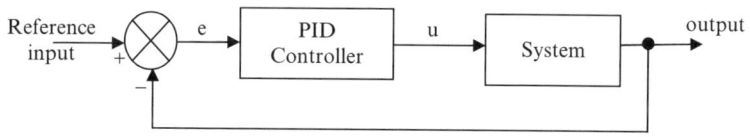

Fig. 8.46. Linear PID feedback control system

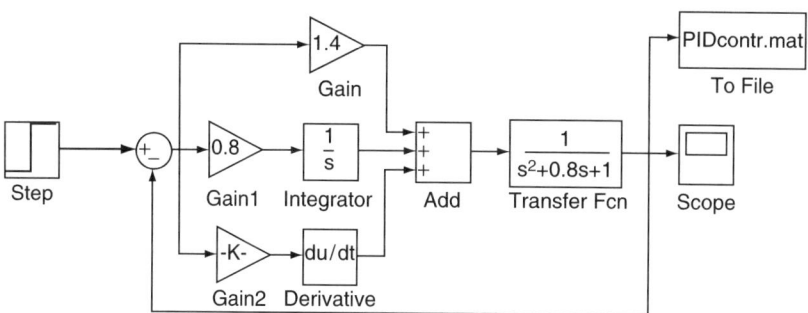

Fig. 8.47. Simulink model for PID controlled system

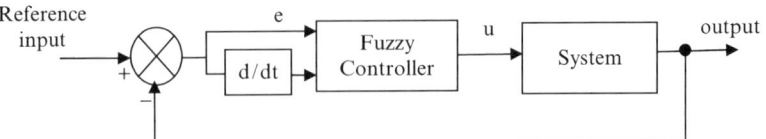

Fig. 8.48. Fuzzy Logic based feedback control system

Input variables = {Negative (N), Zero (Z), and Positive (P)} and

Output variable = {Negative large (NL), Negative small (NS), Zero (Z),

Positive Small (PS) and Positive Large (PL)}.

After selecting appropriate number of input and output variables and their linguistic values, we have to draw the membership function for these linguistic values. The membership function for error, error rate and output variables are shown in Fig. 8.49.

Depending on these input variable values the output variable value is to be decided from the experience encoded in the form of rules. There are nine rules defined as given in the fuzzy associative memory (FAM) Table 8.1. Simulink model for fuzzy controlled system is shown in Fig. 8.50. The three dimension control surface is als shown in Fig. 8.51.

FAM table for fuzzy controller

e de/dt	N	Z	P
N	NL	NS	Z
Z	NS	Z	PS
P	Z	PS	PL

Using Matlab Fuzzy Toolbox, one can easily develop the fuzzy controller with the help of fis editor as shown in Fig. 8.52. Graphically fuzzy rules can be represented as shown in Fig. 8.53.

The performance of PID controller and fuzzy controller are compared as shown in Fig. 8.54 and Table 8.12, and it is found that fuzzy controller is better than PID controller. The performance of fuzzy controller can even be improved by tuning its parameters. Then performance of this controller also compared when there is a delay between the controller and the system of 0.5 s. The fuzzy controller performance is quite good as shown in Fig. 8.55. When there is a variable delay the problem become more sever. These controllers have also been tested for variable delay between controller and system and found that PID controller could not control the system output and it becomes unstable, but fuzzy controller is able to control it as shown in Fig. 8.56.

Response of fuzzy logic controller is almost same as conventional PID controller for linear system but it gives better response for highly nonlinear and delay type system.

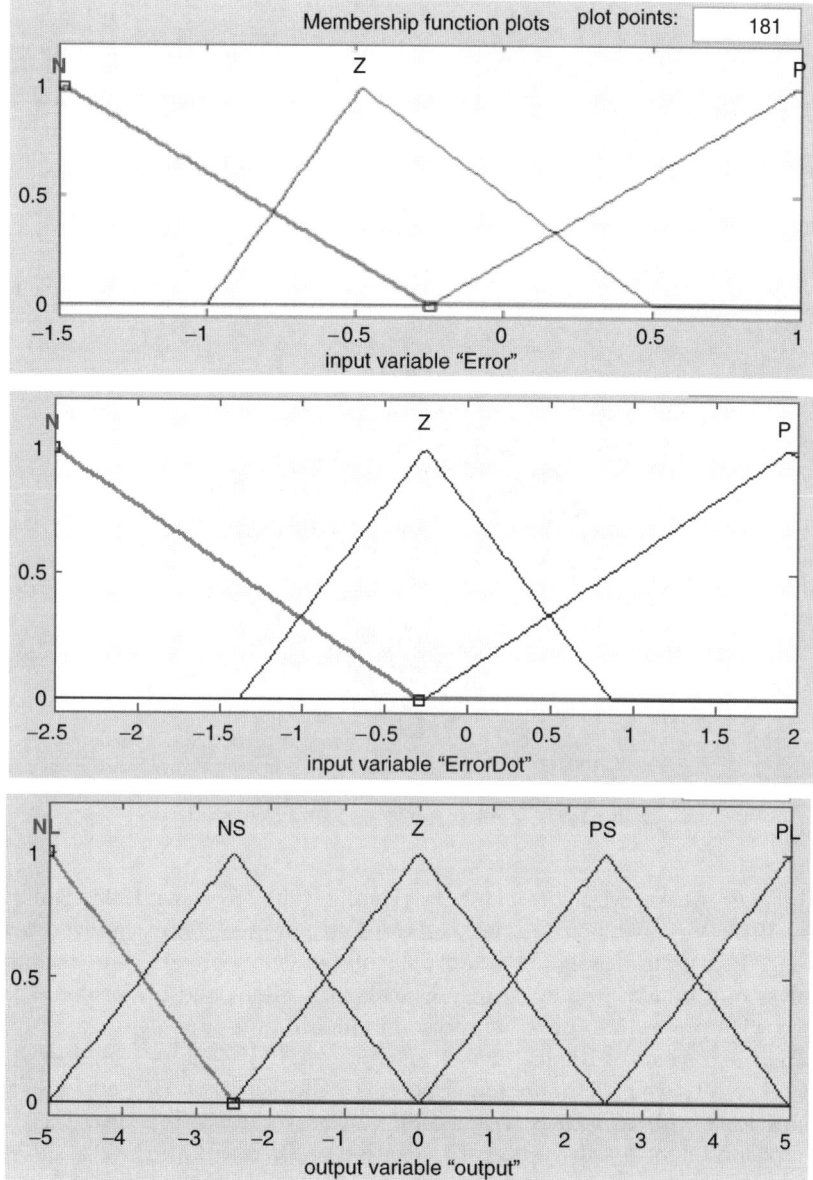

Fig. 8.49. Membership functions for input and output variables

8.3.4 Power System Stabilizer Using Fuzzy Logic

The design and application of power system stabilizers has been the subject of continuing development for many years. Most of the designs of conventional PSS are typically based on analog or digital implementations of lead, lag or

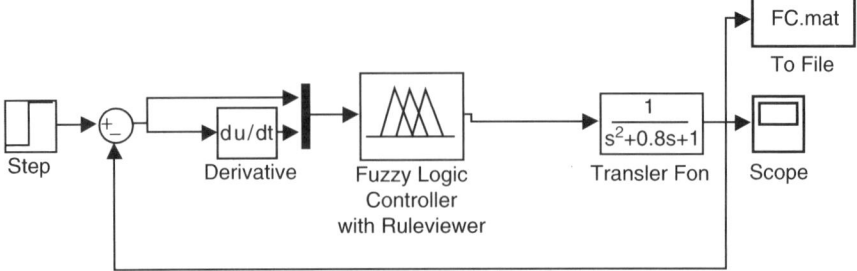

Fig. 8.50. Simulink model of fuzzy logic based controller for second order system

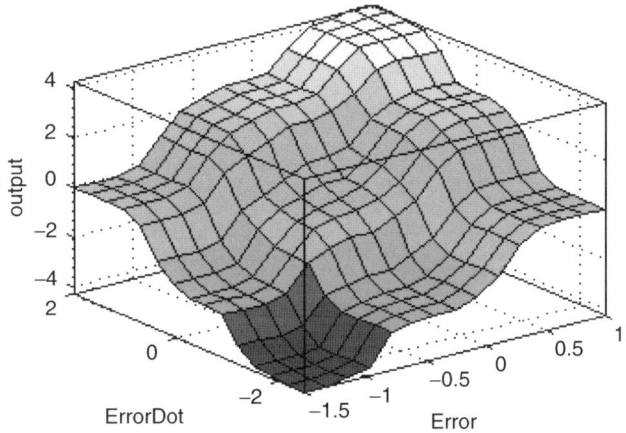

Fig. 8.51. Control surface of fuzzy controller

lead-lag networks. Many researchers (Huang 1991; Kothari 1993; Bollinger et al. 1978; Yu 1990; Srinivas 1981) have used power system stabilizer design via eigen structure assignment or pole placement problem. The technique involves obtaining a frequency response of the systems and the problem associated with power systems when noise inputs are present.

In recent years artificial neural network (ANN) (Guan 1996; Liu 2003) and fuzzy set theoretic approach (Hsu 1990; Gawish 1999; Hosseinzadeh and Kalam 1999; Gibbard 1988; Wang 1995; Ortmeyer 1995; Hassan 1991) have been proposed for power system stabilization problems. Both these techniques have their own advantages and disadvantages. The integration of this approach can give improved results. Fuzzy set is used to represent knowledge in an interpretable manner. Neural networks (Chaturvedi 2004) have the capability of interpolation over the entire range for which they have been trained and provides the capability of adaptability not possessed by fixed parameters devices designed and tuned for one operating conditions. A generalized neuron based PSS (Chaturvedi 2004, 2005) has been proposed to overcome the problems of conventional ANN and combining the advantages of self-optimizing

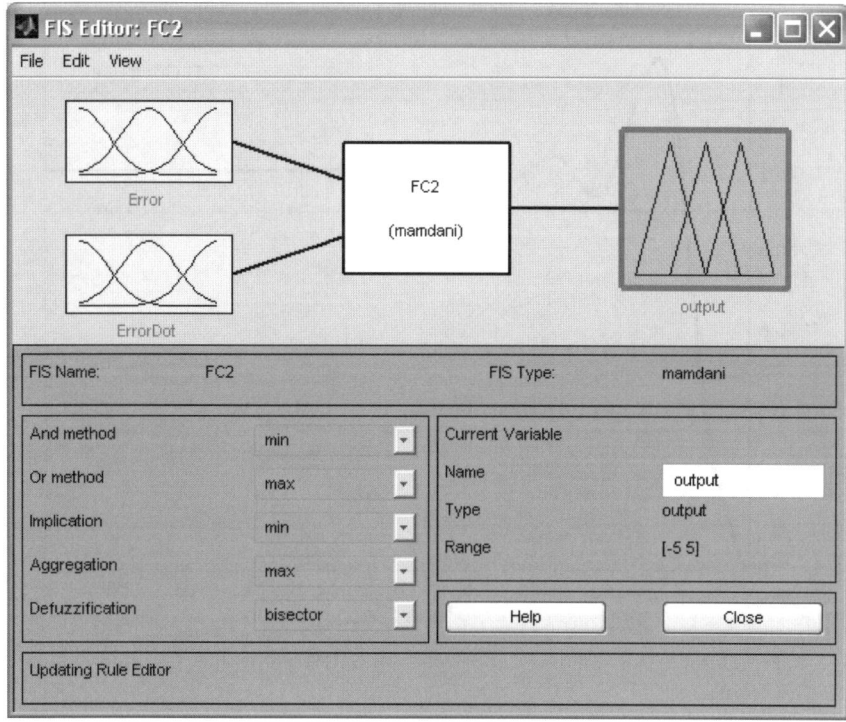

Fig. 8.52. FIS editor of fuzzy tool box of matlab for the development of fuzzy controller

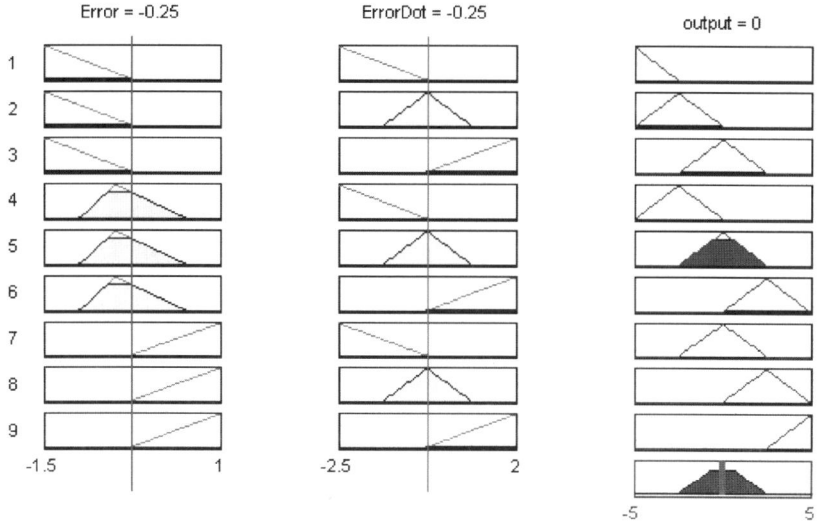

Fig. 8.53. Graphical representation of fuzzy rules

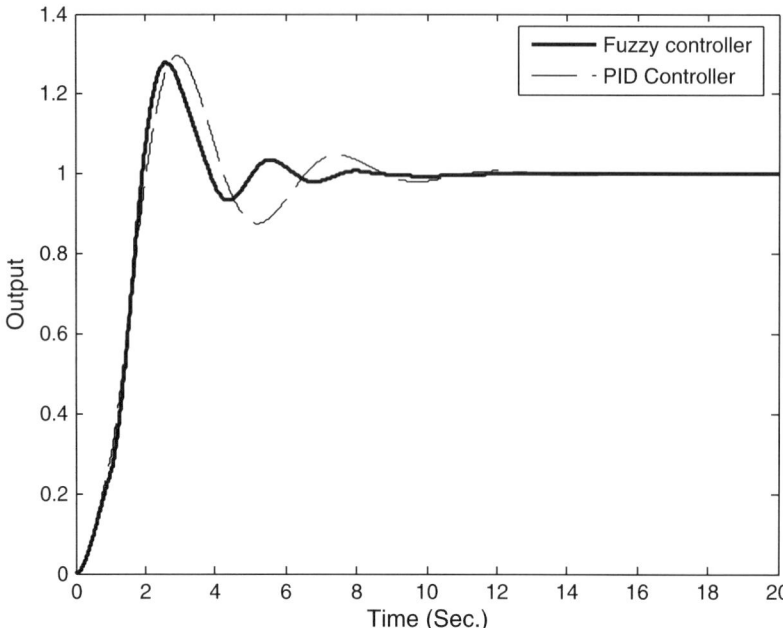

Fig. 8.54. Comparison of fuzzy controller and PID controller for unit change in reference input

Table 8.12. Performance comparison of Fuzzy and PID controller

	PID controller	Fuzzy controller
Max overshoot	1.2955	1.2784
Settling time (with 5% tol.) (s)	7.43 s	4.67 s
Peak time (s)	2.03 s	1.92 s
Rise time (from 10 to 90%) (s)	1.38 s	1.29 s

adapting central strategy and the quick response of the GN, to provide good damping to the power system over a wide operating range. The main issue of ANN stabilizer is its training. To properly train ANN one needs sufficient and accurate data for training, efficient training algorithms and optimal structure of ANN. It is really difficult to get appropriate and accurate training data for real life problems. It is also difficult to select an optimal size of ANN for a particular problem.

To overcome these problems the fuzzy logic controllers are proposed where no numerical training data is required and the operator experience can be used. This makes fuzzy logic controllers very attractive for ill defined systems or systems with uncertain parameters. With the help of fuzzy logic concepts, expert's knowledge can be used directly to design a controller. Fuzzy logic allows one to express the knowledge with subjective concepts such as very

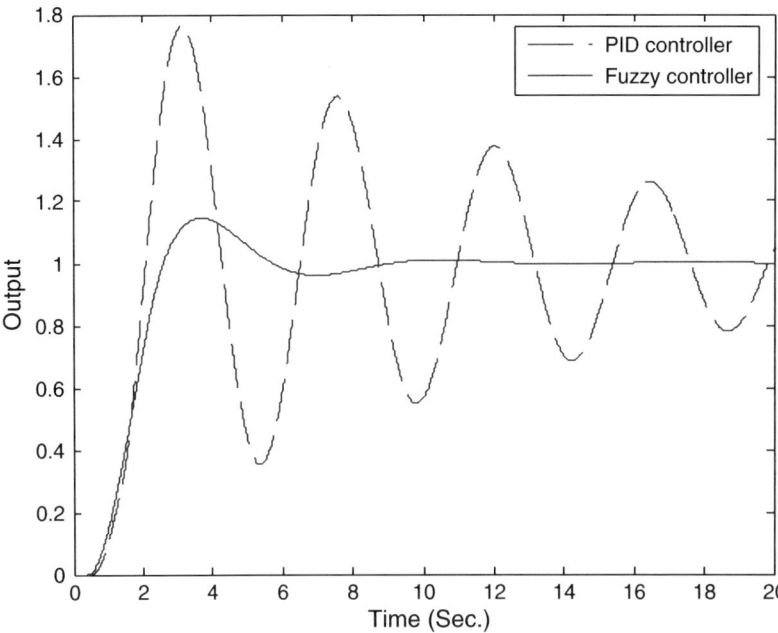

Fig. 8.55. Performance of PID and fuzzy controller when delay in signal between system and controller

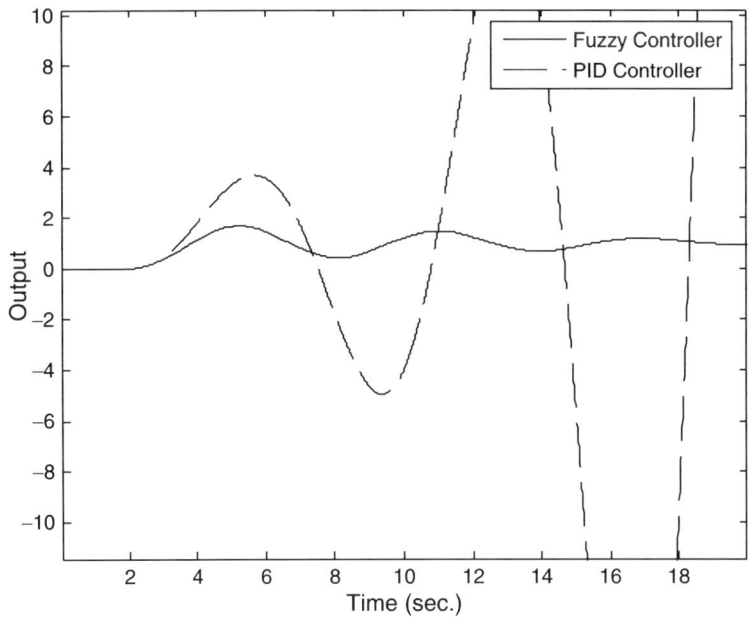

Fig. 8.56. Controllers performance when there is variable time delay in the system

big, too small, moderate and slightly deviated, which are mapped to numeric ranges (Zadeh 1973). Fuzzy control implementation of power system stabilizers has been reported in a number of publications (Hsu 1990; Gawish 1999; Hosseinzadeh 1999; Gibbard 1988; Wang 1995; Ortmeyer 1995; Hassan 1991). Due to its lower computation burden and its ability to accommodate uncertainties in the plant model, fuzzy logic power system stabilizers (FPSS) appear to he suitable for implementing PSSs. FPSSs can he implemented through simple microcomputers with A/D and D/A converters (Hiyama 1993; El-Metwally 1996). The performance of FPSSs depends on the operating conditions of the system, although it is less sensitive than conventional linear PSSs (Parniani 1994).

In this paper, an adaptive polar fuzzy controller has been proposed for the power system stabilizer to reduce the computational burden and time to compute the stabilizing signal. Generalized neuron is used as identifier, which can identify complex and dynamic system through learning, which can easily accommodate non-linearities and time dependencies. Adaptive polar fuzzy PSS can adjust its parameters according to the changes in environment and maintains desired control ability over a wide operating range of power system.

8.3.4.1 Polar Fuzzy Sets

The Polar fuzzy sets were first introduced by Hadipriono and Sun in 1990. Polar fuzzy sets differ from standard fuzzy sets only in their coordinate description. Polar fuzzy sets are defined on a universe of angle, hence are repeating shapes every 2π cycles.

Polar fuzzy in environmental engineering deals with the qualitative information about the pollutions. The pH value of some water samples collected from polluted pond and give these pH reading linguistic labels such as very basic, fairly acidic, etc. The neutral solution has a pH value of 7. The linguistic terms can be built in such a way that a "neutral" (N) corresponds to $\theta = 0\,\text{rad}$, "absolutely basic" (AB) corresponds to $\theta = \pi/2\,\text{rad}$ and "absolutely acidic" corresponds to $\theta = -\pi/2$. Levels of pH varies between 14 and 7 can be labeled as "very acidic" (VA). "acidic" (A) " fairly acidic" (FA) and are represented between $\theta = 0$ and $\pi/2$. The Polar fuzzy set model for these linguistic labels of pH is shown in Fig. 8.57.

The membership functions for the pH value of water is shown in Fig. 8.58. Polar fuzzy is useful in situations that have a natural basis in polar coordinates or in situations where the value of a variable is cyclic.

8.3.4.2 Polar Fuzzy Logic Based Power System Stabilizer

A power systems model consisting of a synchronous machine connected to a constant voltage bus through a double circuit transmission line is shown in Fig. 8.59. The state equations representing the power system and the synchronous machine, governor and the AVR parameter are given in Appendix. The

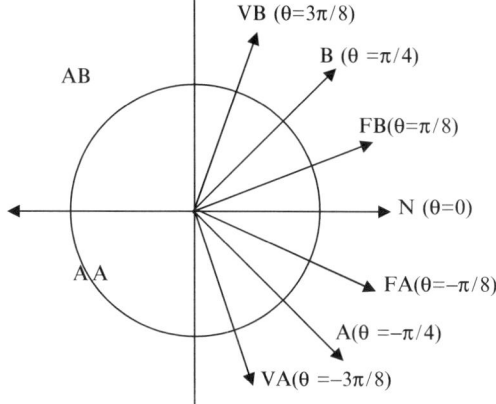

Fig. 8.57. Polar fuzzy set for linguistic label of pH

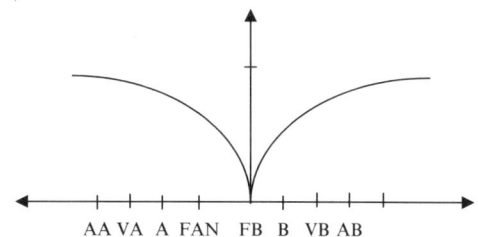

Fig. 8.58. Membership Values for pH

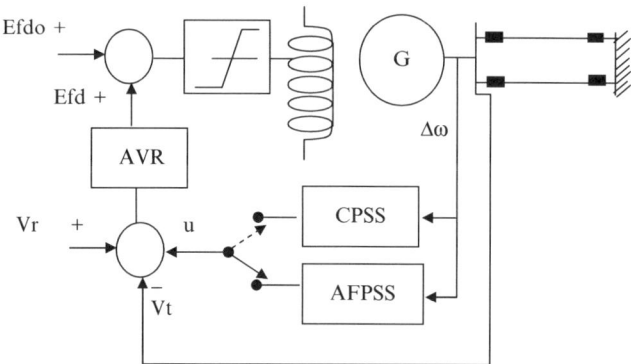

Fig. 8.59. Single machine infinite bus system

control signal generated by the PSS is injected as a supplementary stabilizing signal to the AVR summing point.

The fuzzy logic based PSS with several rules indeed gives a consistently better performance than the CPSS but controller design and mathematical operation with several rules is rather complex and time consuming. Simple rule based stabilizer is proposed depending upon the polar fuzzy information. One

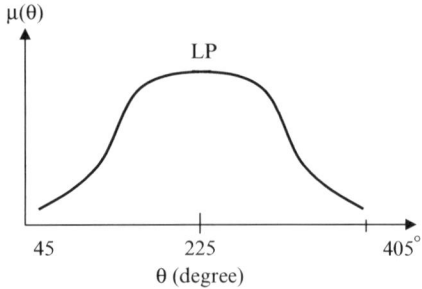

Fig. 8.60. Membership functions for angle

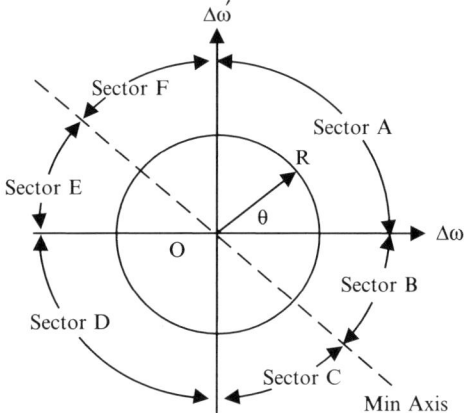

Fig. 8.61. Six sector phase plane

of the alternatives would be to represent the two states of the generator speed deviation $\Delta\omega$ and acceleration $\Delta\dot{\omega}$ in polar form and generate the PSS signal based on the magnitude R and angle θ as shown in Fig. 8.61. The origin "O" is the desired equilibrium point and all the control efforts should be directed to shift the generator state R toward the origin "O" of the phase plane.

Two fuzzy membership functions LN and LP were defined for the input angle θ. The values of these two functions are complimentary to each other, were either 0 or 1 or varied linearly over the operating range as shown in Fig. 8.60. *Fuzzy Inputs: Angle*

LP = [1.0/45 0.9/90 0.5/135 0.2/180 0/225 0.2/270 0.5/3150.9/360 1.0/405]

and

LN = 1 − LP;
= [0.0/45 0.1/90 0.5/135 0.8/180 1.0/225 0.8/270 0.5/315 0.1/360 0.0/405]

Fuzzy Output of Controller Uc:

P = [0/0.5 0.25/0.56 0.50/0.63 0.75/0.69 1.00/0.75 0.75/0.81 0.50/0.88
    0.25/0.94 0.0/1.0];

N = [0/−1 0.25−0.94 0.50−0.88 0.75−0.81 1.00−0.75 0.75−0.69 0.50−0.63
    0.25−0.56 0−0.5];

The overall gain of PSS was linearly increased with magnitude R and deviation in angular speed. The output of the PSS was generated directly by utilizing the values of LN and LP. This effectively means that LN = 0 for $(\alpha < \alpha i$ or $\alpha > 2\pi)$.

And LP = 1 − LN where α_i is tuning parameter. The maximum and minimum value of membership function LN and LP respectively dependent on α_i which is depending on the system requirements sometime it is necessary to rotate the min-max axis by a suitable angle this can be done either by modification of input membership function along the axis of angle θ or by modifying the angle θ directly. The former method is used where upper domain of θ for MFS is fixed to 2π but the problem associated with this is that, when α_i is shifted then change in angle θ at lowest domain is maximum and keep on decreasing towards $\theta = 2\pi$ and finally to zero. So the effect on controller performance is not same for all θ in the domain [θ2π].

In PFPSS there is no need to use two separate input gains for Δω and Δώ because the FLC of PFPSS uses the polar angle of the properly scaled inputs which depends upon the ratio of the scaled inputs and consider only one gain Kacc. The scaling factor Kacc decides as to which variable, speed deviations or acceleration has more weightage in magnitude R(k). In the proposed controller the magnitude of output from the FLC is set to be maximum and minimum at 45° and 135° axis, respectively. The maximum and minimum is fixed at these angles. But due to scaling of Δώ the gain Kacc, all the points in the phase plane are relocated and sometimes system conditions may also require these points to be relocated. Hence there is a need for clockwise or anticlockwise rotation, for better tuning of the PSS. This can be done by adding or subtracting an angle "β" from phase plane angle "θ" of the polar form. The configuration of the proposed polar fuzzy logic controller is shown in Fig. 8.62.

The required control strategy is

a) In sector A (0°–90°) controls from FLC should be large positive as both scaled Δω and Δώ are positive.
b) In Sector B (315°–360°) control signal from FLC should be low positive as scaled Δω is large positive and scaled Δώ is small negative.
c) It sector – C (270°–315°) control signal from FLC should be low negative as scaled Δω is small positive and scaled Δώ is large negative.
d) In should D, E, F all the situation are completely opposite to those in sector A, B and C, respectively.

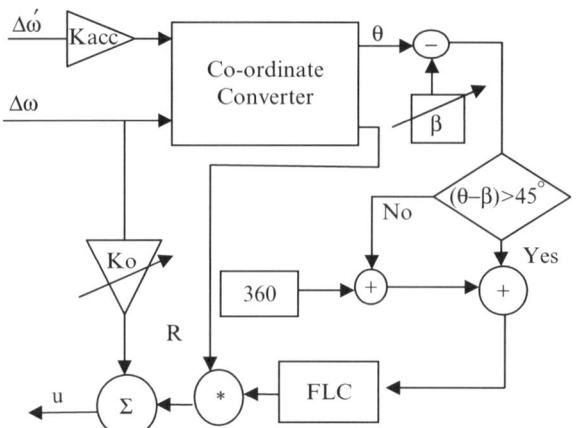

Fig. 8.62. Block diagram of proposed polar fuzzy logic controller

The output of FLC is divided into two linguistic variables variable "P" and "N", which are triangular membership function. So here only two simple rules are generated.

1. If "θ" is LP then Vo is P or mathematically:

R1 = LP × P or

R1 = [1.0/45 0.9/90 0.5/135 0.2/180 0/225 0.2/270 0.5/315 0.9/360 1.0/405]
   × [0/0.5 0.25/0.56 0.50/0.63 0.75/0.69 1.00/0.75 0.75/0.81 0.50/0.88
   0.25/0.94 0.0/1.0];

2. If θ′ is LN then Vo is N, i.e.

R2 = LN × N

R2 = [0.0/45 0.1/90 0.5/135 0.8/180 1.0/225 0.8/270 0.5/315 0.1/360 0.0/405]
   × [0/−1 0.25/−0.94 0.50/−0.88 0.75/−0.81 1.00/−0.75 0.75/−0.69
   0.50/−0.63 0.25/−0.56 0/−0.5];

These can be represented in the matrix form in Tables 8.13 and 8.14.

At an angle of 45° or 405° the value of membership function of LP is maximum and that for LN is minimum so that "Vo" is positive maximum. At an angle of 135° and 315° the value of membership function for both LP and LN is same, so that Vo is minimum (zero). At an angle of 225° the value of membership function LP is minimum and that for LN is maximum so that Vo is negative maximum. The input to FLC is angle θ′ is defined as:

$$\theta' = (\theta - \beta) + 360° \quad \text{for } \theta - \beta < 45°$$
$$\theta' = \theta - \beta \quad\quad\quad\; \text{for } \theta - \beta \geq 45°.$$

Table 8.13. Rule matrix R1

	.5	0.56	.63	0.69	.75	0.81	.88	0.94	1
45	0	0.25	0.5	0.75	1.0	0.75	0.5	0.25	0
90	0	0.25	0.5	0.75	0.9	0.75	0.5	0.25	0
135	0	0.25	0.5	0.5	0.5	0.5	0.5	0.25	0
180	0	0.25	0.5	0.75	0.8	0.75	0.5	0.25	0
225	0	0	0	0	0	0	0	0	0
270	0	0.2	0.2	0.2	0.2	0.2	0.2	0.2	0
315	0	0.25	0.5	0.5	0.5	0.5	0.5	0.25	0
360	0	0.25	0.5	0.75	0.9	0.75	0.5	0.25	0
405	0	0.25	0.5	0.75	1	0.75	0.5	0.25	0

Table 8.14. Rule matrix R2

	−1	−.94	−.88	−.81	−.75	−.69	−.63	−.56	−.5
45	0	0	0	0	0	0	0	0	0
90	0	0.1	0.1	0.1	0.1	0.1	0.1	0.1	0
135	0	0.25	0.5	0.5	0.5	0.5	0.5	0.25	0
180	0	0.25	0.5	0.75	0.8	0.75	0.5	0.25	0
225	0	0.25	0.5	0.75	1.0	0.75	0.5	0.25	0
270	0	0.25	0.5	0.75	0.8	0.75	0.5	0.25	0
315	0	0.25	0.5	0.75	0.5	0.75	0.5	0.25	0
360	0	0.1	0.1	0.1	0.1	0.1	0.1	0.1	0
405	0	0	0	0	0	0	0	0	0

Hence, the output of FLC unit is $u_{FLC} = f1(\theta, \beta)$, and final output $u = f2$ $(u_{FLC}, Ko, R, \omega_i, d\omega_i)$

where

θ – angle in degree,
β – modifier (tunning parameter) in degree,
Ko – multiplier (tunning parameter),
R – magnitude
$\Delta\omega_i$ – angular speed as GN-identifier output, rad/s.
$\Delta\dot{\omega}$ – angular acceleration as GN-identifier output, rad/s^2

8.3.4.3 Generalized Neuron Identifier

The generalized neuron developed earlier is used as the plant identifier using forward modeling is shown in Fig. 8.63. A GN identifier is placed in parallel with the system and has the following inputs:

$$X_i(t) = [\omega\_vector, \dot{\omega}\_vector, u\_vector]$$

where

$\omega\_vector = [\Delta\omega(t), \ \Delta\omega(t\text{-}T), \ \Delta\omega \ (t\text{-}2T)]$

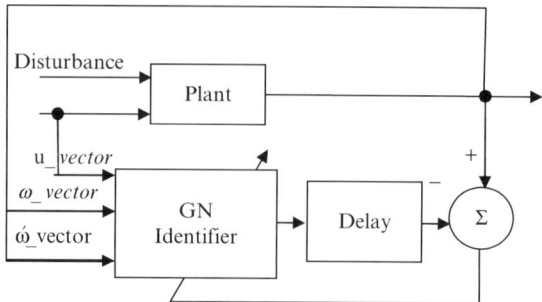

Fig. 8.63. Schematic diagram of proposed GN identifier

$\acute{\omega}\_vector = [\Delta\acute{\omega}\ (t),\ \Delta\acute{\omega}\ (t\text{-}T),\ \Delta\acute{\omega}(t\text{-}2T)]$
$u\_vector = [u(t\text{-}T),\ u(t\text{-}2T),\ u(t\text{-}3T)]$
T is the sampling period,
$\Delta\omega$ is angular speed deviation from synchronous speed, rad s^{-1},
$\Delta\acute{\omega}$ is angular acceleration rad/sec^2 and u is controller output.

The dynamics of the change in angular speed of the synchronous generator can be viewed as a non-linear mapping with the inputs mentioned in (1) and could be mathematically written as:

$$[\Delta\omega_i\Delta\acute{\omega}] = f_i(X_i(t)) \tag{8.1}$$

where f_i is a non-linear function.
Therefore, the GN-identifier for the plant can be represented by a non-linear function F_i.

$$\text{Identifier\_out} = [\Delta\omega_i\Delta\acute{\omega}] = \text{F}_i(X_i(t), W_i(t)) \tag{8.2}$$

where, $W_i(t)$ is the matrix of GN identifier weights at time instant t.

The error between the system and the GN identifier output at a fourth order delay is used as the GN identifier-training signal. The error square is used as the performance index:

$$\text{Ji(t)} = 0.5 * (\text{Delayed\_Identifier\_out} - \text{Desired\_out})^2 \tag{8.3}$$

The weights of the GN identifier are updated during training to minimize the error. After training the GN-identifier represents the plant characteristics reasonably well, i.e.

$$\omega(t+T) = f_i(X_i(t)) \approx \omega_i(t+T) = F_i(X_i(t), W_i(t)),$$

the proposed GN-identifier will be connected to the power system for on-line update of weights. The performance of the GN identifier are shown in Figs. 8.64 and 8.65.

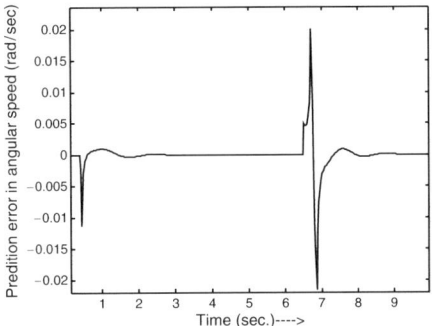

Fig. 8.64. Results of GN identifier during the line removal and reconnection of one line in a double line circuit at P = 0.9 pu and Q = 0.9 pu (lag)

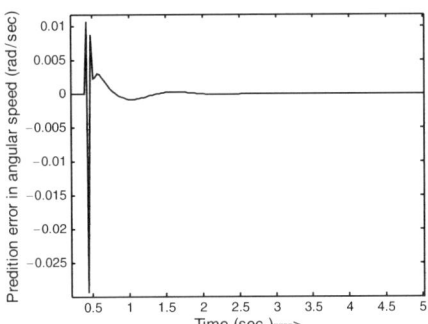

Fig. 8.65. Results of GN identifier during three phase fault of 100 ms at P = 0.5 pu and Q = 0.4 pu (lag)

8.3.4.4 Adaptive Polar Fuzzy PSS

The output of GN-identifier is passed on to the polar fuzzy controller as shown in Fig. 8.66 to make adaptive neuro-fuzzy PSS.

8.3.4.5 Simulation Results and Discussion

The proposed adaptive polar fuzzy logic based power system stabilizer (AF-PSS) is exposed to variety of operating conditions and disturbances to check its performance. The results are compared to CPSS.

A. *Normal load condition*

Under normal operating condition P = 0.9 pu and Q = 0.3 pu lag, the controllers are tested for different situations like 3pf dead short circuit transient fault for 100ms, one line removal from the double circuit line, and change in reference voltage. Simulation results of PFPSS and CPSS are shown in Fig. 8.67.

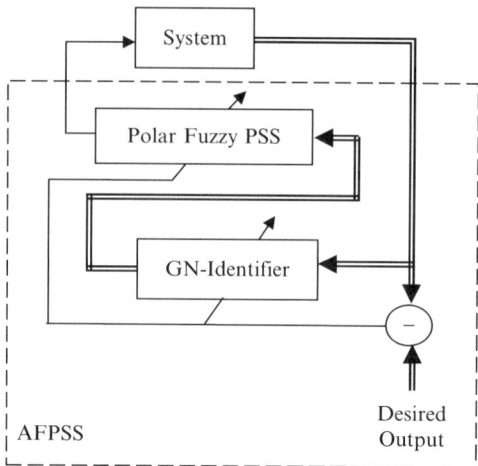

Fig. 8.66. Adaptive polar fuzzy PSS

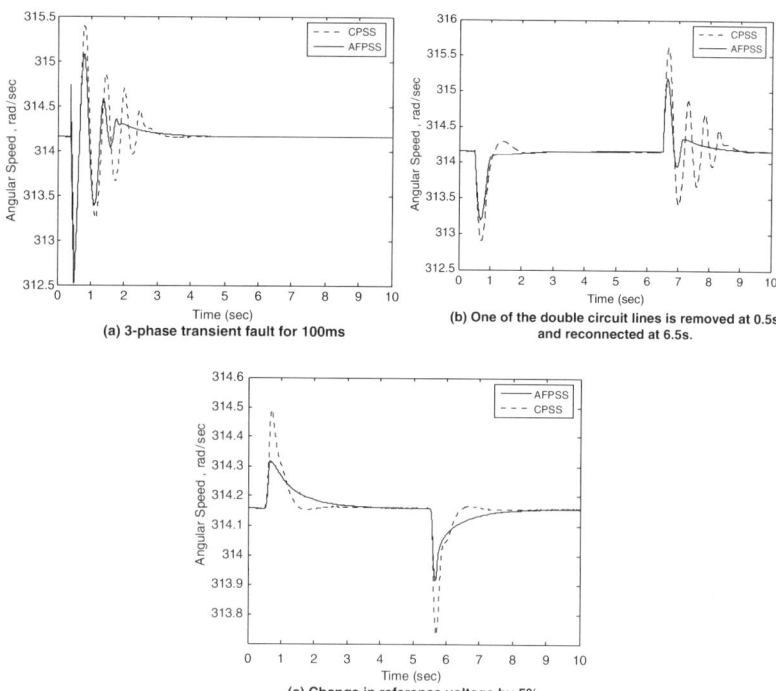

Fig. 8.67. Comparison of AFPSS and CPSS at normal load P = 0.9 and Q = 0.3 (lagging)

B. *Light load condition*

At P = 0.3 pu and Q = 0.3 pu lag and the performance of both PSSs are evaluated and compared as shown in Fig. 8.68.

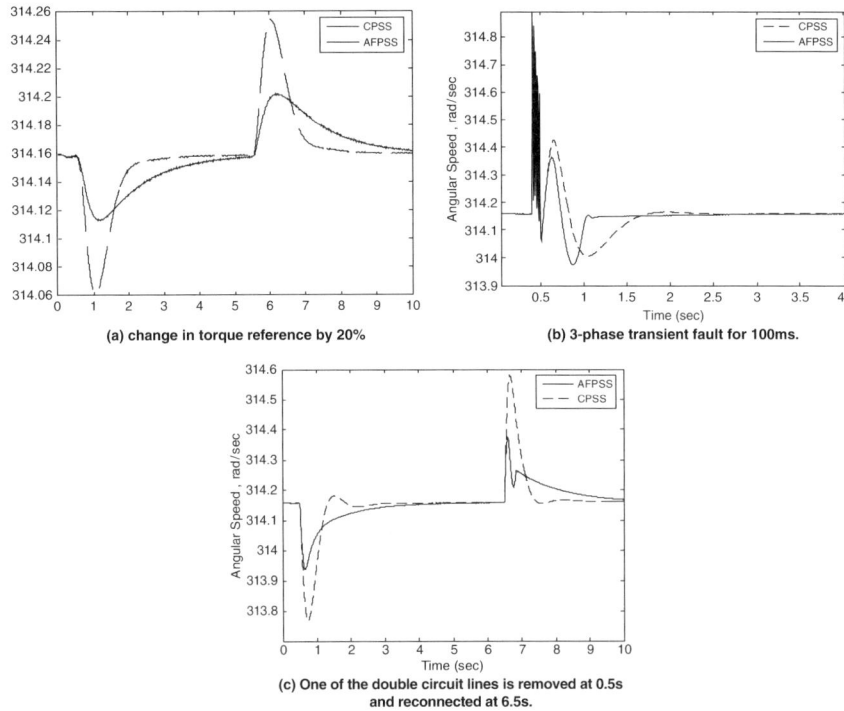

Fig. 8.68. Comparison of AFPSS and CPSS at P = 0.3 and Q = 0.3 (lag)

C. *Heavy Load condition*

The system is operating at P = 1.1 pu and Q = 0.4 pu lag the performance of the AFPSS, CPSS has been compared under these loading conditions and the results are shown in Fig. 8.69.

Under these conditions when there is 3-phase ground fault on one line of double circuit, which was removed after 220 ms by disconnecting the faulty line and reconnected at 5.5 s and it is found that the CPSS could not control it and system became unstable, but the AFPSS worked well and system stability is maintained as shown in Fig. 8.70.

8.3.4.6 Experimentation on Physical Set Up

Behavior of the proposed AFPSS has been further investigated on a physical model in the Power System Research Laboratory at the University of Calgary. The physical model consists of a three-phase 3-kVA micro-synchronous generator connected to a constant voltage bus through a double circuit transmission line. The transmission lines are modeled by six Π sections, each section is equivalent of 50 km length. The transmission line parameters are the equivalent of 1,000 MVA, 300 km and 500 kV.

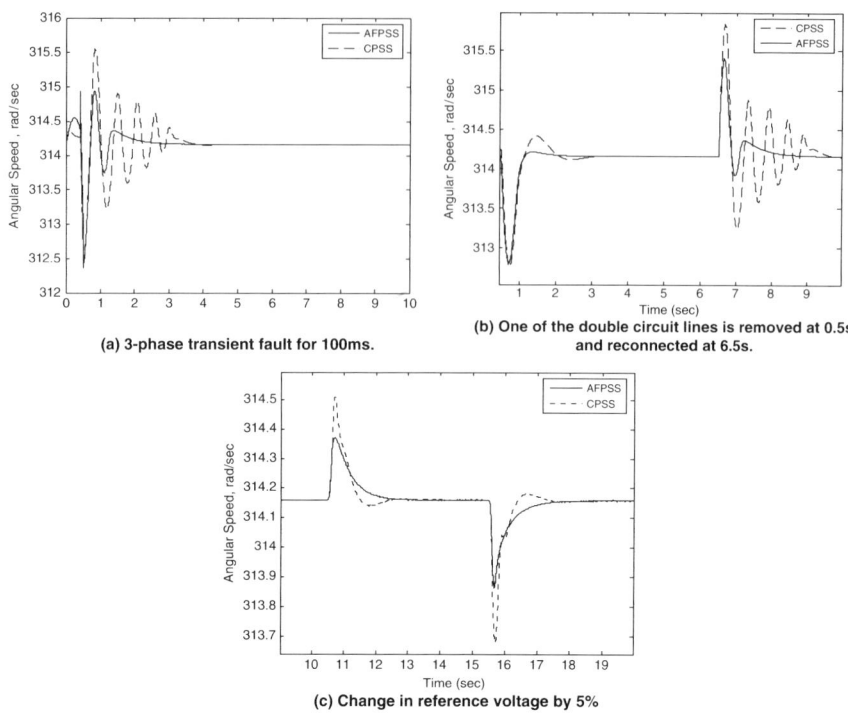

(a) 3-phase transient fault for 100ms.

(b) One of the double circuit lines is removed at 0.5s and reconnected at 6.5s.

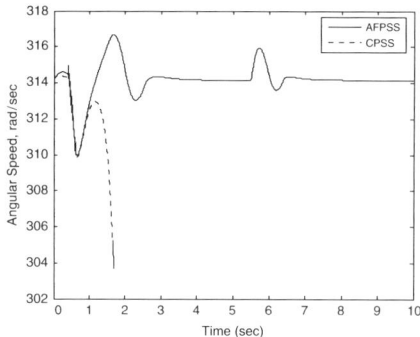

(c) Change in reference voltage by 5%

Fig. 8.69. Comparison of AFPSS and CPSS at P = −1.1 and Q = −0.4

Fig. 8.70. Comparison of AFPSS and CPSS at P = −1.1 and Q = −0.8 when there is 3-phase ground fault on one line of double circuit, which was removed after 220 ms by disconnecting the faulty line and reconnected at 5.5 s

The governor turbine characteristics are simulated using the micro-machine prime mover. It can be achieved by dc motor which is controlled as a linear voltage to torque converter. The Laboratory model mainly consists of the turbine, the generator, the transmission line model, the AVR, DAQ board and Man-machine interface.

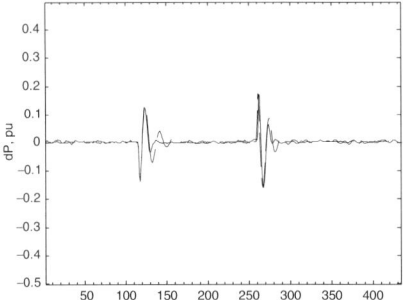

Fig. 8.71. Experimental results of AFPSS and CPSS at torque disturbance at P = 0.67 pu and pf = 0.9 (Lag)

The AFPSS control algorithm is implemented on real time window (RTW) toolbox of Matlab Ver. 6.5 with 50 ms step size. The DAQ board is installed in a personal computer with the corresponding development software. The analog to digital input channel of DSP board receives the input signal and control signal output is converted by the digital to analog converter. CPSS is also implemented on the same RTW of Matlab. The following tests have been performed on the experimental set up to study the performance of the AFPSS and CPSS.

A. *Step change in power reference (Pref)*
   A disturbance of 22% step decrease in reference power was applied and then again increased to the same initial value. The angular speed deviation (Δw) with GNPSS and CPSS is shown in Fig. 8.71. The proposed controller exhibits fast and well-controlled damping as shown by solid line in the figures.
B. *Transient faults*:
   To investigate the performance of the GNPSS under transient conditions caused by transmission line faults various tests have been conducted on the experimental set up.

1. *Single-phase to ground fault test*
   In this experiment, the generator was operated at P = 0.67 pu and 0.9 pf lead. At this operating condition and with both lines in operation, a single-phase to ground (transient) fault was applied in the middle of one transmission line for 100 ms. The system performance is shown in Fig. 8.72. It can be observed that the GNPSS provides faster settling.
2. *Two-phase to ground fault test*
   The two-phase to ground fault test was performed at P = 0.3 pu and 0.8 lagging power factor, and at the middle of one transmission line of double circuit transmission. The results of these experiments shown in Fig. 8.73 are consistently better with the AFPSS.

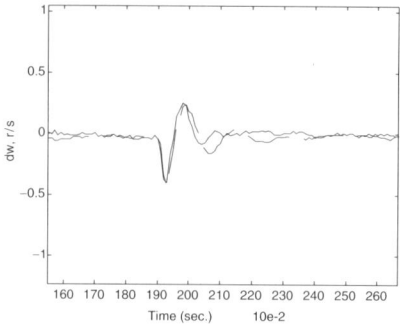

Fig. 8.72. Experimental results of AFPSS and CPSS at single phase to ground fault (transient) at P = 0.67 pu and pf = 0.9 (Lead)

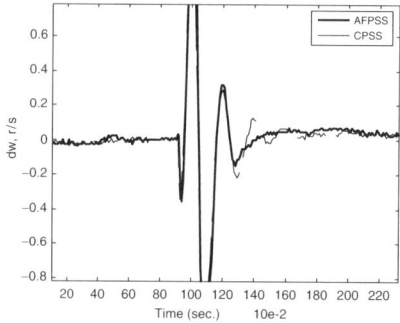

Fig. 8.73. Experimental results of AFPSS and CPSS at two-phase to ground fault (transient) at P = 0.3 pu and pf = 0.8 (Lag)

8.3.4.7 Multi-Machine System

A five-machine power system without infinite bus, that exhibits multi-mode oscillations as shown in Fig. 8.74, is used to study the performance of AFPSS. In this system, generators #1, #2 and #4 are much larger than generators #3 and #5. All five generators are equipped with governors, AVRs and exciters. This system can be viewed as a two area system connected through a tie line between buses #6 and #7. Generators #1 and #4 form one area and generators #2, #3 and #5 form another area. Parameters of all generators, loads and operating conditions are given in the Appendix. Transmission line parameters are given in (Chaturvedi 2004). Under normal operating condition, each area serves its local load and is almost fully loaded with a small load flow over the tie line.

A. *Simulation studies with AFPSS installed on one generator*
   The proposed AFPSS is installed only on generator #3 and CPSSs with the following transfer function are also installed on same generator to compare their performance.

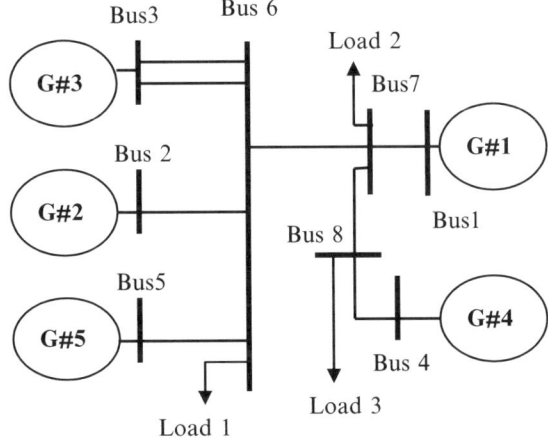

Fig. 8.74. Schematic diagram of a five-machine power system

$$u_{pss} = \frac{k_\delta}{ka} \frac{sT_5}{(1 - sT_5)} \frac{(1 + sT_1)\,(1 + sT_3)}{(1 + sT_2)\,(1 + sT_4)} \Delta P_e(s) \tag{8.4}$$

The following parameters are set for the fixed parameter CPSS for all studies in the multi-machine environment:

$K_\delta = 10, \text{T1} = \text{T3} = 0.19, \text{T2} = \text{T4} = 0.04, \text{ T5} = 10$ for G3, G5

$K_\delta = 25, \text{T1} = \text{T3} = 0.06, \text{ T2} = \text{T4} = 0.01, \text{T5} = 10$ for G1, G2 G4

Speed deviation of generator #3 is sampled at a fixed time interval of 50 ms. The system response is shown in Fig. 8.75 for the operating conditions given in the Appendix. Each part of the figure shows difference in speed between two generators. The results show that the AFPSS unable to do much on the oscillations between Generator #1 and #2 (ω_{12}), but ω_{23} and ω_{31} damped out nicely, because it is installed on Generator #3.

B. *AFPSS installed on three generators*
   In this test, AFPSSs are installed on generators #1, #2 and #3. A 10% step decrease in mechanical input torque reference of generator #3 was applied at 1 s and returns to its original level at 10 s. The simulation results of only AFPSSs and only CPSSs applied at generators #1, #2, and #3 are shown in Fig. 8.76. It is clear from the results that both modes of oscillations are damped out very effectively.

C. *Three phase to ground fault:*
   In this test, a three phase to ground fault is applied at the middle of one transmission line between buses #3 and #6 at 1 s and the faulty line is removed 100 ms later. At 10 s, the faulty line is restored successfully. The AFPSSs are installed on all five generators. The system responses

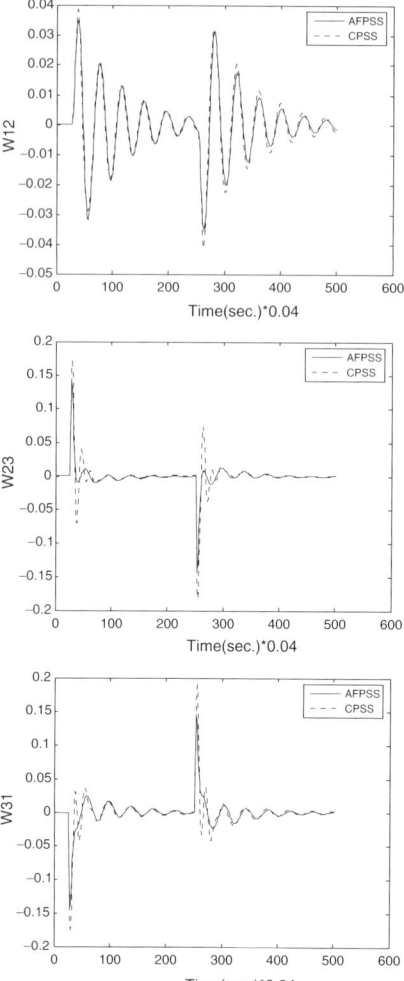

Fig. 8.75. Performance of CPSS and AFPSS when applied to Generator 3 and Reference torque is changed by 10%

are shown in Fig. 8.77. The results with CPSSs installed on the same generators are also shown in the same figures. From the system responses, it can be concluded that although the CPSS can damp the oscillations caused by such a large disturbance; the proposed AFPSS has much better performance.

D. *Coordination between AFPSS and CPSS*
The advanced PSSs would not replace all CPSSs being operated in the system at the same time. Therefore, the effect of the AFPSS and CPSSs work-

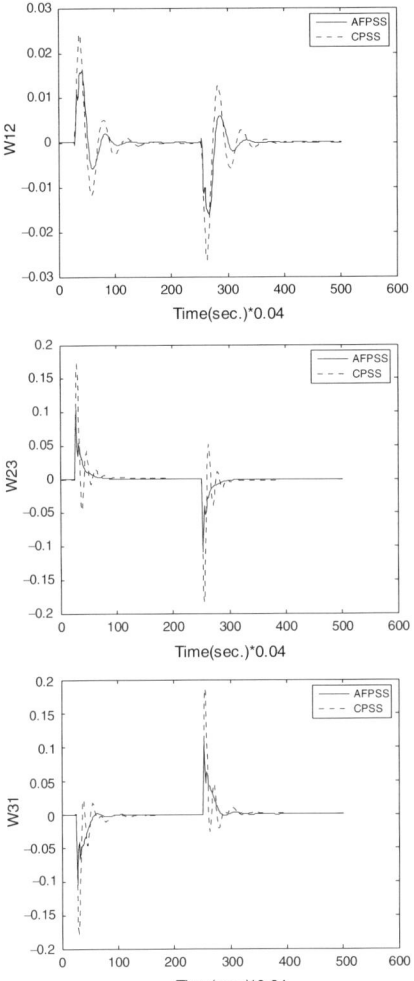

Fig. 8.76. System response under change in Tref with only AFPSS and only CPSS installed on G1, G2, and G3

ing together needs to be investigated. In this test, the proposed AFPSS is installed on generators #1 and #3 and CPSSs on generators #2, #4 and #5. The operating conditions are the same as given in the Appendix. A 0.2 pu step decrease in the mechanical input torque reference of generator #3 is applied at 1 s and returns to its original level at 10 s. The system responses are shown in Fig. 8.78. The results demonstrate that the two types of PSSs can work cooperatively to damp out the oscillations in the system. The proposed AFPSS input signals are local signals. The AFPSS coordinates itself with the other PSSs based on the system behaviour at the generator terminals.

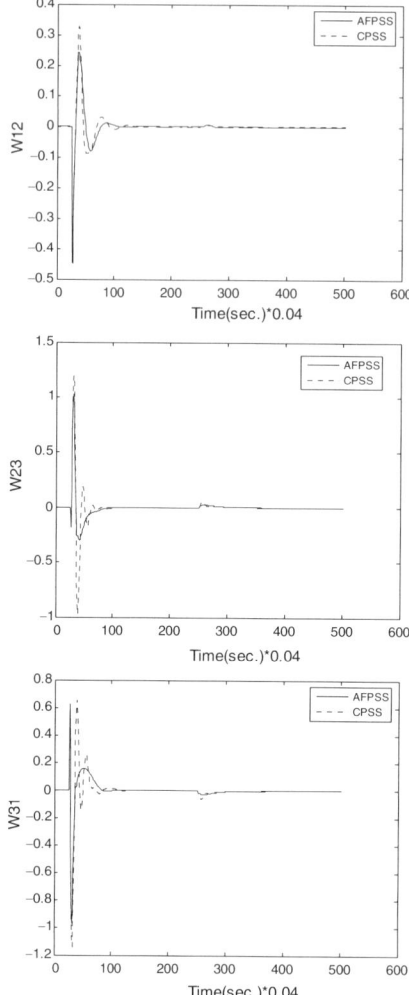

Fig. 8.77. Performance of AFPSS and CPSS when these PSS applied to all the machines and three phase fault is created at the middle of bus 3 and Bus 6

8.3.4.8 Conclusions

AFPSS has been employed to perform the function of a PSS to improve the stability and dynamic performance of the single machine infinite bus as well as a multi-machine power system. Simulation studies described in the chapter show that the performance of the AFPSS is good over a wide range of operating conditions. The effectiveness of AFPSS to damp multi-mode oscillations in a five machine power system provides satisfactory results and can cooperate with other AFPSSs or CPSSs.

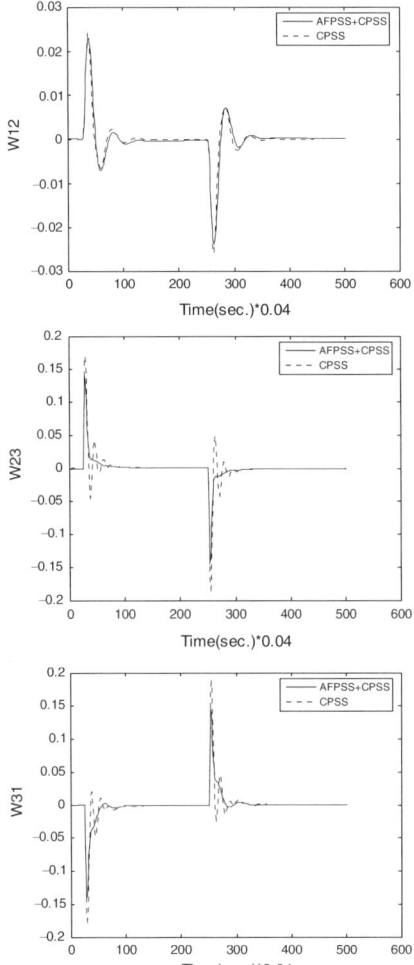

Fig. 8.78. System response with AFPSS at G1, G3 and CPSS on G2, G4, G5, for ±0.1 pu step change in torque reference

8.4 Summary

- In this chapter the Fuzzy logic system is used for the following applications:
  1. Electrical machines modeling
  2. Load frequency control problem
  3. Power system stabilizer problem
      a. Single machine infinite bus system and
      b. Multi- machine power system.
- The effect of different connectives, implications and compositional operators are studied on electrical machines modeling problem.

- The effect of different membership functions and their overlapping are also studied.
- FLC is used to control load frequency problem and the performance is compared with conventional integral controller. It is found that the fuzzy controller works better than conventional controller.
- Direct and indirect adaptive controllers are also discussed and adaptive polar fuzzy power system stabilizer is theoretically developed and experimentally tested on the physical system.
- Polar fuzzy PSS work is extended for multi-machine power system. The co-ordination aspects of fuzzy PSS and conventional PSS are explained.

8.5 Bibliography and Historical Notes

Fuzzy logic offers the option to try to model the non-linearity of the functioning of the human brain when several pieces of evidence are combined to make an inference. There are several publications available on fuzzy aggregation operators, of which a few notable ones are (Kaymak and Sousa 2003; Mendis et al. 2006; Bandemer and Gottwald 1995; Bloch 1996; Combettes 1993; Cubillo et al. 1995; Czogala and Drewniak 1984; Dubois and Prade 1985; 1992; Klir and Yuan 1995; Trillas et al. 1983; Yager 1977, 1981, 1988, 1991, 1996; Radko et al. 2000; Ai-Ping et al. 2006).

The choice of fuzzy implication as well as other connectives is an important problem in the theoretical development of fuzzy logic, and at the same time, it is significant for the performance of the systems in which fuzzy logic technique is employed. There are mainly two ways in fuzzy logic to define implication operators: (1) an implication operator is considered as the residuation of conjunction operator; and (2) it is directly defined in terms of negation, conjunction, and disjunction operators (Demirli and Turksen 1994; Ying 2002).

The different compositional operators are described in (Zeng and Trauth 2005; Wang 2006).

Fuzzy implication rules and generalized modus ponens were first introduced by Zadeh (1975). This area soon became the most active research topic for fuzzy logic researchers. Gains (1976, 1983, 1985) described a wide range of issues regarding fuzzy logic reasoning, including fuzzy implications.

Fukami, Mizumoto and Tanaka proposed a set of intuitive criteria for comparing fuzzy implications.

There is great interest in the area of hierarchical fuzzy control systems. Wang and his colleagues (1993, 1994a) did some work in this area. Similarly, auto tuning PID type controller as well as other types of adaptive fuzzy control could also be considered as hierarchical type control strategies (Wang 1994). A book edited by Terano et al. (1984) contains a chapter that describes several good applications of fuzzy expert systems. Negoita (1985) also wrote a book on fuzzy expert systems.

Pin (1993) implemented fuzzy behavioural approaches to mobile robot navigation which have been implemented using special purpose fuzzy inference chips.

Several approaches to fuzzy object-oriented data models have been proposed in the late 1990s. Cheng and Ke proposed a scheme for translating fuzzy queries into relational data bases in the late 1970s.

The literature dealing with the use of neural networks for learning membership functions or inference rules is rapidly growing; the following are a few relevant references: (Takagi and Hayashi 1991; Wang and Mandel 1992; Jang 1993; Wang 1994).

The overview of the role of fuzzy sets in approximate reasoning were prepared by Dubois and Prade (1991) and Nakanishi et al. (1993).

Most of the fuzzy logic applications are in the area of control systems. There are many research papers published in this area starting with plain fuzzy controller (Antsaklis 1993) to adaptive controller (Akbarzadeh et al. 1997; Amizadeh 1994).

Fuzzy computer hardware is also one of the most growing area. There is long list of publications in computer hardware (Gupta and Yamakawa 1988; Diamond and Kloeden 1994). An important contribution to fuzzy logic hardware is a design and implementation of a fuzzy memory to store one digit fuzzy information by Hirota and Ozawa (1989).

In the area of communication fuzzy set theoretic approach is used as fuzzy adaptive filters to non-linear channel equalization (Wang and Mandel 1993), signal detection (Saade and Schwarzlander 1994).

It is also used for analysis of Fuzzy cognitive maps (Styblinski and Meyer 1991), in robotics (Palm 1992; Nedugadi 1993; Chung and Lee 1994), vision, identification, reliability and risk analysis. Besides this Fuzzy logic is used in medical diagnosis, Ray et al. (2006) used fuzzy PI control application to load-frequency problem.

8.6 Exercises

8.1 Consider a simple first order plant model which is given by

$$\dot{x}(t) = Ax(t) + bu(t)$$

where values of A and B are dependent on system parameters.

$$u(t)\text{–input vector.}$$

Develop a fuzzy logic controller for this simple system and compare the performance with conventional PI controller.

8.2 Explain how is fuzzy controller different to conventional controller?

8.3 Is fuzzy controller non-linear? If yes, explain reasons.

8.4 Explain qualitatively the influence of membership function, cross over points and method of defuzzification on fuzzy controllers.

8.5 Discuss different types of rules used to generate the controller output.

8.6 Generate table which shows all sets of rule relating, e, □e, with u.

8.7 Discuss role of Normalization and scaling for generating set of rules for the knowledge base.

8.8 Described methods for realizing fuzzy PID controllers.

8.9 List out various steps involved in development of fuzzy logic control.

8.10 Distinguish between self organizing and fixed rules controller using fuzzy logic approach.

8.11 Explain the following rule

If PE = NB and CPE = not(NB or NM) and SE = ANY and CSE = ANY
    then   HC = PB

elseif PF = NB or NM and CPE = NS and SE = ANY and CSE = ANY
    then   HC = PM.

(a) Represent the rule using "max" and "min" operators.

(b) Represent the rule in narrative style.

9

Genetic Algorithms

"Genetic algorithms are good at taking large, potentially huge search spaces and navigating them, looking for optimal combinations of things, solutions you might not otherwise find in a lifetime."

Salvatore Mangano
Computer Design, May 1995

9.1 Introduction

Origin with a protozoa (prime unicell animal) to existence of human (most developed living being) in nature as a result of evolution, is the main theme, adopted by genetic algorithms (GA), one of the most modern paradigm for general problem solving. Since the paradigm simulates the strategy of evolution, it is surprisingly simple but powerful, domain free approach to problem solving. GAs are gaining popularity in many engineering and scientific applications due to their enormous advantages such as adaptability, ability to handle non-linear, ill defined and probabilistic problems. As the approach is domain free, it has wide scope of applications and most of the optimization problems can be handled successfully with this approach.

The emergence of massively parallel computers made these algorithms of practical interest. There are various well known programs in this class like evolutionary programs, genetic algorithms, simulated annealing, classifier systems, expert systems, artificial neural networks and fuzzy systems. This chapter discusses a genetic algorithm – which is based on the principle of evolution (survival of fittest). In such algorithms a population of individuals (potential solution) undergoes a sequence of transformations like mutation type and crossover type. These individuals strive for survival; a selection scheme, biased towards fitter individuals, selects the next generation. After some number of generations the program converges to the optimal value.

Genetic algorithm has been applied to various problems in electrical power systems such as generation scheduling (Orero and Irving 1996, 1996a, 1998;

D.K. Chaturvedi: *Soft Computing Techniques and its Applications in Electrical Engineering*, Studies in Computational Intelligence (SCI) **103**, 363–381 (2008)
www.springerlink.com

Huang 1998), Economic load dispatch (Song and Xuan 1998), reactive power optimization (Iba 1994), distribution network planning (Miranda et al. 1994), alarm Processing (Wen et al. 1998), Electrical long term load-forecasting (Chaturvedi et al. 1995) and optimal control problems. Genetic algorithms are more robust than existing directed search methods. Another important property of GA based search methods is that they maintain population of potential solutions – all other methods process a single point of the search space like hill climbing method. Hill climbing methods provide local optimum values and these values depend on the selection of starting point. Also there is no information available on the relative error with respect to global optimum. To increase the success rate in hill climbing method, it is executed for large number of randomly selected different starting points. On the other hand, GA is a multi-directional search maintaining a population of potential solutions and encourages information formation and exchange between these directions. The population undergoes a simulated evolution and at each generation the relatively good solution reproduce, while the relatively bad solutions die out. To distinguish between different solutions we use an objective function which plays the role of an environment.

9.2 History of Genetics

Genetics is a science which deals with the transfer of biological information from generation to generation. Genetics deals with the physical and chemical nature of these informations itself. Geneticists are concerned with whys and hows of these transfer of biological information, which is the basis for certain differences and similarities that are recognized in a group of living organisms. What is the source of genetic variations? How are difference distributed in populations? Why not all variations among living things however are inherited? All these are concern with genetics.

Long before human began to wonder about genetic mechanism, they already operating effectively in nature. Population of plants and animals are now known to have built in potentials for consistency and change that are dependent on genetics. Change that are established through these mechanism over long period of time in a population of living things is called EVOLUTION.

Many potential changes have been accomplished by human interventions in genetic mechanism that now accrue to benefit human beings. By selective breading, domesticated organisms have been made to serve human society increasingly better. Improve quantity and quality of milk, eggs, meat, wool, maize, wheat, rice, cotton and many other sources of food, fiber and shelter at least to the success of human intervention in genetic mechanism.

The mechanism of genetics is entirely based on gene. The gene concept however, had been implicit in model's visualization of a physical element or factor that acts as the foundation of development of a trait. He first postulated the existence of genes from their end effects, as expressed in altered characteristics. The word "allelmorph", shortened to "allele" is used to identify the member of paired genes that control different alternative traits. The gene is characterized as an individual unit of genetic mechanism. Genes replicate themselves and reproduce chromosomes, cells and organisms of their own kind. Gene is the part of chromosome. Some chromosomal genes work together, each making a small contribution to height, weight or intelligence, etc. Genes not only have a basic role in the origin and life of individual organisms, but they also, through variation in gene frequencies, cause change in populations.

Let us have a quick look at the brief history of genetics:

"The fundamental principle of natural selection as the main evolutionary principle long before the discovery of genetic mechanism has been presented by C. Darwin. He hypothesized fusion or blending inheritance, supposing that parental qualities mix together.

This theory was first time objected by Jenkins. He mentioned that there is no selection in homogenous populations. It is simply a nightmare called Jenkins nightmare.

In 1865, Gregor Johann Mandel discovered the basic principles of transference of hereditary factors from parents to offspring and explained the Jenkins nightmare. The Danish biologist Wilhelm Johannsen called these factors genes. It is now known as the genes not only transmitted hereditary traits but also mastermind the entire process of life. The genes are located in the chromosome (thread-like bodies) which are themselves situated in the nucleus of the cell. They are always found in pairs. Chromosomes vary in number according to species. The fruitfly, for example, has 4 pairs or 8 chromosomes in all, and the garden pea has 7 pairs (14 in all), mice have 20 pairs (Lawrence 1991) and humans 23 pairs (Brest et al. 2006).

Genetics was fully developed by Morgan and his collaborators. They proved experimentally that chromosomes are the main carriers of hereditary information, which later proved that Mendelian laws to be valid for all sexually reproducing organisms.

1920s Cetverikov proved that Mendel's genetics and Darwin's theory of natural selection are in no way conflicting and that their marriage yields modern evolutionary theory.

Prof. John Holland of the University of Michigan, Ann Arbor envisaged the concepts of GA algorithms and published a seminal work (Holland 1975).

There after, number of other researchers (Davis 1991; Goldberg and Holland 1989; Michalewiccz 1992) contributed to developing and improve the original GA.

9.3 Genetic Algorithms

The beginnings of genetic algorithms can be traced back to the early 1950s when several biologists used computers for simulations of biological systems (Goldberg and Holland 1989). However, the work done in late 1960s and early 1970s at the University of Michigan under the guidance of John Holland led to genetic algorithms as they are known today. GA vocabulary is being borrowed from natural genetics. The idea behind genetic algorithms is to do what nature does. Genetic algorithms (GAs) are stochastic algorithms whose search methods inspired from phenomena found in living nature. The phenomena incorporated so far in GA models include phenomena of natural selection as there are selection and the production of variation by means of recombination and mutation, and rarely inversion, diploid and others. Most Genetic algorithms work with one large panmictic population, i.e. in the recombination step each individual may potentially choose any other individual from the population as a mate. Then GA operators are performed to obtain the new child offspring; the operators are:

  i. Selection
 ii. Crossover,
iii. Mutation,
 iv. Survival of fittest (Heistermann 1990; Michalewiccz 1992; Muzhlenbein 1989; Holland 1973; Nowack and Schuster 1989).

9.3.1 Selection

As in natural surroundings it holds on average: "the better the parents, the better the offsprings" and "the offspring is similar to the parents". Therefore, it is on the one hand desirable to choose the fittest individuals more often, but on the other hand not too often, because otherwise the diversity of the search space decreases (Braun 1990). GA researchers have developed a variety of selection algorithms to make it more efficient and robust. In the implementation of genetic algorithm the best individuals have been select using roulette wheel with slot sized according to fitness, so that the probability of selection of best strings are more as shown in Fig. 9.1a. Besides the roulette wheel selection, researchers have developed a variety of selection algorithms like proportionate selection, linear rank selection, tournament selection and stochastic remainder selection.

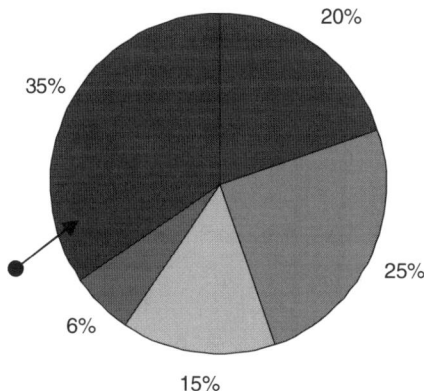

Fig. 9.1a. Roulette Wheel Selection

9.3.1.1 Roulette Wheel Selection

In roulette selection chromosomes are selected based on their fitness relative to all other chromosomes in the population as shown in Fig. 9.1a. One disadvantage of using roulette wheel is that its selective pressure reduces as population converges upon a solution, which reduces the convergence rate and may not allow finding the better solutions.

% Matlab sub-routine for roulette wheel selection

```
function x=roul(num_to_be_sel,popsize,pop,f);
        f=f/sum(f);
        roul_wheel(1)=f(1);
        for i=2:popsize
                roul_wheel(i)=roul_wheel(i-1)+f(i);
        end;
% create n=num_to_be_sel random numbers for roulette
        selection : r(i)
        r=rand(num_to_be_sel,1);
% determine selected strings according to roul_wheel :
sel_str(popsize,lchrom)
        for i=1:num_to_be_sel
          flag=1;
          for j=1:popsize
                if (r(i)<=roul_wheel(j) & flag==1)
                    x(i,:)=pop(j,:);
                    flag=0;
                end;
            end;
        end;
```

9.3.1.2 Tournament Selection

In this process of selection, one parent is selected by randomly comparing other individual in the same population and select with the best fitness. To select the second parent the same process is repeated. It is most popular selection method due to its simplicity (Baker 1987).

9.3.1.3 Linear Rank Selection

In this method the individuals are ordered according to their fitness values (Grefenstette 1986). The individuals of highest fitness are kept on the top and worst on the bottom. Then each individual in the population is assigned a subjective fitness based on linear ranking function as

f(r)=(popsize-rank)(max-min)/(Popsize-1)+min
where popsize – population size
           rank – rank in the current population
           max, min – maximum and minimum subjective fitness determined by the user.

Now this subjective fitness value is assigned to the individual and the selection is done on the basis of roulette wheel spinning. In this selection process the selective pressure is constant and does not change with generation to generation. However, in this process, it is necessary to sort the population according to their fitness values and the individuals of same fitness will not have the same probability of being selected.

9.3.2 Crossover

Obviously, selection alone can not generate better offsprings. To produce better new off springs a crossover operator is required. A crossover operator can be termed loosely as recombination or slice-exchange-merge operator. The most common type of crossover operator mentioned above is called single point crossover. In this operation select two parents and randomly selects a point between two genes to cut both chromosomes into two parts. This point is called crossover point. In crossover operation combine the first part of first parent and second part of second parent to get first offspring. Similarly, combine the first part of second parent and second part of first parent to get second offspring. These offsprings belong to the next population. The crossover operator has three distinct substeps:

  a. Slice each of the parent strings in two substrings.
  b. Exchange a pair of corresponding substrings of parents.
  c. Merge the two respective substrings to form offsprings.

For example, suppose following two binary strings are mated together and undergoes the crossover operation. The strings are 1100000, 0101111. By a random choice the crossover site is fixed at 3 which is shown by a vertical bar. Then the effect of cross over will be as shown in Fig. 9.1b.

To increase the speed of convergence of GA, the population is divided from the middle and two halves (subgroups) are used in group cross over as shown in Fig. 9.1 c. Another type of crossover is multi-point cross over, in which two or more than two sites have been selected and exchange have been done as illustrated in Fig. 9.1d.

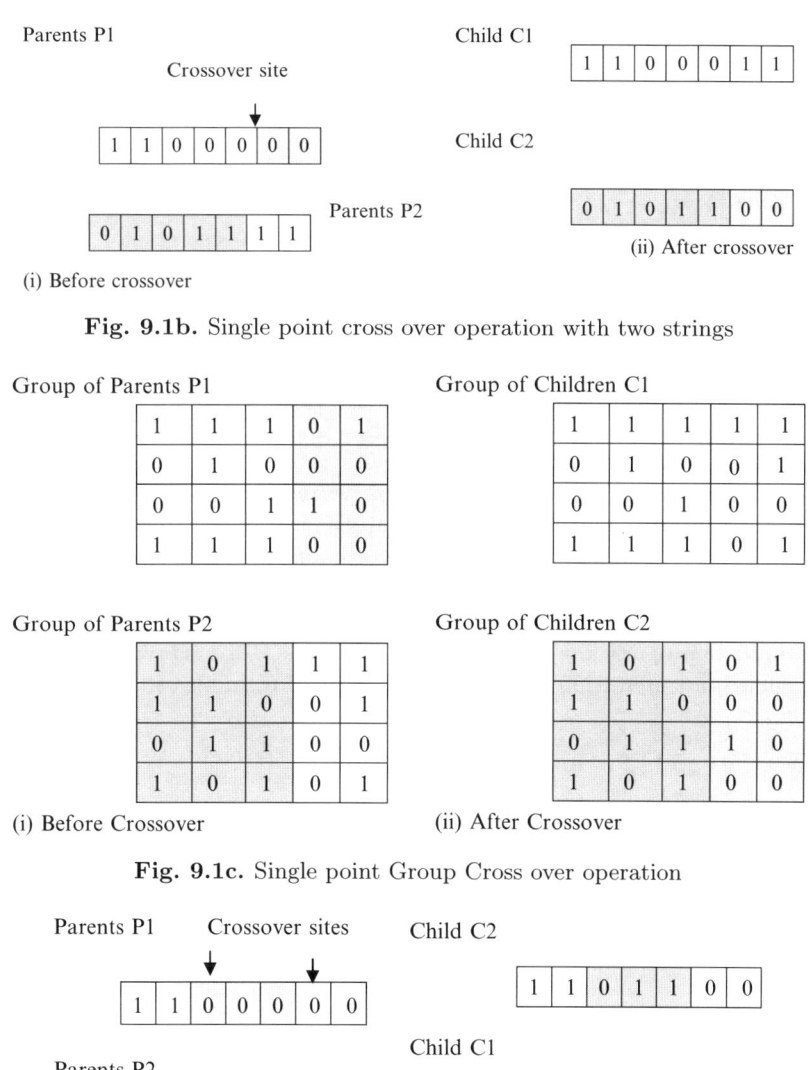

Fig. 9.1b. Single point cross over operation with two strings

Fig. 9.1c. Single point Group Cross over operation

Fig. 9.1d. Multipoint crossover

% Matlab code for single point crossover operation

```
j2=2*i;
j1=j2-1;
a=rand(1);                        % Random number generation.
site=round(a*(lchrom-2)+1); % Random selection of crossover site
temp=sel_str(j1,site:lchrom);   % lchrom - length of cromosome
sel_str(j1,site:lchrom)=sel_str(j2,site:lchrom);
sel_str(j2,site:lchrom)=temp;
```

9.3.3 Mutation

The newly created individuals have no new inheritance information and the
number of alleles is constantly decreasing. This process results in the con-
traction of the population to one point, which is only wished at the end of
the convergence process, after the population works in a very promising part
of the search space. Diversity is necessary to search a big part of the search
space. It is one goal of the learning algorithm to search always in regions not
viewed before. Therefore, it is necessary to enlarge the information contained
in the population. One way to achieve this goal is mutation. The mutation
operator M (chromosome) selects a gene of that chromosome and changes the
allele by an amount called the mutation variance (mv), this happens with
a mutation frequency (mf). The parameter mutation variance and mutation
frequency have a major influence on the quality of learning algorithms. For
binary coded GAs mutation is equivalent of flipping a bit at any particular
position. Since, mutation is to be used sparingly its probability is very low.
The mutation operation may be shown as in Fig. 9.1e. The group mutation
and multipoint mutation may also be performed to improve the results.

% Matlab code for mutation operation

```
% sel_str is now the intermediate pop for mutation
for i=1:popsize
        for j=1:lchrom
                if (flip(pmute)==1)
                        if(sel_str(i,j)==0)
                                sel_str(i,j)=1;
                        else
                                sel_str(i,j)=0;
                        end;
                end;
        end;
end;
function y=flip(prob)
        a=rand(1);
```

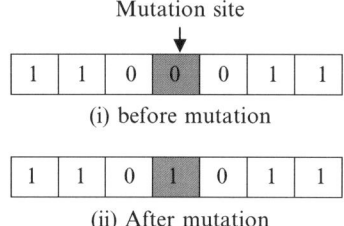

Fig. 9.1e. Single point mutation operation

```
if (a <=prob)
        y=1;
else
        y=0;
end;
```

9.3.4 Survival of Fittest

Further more we only accept an offspring as a new member of population, if it differ enough from the other individuals, that means here its fitness differ from all other individuals at least by some significant amount. After accepting a new individual we remove one of the worst individual (i.e. its fitness value is quite low) from the population in order to hold the population size constant.

To maximize the efficiency of GAs, three inherent parameters of GAs are to be optimized, the mutation probability **Pm**, the crossover probability **Pc**, and the population size **POPSIZE**. For GA parameter optimization several results have been obtained over the last few years. DeJong and Schuster proposed heuristics for an optimal setting of the mutation probability **Pm** (Nowack and Schuster 1989; Schuster 1985), Fogarty and Booker investigated time dependencies of the mutation and the crossover probability respectively (Fogarty 1989), Greffenstette Schaffer and Jong found optimal settings for all three parameters of the GA by experiment (Greffensette 1986; Schaffer et al. 1989; De Jong and Spears 1990). The brief description of these parameters are given below:

Duplicates

Individuals that represent the same candidate solution are known as duplicate individuals. It has been mentioned (Davis 1991) that eliminating duplicates increases the efficiency of a genetic search and reduces the danger of premature convergence.

9.3.5 Population Size

A group of individuals (chromosome) collectively comprise is known as population. Population size is the number of individuals (chromosome) in the population maintained by a GA. As discussed by De Jong and Spears (1990) [30]

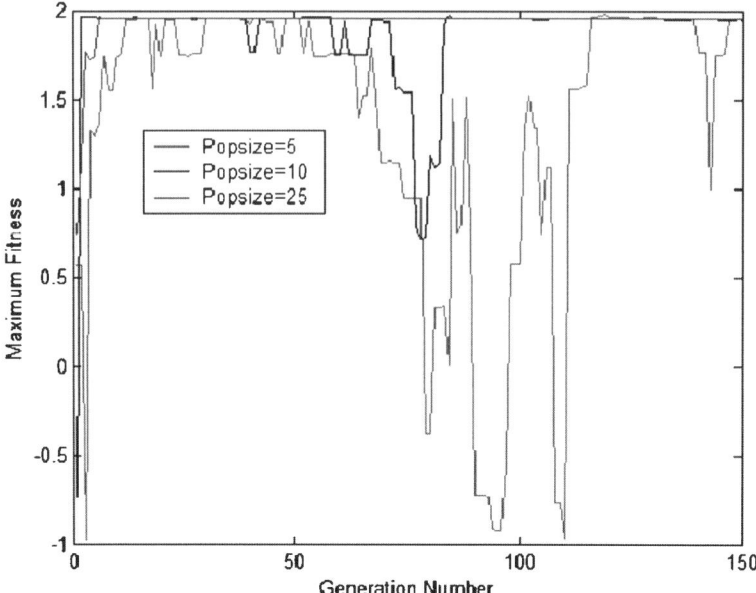

Fig. 9.2. Effect of population size on maximum fitness

that the choice of population size has a strong interacting effect on the results. Smaller population size tends to become homogeneous more quickly and there is a danger of premature convergence upon a suboptimal solution. With large population size the crossover productivity effect is much less dramatic, hence takes longer time to converge upon a solutions.

Usually the population size for GA varying from tens to thousands and it is noted that this parameter is mostly problem dependent. If the problem in hand is simpler then smaller population size can also serve the purpose, but if the problem is complex, large population size is required and it is also necessary to run for large number of generations.

The effect of population size on maximum fitness value of GA is shown in Fig. 9.2. Form the figure it is clear that the GA performance is good for population size 50, 80 and 100. The optimal performance of GA is obtained at popsize equals to 50. The average fitness is also compared for different population sizes as shown in Fig. 9.3.

9.3.6 Evaluation of Fitness Function

The evaluation function of a GA is used to determine the fitness of chromosomes in the population. The binary coded chromosomes also known as a genotype. To find the fitness of binary coded chromosomes, they must be decoded first and then evaluated the fitness. But in case of real coded chromosome which is also called as phenotype and for them, there is no need of decoding is required.

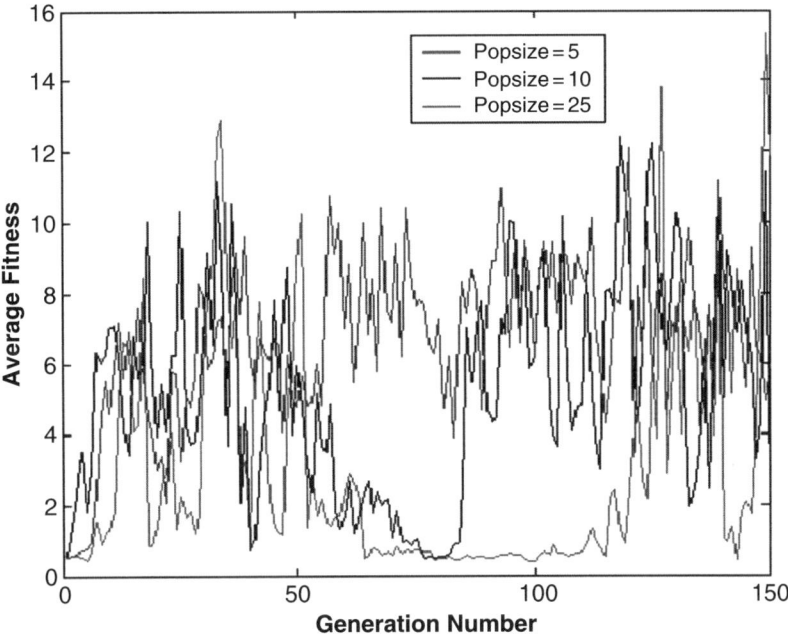

Fig. 9.3. Effect of population size on average fitness

9.4 Effect of Crossover Probability on GA Performance

For better results, it is advisable to select the crossover rate quite large than mutation rate. This is the usual practice to take crossover rate 20 times greater than the mutation rate. Crossover rate generally ranging from 0.25 to 0.95. The effect of crossover probability (pcross) on GA performance in terms of average fitness is shown in Fig. 9.4.

9.5 Effect of Mutation Probability on GA Performance

Schaffer (Mbamalu and Hawary 1993) found experimentally that mutation probability (Pm) is approximately inversely proportional to the population size. Mutation rate generally varying from 0.001 to 0.05. The effect of mutation rate is shown in Fig. 9.5. If the mutation rate is high then there are more fluctuations in the fitness value. On the other hand if the mutation rate is low then it the search area is reduced. Hence, the optimal value of mutation rate is selected for good performance of GA or one can dynamically change its value.

Maximum number of generations

The selection of maximum number of generations is a problem dependent parameter. For complex problems, the maximum number of generations is large enough, so that the results should converge to optimal value (Greffensette 1986).

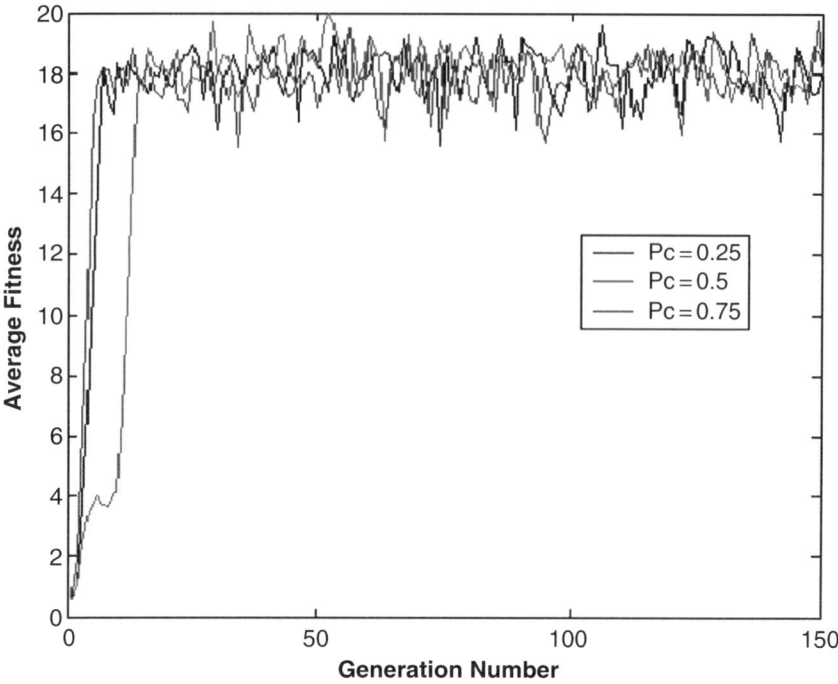

Fig. 9.4. Effect of crossover probability on average fitness

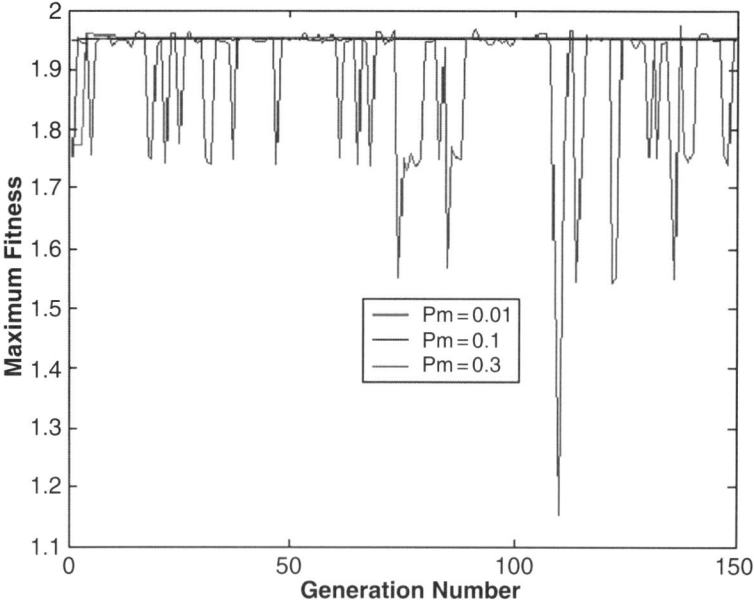

Fig. 9.5. Effect of mutation probability on maximum fitness

Length of chromosome (lchrome)

The value of lchrome is dependent to the precision required and can be calculated with the help of the following expression –

$$2^{lchrome} = (\max parm - \min parm) * 10^r$$

Where, r is number of places after decimal, up to which the precision is required.

Max parm – Upper bound of parameter
Min parm – Lower bound of parameter

9.6 Main Components of GA

A GA (or any evolutionary program) for a particular problem must have the following five components:

1. A genetic representation for potential solutions to the problem (Coding).
2. A way to create an initial population of potential solution.
3. To evaluate rank of a solution define an objective function.
4. To alter the composition of offspring's define genetic operators.
5. Define GA parameters like population size, probabilities of genetic operators, etc.

Coding

In order to solve any problem with genetic algorithms, variables are first coded in some string structures. There are some of the studies in which directly variables values are taken, but most of the GAs work with binary coded variable strings (Fig. 9.6).

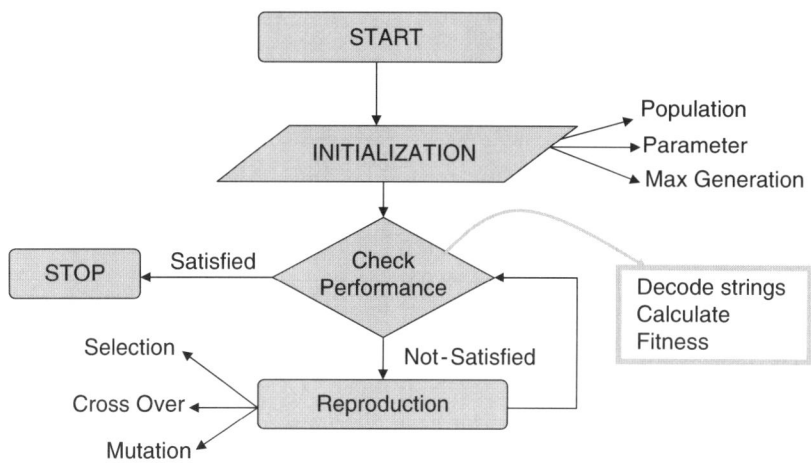

Fig. 9.6. Flow chart of simple genetic algorithm

Example 1. Minimize the surface area of a cylindrical closed end container, with the following constraints:

   i. Volume is $5\,m^3$
  ii. Radius of cylinder is not less than $0.5\,m$ and not more than $1.0\,m$.

Solution

A. *Problem formulation*

It is a double variable optimization problem, which could be modified into a single variable optimization problem.

$$\text{Volume of container V} = \pi\ r^2 l = 5m^3$$
$$L = 5/\pi\ r^2$$
$$\text{Surface area A} = 2\pi\ r^2 + 2\pi r\ l$$
$$= 2\pi\ r^2 + 2\pi\ r\ (5/\pi\ r^2)$$
$$= 6.28\ r^2 + 10/r$$

B. *Implement GA code in MATLAB*
C. *Given*
    Upper bond ub $= 1.0$
    Lower bond lb $= 0.5$
D. *Data preparation*
    i. Size of population popsize $= 30$
   ii. Length of chromosome lchrom is determined as:

$$2^{\text{lchrom}} \geq (\text{ub} - \text{lb}) * 10^{\text{dp}}$$

If the precision required upto four places after decimal dp $= 4$.
Hence, lchrom $= 13$
  iii. Since it is a minimization problem and genetic algorithms evaluate the strings on the basis of fitness function, which is maximization function.
Objective function $=$ C $-$ A.
Where C is a constant, whose value is more than the area A.

$$= 50 - A$$

   iv. Number of maximum generation $= 10$
    v. Cross over probability Pc $= 0.5$
   vi. Mutation probability Pm $= 0.01$
E. Generate initial population

Generate a random binary matrix of size (popsize $\times$ lchrom ($= 30 \times 13$)).

F. Determine the fitness value for each chromosome in the population.
G. If the fitness is equal to some specified value then stop, otherwise perform GA operations (Table 9.1).

Table 9.1. Fitness values in different generations

Generation number	Maximum Fitness value	Minimum Fitness value
0	33.8156	33.2806
1	33.8158	33.3760
2	33.8158	33.3855
3	33.8158	33.4126
4	33.8158	33.4576
5	33.8158	33.4853
6	33.8158	33.5147
7	33.8158	33.6076
8	33.8159	33.5559
9	33.8159	33.6155
10	33.8159	33.6349

Overall maximum fitness value = 33.8159

Hence, the minimum surface area A = 50 − Maximum fitness = 50 − 33.819 = 16.1841 m^2

Alternate method

The calculus method may also be used to solve the above mentioned problem, because there is one variable in the objective function. The function which is to be minimized is

$$\text{Surface area A} = 6.28r^2 + 10/r$$

Differentiate the ara A with respect to r and equate it to zero.

$$\partial A/\partial r = 0$$
$$6.28 * 2\,r - 10/r^2 = 0$$
$$r^3 = 10/(6.28 * 2) = 5/6.28$$
$$r = 0.9268$$

Substitute the value of r in the expression of surface area.

$$A_{min} = 6.28 * (0.9268)^2 + (10/0.9268)$$
$$= 16.1840\,m^2$$

Hence it is very clear that the GA results are quite close to the results calculated from the calculus method. The calculus method is good for small size problems, but if the problem size is large and complex or large number of variables. Then calculus method may give good results.

9.7 Variants

The simplest algorithm represents each chromosome as a bit string. Typically, numeric parameters can be represented by integers, though it is possible to use floating point representations. The floating point representation is natural to evolution strategies and evolutionary programming. The notion of real-valued genetic algorithms has been offered but is really a misnomer because it does not really represent the building block theory that was proposed by Holland in the 1970s. This theory is not without support though, based on theoretical and experimental results. The basic algorithm performs crossover and mutation at the bit level. Other variants treat the chromosome as a list of numbers which are indexes into an instruction table, nodes in a linked list, hashes, objects, or any other imaginable data structure. Crossover and mutation are performed so as to respect data element boundaries. For most data types, specific variation operators can be designed. Different chromosomal data types seem to work better or worse for different specific problem domains.

When bit strings representations of integers are used, gray coding is often employed. In this way, small changes in the integer can be readily effected through mutations or crossovers. This has been found to help prevent premature convergence at so called *Hamming walls*, in which too many simultaneous mutations (or crossover events) must occur in order to change the chromosome to a better solution.

Other approaches involve using arrays of real-valued numbers instead of bit strings to represent chromosomes. Theoretically, the smaller the alphabet, the better the performance, but paradoxically, good results have been obtained from using real-valued chromosomes. A very successful (slight) variant of the general process of constructing a new population is to allow some of the better organisms from the current generation to carry over to the next, unaltered. This strategy is known as *elitist selection*.

Parallel implementations of genetic algorithms come in two flavours. Coarse grained parallel genetic algorithms assume a population on each of the computer nodes and migration of individuals among the nodes. Fine grained parallel genetic algorithms assume an individual on each processor node which acts with neighboring individuals for selection and reproduction. Other variants, like genetic algorithms for online optimization problems, introduce time-dependence or noise in the fitness function.

It can be quite effective to combine GA with other optimization methods. GA tends to be quite good at finding generally good global solutions, but quite inefficient at finding the last few mutations to find the absolute optimum. Other techniques (such as simple hill climbing) are quite efficient at finding absolute optimum in a limited region. Alternating GA and hill climbing can improve the efficiency of GA while overcoming the lack of robustness of hill climbing.

A problem that seems to be overlooked by GA-algorithms thus far is that the natural evolution maximizes mean fitness rather than the fitness of the individual (the criterion function used in most applications).

An algorithm that maximizes mean fitness (without any need for the definition of mean fitness as a criterion function) is Gaussian adaptation, provided that the ontogeny of an individual may be seen as a modified recapitulation of evolutionary random steps in the past and that the sum of many random steps tend to become Gaussian distributed (according to the central limit theorem).

This means that the rules of genetic variation may have a different meaning in the natural case. For instance – provided that steps are stored in consecutive order – crossing over may sum a number of steps from maternal DNA adding a number of steps from paternal DNA and so on. This is like adding vectors that more probably may follow a ridge in the phenotypic landscape. Thus, the efficiency of the process may be increased by many orders of magnitude. Moreover, the inversion operator has the opportunity to place steps in consecutive order or any other suitable order in favour of survival or efficiency. (See for instance (Mahalanbis et al. 1991)).

Gaussian adaptation is able to approximate the natural process by an adaptation of the moment matrix of the Gaussian. Gaussian adaptation may serve as a genetic algorithm replacing the rules of genetic variation by a Gaussian random number generator working on the phenotypic level. Population-based incremental learning is a variation where the population as a whole is evolved rather than its individual members.

9.8 Applications of Genetic Algorithms

GA is not only used for solving optimization problems, but there are number of GA applications as mentioned below:

1. Industrial design by parameterization
2. Scheduling problems such as manufacturing, facility scheduling, allocation of resources, etc.
3. System design
4. Time series prediction
5. Data base mining
6. Control system
7. Artificial life system
8. Various medical applications, such as image segmentation and modeling
9. Combinatorial optimization problems like travelling sales man problem, routing, bin packing, graph partitioning and colouring.
10. Trajectory planning of robots
11. Game playing like chase playing, prisoner's dilemma, etc.
12. Resource allocation problem
13. Graph colouring and partitioning, etc.

9.9 Summary

During the last few decades there has been growing interest in natural process based algorithm. In this chapter, we provided a brief introduction to the field of evolutionary computing and an overview of genetic algorithms (GAs).GA

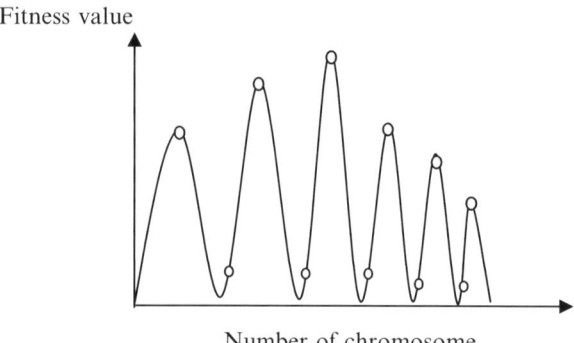

Fig. 9.7. Multiple solutions of GA for single population

describes the behaviour of genetic search. GA has a capability to provide multiple solutions for a given problem as shown in Fig. 9.7. These solutions (fitness value) improve from generation to generation.

9.10 Bibliography and Historical Notes

Some of good books on genetic algorithms are written by Goldberg and Holland (1988, 1989) and Lozano et al. 2006. The practical aspects and applications of genetic algorithms are given in Handbook of Genetic Algorithms edited by Davis (1991). The classical papers on genetic algorithms are given in Genetic Algorithms, edited by Buckles and Petry (1992) and Auger and Hansen (2005a,b) and Srinivas and Patnaik (1994a). Yaochu J. and Branke (2005) wrote a very lucid survey paper on evolutionary optimization. Albert et al. (2005) Hybrid Optimization Approach for a Fuzzy Modelled Unit Commitment Problem. Bath et al. (2007) had optimized the security constrained multi-objective optimal power dispatch.

Genetic programming (GP) is very computationally intensive and so in the 1990s it was mainly used to solve relatively simple problems. But more recently, thanks to improvements in GP technology and to the exponential growth in CPU power, GP produced many novel and outstanding results in areas such as quantum computing, electronic design (Garrison et al. 2006), game playing, sorting, searching (Smith 2002) and many more. GP has also been applied to evolvable hardware as well as computer programs (Eiben and Smith 2003; Fogel 2000; Auger and Hansen 2005a,b).

9.11 Exercises

1. What do you understand by genetic algorithms?
2. How does genetic algorithm work?
3. What do you mean by crossover and muation operations in GA. Write Matlab codes for these operations.

4. Mention different types of crossover operations and compare them.
5. Write Peuso codes for simple GA and implement simple GA using Matlab and study the effect of chromosome length, crossover and muation rates for the minimization of ocnnection length on a printed circuit board.
6. Simulated the above problem with different population size and compare the results.

10

Applications of Genetic Algorithms to Load Forecasting Problem

Evolutionary programs are gaining popularity in many engineering and scientific applications due to their enormous advantages such as adaptability, ability to handle non-linear, ill-defined and probabilistic problems. Specific reference to genetic algorithms (Gas), some parameters that influence the convergence to the optimal value are the population size (popsize), the crossover probability (Pc) and the mutation propability (Pm). Normally these values are prescribed initially and do not vary during the execution of the program, although these parameters greatly affect the performance of GA.

The present chapter deals with the development of an improved genetic algorithm (IGA) by introducing a variation in the values of the parameters like population size (popsize), the crossover probability (Pc) and the mutation propability (Pm). The aim of this variation is to minimize the convergence time. This work presents a method of dynamically varying the parameters of operation of the GA program using fuzzy state theory (FST) so that the final convergence is obtained in a shorter time.

Also, in this chapter a function has been developed and optimized for long-term load forecasting problem using IGA. This technique does not require any previous assumption of a function for load forecasting, further, it does not need any functional relationship between dependent and independent variables. The results obtained by this technique are compared with the data available from central electricity authority (CEA), India to demonstrate the effectiveness of the proposed technique.

10.1 Introduction

During the last few years there has been a growing interest in algorithms that rely on analogies to natural processes. The emergence of massively parallel computers made these algorithms of practical interest. There are various well-known programs in this class like evolutionary programs, genetic algorithms, simulated annealing, classifier systems, artificial neural networks and fuzzy systems, etc.

D.K. Chaturvedi: *Soft Computing Techniques and its Applications in Electrical Engineering*, Studies in Computational Intelligence (SCI) **103**, 383–402 (2008)
www.springerlink.com

This chapter discusses a genetic algorithm – which is based on the principle of evolution (survival of fittest). In such algorithm a population of individuals (potential solution) undergoes a sequence of transformations like mutation type and crossover type. These individuals strive for survival; a selection scheme, biased towards fitter individuals, selects the next generation. After some number of generations the program converges the optimal value. The rate of convergence the optimal value is dependent on the values of various GA parameters and the problem in hand (Deb 1995).

In the present work, it has been suggested that the convergence is not the optimal for any fixed set of GA parameter values for all the problems. Dynamic control of the parameters values should lead to improvement in the performance of the genetic algorithm and converged results should be available in lesser time. An improved genetic algorithm (IGA) program has been developed which controls the GA parameters (i.e. population size (popsize), crossover probability (Pc) and mutation probability (Pm)) during execution using fuzzy set theoretic approach. The IGA developed above has been used to model long term load forecasting problem and compared the results obtained with the load given by CEA.

10.2 Introduction to Simple Genetic Algorithms

Genetic algorithm (GAs) are inspired from phenomena found in living nature. The phenomena incorporated so far in GA models include phenomena of natural selection as there are selection and the production of variation by means of recombination and mutation, and rarely inversion, diploid and others. Most genetic algorithms work with one large population, i.e. in the recombination step each individual may potentially choose any other individual from the population as a mate. Then GA operators are performed to obtain the new individuals (chromosome vectors). The operators are selection, crossover, mutation and survival of fittest (Booker 1987; Braun 1990; De Jong et al. 1990; Heistermann 1990; M{u}hlenbein 1989).

The genetic algorithm involves in finding a mathematical function (expression) in symbolic form, that provides a good, best or perfect fit between a given finite sampling of values of independent variables and the associated values of the dependent variables. Unlike the conventional linear, quadratic or polynomial regression, which merely involve finding the numeric coefficients, GA used for finding both the functional form and the numeric coefficients for modeling the long-term load-forecasting problem.

A GA performs a multidirectional search by maintaining a population of potential solutions and encourages information and exchange between these directions. The population undergoes a simulated evolution: at each generation the relatively "good" solution reproduces, while the relatively "bad" solution dies. To distinguish between different solutions we use an objective (evaluation) function which plays the role of an environment.

During any iteration n, a GA maintains a population of potential solutions (chromosome vectors),

$$P(n) = \{X_1{}^n, X_2{}^n, X_3{}^n, X_4{}^n, \ldots \ldots \ldots \ldots X_{popsize}{}^n\}$$

where popsize – the population size and

$X_i{}^n$ – chromosome vectors (i.e. binary strings or symbolic functions)
n – generation number (varying from $1, 2, 3, \ldots \ldots$ maxgen.)

Each $X_i{}^n$ is decoded and then evaluated to give some measure of its "fitness". Then selecting the more fit individuals to form a new population (next iteration $n+1$). While getting a new population some $X_i{}^n$ directly taken if their expected count is greater than one. Remaining members of $X_i{}^{n+1}$ of new population is obtained by selecting the chromosome vectors from the roulette wheel which is developed according to the decimal part of expected count of $X_i{}^n$ undergoes reproduction by means of crossover and mutation. The expected count is defined as:

$$\text{Expected count } = (popsize/sum(H))^*H;$$

where H is the fitness value of objective function.

In this section, the crossover, mutation operations and population size are described in brief.

10.2.1 Crossover Operation

The crossover operation in the creation of two new individuals (functions) out of each pair of parent individuals (functions) of the current population of individual (Ichalewiccz 1992). An individual can be viewed as a function on chromosome level, e.g.

F1 (x,y) = (x+3)/(y−4) + (4−x)*(7+y)

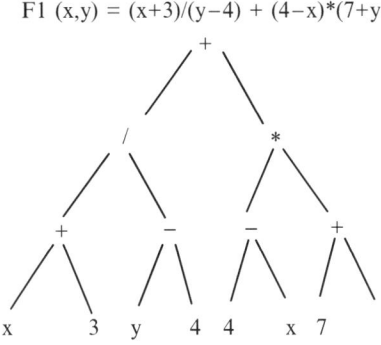

A crossover site is chosen at random and the child chromosome are obtained by exchanging the part of the parent strings to one side of the site as shown below:

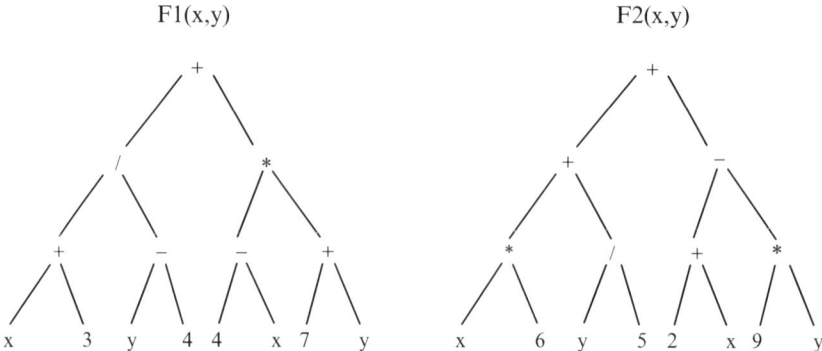

Then the child strings are obtained after crossover as

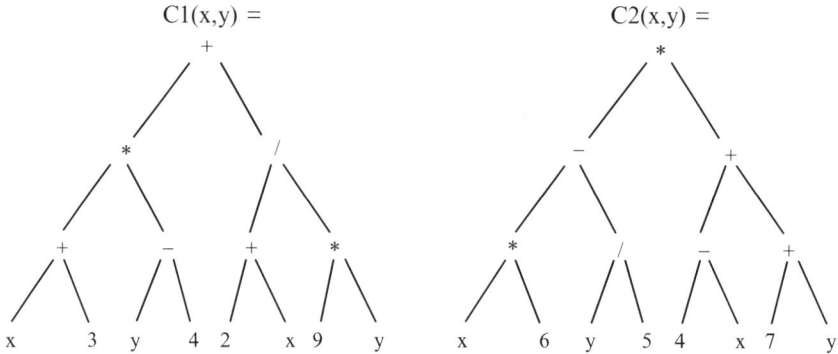

It is generally recommended that the crossover probability should be large in comparison to the mutation propability.

10.2.2 Mutation

The crossover operation results in new offspring having no new inheritance information. This can be corrected by having a mutation operator which would create diversity in the population and which would enable checking of a larger area of the search space, e.g. if a chromosomes is F (x,y) and mutation occurs at the mutation site (m site), then the new string will be F1 (x,y) as shown below:

Mutation arbitrarily alters one or more bit of information of a selected individual (function) by a random change with a probability equal to the mutation probability (Pm). Schaffer et al. (1989) has reported that the mutation probability (Pm) should be approximately inversely proportional to the population size (popsize).

F1 (x,y)   F1' (x,y)

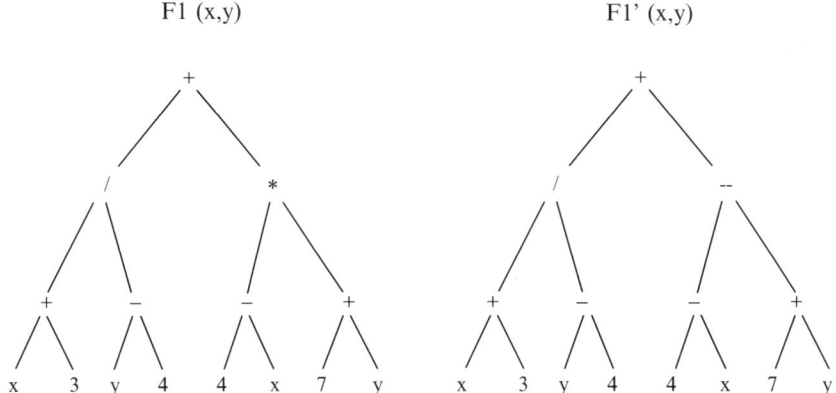

10.2.3 Population Size (Pop Size)

The choice of the population size had a strong interacting effect on the results. Smaller population sizes tend to become homogenous more quickly. With large population sizes the crossover probability effects are much less dramatic.

The choice of the population size is largely dependent on the complexity of the problem. If the problem in hand is simple then a smaller population size is sufficient, but a complex requires a large population size.

10.3 Development of Improved Genetic Algorithm (IGM)

The details of parameters variations and their influence on the optimization process have been studied by many reasearchers (Holland 1973; Fogarty 1981; Goldberg and Holland 1988, 1989; Schaffer et al. 1989; Deb 1995; Chaturvedi et al. 2000; Back 1997; Beyer 2001, 2002; Bingul 2007). In all these studies the objective function is optimized using GA for different set of parameters which are initialized at the time of starting and during optimization those parameters are kept constant.

In the improved genetic algorithm (IGA) there are three important GA parameter:

  i. Crossover probability (Pc),
 ii. Mutation probability (Pm) and
iii. Population size (popsize)

are proposed to be varied dynamically during execution of the program according to the fuzzy knowledge base which has been developed from the experience to maximize the efficiency of GA. These parameters and the basis of their change are described in this section.

10.3.1 Basis of Variation of Pc, Pm and Popsize

The philosophy behind varying these parameters is that the response of the optimization method/procedure depends on the stage of optimization, i.e. a high fitness value may require a relative high population size, a low cross over probability and high mutation probability for further improvement; alternatively, at low fitness values the response would be better with a relatively low population size, a high cross over probability and a low mutation probability. The reason behind it is that at the time of starting high cross over probability (Pc) and low mutation probability (Pm) yield good results, because large number of crossover operation will produce better chromosome vectors whose fitness value are relatively high. This process will continue for some finite number of generations after that the fitness value of each chromosome vector becomes almost same (around 0.9). Beyond that the effect of crossover is not significant due to little variation in the chromosome vectors in that particular population. Hence, at this stage, the population can be diversified by the following means:

1. By increasing the mutation rate of the chromosome vector to inculcate the new characteristics in the existing population.
2. By introducing new chromosome vectors in the existing population whose characteristics are different from the existing chromosomes vector (i.e. by increasing the population size (popsize)).
3. By introducing new chromosome vector in the existing population whose characteristics are different from the existing chromosome vectors and remove the chromosome vector from the existing population whose fitness value H is relatively low (i.e. Keeping the population size (popsize) constant).

Several methods of optimization have been proposed over the past few years; Schuster proposed heuristics for an optimal setting of the mutation probability Pm, Fogarty (1981) and Booker (1987) investigated time dependencies on the mutation and the crossover probability respectively, and Grefenstette (1981) found optimal settings for all three parameters of the GAs by experiment.

In the present work, a fuzzy control of the values of the three parameters is attempted. For this purpose the proposed ranges of these parameters have been divided into low, medium and high membership function, and each is given some membership value. These are presented graphically in Table 10.1.

The parameters are varied based on the value of the fitness function and its variation:

1. For this purpose the best fitness (BF) for each generation is considered.
2. This value is expected to change over generations. If the BF does not change significantly over a number of generations (UN) then this information also is considered to effect changes in the GA parameters.
3. The diversity of population is one of the factors which influence the search for a true optimum.

Table 10.1. Membership functions and range of variables

S. No.	Variable		Linguistic term	Membership functions
1.	Crossover probability	Pc	Low medium High	
2.	Mutation probability	Pm	Low medium High	
3.	Population size	Popsize	Low medium High	
4.	Best fitness	BF	Low medium High	
5.	Number of gen. for unchanged BF	UN	Low medium High	
6.	Variance of Fitness	VF	Low medium High	

The variances of the fitness values of objective functions (VF) of a population is a measure of its diversity and hence, considered as a factor based on which the GA parameters may be changed.

10.3.2 Development of Fuzzy System

The membership functions and membership values for these three quantities are selected based on the experience and ease in computation. The support and overlapping of these membership functions are optimized and final results are presented graphically in Table 10.1.

The knowledge base for modifying the GA parameters is described below:
Fuzzy knowledge base

(a) *For controlling Pc*
  1. If BF is low then Pc is High
  2. If BF is medium or High and UN is Low then Pc is High
  3. If BF is medium or High and UN is Medium then Pc is medium
  4. if UN is High and VF is Low or Medium then Pc is Low
  5. if UN is High and VF is High then PC is Medium

(b) *For controlling Pm*
  6. If BF is Low then Pm is Low
  7. If BF is Medium or high and UN is low then Pm is Low
  8. If BF is Medium or High and UN is Medium then Pm is Medium
  9. If UN is High and VF is Low then Pm is High
  10. If UN is High and VF is Medium or High then Pm is Low.

(c) *For controlling Popsize*
  11. If BF is Medium and UN is High then pop size is Medium
  12. If BF is low and UN is high then popsize is High
  13. If BF is high and UN is low and popsize is Low

10.4 Application of Improved Genetic Algorithm (IGA) to Electrical Load Forecasting Problem

Improved genetic algorithm is best suited for the function generation and optimization of problems like load forecasting. Load forecasting plays an important role in power system planning, designing, operation and control. The load at the various load buses is required to be known a few seconds to several minutes before to plan the generation and distribution schedules contingency analysis and for checking system security (know as very short time load forecasting). For the allocation of the spinning reserve, it would be necessary to predict the load demands at least half an hour to a few hours ahead (know as short term load forecasts). On the other hand preparing to meet the load requirements at the height of the winter or summer season may require a load forecast to made a few days to few weeks in advance. Forecasts with such lead time constitute medium term load forecasts. Finally to plan the growth of the generation capacity, it would be necessary to make "long term" load prediction which may involve a lead time of a few months to a few years. Long term load forecasting of a future demands on a realistic basis is important in power planning.

Major power projects have long gestation periods, which may extend to 10 years more. Therefore, decisions on investment have to be taken in advance for demands, if the energy benefits are to materialize at appropriate time needed. Thus, it is necessary not only to have demand forecast covering a 15–20 years period but only to update the same every 3–5 year in order to fit into the 5 years plan.

Many techniques and approaches have been used for the electrical load forecasting problem in the last two decades (IEEE Report 1980, 1981; Mahalanbis et al. 1991). These are after different in nature and apply different engineering considerations and economic aspects. Traditional approaches are load survey methods, mathematical models like correlation or extrapolation models (linear growth pattern, exponential growth pattern, parabolic growth pattern, or sigmoid growth pattern) combination of both, regression and time series analysis and mathematical models considering economic parameter. These traditional methods for load forecasting may not give sufficiently accurate results. Also these methods are cumbersome and time consuming, as they require a lot of information about variables on which load forecasting depends (Georg 1987). The information regarding variables may be incorrect, improper and insufficient, casing error in forecasting; more the number of such variables, higher may be the error in forecasting. Therefore, a method is required which can forecast the power demand with minimum number of variables giving sufficient accuracy and the method is not quite complex and cumber some. In this respect, Lee and his collogues (Lee et al. 1997) have taken population and GDP as two variables for predicting the long term demand for Korea.

In this section an attempt has been made to develop a function for long term load forecasting from the available data of peak demand using IGA in two steps:

I. *Function development*: the functional relationship is developed between the variables and constants.

II. *Function optimization*: the function obtained in the previous step is to be optimized by selecting the suitable values of coefficients used in the function.

A. *Function Development*

A function or expression is composed of three parts:

I. Variables $(x1, x2, x3 \ldots \ldots \ldots \ldots)$

II. Constant $(k1, k2, k3 \ldots \ldots \ldots \ldots)$ and

III. Operation $(o1, o2, o3 \ldots \ldots \ldots \ldots)$

The operators connect the variables and constants constituting a function. Therefore a function or expression is a string of variables, constant and operators arranged in a proper way as given below:

F (x1, x2..........) = x1 o1 k1 o1 x2 o3 k2 o4 x3 o5 k3........x_n o_n k_n

In the above string, the constants (k1, k2, k3...........k_n) may be real or integers; the operators (o1, o2 o3............o_n) are mathematical operators like {'+', '−', '/', 'exp', 'log','*', }, etc.

The stepwise procedure of function development for load forecasting problem through IGA is given below:

Step-1: *Initialization*: Give the initial values to GA parameters to start GA optimization. In load forecasting problem following initial values for GA parameters have been taken.

Chromosome length (lchrome)   = 22
Population size (popsize)   = 60
Maximum number of generation (maxgen)   = 20
Crossover probability (Pc)   = 0.9
Mutation Probability (Pm)   = 0.005

Step-2: Generate the initial population of function in the form of binary strings equals to the popsize and length of each function string equal to the lchrome.

Step-3: Decode the value of constant as well as operators of each function string and develop the functions for the complete population. A developed function from the corresponding string of population is shown below which representing the change in electrical demand:
00110100001010110011010 $F(x) = (x+1)^*(x/4) + (x-1) - x^{\wedge}3)$
Population string   Developed Function

Step-4: Each string of a population is decoded in the form of a function which predict the change in demand. From this change in electrical demand, the demand as well as the error in predicted load and actual load can be calculated. The objective function developed for long term load forecasting problem uses the error calculated above.

$$\text{Objective function} = 1/(1 + \Sigma \, \text{Error}_i)$$

Where Error_i - is the error of the demand predicted by ith decoded chromosome vector of nth Generation and the demand available from CEA.

Step-5: Performs crossover and mutation operations among selected strings of population according to the probabilities of mutation and crosses over. Also, the fuzzy controller during execution through fuzzy knowledge base controls the crossover and mutation probabilities in IGA. In this step the old population is modified by crossover and mutation operation to produce better strings.

Step-6: Repeat Step 3, 4 and 5 till the fitness value will not reach near unity or the error in forecast will minimize.

b. *Function Optimization*

The function, which is developed in the previous section, is optimized to predict the long term load with better accuracy. In function optimization, the coefficient of variables and constant are optimized using IGA to reduce the error during prediction. While optimizing the functional coefficients and constants using IGA, it was observed that the best fitness (BF) value continuously increases as the generation increase. Very rarely the BF Value of present generation goes below the BF value of the previous generation. If it goes below then IGA select the appropriate value of Pc and Pm from fuzzy knowledge base and immediately recovers the BF value to the same or more than that. On the other hand, in simple GA if the BF value goes below in any generation it could not be recovered and remains same for significant number of generations.

10.5 Results

The simple GA is used to develop the functional relationships between variables and constants. After selection the function for long term load forecasting problem, improved genetic algorithm is used to optimize the functional coefficients of the function generated by the simple GA. For load forecasting problem, the results obtained by improved genetic algorithms (IGA) are compared with the data given by annual power survey (APS) carried out by CEA as shown in Fig. 10.1; which is very close to the actual curve.

10.6 Integrated Fuzzy GA Technique

The fuzzy set theoretic approach has been described in earlier chapters. It is a practical, robust, economical and intelligent alternative for system's modeling and control. However, this technique has a serious set back that without the precise and quality expert knowledge it could not give good results. Most of the time it is difficult to get precise quality expert knowledge due to various reasons like unavailability of domain experts, expert hide some of the information or forget to give, etc. To solve this problem, fuzzy logic is complimented with various other techniques such as ANN, Evolutionary algorithms, etc.

Evolutionary algorithms are one of the suitable tools for optimizing the behaviour of fuzzy systems without the human intervention. There is no problem of local minima as in gradient descent algorithms. Also it provides number of solutions at a time.

Hence, hybridization of Fuzzy systems and GAs is a good choice for developing the adaptive fuzzy systems. The tuning of fuzzy system is done by

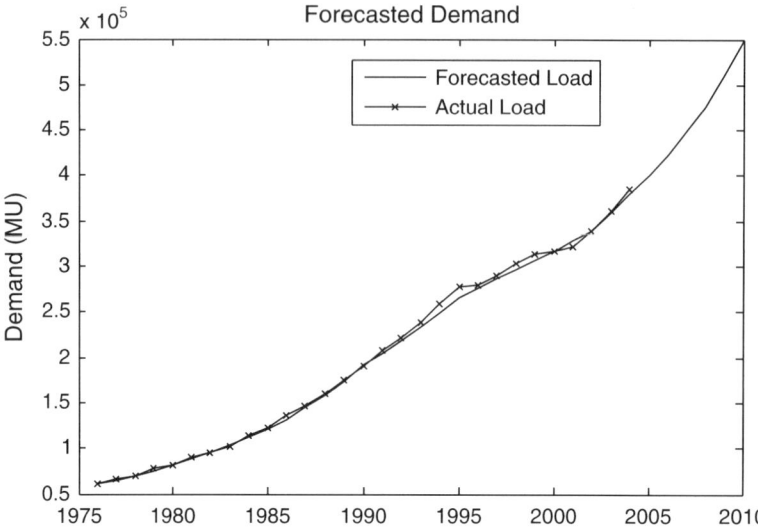

Fig. 10.1. Comparison of forecasted load using GA and actual demand

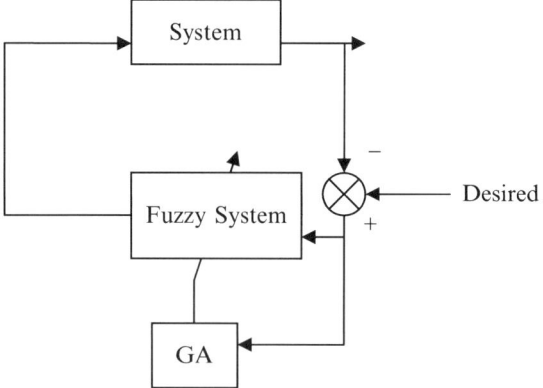

Fig. 10.2. Block diagram of adaptive fuzzy system

optimizing its parameters using GAs. GA performs well in noisy, non-linear and uncertain optimization of fuzzy systems.

Various issues like determining the shape of fuzzy set, its parameters, overlapping and parameter of aggregation operators can be handled with GA optimization. GA optimization includes large search space, non-differentiable objective function involving uncertainty and noise. Due to these characteristics GA is quite suitable for fuzzy system parameter optimization and gives potential solution in parallel and hence easily avoid local optima. The block diagram of this adaptive fuzzy system is shown in Fig. 10.2.

The following steps are used in adaptive fuzzy system parameter optimization using GA:

Step-1: Identify parameters of fuzzy system for optimization using GA. The number of fuzzy system parameters affect the complexity of objective function of GA, which in turn slows down the optimization process.

Step-2: Representation of these parameters (Encoding)

The system parameters should be represented in a proper form, so that GA optimizes them. It could be represented by binary strings (genomes). If there will be large number of parameters then the length of chromosomes should also be large. It will increase the complexity, slow down the convergence and also affect the accuracy of results. In such situations normalized real number representation improves the convergence and accuracy.

Step-3: Initialization

Initial population greatly affect the convergence time.

Step-4: Definition of objective function

The designing of objective/fitness function has large impact on the performance of GA. Let us consider a minimization problem, the objective function could be designed as

1. Fitness function = Max value – functions f(xi)
2. Fitness function = 1/(1 + function)
3. Fitness function = exp(–function)

10.6.1 Development of Adaptive Fuzzy System

The most crucial and important part of fuzzy system is its knowledge base. Knowledge required for fuzzy system can be divided into two parts:

1. Domain knowledge – It is acquired from domain expert.
2. Meta knowledge – This contains the knowledge about the fuzzy membership functions, firing strength of each rule, aggregation operator, defuzzification method, implication and compositional rules.

Based on the above two types of knowledge, fuzzy system can be made adaptive by

a. Fixing the domain knowledge and optimize meta knowledge.
b. Fixing the meeta knowledge and optimize the domain knowledge.
c. Optimizing domain knowledge (coarse tuning) and then optimizing meta knowledge (fine tuning).
d. Optimizing both the knowledge simultaneously.

(a) Optimization of membership functions:

1. In the optimization of membership functions, it is necessary to find the universe of discourse in which fuzzy sets could be defined.
   Universe of discourse $U = [-u + u]$
2. Type of membership function
   Triangular, trapezoidal, Gaussian or generalized membership function, etc.
3. Encode for GA optimization

   Type of membership function and shape must be encoded as shown below:
   Various types of membership functions are used in fuzzy systems like triangular, trapezoidal, Gaussian or one can also use the generalized membership function of four segments. Most often the triangular membership functions are used in fuzzy systems due to its simplicity.

   Case-1 If core $= 0$ and left base $=$ right base then membership function (MF) is symmetrical triangular function.
   Case-2 If core $= 0$ and left base or right base (any one) $= 0$ then right angled triangle MF.
   Case-3 when all these parameters are there it is trapezoidal MF.

Hence, the encoded value as binary strings and length of these strings depend on the accuracy required. The complete membership function will be defined as binary string or real value is shown in Table 10.2.

$$\text{and left base} + \text{core} + \text{right base} <= 1.$$

In four segment generalized membership function as shown in Fig. 10.3, every segment has two attributes with it. First is the length of the segment and second is the angle associated with it.

For segment a, the length is l_1 and angle θ_1. Similarly for b (l_2, θ_2), c (l_3, θ_3) and for d (l_4, θ_4). In this case, segment length is in the range 0–2, but the angle is 0–180°. Both these parameters in different ranges, hence it is necessary to normalize the value of these attributes in the same range from 0 to 1.

Suppose multi-input – single output (MISO) systems, with two inputs as shown in Fig. 10.4. Each input three fuzzy sets (membership functions). Each fuzzy membership function has four segments and each segment has two attributes, which could be represented by say three binary bits. Hence, the total length of chromosome for single fuzzy input variable will be

Table 10.2. Encoding of membership functions

	Left base	Core	Right base
Binary strings	01010	0110	01010
Real values	0.1	0.2	0.1

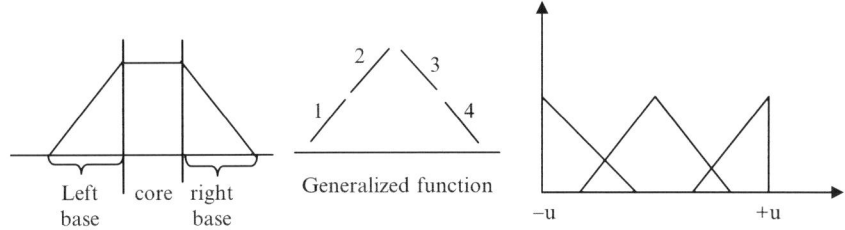

Fig. 10.3. Membership functions and universe of discourse

(a) Fuzzy system

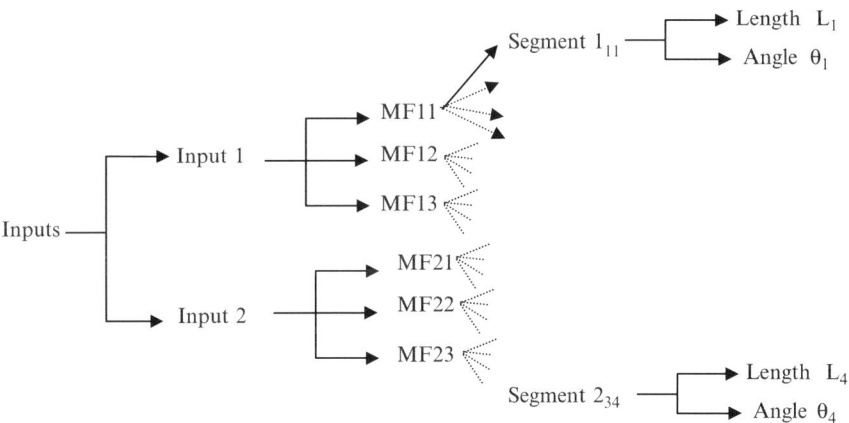

(b) Details of chromosome

Fig. 10.4. Details of optimization parameter of adaptive fuzzy system

Length of chromosome = number of variables × number of fuzzy sets

$$\times \text{ number of segments} \times \text{ number of attributes}$$

$$\times \text{ number of binary bits}$$

$$= 2 \times 3 \times 4 \times 2 \times 3 = 144.$$

If the number of variables is more, GA optimization will be more complicates. Therefore, it is necessary to represent the chromosome in real normalized values, to reduce the length of chromosome. In this case it will be 24, which may be calculated as

Table 10.3. FAM table for fuzzy system

	MF21	MF22	MF23
MF_{11}	P_{o1}	P_{o2}	P_{o3}
MF12	P_{o4}	P_{o5}	P_{o6}
MF13	P_{o7}	P_{o8}	P_{o9}

Length of chromosome = number of variables × number of fuzzy sets

$$\times \text{ number of segments} \times \text{ number of attributes}$$

$$= 2 \times 3 \times 4 \times 2 = 48.$$

(b) Optimization of Rule base

For the above mentioned system the rule base consisting of maximum nine rules which is represented by rule table, also called fuzzy associative memory (FAM) Table 10.3.

The initial population may be randomly generated consisting FAM tables with different output fuzzy sets as given below. Then select any two FAM tables and perform point radius crossover and mutation. Finally check the fitness of each FAM table of child population and prepare the parent population from child population using Darwinian survival of fittest principle. Repeat this process till the good solution is not obtained (Fig. 10.5).

(c) Optimization of rule strength of each rule

The rule strength is the strength of the particular rule which affects the final solution if that rule is fired. Normally, the rule strength is unity i.e. all the rule output equally participate in the final output of fuzzy system. If someone want to reduce the weightage of any rule or wish to increase it then rule strength may be changed. It is also one of the parameter to optimize using GA to get suitable rule strengths of each rule.

(d) Optimization of aggregation parameter

As mentioned in the earlier chapter the aggregation function affects the firing strength of rules which are fired for the given inputs. Normally max and min aggregations operators are used. Some of the researcher used their own aggregation operators, compensatory or averaging operator. Lee and Esbensen (1996) optimize the parameters of these operators to improve upon the results.

Example – Cart Pole Problem

It is a system consisting of an inverted pole on a moving cart. The goal is balance this pole by moving the cart to and fro as shown in Fig. 10.6. This problem is also called inverted pendulum problem.

Parent population

Child population

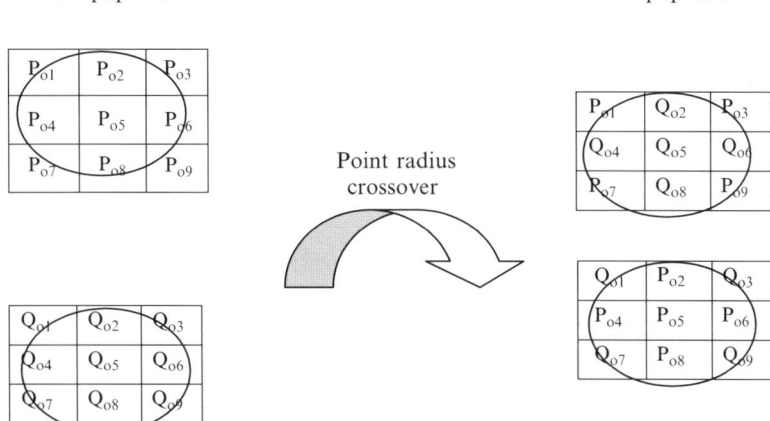

(a) Point radius crossover on FAM tables of parental population.

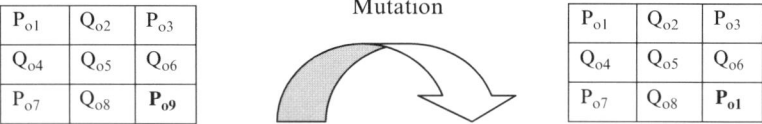

(b) Mutation operation on FAM table.

Fig. 10.5. Crossover and mutation operation to optimize FAM table

The cart can move right or left on rails when a force is exerted on it. This dynamic system is characterized by the following equation

$$\text{Angular acceleration } \dot{\omega} = \frac{g \cdot \sin\theta + \cos\theta \left(\dfrac{-f - ml\omega^2 \sin\theta}{m_c + m} \right)}{l \left(\dfrac{4}{3} - \dfrac{m \cdot \cos^2\theta}{(m_c + m)} \right)}$$

$$\text{Linear acceleration } \dot{v} = \frac{f + ml(\omega^2 \sin\theta - \dot{\omega}\cos\theta)}{(m_c + m)}.$$

where

θ – Angle of pole from the vertical plane (rad)

ω – Angular velocity (rad/s^2)

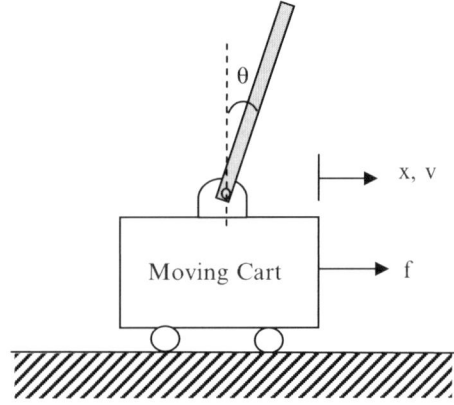

Fig. 10.6. Cart pole problem

x – displacement (m)
v – Linear velocity (m/s^2)
l – half length of pole (m)
m – mass of pole (kg)
m_c – mass of cart (kg)
g – acceleration due to gravity $(9.81\,m/s^2)$

The state variable vector is consisiting of $[\theta\ \omega\ x\ v]$. The aim is to find suitable force F to balance the pole in a satisfactory manner, of regardless of the cart position.

Problems with conventional controller:

i. System is non-linear.
ii. Delay is involved.
iii. Saturation of core due to heavy flux when heavy current flows in the motor of moving cart.
iv. System parameters are time varying.

Due to these problems conventional controller is unable to perform well. Lee and Takagi (1993) used fuzzy controller for this problem.

Diversity in solutions has an important role on the convergence property of GA. Proposed different methods for maintaining diversity by dynamically changing mutation rate. Consider the two extreme case of mutation rate.

1. If mutation rate is zero, it means there is no diversity at all. In this case after some iteration, all the solutions in a particular population are equally good and crossover operation does not affect the fitness. It results the premature convergence of GA. Lower value of mutation rate is normally used for fine tuning of adaptive fuzzy systems and on-line implementation of optimization tool.

2. If mutation rate is one means always there is a random population, which reduces the rate of convergence and leads low average fitness and high diversity. High value of mutation rate explores almost complete search space. The higher values of mutation rate are good if the expert's knowledge is not available or available but incomplete for the development of adaptive fuzzy systems.

Hence, a suitable value of mutation rate is required to quickly get an optimal solution.

Selection of Fitness function for GA optimization

GA performance also depends on the fitness function used in optimization to guide the direction of its search. There is no general way to define fitness function for a problem; however, it is often designed such that the more desirable solutions corresponding to high solutions. Fitness function includes all the parameters which need to be optimized. For example, in a second order system it is desirable to get a fast and accurate response with lower overshoot.

$$\text{Therefore the fitness function} = \int_{to}^{t_f} \frac{1}{e^2 + \zeta^2 + 1} dt$$

where, to initial time

tf – finish time

e – error, i.e. the difference between desired and actual outputs.

ξ – overshoot

$$\text{More general fitness function} = \frac{\zeta}{t_f - to} \int_{to}^{t_f} \frac{1}{K_1 P_1^2 + K_1 P_1^2 + \ldots \ldots K_n P_n^2 + \zeta} dt$$

Ki – is the weightage

10.7 Limitations of GA

a. GA is quite time intensive algorithm for optimization. Even with high speed computers, real time GA implementation is still a challenge. In GA, GA may provide on-line learning capability in intelligent adaptive robots and evolve their performance, on-line GA implementation is still difficult.

b. In GA based adaptive fuzzy systems, GA drastically alters fuzzy knowledge base system architecture and after that it is difficult or no longer identifiable by a human operators. This is important when intuitive human understanding of system is a significant component in its control.

c. Meaningful relation between membership function and fuzzy rules may be lost from human point of view after GA optimization.
d. For more intelligent systems require more complex and autonomous systems. GA-fuzzy integrated technique best suited for this type of applications.

10.8 Summary

The genetic algorithm, which is inspired from the biological genetics, is simple, powerful, domain free and probabilistic approach to general problem solving techniques. It is best suited for the problems like load forecasting for electrical power.

The performance of genetic algorithms can be improved by dynamically varying its operating parameters during execution. The effect of varying the probability of mutation and the probability of crossover is quite significant in controlling the trend of the best fitness value, especially when the best fitness is not very close to its maximum value.

IGA claims to provide near optimal or optimal solution for computationally intensive problems. Therefore, the effectiveness of genetic algorithm solutions should always be evaluated by experimental results.

The graph obtained from developed function through IGA is much closer to the graph obtained by APS carried out by CEA, India. Therefore, it is capable to produce a function for non-linear variations from the available data, which can save a lot of labor and complexity during analysis of any type of non-linear phenomena.

There is a vast area of application in which the GAs can be used. The idea of producing a function for non-linear variation has a vast application area. It can be applied to solve problems of science, engineering and technical fields. Additional features like multi-site crossover and other algorithms for chromosome selection should also be tested with GA optimization.

Exercise

1. Write the merits and demerits of GA and Fuzzy System.
2. Explain the different possibilities of combining GA and Fuzzy systems to get integrated tool.
3. Using GA-Fuzzy systems approach solve traveling salesman problem to minimize the distance traveled for n-cities.

Synergism of Genetic Algorithms and Fuzzy Systems for Power System Applications

11.1 Introduction

The power system of today is a complex interconnected network having four major components – generation, transmission, distribution and loads. Electricity is being generated in large hydro, thermal and nuclear power stations, which are normally located far away from the load centers. Large and long transmission networks are wheeling the generated power from these generating stations to different distribution systems, which ultimately supply the load. The distribution system is that part of the power system which connects the distribution substations to the consumers' service-entrance.

Earlier the utilities were mainly concerned about the optimal dispatch of active power only, but evolvement of competition has also resulted in the optimal dispatch of reactive power. When only total cost is minimized by real power scheduling of available generator in a system, the optimal power flow (OPF) corresponds to *Active Power Dispatch*. Some of the solution techniques successfully used for active power dispatch include classical co-ordination methods based on Lagrangian multiplier approach (Chowdhury and Rahman 1990), Linear programming (LP) based methods (Stott and Hibson 1978; Stott and Marinho 1979), quadratic programming (QP) approach (Nanda et al. 1989), Gradient method using steepest descent technique (Dommel and Tinney 1968) and Newton's methods (Sun et al. 1984; Maria and Findlay 1987). A comprehensive review of various optimization techniques available in the literature is reported in references (Happ 1977; Sasson and Merril 1974; Carpantier 1985). The classical method of optimization is relatively simple, fast and requires less memory space but sometimes it is unable to handle the system constraints effectively and sometimes convergence is not obtained. The LP based method involves approximation in linearizing the objective function and constraints and may result in zigzagging of the solution. Gradient based methods compute the derivative of the function at each step. They require a close initial guess and in general suffer from convergence difficulties and may stuck to local minima.

D.K. Chaturvedi: *Soft Computing Techniques and its Applications in Electrical Engineering*, Studies in Computational Intelligence (SCI) **103**, 403–477 (2008)
www.springerlink.com © Springer-Verlag Berlin Heidelberg 2008

The GAs (genetic algorithms) have been applied to solve unit commitment problem (Kazarlis et al. 1996), Optimal reactive power dispatch (ORPD) problem (Singh et al. 1993; Swarup et al. 1996; Lee et al. 1997) and for economic load dispatch problem (Walters and Sheble 1993; Sheble and Britigg 1995; Chen and Chang 1995; Orero and Irving 1996; Achyuthakan 1997). Miranda et al. (1996) have provided a survey of three branches of evolutionary programming (EP), genetic algorithms (GAs) and discusses their relative merits and demerits.

The superiority of GA methods in handling continuously non-differentiable objective has been given in (Walters and Sheble 1993; Achyuthakan 1997). For better results and faster convergence, conventional GA models have been modified by including new operators such as elitism, shuffle in reproduction, multi-point or uniform crossover and creep mutation. Considering three added features, a refined GA is used to solve economic load dispatch (ELD).

A pyramid genetic algorithm (PGA) has been used in Lee et al. (1997) for voltage profile optimization. The PGA can analytically determine the bound values of mutation and crossover probabilities, which are otherwise, chosen by experience. The GA-Fuzzy approach presented in this paper is developed to get above mentioned advantages by varying crossover and mutation probabilities throughout the generations by fuzzy-rule base.

11.2 Transmission Planning, Pricing and Structure/Models of Indian Power Sector

A bibliographical survey of power system wheeling under deregulated environment is presented by Sood et al. (2002). A lot of literature is available for the issue of transmission open access. Christie and Anjan Bose (1996) discussed the complete deregulation scenarios and technical issues related to operation and control of the system. Various aspects of pricing of transactions in open access are discussed by Silva et al. (1998), David (1998), Arriaga et al. (1995b) and Vojdani et al. (1996).

1. *Transmission planning*

A genetic algorithm based dynamic transmission planning methodologies are formulated by Rudnick et al. (1996) and Lima et al. (1998) to determine the economically adapted transmission system in open access. But since the transmission planning is a complex, nonlinear and dynamic problem, a simple GA method is not suitable. Object oriented software for transmission planning in open access is proposed by Handschim et al. (1998). Raga et al. (2005) have presented a multi-criteria formulation (i.e. investment costs, operational cost and the expected energy not supplied) for multiyear dynamic transmission expansion planning problems. The solution algorithm adapts an interactive decision-making approach that starts at a non dominated solution of the problem.

2. *Transmission pricing methods*

Transmission pricing has been discussed in a detail in the literature. Happ (1994) has presented computational procedure and data requirements for embedded cost methods, incremental cost methods and marginal costs methods.

Caramanis et al. (1986) has presented new wheeling rates for buying and selling. A computer program WRATES is developed by Caramanis et al. (1989) is used to provide the practical means of computing the marginal cost of wheeling. This analysis requires a load flow program integrated with constraints and economic load dispatch simulation.

First extensive computations of marginal cost of wheeling and rates based on marginal costs are carried out by Merrill and Erickson (1989). Different methodologies for costing of transmission services have been developed and are reported by Shirmohammadi and Thomas (1991) and Shirmohammadi et al. (1991). The theory to evaluate optimal wheeling rates for the case of bus to bus wheeling is developed in (Lo and Zhu 1993). It is based on the marginal cost theory which has been used for electricity pricing. On the basis of the equitable sharing of the benefits arising from wheeling transactions among the wheeler, the power seller and the buyer, this approach has the advantage over others. It avoids the direct evaluation of the network maintenance cost and the quality of supply cost components.

Several methodologies have been reported for cost of wheeling. A nonlinear optimization program with linear constraints is developed by Li et al. (Li and David 1993) to calculate the amount of wheeled energy and wheeling price solved by gradient projection method.

The principles and practices of a new methodology for wheeling rate evaluation without assuming the existence of the spot price based market place is describe by Lo and Zhu (1994). Li et al. (1994) have used a wheeling rate based on marginal cost pricing and implemented using the modification of the optimal power flow.

A load flow based model for calculating the various cost components is presented by Kovacs and Leverett (1994). A separate pricing of transmission and distribution services is proposed by Farmer et al. (1995). SRMC and LRMC based models are proposed by Lima and Pereira et al. (1995) for allocating transmission cost among users of centralized transmission service. A novel approach that alleviates the inherent shortcomings of SRMC based pricing and maintains the economic efficiency of the price signals are proposed by Farmer et al. (1995). But the effect of security analysis has not been taken care while considering the optimal conditions.

Lima (1996) has proposed load-flow based Megawatt-mile, Modulus, Zero counter complex calculations and greater data and provides no incentive to users.

Pereira et al. (1996) have presented a method for evaluating an optimal set of transmission prices to be charged for use of a transmission system on a time-of-use basis. Prices are calculated by maximizing the global benefit of using the transmission system that allocates both capacity and operational cost.

A methodology to allocate the cost of transmission network facilities to wheeling transactions in decentralized power systems using Game theory is proposed by Tsukamoto et al. (1996) and Ferrero et al. (1998). The concept of game theory is employed to deal with the conflicts in a deregulated power system.

Wakefield et al. (1997) have presented transmission costing framework and its application for analyzing the transmission costing issues. Zobian and Illic (1997) have proposed a methodology for allocating transmission cost among users of a centralized transmission service. The share of each participant is proportional to its impact on system transmission investment requirements. This allocation rule provides incentives for all participants to remain in the pool and ensures revenue reconciliation. Yu and David (1997) have proposed an approach which distinguishes between operating and embedded costs and have developed separate methods in respect of each of these components.

In (1999) a method for long run marginal cost (LRMC) based pricing in multi-area interconnected system, based on the incremental use of each area's transmission network at times of peak flow, is proposed.

In (Muchayi Maxwell and El-Hawary 1999) unlike other methods which use only the variation of fuel cost for generation to estimate the rate structures, the proposed pricing algorithm incorporates the optimal allocation of transmission system operating costs based on time-of-use pricing. The transmission costs are obtained by assigning a price k to each unit of power flow in the network.

In (Moya 2002), a model of marginal adequacy costs is developed in order to reflect the influence that any nodal load has on system static security. An adequacy cost function is defined, making use of the load that must be theoretically withdrawn at each node in order to re-establish power flows on transmission elements, after any static contingency of a predefined set occurs.

Chen et al. (2002) have presented a method to provide a detailed description of each nodal price, by breaking down each nodal price into a variety of parts corresponding to the concerned factors, such as generations, transmission congestion, voltage limitations and other constraints or elements.

Gang et al. (2005) have proposed a transmission and wheeling pricing method based on the monetary flow tracing along power flow paths: the monetary flow–monetary path method. Active and reactive power flows are converted into monetary flows by using nodal prices. The method introduces an uniform measurement for transmission service usages by active and reactive powers.

Gil et al. (2006) presents an approach for the allocation of transmission network costs by controlling the nodal electricity prices. The proposed approach introduces generation and nodal injection penalties into the traditional economic dispatch so as to create nodal price differences that recover the required transmission revenue from the resulting congestion rent.

Galetovic and Montecinos (2006) describes the new method used in Chile to allocate transmission charges among generating companies and customers. They show that the new Chilean transmission charge scheme is a hybrid based on marginal cost pricing, identification of use through economic benefits and flow identification methods, and last, a postage stamp to redistribute almost all the charges that customers have to pay.

Sedaghati (2006) has proposed a novel method for allocation of the fixed cost of the transmission systems to agents using facilities. In (Verma and Gupta 2006), a nonlinear optimization problem has been formulated to maximize the social welfare in the open power market using a unified power flow controller (UPFC).

3. *Market structures/models*

A Poolco model suitable for power system planning and decomposing spot prices to reveal components caused by congestion is presented by Finny et al. (1997).

Illic and Prasad et al. (2003) provided simulation-based demonstrations of hybrid electricity market that combines the distributed competitive advantages of centralized markets.

Ren et al. (2004, 2004a) compared the quantitative behavior of the two markets, i.e. pay as bid and marginal pricing, assuming that generators submit the best strategic offers that correspond to the specified pricing method. In Part I of their two-part study, assuming that the system marginal costs for pay-as-bid (PAB) and marginal pricing (MP) are random with known probability density functions, they develop generator strategic offers by maximizing the corresponding expected values of the generator profits over the offer parameters. In Part II relations are established between the system marginal costs (SMCs) for each market type and a common random demand, thus allowing the two markets to be compared through the expected values and variances of the individual generation profits and of the consumer payments.

Competitive markets for electricity determine either a uniform marginal price (UMP), a set of nodal marginal prices (NMPs), or a smaller set of zonal marginal prices (ZMPs). Ding and David (2005) prove that, the UMP or ZMP models (a) do not affect the total economic surplus, (b) redistribute the surplus among generators and loads at the different nodes, and (c) give perverse incentives for generation expansion.

Fleten and Erling (2005) have proposed a stochastic linear programming model for constructing piecewise-linear bidding curves to be submitted to Nord Pool, which is the Nordic power exchanger. They have considered the case of a price-taking power marketer who supplier electricity to price-sensitive end users.

Plazas et al. (2005) considers a profit-maximizing thermal producer that participates in a sequence of spot markets, namely, day-ahead, automatic generation control (AGC), and balancing markets. The producer behaves as a price-taker in both the day-ahead market and the AGC market but as a potential price-maker in the volatile balancing market.

Li and Mohammad (2005) describes a method for analyzing the competition among transmission-constrained generating companies (GENCOs) with incomplete information. Each of GENCO models and its opponents' unknown information with specific types for transforming the incomplete game into a complete game with imperfect information.

Ongasakul and Chayakulkheeree (2006) have proposed a coordinated fuzzy constrained optimal power dispatch (CFCOPD) algorithm for bilateral contract, balancing electricity and ancillary services markets.

Bompard et al. (2006) has presented comprehensive approach to evaluate the performance of the electricity markets with network representation in presence of bidding behavior of the producers in a pool system. A supply function strategic bidding model for the producers is introduced, and then different scenarios in terms of bidding behavior and network constraints are studied and compared on the basis of a set of microeconomic metrics.

Philpot and Erling (2006), present a model of a purchaser of electricity in Norway, bidding into a wholesale electricity pool market that operates a day ahead of dispatch.

Olmos and Neuhoff (2006) have proposed an algorithm and apply it to the European electricity network to identify a balancing point that reduces market power of generation companies and is well connected. Market-level data or detailed information about demand is not required.

4. *Congestion management*

Congestion management is one of the major tasks performed by system operators (SOs) to ensure the operation of transmission system within operating limits. In the emerging electric power market, the congestion management becomes extremely important and it can impose a barrier to the electricity trading. Kumar et al. (2005) presented bibliographical survey of papers/literature on congestion management issues in the deregulated electricity markets.

A study of congestion management based on congestion pricing is proposed by Glavitsch and Fernando (1998).

Singh et al. (1998) studied the management of costs associated with transmission constraints (i.e. transmission congestion costs) in a competitive electricity market. The paper examines two approaches for dealing with these costs. The first approach is based on a nodal pricing framework and forms the basis of the so-called pool model. The second approach is based on cost allocation procedures proposed for the so-called *bilateral* model. An advanced analytical method for secure and efficient operation of power system is proposed by Shirmohammadi et al. (1998).

A congestion problem formulation should take into consideration interactions between intra-zonal and inter-zonal flows and their effects on power systems. It is perceived that phase-shifters and tap transformers play vital preventive and corrective roles in congestion management. These control devices help the ISO mitigate congestion without re-dispatching generation away from preferred schedules. In Ref. (2000) a procedure is introduced for minimizing the number of adjustments of preferred schedules to alleviate congestion and apply control schemes to minimize interactions between zones while taking contingency-constrained limits into consideration.

Service identification and congestion management are important functions of the ISO in maintaining system security and reliability. In Fu and John et al. (2001), a combined framework for service identification and congestion management is proposed. Verma et al. (2001) presents the development of simple and efficient models for suitable location of unified power flow controller (UPFC), with static point of view, for congestion management.

Gan and Donald et al. (2002) briefly review the New England power system (NEPOOL) locational pricing proposal being implemented. Two new approaches for locational market power screening are presented. The first one is based on a zonal network model and the second is based on a nodal transmission model.

The paper by Bompard et al. (2003) briefly reviews the congestion management (CM) schemes and the associated pricing mechanism used by the independent grid operators (IGOs) in five representative schemes. These are selected to illustrate the various CM approaches in use: England and Wales, Norway, Sweden, PJM, and California. They develop a unified framework for the mathematical representation of the market dispatch and redispatch problems that the IGO must solve in CM.

Kristiansen (2004) gives an overview of the current practice for congestion management, transmission pricing, and area price hedging in the Nordic region. Transmission congestion in the Nordic region is managed by using the area price model and counter trade. In Kumar et al. (2004), a new zonal/cluster-based congestion management approach has been proposed. The zones have been determined based on lines real and reactive power flow sensitivity indexes also called as real and reactive transmission congestion distribution factors. The generators in the most sensitive zones, with strongest and nonuniform distribution of sensitivity indexes, are identified for rescheduling their real power output for congestion management.

A new congestion management system is proposed in Mendez and Hugh (2004), applied under nodal and zonal dispatches with implementation of fixed transmission rights (FTR) and flow gate rights (FGR). The FTR model proves to be especially suitable for congestion management in deregulated centralized market structures with nodal dispatch, while the FGR is suitable for decentralized markets.

In the paper (Aguado et al. 2004), authors deal with the operation of power systems consisting of several interconnected electricity markets. They proposed an alternative approach to inter-regional trade that avoids the flaws of forward markets with explicit auctioning of interconnections capacities. They proposed the integration of a forward market with a balancing (spot) market for inter-regional exchanges based on nodal pricing.

Alomoush (2005) presents some performance indices to compare different dispatch options, where it proposes to use some congestion and system utilization measures. These measures are used in the paper to indicate level of system usage and congestion severity under different dispatch scenarios, and may enable the system operator or the qualified dispatch decision-making entity to decide which dispatch, among different dispatch scenarios, is the optimal.

The paper by Conejo et al. (2006) addresses the congestion management problem avoiding offline transmission capacity limits related to stability. These limits on line power flows are replaced by optimal power flow-related constraints that ensure an appropriate level of security, mainly targeting voltage instabilities, which are the most common source of stability problems.

11.3 GA-Fuzzy System Approach for Optimal Power Flow Solution

The present day power system is a very large and integrated power system comprising of several generators and buses. Recent trends of deregulation of power system have resulted in increased competition in the area of generation, transmission and distribution of power. The problem of economic operation of power system had emerged when it was required to operate two or more units to meet economically the demand when net generation exceeds the demand.

In the recent past, methods using genetic algorithms (GAs) (Goldberg 1989) have become popular to solve the optimization problems mainly because of its robustness in finding optimal solution and ability to provide near optimal solutions close to global minima. GAs are search algorithms based on the mechanics of natural selection and natural genetics. The performance of GA can also be improved by introducing new problem specific genetic operators. In Maha et al. (2006) a new genetic operator named *pluck* is introduced that incorporates a problem specific knowledge in population generation and leads to a better channel utilization in mobile computing problem. GAs are different from other optimization methods in the following ways:

- GAs search from population of several points, not a single individual point in the population.
- GAs have inherent parallel computation ability.
- GAs use pay off information (objective function) and not derivatives or auxiliary knowledge.
- GAs use probabilistic transition rules, so they can search a complicated and uncertain area to find the global optimum.

The basic idea in GA is to maintain a population of chromosomes that evolves through a process of competition and controlled variation. Simple forms of GAs performance largely depend on the appropriate setting of genetic parameters namely crossover probability and mutation probability. It has been observed that after few generations, the fitness value of each chromosome becomes almost equal to other chromosomes from the same population. The effect of crossover beyond this stage becomes insignificant due to very small variation in the chromosomes in a particular population. Therefore, it is difficult to find optimal settings for these parameters.

The techniques developed to set these parameters are classified by Eiben and Smith (2003) as parameter tuning and parameter control. For parameter tuning, the parameter values are set in advance (before the run) and are kept constant during the whole execution of the algorithm. In parameter control techniques, parameters are initialized at the start of execution and are allowed to change during the run. Parameter control techniques are classified mainly into three groups based on the type of change they introduce:

- *Deterministic*: the parameter value is updated according to some deterministic rule without using any feedback from the population. The deterministic mutation rate schedule implementation proposed in Smith and Fogarty, (1997) has successful results for hard combinatorial problems.
- *Self adaptive*: the parameter is evaluated and updated by the evolutionary algorithm itself by encoding the parameters into the chromosomes and undergo mutation and recombination. The basic idea is that better parameter values will survive in the population since they belong to the surviving individuals. Bäck (1993) refers to this approach as on-line learning. In their work, they propose a self adaptation mechanism of a single mutation rate per individual.
- *Individually adaptive*: the parameter value is updated based on some feedback (usually fitness values of individuals) from the population. Srinivas and Patnaik. (1994) has proposed this approach by giving mutation rate adaptation rule in the form of following equations:

$$p_m = k_2(f_{\max} - f)/(f_{\max} - f_{avg}), \ f \geq f_{avg}$$
$$p_m = k_4, \ f < f_{avg}$$

where

f = fitness value of the individual,
$f_{\max}$ = best fitness value of the current generation,
f_{avg} = average fitness value of the current generation,
constants k_2 and $k_4 = 0.5$.

In an adaptive GA-Fuzzy algorithm developed in present research work has two important parameters namely, crossover probability (P_c) and mutation probability (P_m). They are varied dynamically during the execution of the program according to a fuzzy knowledge base which has been developed from experience to maximize the efficiency of GA.

11.3.1 OPF Problem

The optimal power flow problem is concerned with optimization of steady state power system performance with respect to an objective function f, subject to numerous constraints. For optimal active power dispatch, the objective function f is the total generation cost as expressed below:

$$\min\ f = \sum_{i=1}^{N_g}(a_i + b_i P_{gi} + c_i P_{gi}^2) \tag{11.1}$$

where

N_g = total number of generation units,
a_i, b_i and c_i = cost coefficients of generating unit,
P_{gi} = real power generation of ith unit i = $1, 2, \ldots . N_g$

subject to following constraints:
  Equality constraints as

$$P_{gi} - P_{di} - \sum_{j=1}^{N}|V_i||V_j||Y_{ij}|\cos(\delta_i - \delta_j - \theta_{ij}) = 0 \tag{11.2}$$

and

$$Q_{gi} - Q_{di} - \sum_{j=1}^{N}|V_i||V_j||Y_{ij}|\sin(\delta_i - \delta_j - \theta_{ij}) = 0 \tag{11.3}$$

Inequality constraints as

$$P_{gi}{}^{min} \le P_{gi} \le p_{gi}{}^{max} \tag{11.4}$$

$$Q_{gi}{}^{min} \le Q_{gi} \le Q_{gi}{}^{max} \tag{11.5}$$

$$V_i^{min} \le V_i \le V_i^{max} \tag{11.6}$$

$$t_k{}^{min} \le t_k \le t_k{}^{max} \tag{11.7}$$

$$\delta_{gi}{}^{min} \le \delta_{gi} \le \delta_{gi}{}^{max} \tag{11.8}$$

$$line\_flow_1 \le line\_flow_1^{max} \tag{11.9}$$

$$Q_{cm}{}^{min} \le Q_{cm} \le Q_{cm}{}^{max} \tag{11.10}$$

where, N = Total number of buses,
  N_T = Total number of tap changing transformers,
  Q_{cm} = m^{th} shunt capacitor/reactor compensations,
  N_l = Total number of lines,
  N_c = Total number of shunt capacitors
  i and $j = 1, 2, \ldots . N$,
  $k = 1, 2, \ldots .. N_T$,
  $l = 1, 2, \ldots . N_l$,
  $m = 1, 2, \ldots . N_c$,
  P_{gi} and Q_{gi} = real and reactive power generation at bus i,
  P_{di} and Q_{di} = real and reactive power demands at bus i,
  $|V_i|$ and $|V_j|$ = voltage magnitudes at bus i and j respectively,
  δ_i and δj = voltage angles at bus i and j,

$Y_{ij} = |Y_{ij}| \angle \theta_{ij} =$ admittance matrix,
$t_k =$ tap setting of kth transformer,
$line\_flow_l =$ line flow at l^{th} line

11.3.2 Synergism of GA-Fuzzy System Approach

At the starting stage, high crossover probability and low mutation probability yield good results, because a large number of crossover operations produce better chromosomes for a finite number of generations, after that the fitness value of each chromosome vector becomes almost equal. Beyond this the effect of crossover is insignificant due to little variation in the chromosome vectors in that particular population. At later stages, increasing the mutation rate of the chromosomes inculcates new characteristics in the existing population and therefore diversifies the population.

Therefore, philosophy behind varying Pc and Pm is that the response of the optimization procedure depends largely on the stage of optimization, i.e. a high fitness value may require relatively low crossover and high mutation probabilities for further improvement, alternatively, at low fitness values the response would be better with relatively high crossover and low mutation probabilities.

Schuster (1985) proposed heuristics for optimal setting of the mutation probability (Pm). Fogarty, (1981) and Booker (1987) investigated time dependencies on the mutation and crossover probabilities respectively. Grefenstette, (1981) and Schaffer (1981) found optimal settings for all these parameters of the GA by experiment.

In this work, a GA-Fuzzy approach is used in which ranges of parameters – crossover probability (Pc) and mutation probability (Pm) have been divided into LOW, MEDIUM and HIGH membership functions.

The GA parameters (Pc and Pm) are varied based on the fitness function values as per the following logic:

The value of the best fitness for each generation (BF) is expected to change over a number of generations, but if it does not change significantly over a number of generations (UN) then this information is considered to cause changes in both Pc and Pm.

The diversity of a population is one of the factors, which influences the search for a true optimum. The variance of the fitness values of objective function (VF) of a population is a measure of its diversity. Hence, it is also considered as another factor on which both Pc and Pm may be changed.

The membership functions and membership values for these three variables (BF, UN and VF) are selected after several trials to get optimum results.

11.3.3 GA-Fuzzy System Approach for OPF Solution (GAF-OPF)

Figure 11.1 is a diagrammatic representation of an approach to incorporate fuzzy logic to find GA based OPF solution.

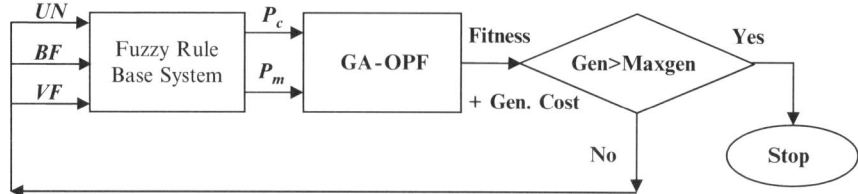

Fig. 11.1. Implementation of fuzzy system to GA for OPF solution

Therefore, this approach may be divided broadly in two parts namely GA-OPF and fuzzy rule base system (for controlling the GA parameters Pc and Pm dynamically during execution).

(A) GA technique for OPF

In GA-OPF, GA is used as a search technique for optimization of power flow in different lines of the power system. The GA requires the evaluation of the so-called fitness function (FF) to assign a quality value to every solution produced. Movement in a GA is accomplished using three primary operations: Parent reproduction, crossover and mutation. The details of important operations during solution of GA-OPF are as follows:

1. *Encoding*

Binary coded strings having 1s and 0s are used for building chromosomes through random process. The randomly generated chromosomes represent binary coded values of controllable variables e.g. power generation at all generator (PV) buses other than slack bus, the voltage magnitude at all PV buses, tap settings of variable tap transformers and shunt capacitor/reactor compensations.

The bits of each chromosome are separated out for different control variables and are converted into equivalent decimal values by the following formula:

$$X_i = X_i^{min} + dec_i(b_1 b_2 \ldots\ldots)_2 \times ((X_i^{max} - X_i^{min})/(2^{bits\_reqd}i - 1)) \quad (11.11)$$

where,

$dec_i(b_1\ b_2\ldots\ldots)_2$ =decimal values of bits corresponding to *ith* control variable,
X_i^{min} = minimum generation value of *ith* control variable,
X_i^{max} = maximum generation value of *ith* control variable,
$bits\_reqd$ = Total number of bits required to represent *ith* control variable.

Load flow using Newton–Raphson method is run for set of control variables values belonging to each chromosome. If load flow converges and slack bus generation obtained from load flow solution is within specified limits then chromosome is included to complete initial population. Otherwise, a new chromosome is generated according to same procedure and checked again.

2. *Fitness function evaluation*

GAs are usually designed so as to maximize the fitness function (FF), which is a measure of quality of each candidate solution. The objective of the OPF problem is to minimize the total generation cost including power flow constraint for each line and other equality and inequality constraints stated above. In proposed GA-Fuzzy approach, penalty index (pen_index_i) for each generated chromosome is calculated for lines having power overflows $(over\_flow_l)$, based on respective penalty factors (p_l) as follows:

$$pen\_index_i = \sum_{l=i}^{n_i} p_l * overflow_1 \qquad (11.12)$$

and fitness function is modified to keep line flows under limits as:

$$FF_{i=}\{A/(1+cost_i)\}e^{-(k^* pen\_index)_i} \qquad (11.13)$$

Where as

$i = 1$ to population size,
$n_l =$ total number of lines in system,
$l = 1$ to n_l,
$over\_flow_l =$ overflow in l^{th} line, if any otherwise zero,
$pen\_index_i =$ penalty index for i^{th} chromosome,
$FF_i =$ fitness value of function for i^{th} chromosome,
A *and* $k =$ large numerical constant,
$cost_i =$ cost corresponding to i^{th} chromosome.

3. *GA operators*

As a next step in solution finding process, GA operators – *Reproduction, Crossover* and *Mutation* are applied in above sequence for each generation. The reproduction operator selects a chromosome string from the previous generation based on the string's fitness and its probability of propagation to the next generation. In the reproduction operator a stochastic remainder selection is used instead of simple Roulette wheel. In simple Roulette wheel selection, there is no guarantee that the best strings would be selected. To overcome this problem the stochastic selection is used in this work. Selection continues until the population of the next generation is filled. The crossover and mutation operators work in conjunction with selection similarly as in simple GA. The values for P_c and P_m are assigned respectively for first generation, then after these values are determined by fuzzy rule base for the successive generations.

After crossover and mutation, load flow using Newton-Raphson method is run. If load flow converges and slack bus generation obtained from load flow solution is within specified limits then chromosome is included to valid population. For any generation, the minimum generation cost amongst all valid chromosome and corresponding generation pattern is stored in variable C_{min}.

For first generation, value of C_{min} is stored in another variable C_{min_gen} representing generation minimum cost, and for successive generations if $C_{min} <$ C_{min_gen} then C_{min_gen} is replaced by C_{min} otherwise C_{min_gen} of previous generation is reconsidered. The process continues till last generation.

11.3.3.1 Fuzzy System for Controlling Crossover and Mutation Probability

The best fitness (BF) for each generation, number of generations for unchanged BF (UN) and variance of fitness values of objective functions (VF) for population of each generation are computed. These variables values are fed as input to fuzzy rule base system, as shown in Fig. 11.1.

Fuzzy rule base for GA-fuzzy approach

The GA parameters (PC, Pm) in GA-Fuzzy algorithm are varied based on fuzzy rules base as mentioned in earlier chapter for the solution of optimal power flow (OPF).

11.3.4 Test Results

GA-OPF and GA-Fuzzy OPF proposed here are tested by solving various test systems. These systems are 26-bus system (Saadat 2002), 6-bus system (Osman et al. 2004), IEEE 30-bus system and modified IEEE 30-bus system (Lee et al. 1985; Lai et al. 1997). The data for all the above systems are given in Appendix C, D, E, F respectively. The test examples have been run on 1.7 GHz Celeron with 128 MB RAM PC.

11.3.4.1 6-Bus System

Osman et al., (2004) have developed a modified co-evolutionary genetic algorithm (M-COGA) and compared the results with classical economic dispatch and standard flow (ED+LF), Weber (1997) and simulated annealing (OPFSA) (2003) on a 6-bus system. The proposed GA-Fuzzy OPF and GA-OPF are tested using the GA parameters given below:

Population size:	50,
Maximum no. of Generations:	200,
Selection operator:	Stochastic remainder,
Initial crossover probability:	0.9,
Initial mutation probability:	0.01

The voltage magnitude limits, active and reactive power limits and line flows limits are taken same as in references (Osman et al. 2004; Web 1997). All the lines have power flow limit of 100 MVA, except line 4–5 whose limit is 50 MVA. The values of P_c and P_m changes from 35th generation and remain constant after 71st generation ($P_c \approx 0.5676$ and $P_m \approx 0.0665$), as shown in Fig. 11.2b. It is observed that convergence of GA-Fuzzy OPF is better than GA-OPF as shown in Fig. 11.2a. The results are tabulated in Table 11.1. The results highlight the goodness of this solution technique having minimum generation cost while satisfying all constraints. Load flow solution and lines flows are given in Table 11.2.

In ED + LF method though the cost is low but losses are more as compare to GA-Fuzzy OPF and also there are certain limit violations.

11.3.4.2 26 Bus System

The 26 bus system has 46 branches, 6 generators and 7 variable tap transformers (Saadat 2002). The OPF problem has been solved GA-OPF and GA-Fuzzy OPF. The performance of the method proposed by Sadaat (2002) and GA-OPF are compared with GA-Fuzzy OPF. GA-OPF and GA-Fuzzy OPF are compared for same initial population and following GA parameters in Table 11.3.

In GA-Fuzzy OPF approach, P_c and P_m are dynamically changed during execution and governed by fuzzy rules as shown in Fig. 11.3b.

For GA-OPF and GA-Fuzzy OPF, transformers tap settings are assumed to vary within a range of ±10% of rated values. The lower voltage magnitude limits for all buses are 0.9 p.u. whereas the upper limits for PV buses are 1.1 p.u. and for remaining buses including the slack bus the limit is 1.025 p.u..

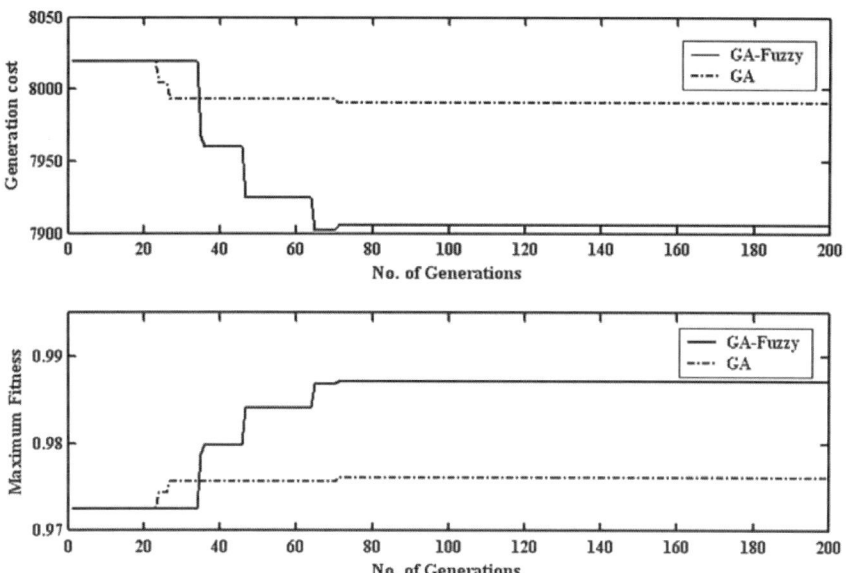

Fig. 11.2a. Convergence of generation cost & max. fitness for GA-OPF & GA-Fuzzy OPF for 6-bus. Generation cost is in $/h

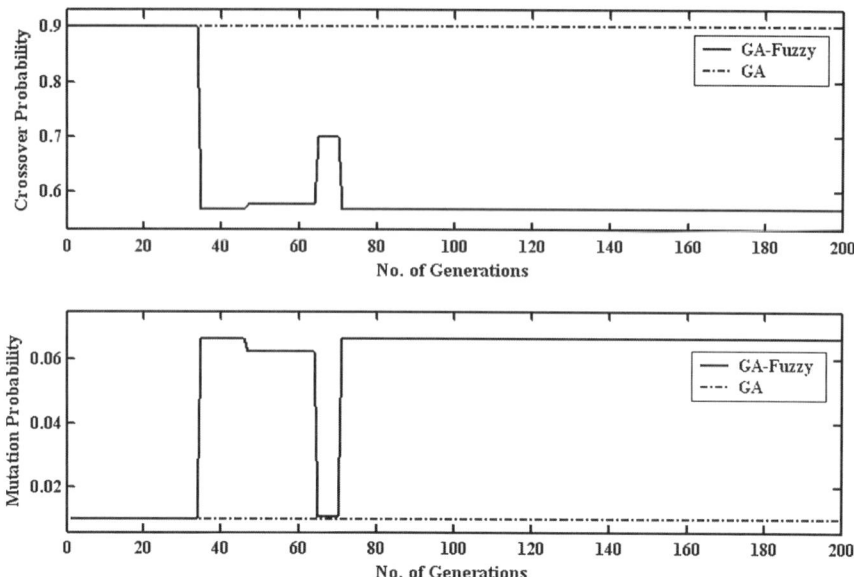

Fig. 11.2b. Crossover and mutation probabilities variations for GA-OPF & GA-Fuzzy OPF for 6-bus system

Table 11.1. Comparison of different OPF methods for 6-bus system

	Classical optimization methods		Non-classical optimization methods			
	ED + LF	Weber [23]	OPFSA [24]	M-COGA [22]	GA-OPF	GA-Fuzzy OPF
Unit 1 (MW)	99.74	160.39	131.80	152.3252	108.466	140.865
Unit 2 (MW)	216.17	133.39	190.98	151.6563	235.337	188.025
Unit 3 (MW)	50.00	143.00	109.15	118.0913	130.938	100.244
Unit 4 (MW)	250.00	169.00	178.24	187.0893	134.262	180.205
Cost ($/h)	7,860	8,062	7,938	7,987.1764	7,990.2795	7,905.9163
Losses (MW)	15.91	5.38	6.33	9.2088	9.003	9.33
Violating quantities	2	0	0	0	0	0

As shown in Fig. 11.3b, the values of P_c and P_m in GA-Fuzzy OPF change from 8th generation and remain constant after 14th generation ($P_c. \approx 0.56759$ and $P_m \approx 0.06654882$). It is evident from Fig. 11.3a and comparison of methods tabulated in Table 11.4, that GA-Fuzzy OPF has better convergence rate and results least generation cost amongst the three methods. Transformer tapings and voltage magnitudes (Table 11.5) are also found to be within limits.

Table 11.2. Load flow solution and lines flows of 6-bus system using GA-Fuzzy OPF

Bus	Voltage (pu)	Angle(degrees)	Load		Generation	
			MW	MVAr	MW	MVAr
1	1.001	0.000	100	20	140.856	8.357
2	1.017	1.338	100	20	188.025	15.495
3	1.01	−5.495	100	20	100.244	95.456
4	1.00	−1.300	100	20	180.205	11.556
5	0.977	−3.489	100	50	0.0	0.0
6	0.981	−5.664	100	10	0.0	0.0

From bus	To bus	Line flow (MVA)
1	2	32.176
2	4	56.43
1	5	72.725
3	5	52.314
4	5	49.347
3	6	31.607
4	6	87.413

Table 11.3. GA parameters

Population size	30
Maximum generation	100
Initial crossover probability	0.9
Initial mutation probability	0.01
Selection operator	Stochastic remainder

11.3.4.3 IEEE 30-Bus System

The proposed GA-Fuzzy OPF is also applied to IEEE 30 bus system. Two sets of generator cost curves are used to illustrate the robustness of the technique. In case (i) a quadratic cost curve (Alsac and Stott 1974; Yuryevich and Wong 1999) is taken. In case (ii), some of the cost curves are replaced with quadratics plus sine components [YUR99]. Therefore in case (ii), there are many local optimum solutions for the dispatch problem and as a result steepest descent (SD) method cannot determine the global optimum solution. The problem is therefore well suitable for validating the proposed algorithm. The GA-OPF and GA-Fuzzy OPF are compared for IEEE 30-bus system for same parameters as 26 bus system discussed earlier except a mutation probability (= 0.005), population size (= 50) and maximum number of generations (= 50).

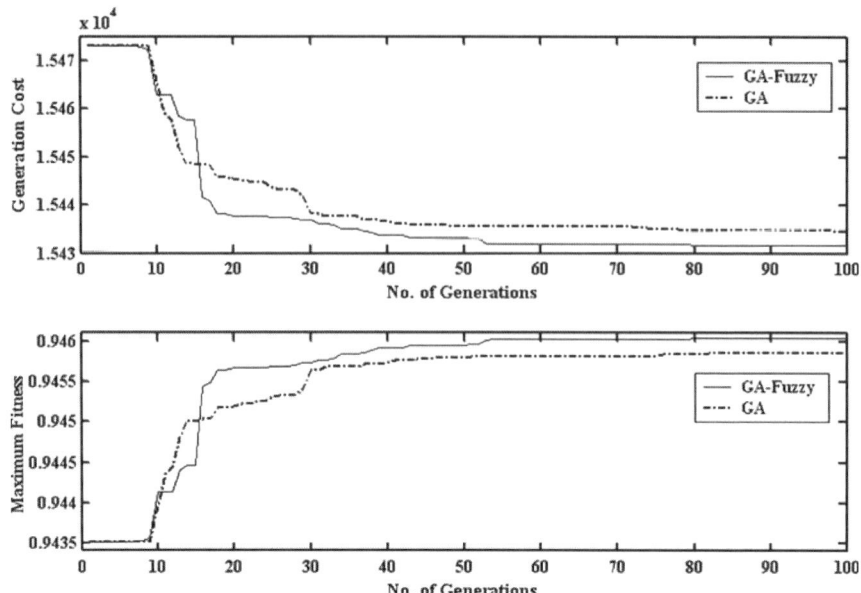

Fig. 11.3a. Convergence of generation cost and max. fitness for GA-OPF & GA-Fuzzy OPF for 26 Bus system. Generation cost is in $/h

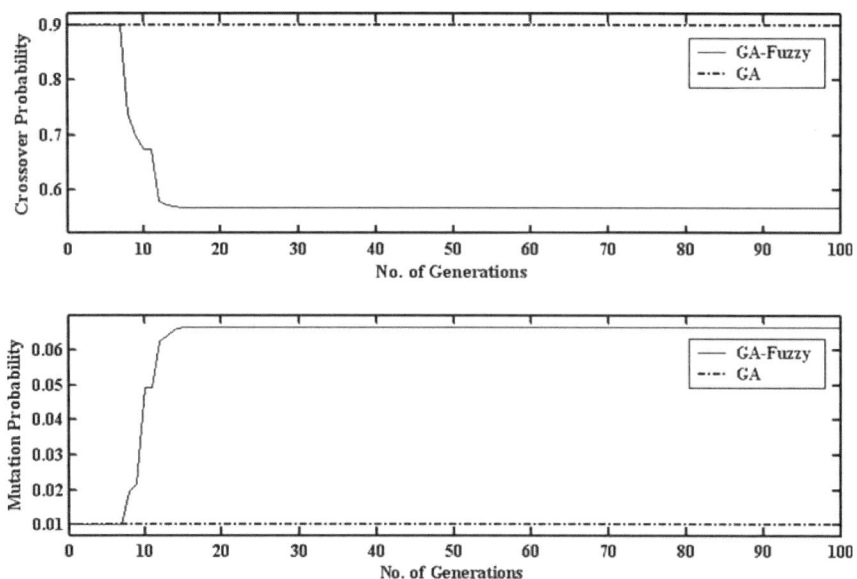

Fig. 11.3b. Crossover and mutation probabilities variations for GA-OPF & GA-Fuzzy OPF for 26 bus system

Table 11.4. Comparison of different OPF methods for 26 bus system

	Generation		
	Sadaat-OPF (in MW)	GA-OPF (in MW)	GA-Fuzzy OPF (in MW)
Bus no. 1	447.611	444.703	449.642
Bus no. 2	173.087	170.968	162.317
Bus no. 3	263.363	258.495	264.086
Bus no. 4	138.716	135.239	139.932
Bus no. 5	166.099	181.525	173.9
Bus no. 26	86.939	83.939	85.924
Gen.Cost ($/h)	15447.72	15434.67	15431.69
Losses (MW)	12.8	11.869	12.8

Table 11.5. Load flow solution and transformer tap settings of 26 bus system using GA-Fuzzy OPF

Bus no.	Voltage magnitude (in p.u.)	Angle (in degrees)	Load	
			MW	MVAr
1	1.025	0	51	41
2	1.025	−0.239	22	15
3	1.074	−0.47	64	50
4	0.91	−2.138	25	10
5	1.026	−1	50	30
6	0.994	−2.771	76	29
7	0.992	−2.388	0	0
8	0.992	−2.258	0	0
9	0.979	−4.476	89	50
10	0.976	−4.334	0	0
11	0.992	−2.798	25	15
12	0.987	−3.325	89	48
13	0.991	−1.12	31	15
14	0.984	−2.338	24	12
15	0.978	−3.156	70	31
16	0.97	−3.907	55	27
17	0.974	−4.563	78	38
18	1.004	−1.872	153	67
19	0.976	−6.075	75	15
20	0.967	−4.777	48	27
21	0.965	−5.452	46	23
22	0.963	−5.363	45	22
23	0.956	−6.428	25	12
24	0.949	−6.726	54	27
25	0.955	−6.329	28	13
26	1.015	−0.324	40	20
Total			1263.00	637.00

Table 11.5. (*Continued*)

Transformer tap settings						
Line 2–3	Line 2–13	Line 3–13	Line 4–8	Line 4–12	Line 6–19	Line 7–9
0.98	1.000	1.080	0.932	0.90	0.983	0.9903

Table 11.6. Best and worst solutions for GA-Fuzzy OPF for IEEE 30 bus system (quadratic cost curve)

	Worst solution (\$/h)	Best solution (\$/h)	% Difference
EP [36]	805.61	802.62	0.371147
GA-Fuzzy OPF	802.32054	802.00031	0.03991

Case (i) Quadratic Cost Curve

In this case the unit cost curves are represented by quadratic functions. The program is tested for 100 different runs. The generation costs of 802.32054 \$/h and 802.00031 \$/h are obtained for worst and best solutions, respectively (0.03991% difference), through GA-Fuzzy OPF. It shows the consistency in the results and better performance of the proposed method than evolutionary programming (EP) OPF for the same number of runs (Table 11.6).

As shown in Fig. 11.4b, the values of P_c and P_m change from 2nd generation and remain constant after 15th generation ($P_c. \approx 0.5676$ and $P_m \approx 0.0666$) for the best solution. The solutions obtained from other GA and non-GA techniques available in literature (Roa and Pavez-lazo 2003; Alsac and Stott 1974; Abido 2002; Paranjothi and Anburaja 2002; Yuryevich and Wong 1999) are compared in Table 11.7. The load flow and transformer tap settings for best solution are provided in Table 11.10. It is observed that in GA-Fuzzy OPF a better convergence rate is obtained (as in Fig. 11.4a) and a minimum generation cost is also achieved in GA-Fuzzy OPF (as in Table 11.7).

Case (ii) Quadratic Cost Curve with Sine Components

In this case, a sine component is added to the quadratic equation cost of the generators at buses 1 and 2 to reflect the valve-point loading effects. The values of cost coefficients are given in Table 11.8.

The cost curves of other generators are taken same as in case (i). The algorithm is tested for 100 different runs. The generation costs of the worst and the best solutions are 924.3387 and 921.3506 \$/h, respectively (0.323% difference). As per Table 11.9, percentage difference between worst and best solution for GA-Fuzzy OPF is less than evolutionary programming (EP) based OPF. Therefore, GA-Fuzzy approach is found to be superior in solving OPF for cost curve with sine components for same number of runs.

As per Fig. 11.6b values of P_c and P_m vary from 4th generation onwards till 50th generation ($P_c. \approx 0.57292$ and $P_m \approx 0.06392$) for best solution. The

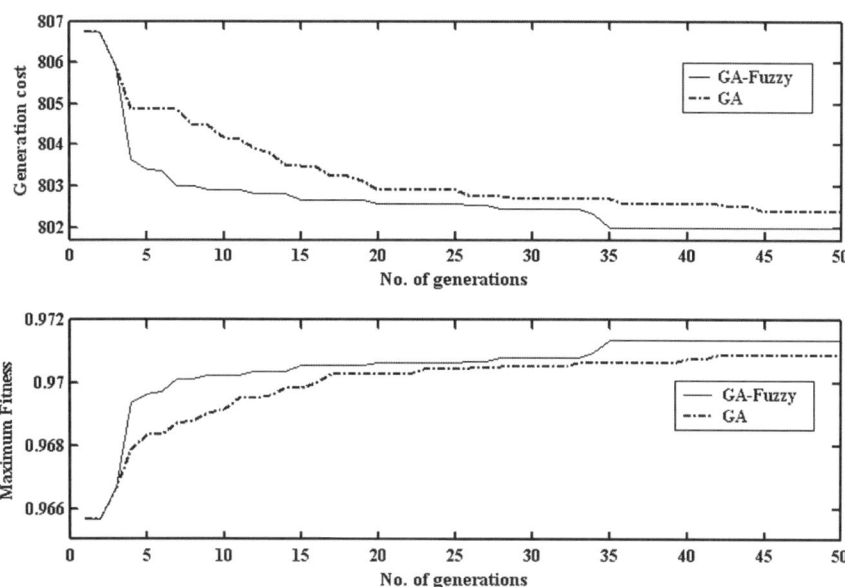

Fig. 11.4a. Convergence of generation cost and max. fitness for GA-OPF & GA-Fuzzy OPF for IEEE 30 bus system (case i). Generation cost is in $/h

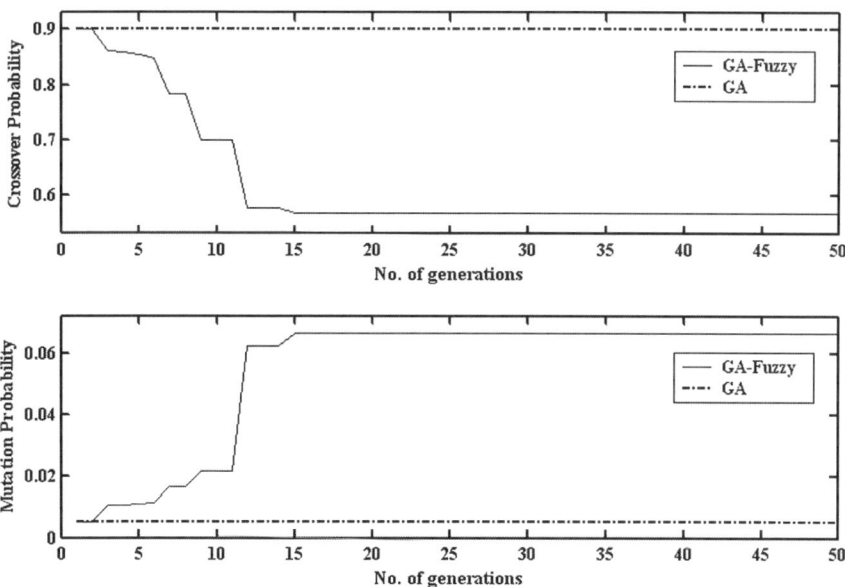

Fig. 11.4b. Crossover and mutation probabilities variations for GA-OPF & GA-Fuzzy OPF for IEEE 30 bus system (case ii)

Table 11.7. Comparison of different OPF methods for IEEE 30 bus system

Quantity	Proj. Gradient [33]	OPFSA [31]	Tabu-Search [34]	GA [35]	RGA [35]	Decoupled OPF by RGA [35]	EP [36]	EGA [37]	GA OPF	GA-Fuzzy OPF
P_1	138.56	173.15	176.04	170.1	174.4	173	173.848	176.2	175.6439	174.9664
P_2	57.56	48.54	48.76	53.9	46.8	48.7	49.998	48.75	48.941	50.35294
P_5	24.56	19.23	21.56	20.6	22.0	26.8	21.386	21.44	21.1765	21.45098
P_8	35.00	12.81	22.05	18.8	23.9	17.7	22.630	21.95	22.647	21.17647
P_{11}	17.93	11.64	12.44	12.0	11.0	10.5	12.928	12.42	12.431	12.66667
P_{13}	16.91	13.21	12.00	17.7	14.5	16.3	12.000	12.02	12.00	12.10980
(in MW)										
Gen.Cost ($/h)	813.74	799.45	802.29	805.94	804.02	806.86	802.62	802.06	802.3841	802.0003

Table 11.8. Generator cost coefficients for case (ii)

Bus	$P_G{}^{min}$ (in MW)	$P_G{}^{max}$ (in MW)	Cost coefficients				
			a	b	c	d	e
1	50	200	150	2.00	0.0016	50.00	0.0630
2	20		25	2.50	0.0100	40.00	0.09890
		80					

Generation cost function: $\cos t_i = a_i + b_i P_{g_i} +_i c_i P_{g_i}^2 + |d_i \sin(e_i(P_{g_i}^{min} - P_{g_i}))|$

Table 11.9. Best and worst solutions for GA-Fuzzy OPF for IEEE 30 bus (quadratic cost curves with sine components)

	Worst solution (\$/h)	Best solution (\$/h)	% Difference
EP [36]	926.68	919.89	0.7327
GA-Fuzzy	924.336729	921.350629	0.323

solution details including load flow, transformer tap settings and line flows are provided for best solution in Table 11.10.

The line flows obtained in this case are within the limits and other constraints are also satisfied. Again GA-Fuzzy OPF proves to be consistently superior to GA-OPF due to faster convergence and lesser generation cost, as shown in Fig. 11.5a.

11.3.4.4 Modified IEEE 30-Bus System

The original IEEE 30-bus network consists of 6 generator buses, 21 load buses and 41 lines, of which 4 lines (6, 9),(6, 10),(4, 12) and (28, 27) are under-load-tap-setting transformer lines. In modified IEEE 30-bus system buses 10, 12, 15, 17, 20, 21, 23, 24 and 29 have been selected as shunt capacitor/reactor compensation buses. The apparent power flow limit in line (8, 28) is taken as 12 MVA.

The GA-OPF and GA-Fuzzy OPF are compared as shown in Fig. 11.6. for this system for same parameters as for 6-bus system discussed earlier except crossover probability ($=0.95$), mutation probability ($=0.005$), population size ($=50$) and maximum number of generations ($=50$).

Hence for best case solution, the changes in values of P_c and P_m start from 4th generation and till 50th generation ($P_c. \approx 0.67479$ and $P_m \approx 0.04901$), as shown in Fig. 11.7a, and b. The OPF solutions obtained from other GA technique based on dynamical hierarchy of the coding system and non-GA technique based on gradient projection method (GPM) are available in literature (Lee et al. 1985), respectively. These results are compared along with other methods in Table 11.11.

It indicates minimum generation cost obtained due to optimal values of controllable variables, i.e. active power generations, generator bus voltages,

Table 11.10. Load flow solution and transformer tap settings of IEEE 30 bus system using GA-Fuzzy OPF

Bus	Voltage in p.u.		Angle in degrees		Generation				Load	
	Case (i)	Case (ii)	Case (i)	Case (ii)	MW Case (i)	MW Case (ii)	MVAr Case (i)	MVAr Case (ii)	MW	MVAr
1	1.05	1.05	0	0	174.9664	199.672	−6.562	−9.126	0	0
2	1.034	1.034	−3.608	−4.335	50.35294	20	22.356	39.212	21.7	12.7
3	1.022	1.016	−5.675	−5.875	0	0	0	0	2.4	1.2
4	1.016	1.008	−6.817	−7.062	0	0	0	0	7.6	1.6
5	1.006	1.006	−10.509	−11.007	21.45098	22.275	30.372	31.548	94.2	19
6	1.008	1.005	−7.944	−8.232	0	0	0	0	0	0
7	0.999	0.998	−9.529	−9.904	0	0	0	0	22.8	10.9
8	1.003	1.003	−8.154	−8.431	21.17647	23.725	18.89	25.987	30	30
9	1.029	1.018	−10.152	−10.121	0	0	0	0	0	0
10	1.021	1.027	−12.059	−11.951	0	0	0	0	5.8	2
11	1.071	1.051	−8.783	−8.483	12.66667	14.706	21.737	16.89	0	0
12	1.018	1.038	−11.139	−11.135	0	0	0	0	11.2	7.5
13	1.048	1.048	−10.228	−10.144	12.1098	13.427	22.635	8.075	0	0
14	1.005	1.023	−12.096	−12.07	0	0	0	0	6.2	1.6
15	1.002	1.019	−12.242	−12.188	0	0	0	0	8.2	2.5
16	1.012	1.026	−11.832	−11.764	0	0	0	0	3.5	1.8
17	1.013	1.021	−12.214	−12.111	0	0	0	0	9	5.8
18	0.997	1.009	−12.917	−12.825	0	0	0	0	3.2	0.9
19	0.996	1.007	−13.113	−13.005	0	0	0	0	9.5	3.4
20	1.002	1.011	−12.911	−12.801	0	0	0	0	2.2	0.7
21	1.009	1.015	−12.535	−12.43	0	0	0	0	17.5	11.2
22	1.009	1.016	−12.524	−12.422	0	0	0	0	0	0
23	0.996	1.01	−12.707	−12.652	0	0	0	0	3.2	1.6
24	0.997	1.008	−12.956	−12.913	0	0	0	0	8.7	6.7
25	1.001	1.015	−12.759	−12.826	0	0	0	0	0	0
26	0.983	0.997	−13.193	−13.248	0	0	0	0	3.5	2.3
27	1.012	1.028	−12.361	−12.5	0	0	0	0	0	0
28	1.002	0.999	−8.433	−8.71	0	0	0	0	0	0
29	0.992	1.008	−13.619	−13.72	0	0	0	0	2.4	0.9
30	0.98	0.996	−14.522	−14.595	0	0	0	0	10.6	1.9
Total					292.7233	293.806	109.43	112.585	283.4	126.20

	Gen. Cost ($/h)	Losses (MW)	Transformer tap setting			
			Line 6–9	Line 6–10	Line 4–12	Line 28–27
Case (i)	802.0003	9.494	1.0032	0.9645	1.0161	0.9645
Case (ii)	921.3506	10.406	1.0355	0.9129	0.9452	0.9452
1–2	117.211	139.6258				
1–3	58.3995	61.0022				
2–4	34.0758	30.7947				
2–5	63.7783	61.7539				
2–6	45.3399	41.6185				
3–4	54.5622	56.8812				
4–6	50.2703	48.6008				
4–12	30.5889	33.3706				
5–7	14.1355	16.1981				
6–7	33.9924	35.221				

Table 11.10. (*Continued*)

Lines	Line flow (in MVA)	
	Case (i)	Case (ii)
6–8	13.6882	9.2172
6–9	22.4033	27.0031
6–10	14.6187	20.0154
6–28	16.5409	16.0689
8–28	3.3685	3.0049
9–11	24.1764	21.6919
9–10	32.7929	31.3223
10–20	11.0315	9.8192
10–17	9.861616	7.33966
10–21	18.96153	18.074
10–22	9.0741	8.5101
12–13	24.9376	15.5102
12–14	7.6911	8.1147
12–15	17.4525	18.8445
12–16	6.34027	7.723838
14–15	1.2313	1.5983
15–18	5.4066	6.18
15–23	4.5343	5.358
16–17	3.2983	3.756
18–19	2.3627	2.8022
19–20	8.5117	7.3363
21–22	2.0887	3.1629
22–24	6.9397	5.7964
23–24	1.4447	1.8148
24–25	1.3934	1.8788
25–26	4.2647	4.2626
25–27	5.633	6.1149
27–29	6.4154	6.4095
27–30	7.2897	7.2825
28–27	19.7428	20.07
29–30	3.7542	3.7525

transformer taps and shunt capacitors/reactive compensations. The convergence of GA-Fuzzy OPF is better than GA-OPF as evident from Fig. 11.7a. The apparent power flows at line (8, 28) is 3.034 MVA and 3.317 MVA for GA-OPF and GAF-OPF respectively. The load flow solution for best solution is provided in Table 11.12.

GA-Fuzzy OPF is run for 100 different runs with different initial populations on above system. The convergence graphs for generation costs and maximum fitness in best and worst cases are shown in Fig. 11.8, which are converging very close to each other. The total generation cost obtained in worst case is 801.1601 $/h. Therefore, GA-Fuzzy OPF gives consistently good results as percentage deviation between best case and worst case generation costs is $\approx 0.089\%$, which is a very small variation.

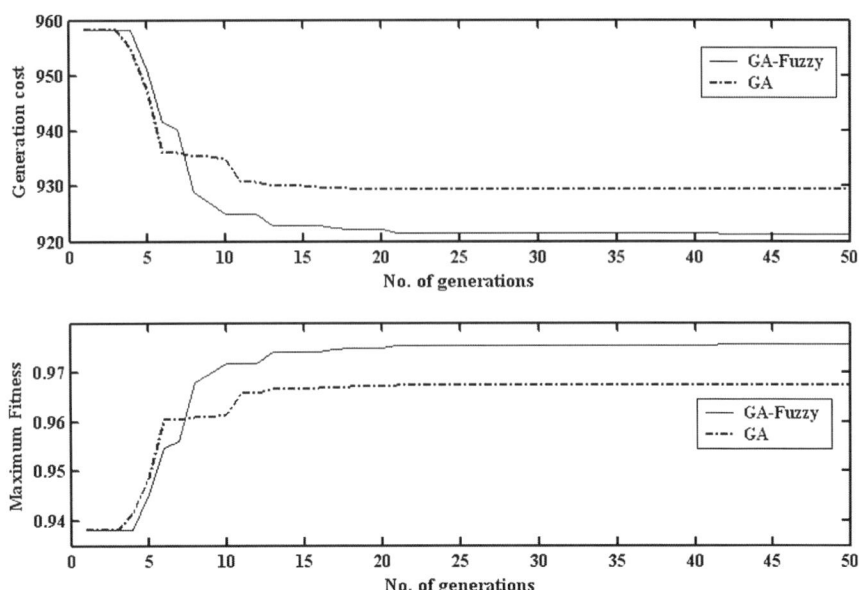

Fig. 11.5a. Convergence of generation cost and Max. fitness for GA-OPF & GA-Fuzzy OPF for IEEE 30 bus system (case ii). Generation cost is in $/h

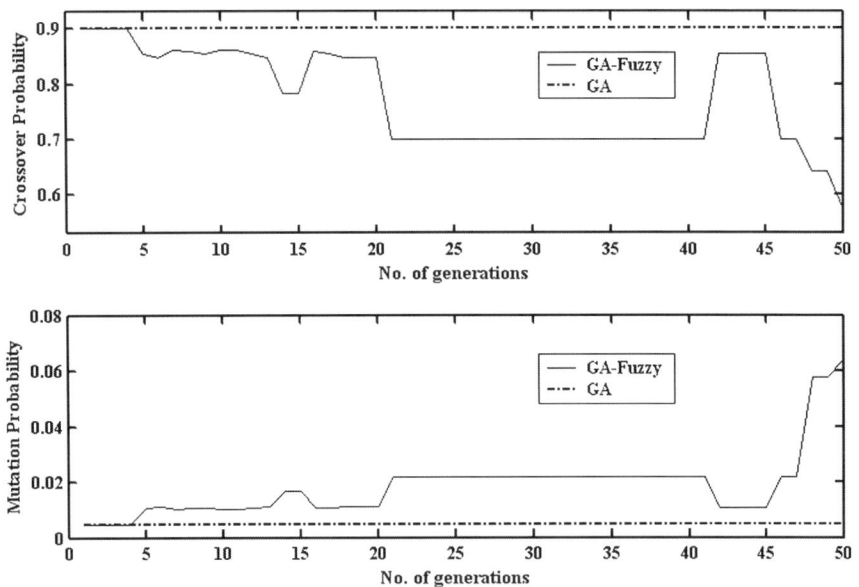

Fig. 11.5b. Crossover and mutation probabilities variations for GA-OPF & GA-Fuzzy OPF for IEEE 30 bus system (case ii)

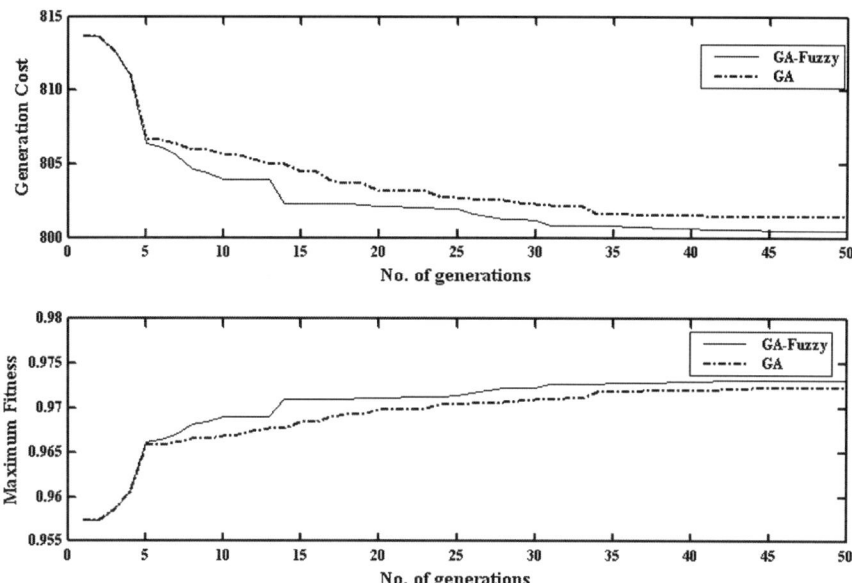

Fig. 11.6a. Convergence of generation cost & max. fitness for GA-OPF & GA-Fuzzy OPF for modified IEEE 30-bus system

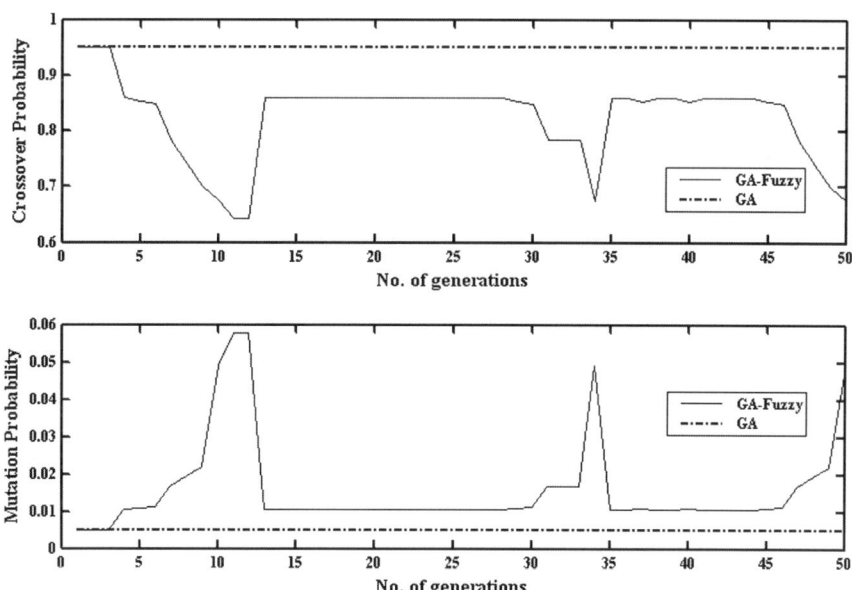

Fig. 11.6b. Crossover and mutation probabilities variations for GA-OPF & GA-Fuzzy OPF for modified IEEE 30-bus system

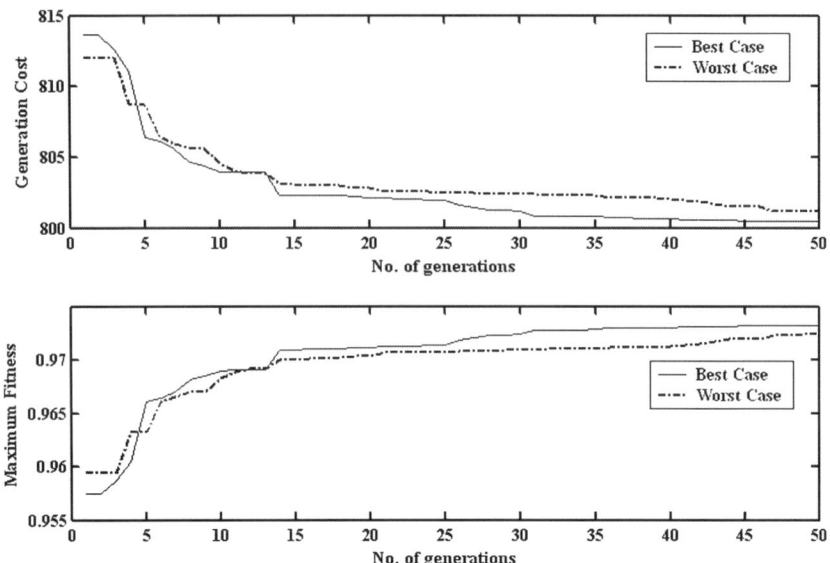

Fig. 11.7. Convergence of generation cost & max. fitness for best and worst cases for modified IEEE 30-bus system using GA-Fuzzy OPF

Table 11.11. Comparison of different OPF methods for modified IEEE 30-bus system

Bus	Active power generations (in MW)						Generat-ion cost ($/h)	Real power losses (MW)
	1	2	5	8	11	13		
Lee et al. [25]	187.219	53.781	16.955	11.288	11.287	13.355	804.853	10.485
Lai et al. [26]	177.7594	48.722	21.454	20.954	11.768	12.052	800.805	9.309
GA-OPF	175.462	50.118	22.000	20.686	11.882	12.439	801.447	9.187
GA-Fuzzy OPF	174.886	48.941	21.176	22.647	12.588	12.000	800.442	8.838

Bus	Generator bus voltages (in p.u.)					
	1	2	5	8	11	13
Lee et al. [25]	1.1	1.08	1.03	1.04	1.08	1.08
Lai et al. [26]	1.081	1.063	1.034	1.038	1.1	1.055
GA-OPF	1.073	1.052	1.03	1.04	1.076	1.061
GA-Fuzzy OPF	1.081	1.063	1.031	1.039	1.095	1.07

Table 11.11. (*Continued*)

Line	Transformer tap settings			
	(6,9)	(6,10)	(4,12)	(28,27)
Lee et al. [25]	1.072	1.07	1.032	1.068
Lai et al. [26]	1.0	0.975	0.975	1.0
GA-OPF	1.016	1.0419	1.087	1.0097
GA-Fuzzy OPF	0.99032	0.98387	0.99032	0.96451

Bus	Shunt capacitor/reactor compensations (in MVAr)								
	10	12	15	17	20	21	23	24	29
Lee et al [25]	0.692	0.046	0.285	0.287	0.208	0.000	0.330	0.938	0.269
Lai et al. [26]	0.1	0.7	1.9	2.4	1.5	2.2	4.7	4.7	2.4
GA-OPF	3.033	2.544	4.618	4.266	4.736	0.528	2.476	4.442	4.194
GA-Fuzzy OPF	3.982	0.02	4.149	4.99	4.432	4.354	4.54	4.687	2.097

11.3.5 Conclusions

The proposed GA-Fuzzy OPF has also been tested in different test systems as indicated earlier. It has shown better results in terms of convergence, consistency in different runs and minimum generation cost as compared to simple GA-OPF and the other techniques. These advantages are mainly due to the changes in crossover and mutation probabilities values which are governed by a set of fuzzy rule base, although they are stochastic in nature. The variations in above GA parameters governed by fuzzy rule base have resulted in lesser generation costs with high convergence rates than other GA and non GA-OPF variants tested for 26-bus, 6-bus, IEEE 30-bus and modified IEEE 30-bus systems (Figs. 11.9–11.12).

In order to demonstrate the real potential of such technique, the proposed GAF-OPF is successfully tested on IEEE 30 bus system for quadratic cost curve with sine components also. The results obtained are compared with EP based OPF with greater satisfaction. This proves the superiority of the proposed GA-Fuzzy OPF method to the gradient-based conventional and other GA variants for finding OPF solution.

11.4 Transmission Pricing Model Under Deregulated Environment

11.4.1 Introduction

Several methods are developed for allocation of the costs embedded in the system to various transactions (embedded cost based pricing) and those incurred by system from one additional transaction (incremental

Table 11.12. Load flow solution for modified IEEE 30-bus system using GA-Fuzzy OPF

Bus	Voltage in p.u.	Angle in degrees	Generation		Load	
			MW	MVAr	MW	MVAr
1	1.081	0	174.886	5.021	0	0
2	1.063	−3.279	48.941	30.435	21.7	12.7
3	1.05	−5.273	0	0	2.4	1.2
4	1.05	−6.347	0	0	7.6	1.6
5	1.034	−9.588	21.176	29.221	94.2	19
6	1.041	−7.361	0	0	0	0
7	1.03	−8.761	0	0	22.8	10.9
8	1.039	−7.568	22.647	36.845	30	30
9	1.049	−9.574	0	0	0	0
10	1.025	−11.429	0	0	5.8	2
11	1.095	−8.414	12.588	17.344	0	0
12	1.032	−10.671	0	0	11.2	7.5
13	1.07	−9.736	12	10.337	0	0
14	1.021	−11.649	0	0	6.2	1.6
15	1.02	−11.883	0	0	8.2	2.5
16	1.023	−11.319	0	0	3.5	1.8
17	1.022	−11.673	0	0	9	5.8
18	1.012	−12.495	0	0	3.2	0.9
19	1.01	−12.661	0	0	9.5	3.4
20	1.014	−12.451	0	0	2.2	0.7
21	1.016	−11.968	0	0	17.5	11.2
22	1.016	−11.95	0	0	0	0
23	1.015	−12.378	0	0	3.2	1.6
24	1.007	−12.42	0	0	8.7	6.7
25	1.014	−12.173	0	0	0	0
26	0.996	−12.596	0	0	3.5	2.3
27	1.026	−11.749	0	0	0	0
28	1.035	−7.814	0	0	0	0
29	1.008	−13.027	0	0	2.4	0.9
30	0.996	−13.878	0	0	10.6	1.9

cost based pricing). In (Sood 2003), an evolutionary programming based SRMC method is proposed and several embedded cost based methods in (Uttar Pradesh 2004) are proposed for Indian system. But their results are obtained for different transmission subsystems. The methodologies for determination of transmission pricing should be so designed that basic goals of transmission pricing can be achieved. Therefore, the methodology can be designed on the basis of marginal cost or embedded cost or a composite cost, i.e. the combination of marginal and embedded cost.

In this chapter, marginal cost method is used and tested on modified IEEE-30 bus system. Embedded cost allocation methods, e.g. Postage Stamp and

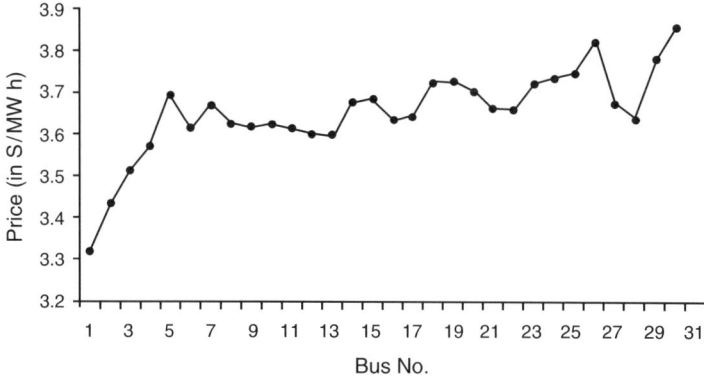

Fig. 11.8. Real marginal prices for modified IEEE 30 Bus System

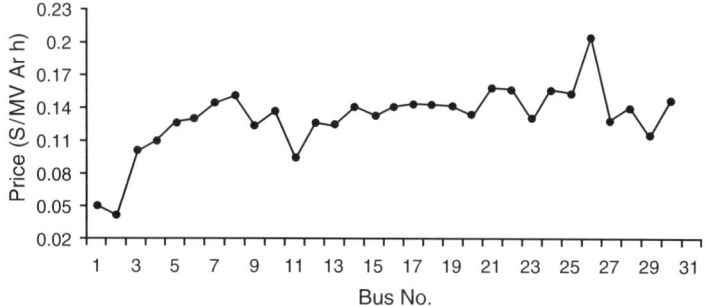

Fig. 11.9. Reactive marginal prices for modified IEEE 30 Bus System

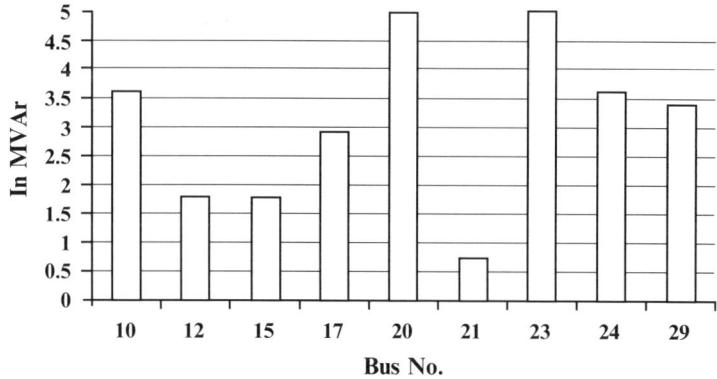

Fig. 11.10. Optimal value of shunt capacitor value for modified IEEE 30 – bus system

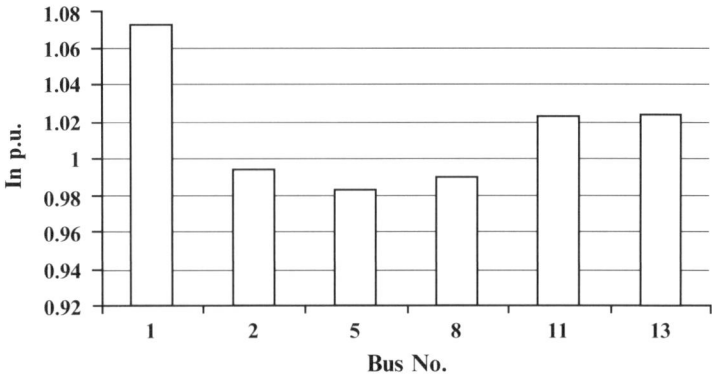

Fig. 11.11. Generator bus voltages for modified IEEE 30 – bus system

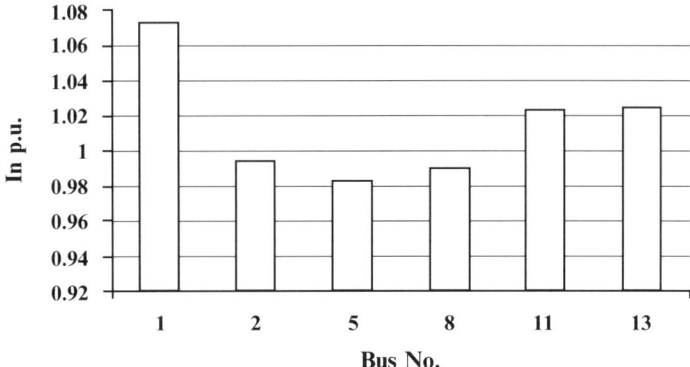

Fig. 11.12. Optimal real power generation for modified IEEE 30 – bus system

MW-Mile methods are also tested and analyzed on Indian UPSEB 75 bus system. A new variant of MW-Mile is proposed and analyzed. Finally, a hybrid type marginal cost based transmission pricing model is proposed for Indian transmission system with pool, bilateral and multilateral transactions. In this model, supplementary/complementary charges left as unrealized revenue after applying marginal cost method are allocated using the MW-Mile methods. This model is tested on Indian UPSEB 75 bus system.

11.4.2 Marginal Cost Based Transmission Pricing Method

In this section the marginal cost based transmission pricing method is analyzed and tested, which dispatches the pool in combination with privately negotiated bilateral and multilateral wheeling contracts, with maximization of social benefit with all system constraints. In the method, all scheduled firm transactions are considered to be added to the system. The method is based on GA-Fuzzy optimization technique, which has been described earlier. The losses taking place in transmission network due to transactions as well as pool are considered to be supplied from the pool itself. They are not supplied by transactions generators or cope up with transaction loss supply contracts which are complex to setup and coordinate.

As the process of developing suitable transmission pricing methodologies in India is in initial stages, hence following facts are considered for application of pricing method to modified IEEE 30 bus system and Indian UPSEB 75 bus system.

1. All the pool generators are required to bid their generation cost characteristics to the pool along with maximum generation.
2. There are no non-firm bilateral transactions.
3. The active and reactive power of pool loads are known from load forecasting and kept constant during optimization. Therefore, there is no bidding from pool demands.
4. The other costs of system like maintenance and different overheads, etc. are not being included in proposed model, which should be considered independently.

1. *Mathematical formulation*

Let n = number of buses in a system
$nfbt$ = number of schedule firm bilateral transactions
$nfmt$ = Groups of schedule firm multilateral transactions

Firm bilateral transaction load component at jth bus $(Pd_j^{fb}) = \sum_{i=1}^{n} FBT_{ij}$

$$(11.14)$$

where FBT_{ij} = Firm bilateral transactions delivered at the jth load bus from the ith generator bus.

Generation at ith bus for firm bilateral transactions $(Pg_j^{fb}) = \sum_{j=1}^{n} FBT_{ij}$

$$(11.15)$$

For a firm bilateral transaction fbt from ith to a jth bus

$$Pg_i^{fbt} = Pd_j^{fbt} \qquad (11.16)$$

where, Pg_i^{fbt} and Pd_j^{fbt} are real power generation and demand for a firm bilateral transaction fbt, at i^{th} and j^{th} bus, respectively.

Vector of real power demand from firm bilateral transactions may be written as

$$Pd^{fb} = FBT^T \times U = \left\{ Pd_j^{fb}; j = 1, 2, \ldots \ldots, n \right\} \qquad (11.17)$$

where FBT = matrix of firm bilateral transactions delivered at the jth load bus from the ith generator bus.

U = Unity vector of dimension n.

Vector of real power generation from firm bilateral transactions also be written as

$$Pg^{fb} = FBT^T \times U = \left\{ Pd_i^{fb}; i = 1, 2, \ldots\ldots, n \right\} \tag{11.18}$$

In case of a multilateral transaction, there are many generation points (at least more than one), similarly there are many load points (at least more than one).

Let PMT^k = size of kth group of multilateral transaction, i.e. total power that has to be transferred from generation points to the load points of a kth group of multilateral transaction.

ngk = number of generation points for a kth group
ndk = number of demand points for a kth group

Real power demand of kth multilateral transaction = Pd_j^{mk} where $k = 1, 2, \ldots.n$.

Real power generation from kth multilateral transaction = Pg_i^{mk} where $i = 1, 2, \ldots.n$.

For k^{th} group of multilateral transaction with total power transfer PMT^k is

$$\sum_{i=1}^{n} Pg_i^{mk} \sum_{j=1}^{n} Pd_j^{mk} = PMT^k \tag{11.19}$$

Total generation at ith bus due to $nfmt$ groups of multilateral transactions is

$$Pg_i^m = \sum_{k=1}^{nfmt} Pg_i^{mk} \tag{11.20}$$

Total demand at jth bus due to $nfmt$ groups of multilateral transactions is

$$Pd_j^m = \sum_{k=1}^{nfmt} Pd_j^{mk} \tag{11.21}$$

Generation vector of all firm multilateral transaction groups may be written as

$$Pg^m = \{ Pg_i^m; i = 1, 2, \ldots., n \} \tag{11.22}$$

Demand vector of all firm multilateral transaction groups may be written as

$$Pd^m = \{ Pd_j^m; j = 1, 2, \ldots., n \} \tag{11.23}$$

The gencos participating in the pool bid their cost function and maximum generation, which they want to deliver to the pool. After optimization of social benefit generations at power pool generation buses are known.

Let the vector of pool real power generation

$$Pg^p = \{Pg_i^p; i = 1, 2, \ldots\ldots, n\} \tag{11.24}$$

Vector of pool real power demand

$$Pd^p = \{Pd_j^p; j = 1, 2, \ldots\ldots, n\} \tag{11.25}$$

Let the vectors of the total real power demand and generation be

$$Pd^T = \{Pd_j^T; j = 1, 2, \ldots\ldots\ldots, n\} \tag{11.26}$$

$$Pg^T = \{Pg_i^T; i = 1, 2, \ldots\ldots\ldots, n\} \tag{11.27}$$

From equations (11.21), (11.23) and (11.25)

$$Pd^T = Pd^p + Pd^{fb} + Pd^m \tag{11.28}$$

Similarly, from equations (11.18), (11.22) and (11.24)

$$Pg^T = Pg^p + Pg^{fb} + Pg^m \tag{11.29}$$

All firm transactions are ready to pay the system marginal price and they do not bid.

The load point of the transaction and pool may have reactive power component in addition to real power.

Let Qd^p and Qd^{fb} be the vector of the reactive power demand due to pool and firm bilateral transaction, respectively.

$$Qd^p = \{Qd_j^p; j = 1, 2, \ldots..n\} \tag{11.30}$$

$$Qd^{fb} = \{Qd_j^{fb}; j = 1, 2, \ldots..n\} \tag{11.31}$$

In the combined power pool transaction dispatched, gencos supplying the loads by transactions may also participate in the pool. Therefore, all such gencos in combinations may meet the reactive power requirements at all the buses of the system. It means that the power balance equation (11.16) for bilateral transactions and (11.19) for multilateral transaction is not necessary for reactive power. However for specific situation when a genco is not participating in pool and it is supplying loads by a transaction and the reactive power requirement of the load is to be supplied by the genco of the transaction only, then these equations for reactive power are also valid.

It is better to supply the reactive power as per requirement of the system, rather than supplying the reactive power at the generation point of a transaction equal to reactive power of load at the load point of the transaction. All generators are paid and loads are charged for the reactive power accordingly.

2. *Objective function and constraints*

The objective function for the optimization problem is to minimize the overall costs of active and reactive power generation with the capital investment of capacitor. Based on the assumption of constant loads, to minimize the total cost is equivalent to maximize the social benefits. Therefore, suggested objective function to maximize social benefit is given as follows:

$$\min \sum_{j=1}^{ng} [C_i(Pg_i) + C_i(Qg_i)] \sum_{j=1}^{ncap} C_{cj}(Qc_j) \tag{11.32}$$

Let the active power generation cost curve bid of the generator at *ith* bus $= C_i(Pg_i)$

Reactive power generation cost of generator at *ith* bus $= C_i(Qg_i)$

Equivalent production cost of *jth* capacitor $= C_{cj}(Qc_j)$

where, $j = 1, 2, \ldots .. ncap$, as $ncap =$ Total number of capacitors operating in the system

$ng =$ Total number of pool generators.

It is seen that GA-Fuzzy OPF technique works successfully for non-linear active generation cost curves. Therefore, proposed model is also capable of handling all types functions such as linear, quadratic, non-linear, convex or non-convex, continuous or discontinuous, etc. used for representing active power generation cost curve bids function in (11.32). For sake of simplicity cost curves for active power generation are modeled by following quadratic function:

$$C_i(Pg_i) = a + Pg_i + cPg_i^2 \tag{11.33}$$

Guo et al. (2004) have used equation for reactive power generation cost of the same form of quadratic equation as (11.33) but with different a, b and c coefficients. Another form introduced in (Lamont and Fu 1999) and used in (Dai et al. 2001) is based on opportunity cost.

The equivalent production cost for capital investment return of capacitors in (11.32) can be expressed as their depreciated rate (the life span of capacitors is 15 years) as follows:

$$C_{cj}(Qc_j) = Qc_j \times \$11600/MVAr \div (15 \times 365 \times 24 \times h)\text{h}$$
$$= Qc_j \times \$13.24/(100 \ MVArh) \tag{11.34}$$

where h represents the average usage rate of capacitors taken as 2/3. Qc_j is in per unit on 100 MVA base. Equation (11.34) is a linear cost function with the slope of $dC_{cj}(Qc_j)/dQc_j = \$13.24/(100 \ MVArh)$ representing approximately the capacitor investment impacts on reactive pricing.

The equality constraints are load flow equations:

$$g(V, \delta) = 0 \qquad (11.35)$$

where

$$g(V, \delta) = \left\{ \begin{array}{l} Pg_i - Pd_i - P_i(V, \delta) \Rightarrow For\ each\ PV\ and\ PQ\ bus\ except\ slack\ bus \\ Qg_i - Qd_i - Q(V, \delta) \Rightarrow For\ each\ PQ\ bus\ only \end{array} \right\}$$

where

P_i = active power injection into ith bus
Q_i = reactive power injection into ith bus
Pd_i = active load on ith bus
Qd_i = reactive load on ith bus
Pg_i = active generation on ith bus
Qg_i = reactive generation on ith bus

The inequality constraints are:

- Active power generation Pg_i at PV buses

$$Pg_i^{\min} \leq Pg_i \leq Pg_i^{\max} \qquad (11.36)$$

where $Pg_i^{\min}$ and $Pg_i^{\max}$ are respectively minimum and maximum value of active power generation at ith PV bus.
- Reactive power generation Qg_i at PV buses

$$Qg_i^{\min} \leq Qg_i \leq Qg_i^{\max} \qquad (11.37)$$

where $Qg_i^{\min}$ and $Qg_i^{\max}$ are respectively minimum and maximum value of reactive power generation at ith PV bus.
- Reactive power output limit of capacitor

$$0 \leq Qc_j \leq Qc_j^{\max} \qquad (11.38)$$

where $Qc_j^{\max}$ is maximum value of output of capacitor at jth bus.
- Voltage magnitude V of each PV and PQ bus

$$V_i^{\min} \leq V_i \leq V_i^{\max} \qquad (11.39)$$

where, $V_i^{\min}$ and $V_i^{\max}$ are respectively minimum and maximum voltage at ith bus
- Phase angle δ of voltage at all the buses.

$$\delta_i^{\min} \leq \delta_i \leq \delta_i^{\max} \qquad (11.40)$$

where, $\delta_i^{\min}$ and $\delta_i^{\max}$ are respectively minimum and maximum allowed value of voltage phase angle at ith bus

- Transmission power limit

$$S_{ij} \leq S_{ij}^{\text{max}} \tag{11.41}$$

where, S_{ij}^{max} is the maximum rating of transmission line connecting bus i and j.

Based on the above mathematical model the corresponding Lagrangian function of this optimization problem takes the form:

$$
\begin{aligned}
L = &\sum_{i=1}^{ng}[C_i(Pg_i) + C_i(Qg_i)] + \sum_{j=1}^{ncap} C_{cj}(Qc_j) - \sum_{i=1}^{n}\lambda_{pi}[Pg_i - Pd_i - P_i(V,\delta)] \\
&- \sum_{i=1}^{n}\lambda_{qi}[Qg_i - Qd_i - Q_i(V,\delta)] + \sum_{i=1}^{ng}\mu_{pi,\min}(Pg_i^{\min} - Pg_i) \\
&+ \sum_{i=1}^{ng}\mu_{pi,\max}(Pg_i - Pg_i^{\max}) + \sum_{i=1}^{ng}\mu_{qi,\min}(Qg_i^{\min} - Qg_i) \\
&+ \sum_{i=1}^{ng}\mu_{qi,\max}(Qg_i - Qg_i^{\max}) + \sum_{j=1}^{ncap}\mu_{cj,\max}(Qc_j - Qc_j^{\max}) \\
&+ \sum_{i=1}^{n}\sum_{\substack{i=1 \\ j \neq 1}}^{n}\eta_{ij}(S_{ij} - S_{ij}^{\max}) + \sum_{i=1}^{n}\upsilon_{i,\min}(V_i^{\min} - V_i) + \sum_{i=1}^{n}\upsilon_{i,\max}(V_i - V_i^{\max})
\end{aligned}
$$

According to the theory of microeconomics, the marginal prices for active and reactive power on ith bus are λ_{pi} and λ_{qi}, respectively, in the above Lagrangian function and are taken as the corresponding spot prices in electricity markets. Similar to vector λ, the vectors μ, η and υ contain marginal change in cost with respect to the corresponding constraints. The elements of vectors μ, η and υ respectively are different than zero only in case that the corresponding constraints are active.

Optimization of (11.32), with power flow relations included as equality constraints (11.35), inequality constraints (11.36) to (11.41) along with generation bidding constraints GA-Fuzzy approach. All the control variables, e.g. V at PV bus and tap ratio of tap setting transformers are also taken care in this optimization process. GA-Fuzzy approach does not provide Lagrange multipliers required for determination of SRMC (short run marginal cost) during optimization process directly. Therefore, in the proposed model method used to determine LMP (locational marginal prices) and hence SRMC is explained in the next section. A solution to this optimization problem provides the pool demands Pd_i^p and pool generations Pg_i^p.

1. *Method for determination of LMP and SRMC*

The optimization problem is solved, if the following equations of optimality are satisfied.

$$\frac{\partial L}{\partial Pg_i} = \frac{\partial C_i(Pg_i)}{\partial Pg_i} - \lambda_{pi} = 0 \quad i = 1, \ldots \ldots ng \tag{11.42}$$

$$\frac{\partial L}{\partial Qg_i} = \frac{\partial C_i(Qg_i)}{\partial Qg_i} - \lambda_{qi} = 0 \quad i = 1, \ldots \ldots, ng \tag{11.43}$$

$$\frac{\partial L}{\partial \delta_i} = \sum_{j=1}^{n} \left[\lambda_{pj} \frac{\partial P_j}{\partial \delta_i} \right] + \sum_{j=1}^{n} \left[\lambda_{qj} \frac{\partial Q_j}{\partial \delta_i} \right] = 0$$

$$= \left(\lambda_{ps} \frac{\partial P_s}{\partial \delta_i} + \sum_{\substack{j=1 \\ j\neq s}}^{ng+nload} \lambda_{pj} \frac{\partial P_j}{\partial \delta_i} \right)$$

$$+ \left(\lambda_{qs} \frac{\partial Q_s}{\partial \delta_i} + \sum_{\substack{j=1 \\ j\neq s}}^{ng} \lambda_{qj} \frac{\partial Q_j}{\partial \delta_i} + \sum_{\substack{j=1 \\ j\neq s}}^{nload} \lambda_{qj} \frac{\partial Q_j}{\partial \delta_i} \right)$$

$$= \left(\lambda_{ps} \frac{\partial P_s}{\partial \delta_i} + \lambda_{qs} \frac{\partial Q_s}{\partial \delta_i} + \sum_{\substack{j=1 \\ j\neq s}}^{ng} \lambda_{qj} \frac{\partial Q_j}{\partial \delta_i} \right)$$

$$+ \left(\sum_{\substack{j=1 \\ j\neq s}}^{ng+nload} \lambda_{pj} \frac{\partial P_j}{\partial \delta_i} + \sum_{\substack{j=1 \\ j\neq s}}^{nload} \lambda_{qj} \frac{\partial Q_j}{\partial \delta_i} \right) \qquad (11.44)$$

where $i = 1, 2, \ldots.(ng + nload)$ and $i \neq s$

$$\frac{\partial L}{\partial V_i} = \sum_{j=1}^{n} \left[\lambda_{pj} \frac{\partial P_j}{\partial V_i} \right] + \sum_{j=1}^{n} \left[\lambda_{qj} \frac{\partial Q_j}{\partial V_i} \right] = 0$$

$$= \left(\lambda_{ps} \frac{\partial P_s}{\partial V_i} + \sum_{\substack{j=1 \\ j\neq s}}^{ng+nload} \lambda_{pj} \frac{\partial P_j}{\partial V_i} \right) + \left(\lambda_{qs} \frac{\partial Q_s}{\partial V_i} + \sum_{\substack{j=1 \\ j\neq s}}^{ng} \lambda_{qj} \frac{\partial Q_j}{\partial V_i} + \sum_{j=1}^{nload} \lambda_{qj} \frac{\partial Q_j}{\partial V_i} \right)$$

$$= \left(\lambda_{ps} \frac{\partial P_s}{\partial V_i} + \lambda_{qs} \frac{\partial Q_s}{\partial V_i} + \sum_{\substack{j=1 \\ j\neq s}}^{ng} \lambda_{qj} \frac{\partial Q_j}{\partial V_i} \right) + \left(\sum_{\substack{j=1 \\ j\neq s}}^{ng+nload} \lambda_{pj} \frac{\partial P_j}{\partial V_i} + \sum_{\substack{j=1 \\ j\neq s}}^{nload} \lambda_{qj} \frac{\partial Q_j}{\partial V_i} \right)$$

$$(11.45)$$

where $i = 1, 2, \ldots..nload$ *and* $i \neq s$

$$\frac{\partial L}{\partial \lambda_{pi}} = P_i(V, \delta) - Pg_i + Pd_i = 0 \quad (i = 1, \ldots \ldots.n) \qquad (11.46)$$

$$\frac{\partial L}{\partial \lambda_{qi}} = Q_i(V, \delta) - Qg_i + Qd_i = 0 \quad (i = 1, \ldots \ldots.n) \qquad (11.47)$$

where $n =$ *total no. of buses*
$s =$ *slack bus*
$ng =$ *total no. of generator buses*
$nload =$ *total no. of load buses*

Equations. (11.45) and (11.46) can be expressed in matrix form as follows:

$$
\begin{bmatrix}
\lambda ps\dfrac{\partial P_s}{\partial \delta_i} + \lambda_{qs}\dfrac{\partial Q_s}{\partial \delta_i} + \displaystyle\sum_{\substack{j=1\\j\neq s}}^{ng} \lambda_{qj}\dfrac{\partial Q_j}{\partial \delta_i} \quad i=1,\dots(ng+nload) \\
\hline
\lambda ps\dfrac{\partial P_s}{\partial V_i} + \lambda_{qs}\dfrac{\partial Q_s}{\partial V_i} + \displaystyle\sum_{\substack{j=1\\j\neq s}}^{ng} \lambda_{qj}\dfrac{\partial Q_j}{\partial V_i} \quad i=1,\dots(nload)
\end{bmatrix}
$$

$$
+\begin{bmatrix}
\dfrac{\partial P_j}{\partial \delta_i} \begin{matrix} j=1,\dots(ng+nload)\\ i=1,\dots(ng+nload)\\ i \text{ and } j\neq s \end{matrix} & \Bigg\| & \dfrac{\partial Q_j}{\partial \delta_i} \begin{matrix} j=1,\dots nload\\ i=1,\dots(ng+nload)\\ i \text{ and } j\neq s \end{matrix} \\
\hline
\dfrac{\partial P_j}{\partial V_i} \begin{matrix} j=1,\dots(ng+nload)\\ i=1,\dots(ng+nload)\\ i \text{ and } j\neq s \end{matrix} & \Bigg\| & \dfrac{\partial Q_j}{\partial V_i} \begin{matrix} j=1,\dots nload\\ i=1,\dots(ng+nload)\\ i \text{ and } j\neq s \end{matrix}
\end{bmatrix}
\begin{bmatrix}
\lambda_{pj} \begin{matrix} j=1,\dots(ng+nload)\\ j\neq s \end{matrix} \\
\hline
\lambda_{qj} \begin{matrix} \\ j=1,\dots(nload)\\ j\neq s \end{matrix}
\end{bmatrix}
$$

$$
=\begin{bmatrix} 0 \\ - \\ 0 \end{bmatrix}
$$

It can also be expressed as:

$$
\begin{bmatrix}
\lambda_{ps}\dfrac{\partial P_s}{\partial \delta_i} + \lambda_{qs}\dfrac{\partial Q_s}{\partial \delta_i} + \displaystyle\sum_{\substack{j=1\\j\neq s}}^{ng} \lambda_{qj}\dfrac{\partial Q_j}{\partial \delta_i} \quad i=1,\dots(ng+nload) \\
\hline
\lambda_{ps}\dfrac{\partial P_s}{\partial V_i} + \lambda_{qs}\dfrac{\partial Q_s}{\partial V_i} + \displaystyle\sum_{\substack{j=1\\j\neq s}}^{ng} \lambda_{qj}\dfrac{\partial Q_i}{\partial V_i} \quad i=1,\dots nload
\end{bmatrix}
$$

$$
+[J]^T \begin{bmatrix}
\lambda_{pj} \begin{matrix} j=1,\dots(ng+nload)\\ j\neq s \end{matrix} \\
\hline
\lambda_{qj} \begin{matrix} j=1,\dots nload\\ j\neq s \end{matrix}
\end{bmatrix} = \begin{bmatrix} 0 \\ \hline 0 \end{bmatrix}
$$

where $J = $ *Jacobian obtained from N-R load flow method for final optimized results.*

$$
\begin{bmatrix}
\lambda_{pj} & j=1,\dots(ng+nload) \\
\hline
\lambda_{qi} & j=1,\dots nload
\end{bmatrix} = -\left([J]^T\right)^{-1}
$$

$$
\times \begin{bmatrix}
\lambda_{ps}\dfrac{\partial P_s}{\partial \delta_i} + \lambda_{qs}\dfrac{\partial Q_s}{\partial \delta_i} + \displaystyle\sum_{\substack{j=1\\j\neq s}}^{ng} \lambda_{qj}\dfrac{\partial Q_j}{\partial \delta_i} \quad i=1,\dots(ng+nload) \\
\hline
\lambda_{ps}\dfrac{\partial P_s}{\partial V_i} + \lambda_{qs}\dfrac{\partial Q_s}{\partial V_i} + \displaystyle\sum_{\substack{j=1\\j\neq s}}^{ng} \lambda_{qj}\dfrac{\partial Q_i}{\partial V_i} \quad i=1,\dots nload
\end{bmatrix} \tag{11.48}
$$

Equation. (5.30) can be written for slack bus as:

$$\lambda_{ps} = \frac{\partial C_s(Pg_s)}{\partial Pg_s} \qquad (11.49)$$

and (11.43) can be written for slack and PV buses respectively as:

$$\lambda_{qs} = \frac{\partial C_s(Qg_s)}{\partial Qg_s} \qquad (11.50)$$

$$\lambda_{qi} = \frac{\partial C_i(Qg_i)}{\partial Qg_i} \quad i = 1, \ldots \ldots ng \qquad (11.51)$$

Therefore, real (λ_p) and reactive (λ_q) marginal prices for slack bus, PV buses and PQ buses are obtained solving (11.48)–(11.51).

Short run marginal cost (SRMC) of real power wheeling PWC_{ij} and reactive power wheeling QWC_{ij} for transaction from bus i to j are calculated by following equations:

$$PWC_{ij} = PW_{ij}x \ (\lambda_{pj} - \lambda_{pi}) \qquad (11.52)$$
$$QWC_{ij} = QW_{ij}x \ (\lambda_{qj} - \lambda_{qi}) \qquad (11.53)$$

where, PWC_{ij} and QWC_{ij} are real power and reactive power to be wheeled from bus i to j, respectively.

3. *Algorithm for marginal cost transmission pricing method*

Step 1 All system voltages and pool loads are set to initial conditions. All feasible (scheduled) firm transactions are added to the system.

Step 2 For active power generation cost, reactive power generation cost of all pool generators and capacitor reactive power support cost, the optimization of objective function (11.32) is carried out satisfying all constraints (11.35)–(11.41) using GA-Fuzzy approach. The inequality constraints of tap setting transformers are also considered in this optimization process.

Step 3 After the optimization, voltages, tap settings, capacitors reactive supports and pool generations are obtained.

Step 4 Marginal costs for both real and reactive power at all buses are calculated using (11.48–11.50).

Step 5 SRMC of wheeling for bilateral transactions are calculated using (11.52) and (11.53), respectively.

Step 6 The amount to be paid by each demand and amount to be received by each genco is determined based on marginal cost. Similarly, multilateral transaction is treated.

Step 7 The Marginal network revenue is determined based on total payments and receipts.

4. *Application of marginal cost transmission pricing method*

The results of method tested for modified IEEE 30 bus system are presented here. The data and single line diagram of this system is given in Appendix F.

The calculations shown in Tables 11.13–11.17 indicates that due to implementation of marginal prices (i.e. nodal prices), marginal network revenue of 40.301905 $/h is obtained.

Table 11.13. Revenue received from Pool demand

Bus no.	Real demand (MW)	λ_{pi} ($/MW h)	Revenue ($/h)	Reactive demand (MVAr)	λ_{qi} ($/MVAr h)	Revenue ($/h)
1	0	3.31921	0	0	0.049762	0
2	21.7	3.435997	74.56113	12.7	0.042547	0.540345
3	2.4	3.513915	8.433397	1.2	0.101652	0.121983
4	7.6	3.570239	27.13382	1.6	0.110167	0.176267
5	94.2	3.690331	347.6292	19	0.127005	2.413096
6	0	3.612632	0	0	0.129748	0
7	22.8	3.66913	83.65617	10.9	0.144936	1.579805
8	30	3.626385	108.7916	30	0.150447	4.513415
9	0	3.616505	0	0	0.123827	0
10	5.8	3.621814	21.00652	2	0.13621	0.27242
11	0	3.61415	0	0	0.094843	0
12	11.2	3.599261	40.31173	7.5	0.126418	0.948136
13	0	3.598323	0	0	0.123478	0
14	6.2	3.676129	22.792	1.6	0.141555	0.226487
15	8.2	3.685928	30.22461	2.5	0.134784	0.336961
16	3.5	3.634265	12.71993	1.8	0.141915	0.255447
17	9	3.64232	32.78088	5.8	0.143435	0.831922
18	3.2	3.723113	11.91396	0.9	0.142798	0.128519
19	9.5	3.728377	35.41958	3.4	0.142602	0.484848
20	2.2	3.704354	8.149579	0.7	0.13277	0.092939
21	17.5	3.662132	64.08731	11.2	0.159024	1.78107
22	0	3.659034	0	0	0.156488	0
23	3.2	3.722763	11.91284	1.6	0.12908	0.206529
24	8.7	3.736867	32.51075	6.7	0.158234	1.060166
25	0	3.746257	0	0	0.152652	0
26	3.5	3.822748	13.37962	2.3	0.203776	0.468684
27	0	3.674906	0	0	0.128106	0
28	0	3.640752	0	0	0.140059	0
29	2.4	3.783965	9.081516	0.9	0.116025	0.104422
30	10.6	3.858995	40.90535	1.9	0.147013	0.279325
Total			1037.401	Total		16.82278

Table 11.14. Expenditure for generation

Bus no.	Real generation (MW)	λ_{pi} ($/MW h)	Expenditure ($/h)	Reactive generation (MVAr)	λ_{qi} ($/MVArh)	Expenditure ($/h)
1	174.961	3.31921	580.7323	11.902	0.049762	0.592267
2	47.529	3.435997	163.3095	15.599	0.042547	0.663691
5	21.176	3.690331	78.14645	36.06	0.127005	4.5798
8	24.51	3.626385	88.8827	34.885	0.150447	5.248344
11	12.039	3.61415	43.51075	15.297	0.094843	1.450813
13	12.329	3.598323	44.36372	21.845	0.123478	2.697377
Total			998.9454	Total		15.23229

Table 11.15. Revenue received from Bilateral transactions

Transaction no.	From bus	To bus	Size (MW)	SRMC ($/MW h)	Revenue Received ($/h)
1	9	13	5	−0.018182	−0.09091
2	22	25	5	0.087223	0.436115
Total					0.345205

Table 11.16. Revenue received from multilateral transactions

Bus no.	MW	λ_{pi} ($/MW h)	Expenditure ($/h)	Bus no.	MW	λ_{pi} ($/MW h)	Revenue received ($/h)
6	4	3.612632	14.450528	11	2	3.61415	7.2283
7	2	3.66913	7.33826	13	3	3.598323	10.794969
				14	1	3.676129	3.676129
Total			21.788788	Total			21.699398

Table 11.17. Summary of results

S. no.		In ($/h)
1	Revenue received from pool real demand	1,037.401
2	Revenue received from pool reactive demand	16.82278
3	Revenue received from bilateral transactions	0.345205
4	Revenue received from multilateral transactions	−0.08939
5	Expenditure for real generation	998.9454
6	Expenditure for reactive generation	15.23229
7	*Total revenue*	*1,054.479595*
8	*Total expenditure*	*1,014.17769*
9	*Marginal network revenue*	*40.301905*

11.4.3 Postage Stamp Method

It is a simplest method of transmission pricing and makes no distinction between transaction with regard to the power flow path, supply or delivery points, or the time when it takes place.

The results of this method tested for Indian UPSEB-75 bus system are presented in Table 11.18 in this section (Fig. 11.13–11.17). Single line diagram and transmission ARR (annual revenue requirement) data is given Appendix G.

Table 11.18. Embedded cost allocation for Indian UPSEB 75-bus system using postage stamp method

Transactions	Rs. lakh/h
Bilateral T1	0.211683854
Bilateral T2	0.190515469
Bilateral T3	0.15876289
Bilateral T4	0.105841927
Bilateral T5	0.088589693
Bilateral T6	0.034504468
Bilateral T7	0.042336771
Bilateral T8	0.02857732
Bilateral T9	0.031752578
Bilateral T10	0.148178698
Bilateral T11	0.052920963
Bilateral T12	0.465704479
Multilateral	2.937854367
Pool	1.264408828
Total ARR	5.761632306

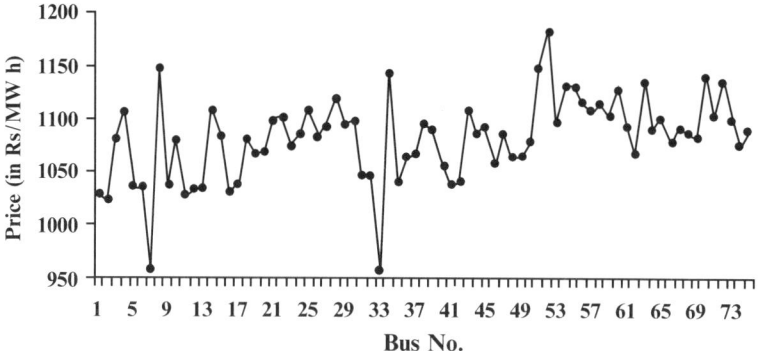

Fig. 11.13. Real marginal price for Indian UPSEB 75-bus system

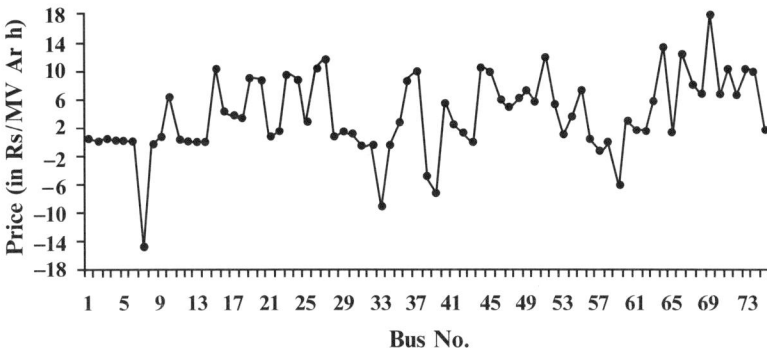

Fig. 11.14. Reactive marginal price for Indian UPSEB 75-Bus System

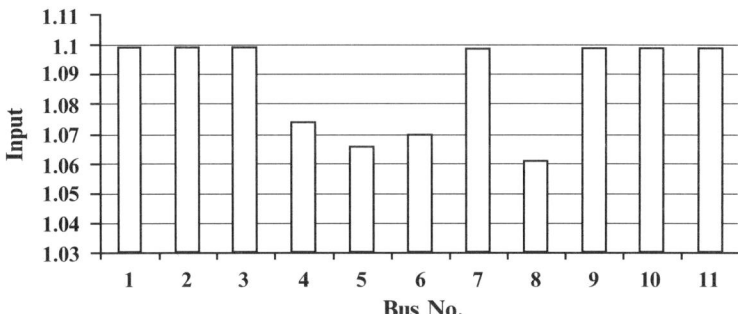

Fig. 11.15. Generator bus voltage for Indian UPSEB 75-Bus System

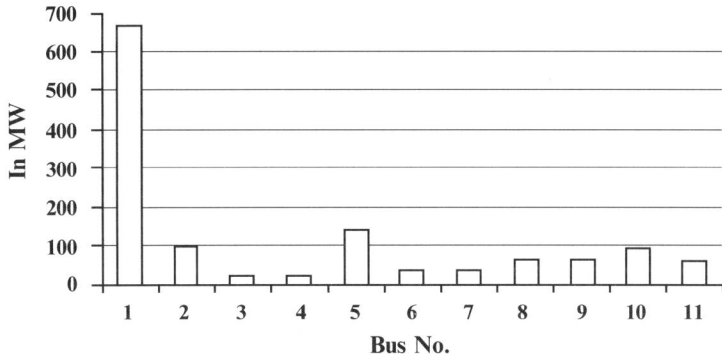

Fig. 11.16. Optimal values of real power generation for Indian UPSEB 75 Bus System

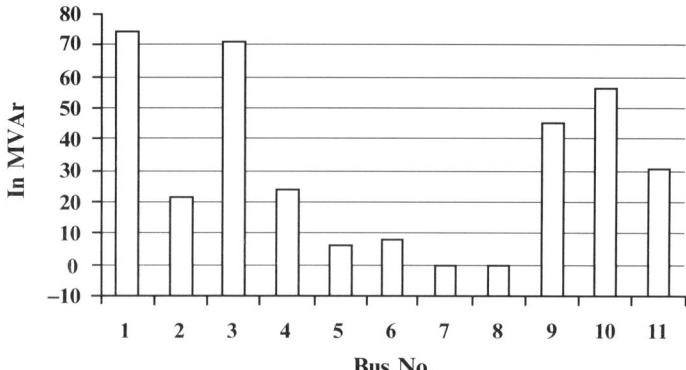

Fig. 11.17. Optimal values of reactive power generation for Indian UPSEB 75 Bus System

11.4.4 MW Mile Methods

It requires the accurate load flow results to compute the power flow in the lines. Once the power flow in each line is known, system usage index for each transaction is calculated. The transmission charge is then proportional to the transmission usage by individual transaction. The system usage index for each transaction is calculated by following relation:

$$UI_{T_i} = \sum_j \left[\frac{P_{j;T_i} * (L_j * F'_j)}{\left(\sum_i P_{j;T_i} + P_{j;pool}\right)} \right] \tag{11.54}$$

UI_{T_i} = Price charged for transaction T_i in \$ (System Usage Index)
P_{j,T_i} = Incremental loading of line j due to transaction (bilateral/multilateral) T_i, MW.
$P_{j;pool}$ = Loading of line j due to pool transactions, MW.
L_j = Length of the line j, mile.
F'_j = Cost of the line per unit length, \$/Mile.

(i) *Procedure to calculate system usage index*
   Step 1: Find the cost of the line by multiplying the unit cost of the line by the line lengths $(L_j^* F'_j)$.
   Step 2: Find the base case power flow on all lines, which can be obtained using an OPF.
   Step 3: Find the new load flow solution with each transaction T_i and hence the power flows on each line.
   Step 4: Calculate the incremental power flows in each line caused by the transaction T_i.
   Step 5: Calculate each line usage due to transaction T_i by multiplying incremental line flows obtained in Step 4 and cost of per unit length of line in Step 1, i.e. $P_{j,T_i}^*(L_j^* F'_j)$, where j is any line.

Step 6: Find the total system usage by transaction T_i, i.e. $\sum_j P_{j,T_i}^* (L_j^* F_j')$.

Step 7: The system usage UI_{T_i} (System Usage Index) of each transaction T_i is calculated for proportional allocation of ARR given by equation (11.54).

Step 8: Calculate the proportional allocation of ARR to transaction T_i.

(ii) *Proposed methods for proportional allocation of ARR*

Let

UI_{T_i} = system usage index of any transaction T_i(bilateral/multilateral) as given by (11.54)

UI_{pool} = system usage index due to pool transactions, as given by (11.55)

$$UI_{pool} = \sum_j \left[\frac{P_{j;pool} * (L_j * F_j')}{\left(\sum_i P_{j;T_i} + P_{j;pool}\right)} \right] \tag{11.55}$$

$UI_{combined_i}$ = system usage index due to all transactions taken simultaneously, i.e. bilateral+multilateral (if any), as given by (5.44)

$$UI_{combined} = \sum_j \left[\frac{P_{j;\sum T_i} * (L_j * F_j')}{(P_{j;\sum T_i} + P_{j;pool})} \right] \tag{11.56}$$

where $P_{j;\sum T_i}$ = Incremental loading of line j due to all transactions taken simultaneously, i.e. bilateral+multilateral (if any), MW.

The ARR allocation can be done by two possible methods discussed below:

Method-1 (When all transactions are considered independently)

Transmission charges paid for transaction $T_i (R_{T_i})$

$$= ARR \times \frac{UI_{T_i}}{\left(\sum_i UI_{T_i} + UI_{pool}\right)} \tag{11.57}$$

Transmission charges paid for pool transactions $(R_{pool}) = ARR -$ Transmission charges paid for transactions T_i, *i.e. (Bilateral and Multilateral, if any)*

$$= ARR \times \frac{UI_{pool}}{\left(\sum_i UI_{T_i} + UI_{pool}\right)} \tag{11.58}$$

In this method ARR is shared by all transactions (bilateral, multilateral and pool) on the basis of their respective system usage. The system usage is measured here in terms of system usage index. Whenever another bilateral or multilateral transaction takes place, the ARR is redistributed among all transactions according to new and lesser system usage index values. Therefore, charges paid by each transaction become less compared to earlier case (i.e. when new transaction did not take place). This method gives incentive to all old transactions whenever new transaction takes place in the system.

This method suffers from a major drawback whenever two or more than two transactions take place simultaneously. In that case it charges higher than actual values (transactions are taken simultaneously) for bilateral and multilateral transactions. Therefore, pool transactions have advantage of paying lesser amount of charges. The reason of this drawback is that combined usage index ($UI_{combined}$) of transactions, i.e. (bilateral + multilateral, if any) is less than sum of usage indexes ($\sum_i UI_{T_i}$) of transactions, i.e. (bilateral + multilateral, if any) treating each of them independently. This is due to difference in actual value of power flow in each line (considering all transactions taking place simultaneously) and algebraic sum of power flow in each line due to bilateral and multilateral transactions (if any) independently.

Method-2 (When all transaction are considered simultaneously)

Allocation of transmission charges paid for bilateral and multilateral transactions simultaneously

$$(R_{combined}) = ARR \times \frac{UI_{combined}}{(UI_{combined} + UI_{pool})} \tag{11.59}$$

Transmission charges paid for transaction $T_i (R_{T_i}) = R_{combined} \times \dfrac{UI_{T_i}}{\sum\limits_i UI_{T_i}}$

$$\tag{11.60}$$

Transmission charges paid for pool transactions $(R_{pool}) = ARR - \sum\limits_i R_{T_i}$

$$\tag{11.61}$$

In this method, collective charges for all bilateral and multilateral transactions (if any) are calculated. Then individual contribution to collective charges for each transaction is calculated on the basis of system indexes of transactions (while considering all transactions independently). Therefore, drawback of method-1 is rectified in this method. This method is more transparent in nature than method-1.

(iii) *Application of proposed MW-Mile methods*

The results obtained for both the methods on Indian UPSEB 75-Bus system are given in Table 11.19, whereas system data and line data are given in Appendix G. System usage indices for both the methods are given in Table 11.20. The results reveal that due to effect of all the transactions taking place simultaneously in method-2 the charges allocated to bilateral and multilateral transactions are lesser as compared to method-1.

Table 11.19. Embedded cost allocation for Indian UPSEB 75-bus system using MW-mile methods

Transactions	Method-1 (Rs. lakh/h)	Method-2 (Rs. lakh/h)
Bilateral T1	0.132874487	0.073341923
Bilateral T2	0.226654476	0.125105094
Bilateral T3	0.042219114	0.023303428
Bilateral T4	0.056802223	0.031352778
Bilateral T5	0.058398303	0.032233757
Bilateral T6	0.015925418	0.008790256
Bilateral T7	0.023775288	0.013123101
Bilateral T8	0.012272944	0.006774222
Bilateral T9	0.195182687	0.107733801
Bilateral T10	0.12254473	0.06764027
Bilateral T11	0.052171774	0.028796937
Bilateral T12	0.498692644	0.275260348
Multilateral	2.227124232	1.229292226
Pool	2.096993986	3.738884167
Total ARR	*5.761632306*	*5.761632306*

Table 11.20. System usage indexes for transactions when all transactions are taking place independently

S. no.	Transaction	System index
1	Bilateral T1	0.001286717
2	Bilateral T2	0.002194855
3	Bilateral T3	0.000408837
4	Bilateral T4	0.000550056
5	Bilateral T5	0.000565512
6	Bilateral T6	0.000154217
7	Bilateral T7	0.000230233
8	Bilateral T8	0.000118848
9	Bilateral T9	0.001890091
10	Bilateral T10	0.001186687
11	Bilateral T11	0.000505216
12	Bilateral T12	0.004829192
13	Multilateral	0.021566811
14	Pool	0.020306668
	Total	*0.05579394*
	(Bilateral+Multilateral+Pool)	

When all bilateral and multilateral transactions are taking place simultaneously

S. No.	Transaction	System index
1	Bilateral+multilateral	0.032869084
2	Pool	0.02668025
	Total (Bilateral+Multilateral+Pool)	*0.059549334*

11.4.5 Hybrid Deregulated Transmission Pricing Model

To facilitate efficient competition in generation, the transmission utility, i.e. Transco (which shall continue to operate as monopoly) is obliged to provide full access to the transmission facilities in a non-discriminatory manner. In order for Transco to operate viably, the charges should be sufficient to cover Transco's revenue requirement. It is noted in (Tabors 1994) and findings of study team report (Echauz and Vachtsevanos 1994) discussed that in a regulated environment such as in electric transmission business, marginal cost based pricing provides an efficient economic and engineering solution to developing a tariff structure.

However, it has been observed that relying solely on this marginal pricing does not generate sufficient revenue for the transmission utility, and the common solution is to establish supplementary charges which when added to the marginal network income would equal to the total network cost. This would mean that a composite cost paradigm may be implemented, based on embedded costs and marginal costs to reflect transmission pricing based on actual costs of existing network facilities, as well as the operation cost. After identifying need of supplementary charges, a brief discussion on the method of supplementary charges allocation and application of the hybrid model to the Indian UPSEB 75-Bus system are in following section.

1. *Method of supplementary charge allocation*
   The allocation of supplementary charges creates additional challenge as how to allocate the charge among transmission users in an equitable manner and to ensure that it does not distort the economic signals provided by marginal pricing. Probably the most popular method is linking the charge with the actual use of the system by the user. In the MW-Mile methodology the actual use of transmission facilities is expressed, conceptually, by a product of power due to a particular transaction times the distance this power travels in the network. Therefore in this hybrid model supplementary charges are allocated on the basis of two MW-Mile methods already explained in earlier section.

2. *Application of proposed Hybrid transmission pricing model to Indian UPSEB 75-Bus system*
   The results of proposed model tested for Indian UPSEB 75-Bus system are presented here.

The results are obtained for marginal prices, generator bus voltages, real power generations and reactive power generations after applying algorithm for marginal cost based transmission pricing method. The calculations shown in Tables 11.21–11.25 that due to implementation of marginal prices (i.e. nodal prices), marginal network revenues of 112,954.4 Rs/h is obtained.

Table 11.21. Revenue received from pool demand

Bus no.	Real demand (MW)	λ_{pi} Case-II (Rs/MW h)	Revenue Case-II (Rs/h)	Reactive demand (MVAr)	λ_{qi} Case-II (Rs/MVAr h)	Revenue Case-II (Rs/h)
1	0	1028.018	0	0	0.558	0
2	0	1023.905	0	0	0.086	0
3	0	1082.344	0	0	0.473	0
4	0	1108.433	0	0	0.162	0
5	0	1034.929	0	0	0.013	0
6	0	1034.556	0	0	0.039	0
7	0	960.118	0	0	−14.59	0
8	0	1146.357	0	0	−0.424	0
9	0	1036.89	0	0	0.636	0
10	0	1080.348	0	0	6.428	0
11	0	1027.98	0	0	0.368	0
12	27	1033.528	27905.26	0	0	0
13	12	1034.753	12417.04	0	0	0
14	0	1109.193	0	0	0	0
15	0	1085.629	0	0	10.432	0
16	0	1030.641	0	0	3.998	0
17	0	1036.924	0	0	3.731	0
18	0	1082.531	0	0	3.335	0
19	0	1067.155	0	0	9.05	0
20	56.37	1069.194	60270.47	1.06	8.928	9.46368
21	0	1098.966	0	0	0.791	0
22	0	1102.138	0	0	1.23	0
23	0	1075.28	0	0	9.573	0
24	27.95	1086.46	30366.56	7.66	8.927	68.38082
25	0	1110.085	0	0	2.591	0
26	0	1084.315	0	0	10.519	0
27	106	1094.203	115985.5	7.83	11.769	92.15127
28	0	1118.675	0	0	0.669	0
29	0	1094.279	0	0	1.201	0
30	0	1098.456	0	0	0.944	0
31	0	1045.211	0	0	−0.742	0
32	18.11	1045.804	18939.51	11.59	−0.419	−4.85621
33	0	959.137	0	0	−9.132	0
34	0	1146.321	0	0	−0.422	0
35	0	1037.723	0	0	2.895	0
36	0	1063.1	0	0	8.97	0
37	0	1067.299	0	0	9.852	0
38	0	1096.058	0	0	−5.066	0
39	0	1089.664	0	0	−7.35	0
40	0	1055.626	0	0	5.665	0
41	0	1037.16	0	0	2.311	0
42	112.5	1039.465	116939.8	−294.7	1.173	−345.683

Table 11.21. (*Continued*)

Bus no.	Real demand (MW)	λ_{pi} Case-II (Rs/MW h)	Revenue Case-II (Rs/h)	Reactive demand (MVAr)	λ_{qi} Case-II (Rs/MVAr h)	Revenue Case-II (Rs/h)
43	0	1109.193	0	0	0	0
44	0	1087.5	0	0	10.409	0
45	0	1093.023	0	0	9.998	0
46	0	1058.116	0	0	5.942	0
47	34.55	1087.35	37567.94	4.38	4.748	20.79624
48	0	1063.089	0	0	6.057	0
49	25.72	1062.683	27332.21	15.8	7.055	111.469
50	2.1	1077.431	2262.605	9.2	5.617	51.6764
51	57.75	1149.66	66392.87	0.62	12.028	7.45736
52	14.27	1183.529	16888.96	−23.36	5.45	−127.312
53	12.63	1096.113	13843.91	0.33	1.16	0.3828
54	21.95	1131.157	24828.9	17.02	3.673	62.51446
55	14.23	1129.993	16079.8	2.81	7.392	20.77152
56	0	1114.006	0	0	0.362	0
57	52.78	1107.552	58456.59	18.53	−1.514	−28.0544
58	54.19	1112.947	60310.6	11.29	−0.335	−3.78215
59	21.89	1103.845	24163.17	11.01	−6.318	−69.5612
60	24.2	1128.782	27316.52	2.44	2.963	7.22972
61	56.5	1092.145	61706.19	6.58	1.651	10.86358
62	17.18	1067.086	18332.54	7.41	1.461	10.82601
63	58.01	1135.825	65889.21	5.31	5.95	31.5945
64	56.79	1090.578	61933.92	13.33	13.518	180.1949
65	47.84	1100.608	52653.09	12.81	1.361	17.43441
66	31.74	1077.662	34204.99	15.18	12.384	187.9891
67	0	1090.699	0	0	8.145	0
68	42.87	1087.015	46600.33	33.6	6.571	220.7856
69	55.94	1081.533	60500.96	32.53	18.014	585.9954
70	23.34	1140.459	26618.31	2.3	6.715	15.4445
71	0	1102.819	0	0	10.169	0
72	52.52	1135.953	59660.25	11.76	6.586	77.45136
73	37	1099.594	40684.98	4.46	10.297	45.92462
74	18	1074.16	19334.88	8.87	9.803	86.95261
75	0	1089.873	0	0	1.252	0
Total			*1306388*	*Total*		*1344.501*

As Transco's total revenue requirement is 576,163.2306 Rs/h, therefore supplementary charges of $(576, 163.2306–112, 954.4 = 463, 208.8306$ Rs/h) can be realized by MW-Mile based supplementary charge allocation method.

Table 11.22. Expenditure for Generation

Bus No.	Real Generation (MW)	λ_{pi} (Rs/MW h)	Expenditure (Rs/h)	Reactive Generation (MVAr)	λ_{qi} (Rs/MVArh)	Expenditure (Rs/h)
1	669.08	1028.018	687826.3	74.39	0.558	41.50962
2	100	1023.905	102390.5	21.46	0.086	1.84556
3	20	1082.344	21646.88	70.59	0.473	33.38907
8	20	1108.433	22168.66	24.05	0.162	3.8961
5	140	1034.929	144890.1	6.36	0.013	0.08268
6	36.3	1034.556	37554.38	8.14	0.039	0.31746
7	33.72	960.118	32375.18	0	−130.59	0
8	60	1146.357	68781.42	0	−0.424	0
9	60	1036.89	62213.4	45.44	0.636	28.89984
10	90	1080.348	97231.32	56	6.428	359.968
11	60	1027.98	61678.8	30.44	0.368	11.20192
12	–	–	–	115.71	0	0
13	–	–	–	28.99	0	0
14	–	–	–	14.42	0	0
15	–	–	–	35	10.432	365.12
Total			*1338757*	*Total*		*846.2303*

Table 11.23. Revenue received from bilateral transactions

Transaction no.	From bus	To bus	Size (MW)	SRMC (Rs/MW h)	Revenue Received (Rs/h)
1	2	50	200	53.426	10, 685.2
2	3	55	180	47.649	8, 576.82
3	4	37	150	−41.134	−6, 170.1
4	5	20	100	34.265	3, 426.5
5	6	52	83.7	148.973	12, 469.04
6	7	62	32.6	106.968	3, 487.157
7	8	57	40.0	−38.805	−1, 552.2
8	9	74	27.0	37.27	1, 006.29
9	10	60	30.0	48.434	1, 453.02
10	11	54	140.0	103.177	14, 444.78
11	16	48	50.0	32.448	1, 622.4
12	75	73	440.0	9.721	4, 277.24
Total					*53, 726.15*

Allocation of supplementary charges

Finally, in order to complete realization of Transco's revenue requirement, allocation of supplementary charges by both the MW-Mile methods is tabulated in Table 11.26. Again, method-2 will be preferred over method-1

Table 11.24. Revenue received from multilateral transactions

Bus no.	MW	λ_{pi} (Rs/MW h)	Expenditure (Rs/h)	Bus No.	MW	λ_{pi} (Rs/MW h)	Revenue Received (Rs/h)
12	1,273	1,033.528	1,315,681	24	100	1,086.46	108,646
13	898.7	1,034.753	929,932.5	25	211	1,110.085	234,227.9
14	150.0	1,109.193	166,379	27	100	1,094.203	109,420.3
15	454.0	1,085.629	492,875.6	28	227	1,118.675	253,939.2
				30	126	1,098.456	138,405.5
				34	141	1,146.321	162,433.7
				39	170	1,089.664	185,242.9
				42	1,000	1,039.465	1,039,465
				46	156	1,058.116	165,066.1
				56	144	1,114.006	160,416.9
				67	200	1,090.699	218,139.8
				71	200	1,102.819	220,563.8
Total			*2,904,868*	Total			*2,995,967*

Net Revenue received = 91,099

Table 11.25. Summary of results

S. no.		(Rs/h)
1	Revenue received from pool real demand	1,306,388
2	Revenue received from pool reactive demand	1,344.501
3	Revenue received from bilateral transactions	53,726.15
4	Revenue received from multilateral transactions	91,099
5	Expenditure for real generation	1,338,757
6	Expenditure for reactive generation	846.2303
7	*Total Revenue*	*1,452,557.6*
8	*Total Expenditure*	*1,339,603.2*
9	*Marginal Network Revenue*	*112,954.4*

because it is more transparent. Moreover, in deregulated competitive business environment method-2 encourages more bilateral and multilateral transactions by charging lesser supplementary charges.

11.4.6 Conclusion

A SRMC based marginal pricing method using GA-Fuzzy technique is developed and tested on IEEE 30-bus system while optimizing real and reactive generation costs and capacitor reactive support cost. This method enables to calculate reactive power wheeling charges also. In category of embedded cost allocation methods – Postage Stamp allocation method and two MW-Mile methods are employed to determine embedded costs revealed that MW-Mile

Table 11.26. Supplementary charges allocation for Indian UPSEB 75-bus system using MW-mile methods

Transaction	Method-1 (in lakh Rs h)	Method-2 (in lakh Rs h)
Bilateral T1	0.106824984	0.058963537
Bilateral T2	0.182219829	0.100578771
Bilateral T3	0.0339422	0.01873487
Bilateral T4	0.045666393	0.025206201
Bilateral T5	0.046949571	0.025914469
Bilateral T6	0.012803304	0.007066962
Bilateral T7	0.019114255	0.010550379
Bilateral T8	0.009866922	0.005446185
Bilateral T9	0.156917909	0.086613025
Bilateral T10	0.098520359	0.054379684
Bilateral T11	0.041943715	0.023151418
Bilateral T12	0.400926047	0.221296713
Multilateral	1.790505797	0.988294604
Pool	1.685887022	3.005891488
Supplementary charges	*4.632088306*	*4.632088306*

(method-2) is best among all the three methods tested for Indian UPSEB-75 bus system.

Finally, a hybrid type marginal cost based deregulated transmission pricing model is proposed and tested for Indian UPSEB 75-bus system with pool, bilateral and multilateral transactions. In this supplementary charges are allocated by MW-mile methods. Therefore, a complete framework for transmission pricing is designed and implemented on Indian system.

11.5 Congestion Management Using GA-Fuzzy Approach

11.5.1 Introduction

Congestion is a consequence of various network constraints characterizing a finite network capacity that may limit the simultaneous delivery of power from an associated set of power transactions (Singh et al. 1998). The network constraints include thermal limits, voltage/VAR requirements and the stability considerations. Among all the constraints, thermal limits are the most frequently considered factor in determining network capacity.

In a deregulated electricity market, the task of ISO (Independent System Operator) is to ensure that contracted power transactions are carried out reliably. However, due to the large number of transactions that take place simultaneously, transmission networks may easily get congested. Congestion may result in preventing new contracts, unfeasibility in existing and new contracts, additional outages and damages to system components.

Managing congestion to minimize the restrictions of the competitive market has become the central activity of systems operators. It has been observed that the unsatisfactory management of transactions could increase the congestion cost which is an unwanted burden on customers. For different power market structures, the approach to manage congestion may vary. A number of methods dealing with congestion management in deregulated electricity markets have been discussed earlier. Hogan (1992) proposed the contract network and nodal pricing approach using the spot pricing theory for pool type market. Chao and Peck (1996) proposed an alternative approach which is based on parallel markets for link based transmission capacity rights and energy trading under a set of rules defined and administered by the System Operator (SO).

A congestion management approach after the deregulation of the Slovenian power system is presented in Grgic et al. (2001, 2002). The method is based on countertrade method where the system operator, based on technical and economic data, decides the optimal redispatch that eliminates congestion.

Singh and David (2003) has proposed dynamic security constrained congestion management in an unbundled electric power system. The different zones have been determined based on lines real and reactive transmission congestion.

Several optimal power flow (OPF) based congestion management schemes for multiple transactions also have been proposed. An approach using the minimum total modification to the desired transactions for relieving congestion is presented. A variant of this least modification approach used a weighting scheme with the weights being the surcharges paid by the transactions for transmission usage in the congestion-relieved network. Marginal cost signals were used for generators to manage congestion. A similar approach is proposed in (Singh et al. 1998), where the congestion cost is bundled with marginal cost at each bus in pool model and a congestion cost minimization is adopted in bilateral model.

Fu and Lamout (2001) has proposed the objective function consisting of congestion cost and service costs. A new mechanism of congestion management in multilateral transaction networks has been developed based on physical flows.

There are two broad paradigms that may be employed for congestion management. The first method includes actions like outage of congested lines or operation of transformer taps, phase shifters or FACTS devices. These means are termed as cost-free only because the marginal costs (and not the capital costs) involved in their usage are nominal.

The *not-cost-free* means include:

(1) *Rescheduling generation*
    Here, system operator re-dispatches power generation in such a way, that resulting power flows does not overload any line. Every generation unit can bid an increase or decrease of its production in a similar manner

as this is done on a balancing market, while the responsibility of system operator is to select bids in efficient way. Somehow, counter trade approach based congestion management can be viewed as simplified optimal power flow problem, where optimization variables are re-dispatch of the active power production and criteria function is minimum of the costs related to this active power re-dispatch.

(2) *Prioritization and curtailment of loads/transactions*

A parameter termed as willingness-to-pay-to-avoid-curtailment was introduced in the objective function. This can be an effective instrument in setting the transaction curtailment strategies which may then be incorporated in the optimal power flow framework.

In this chapter, countertrade congestion management on GA-Fuzzy based OPF formulations incorporating (1) and hybrid type, i.e. both ((1) and (2)) above are presented and tested. The function of above OPF based models is to modify system dispatch to ensure secure and efficient system operation based on the existing operating condition. It would use the dispatchable resources (i.e. real and reactive power generations and capacitor reactive supports) and controls (i.e. transformer tappings) subject to their limits and determine the required curtailment of transactions to ensure uncongested operation of the power system. A new load curtailment scheme for pool loads is proposed where all connected loads are divided into three different groups depending on their *willingness to pay* up to certain load curtailment value.

11.5.2 Transmission Congestion Penalty Factors

A concept of transmission congestion penalty factors is developed and implemented to control line overflows in proposed GA-Fuzzy approach for congestion management. Transmission congestion penalty factor for each transmission line is computed which can adopt a suitable value depending upon amount of power flow (in MVA) above/below the maximum limit. Therefore, the congested line/lines and lines near to congested line/lines have higher values of transmission congestion penalty factors than other lines in the system. These transmission congestion penalty factors are helpful in deciding appropriate re-dispatchment of dispatchable resources. The procedure for determining transmission congestion penalty factors is explained in next section.

1. *Procedure to determine transmission congestion penalty factors*

A base case situation is considered for congestion management. This base case refers to optimal settings of real power generation schedule, transformer tap settings and capacitor reactive support settings under normal state and with these settings now system is subjected to congestion (with one/more than one line limits is/are violated).

The following steps are followed to compute these penalty factors.

Step 1. Load flow solution and line flows ($S_{ij-base}$) are obtained for base case.

Step 2. Set the line limits in congestion case (S_{ij-M}).

Step 3. GA-Fuzzy approach as described earlier, is used to generate population of different generation schedules satisfying equality and non-equality constraints (except line flows limits).

Step 4. Line flows (S_{ij-tr}) are calculated for each such generation schedule and line penalty factors (P_{ij}, where i and j denote bus numbers between which transmission line is connected) are calculated according to Fig. 11.18.

Step 5. Another parameter, *line_flow_sum* representing cumulative effect of penalty factors and transmission line flows in congestion is computed as follows:

$$line\_flow\_sum = \sum_{l=1}^{n_l} P_{ij} * S_{ij-tr}$$

where n_l = no. of transmission lines.

These new types of transmission congestion penalty factors have two advantages. First, separate slope for penalty factor of each transmission line is determined depending upon power overflow above rated line flow value of that transmission line. It means that line with lesser power overflow will have lower value of slope, and thus will result small value of penalty factor. Similarly, it is understood that line with comparatively higher power overflow will have higher value of penalty factor. This adaptive feature is helpful in finding right solution (optimal values of control parameters, e.g. real power genera-

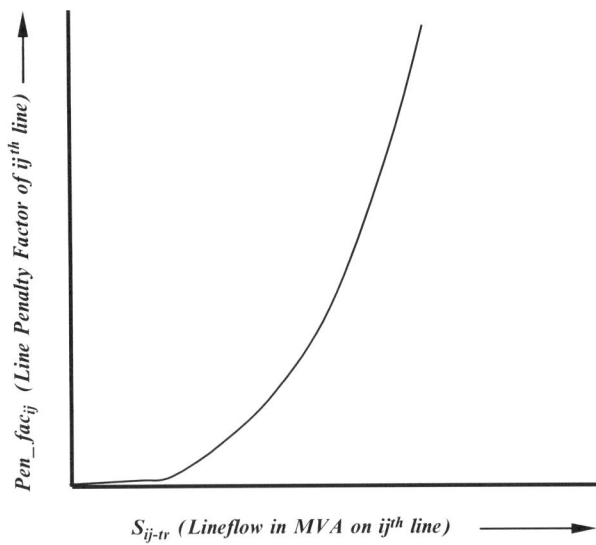

Fig. 11.18. Graphical representation of penalty factors as straight lines

tion, transformer tapping and capacitors values) by search techniques such as GA. Secondly, only single logic mentioned in step-4 works for determining these congestion penalty factors based on magnitude of power overflow in the line/lines. Therefore, no difficulty arises in choosing suitable values of penalty factors.

11.5.3 Proposed Methods for Congestion Management

Three methods are proposed with different objectives using GA-Fuzzy optimal approach and are explained below:

Method-1. Objective of minimization of line overflows only.

Method-2. Objective of minimization of line overflows along with (real power generation + reactive generation) redispatch cost and change in capacitor support cost.

Method-3. Objective of minimization of line overflows along with (real power generation + reactive generation) redispatch cost, change in capacitor support cost and load curtailment.

Mathematical functions representing redispatch cost of real power generation, reactive power generation and change in capacitor support cost are given below. The *real power redispatch cost* $C_{adj}(\Delta P_{g,k-m})$ is computed by adjusting generation of each generating unit less or more than base case value, with the help of *adjustment bids characteristics curves* shown in Fig. 11.19. These curves are decided by special *adjustment bids* $C_{adj,Pg,k-m}$ invited from all the generator units for generating power less or more than base case values. Therefore, *real power redispatch cost* ca be expressed as:

$$C_{adj}(\Delta P_{g,k-m}) - C_{adj,Pg,k-m} * \Delta P_{k-m} \; \$/hr \qquad (11.62)$$

The reactive power cost of generator is also called opportunity cost Dai (2001). The reactive power output of a generator will reduce its active power generation capability which can serve at least as spinning reserve, and the corresponding implicit financial loss to generator is modeled as an opportunity cost. Therefore *reactive power redispatch cost* $C_{adj}(\Delta Q_{g,k-m})$ of generator as defined by Kumar (2004) is:

$$C_{adj}(\Delta Q_{g,k-m}) = \lfloor C_{pg}(S_{G,\max,k-m})$$
$$- C_{pg}(\sqrt{S_{G,\max,k-m}^2 - \Delta Q_{g,k-m}^2})\rfloor kprofit \; \$/h \qquad (11.63)$$

where $C_{pg}(P_{G,k-m}) = a_k + b_k P_{G,k-m} + c_{km} P_{G,k-m}^2$

i.e. the cost of active power generation is modeled by above quadratic function. Where a_k, b_k and c_k are costs coefficients of *kth* generator and $S_{G,\max,k-m}$ is the nominal maximum apparent power of generation and *kprofit* is the profit rate of active power generation taken between 5 and 10% [DAI01].

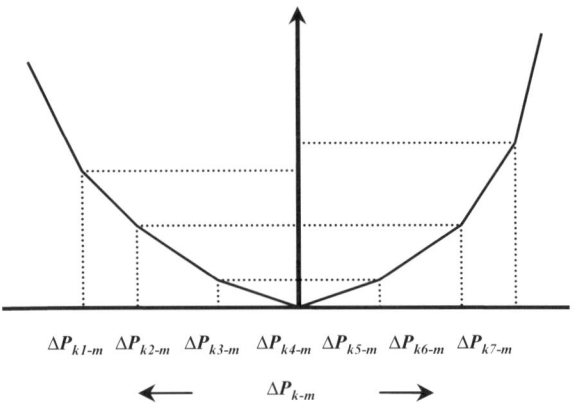

$\Delta P_{k1\text{-}m}$ $\Delta P_{k2\text{-}m}$ $\Delta P_{k3\text{-}m}$ $\Delta P_{k4\text{-}m}$ $\Delta P_{k5\text{-}m}$ $\Delta P_{k6\text{-}m}$ $\Delta P_{k7\text{-}m}$

$\longleftarrow$ $\Delta P_{k\text{-}m}$ $\longrightarrow$

Fig. 11.19. Adjustment bid characteristic representing cost function of the change of active power production at the kth generator

The equivalent cost for return on the capital investment of the capacitors, which is expressed as their depreciation rates (the life span of capacitors is assumed as 15 years) is computed as

$$C(Q_{C,kc-m}) = Q_{C,kc-m} \frac{(\$11600/M\mathrm{var})}{(15{*}365{*}24{*}h)\ hour}$$

$$= Q_{C,kc-m}{*}\$13.24/(100\ M\ \mathrm{var}\ hour) \qquad (11.64)$$

where h is the average usage rate of capacitors taken as $2/3$. Equation (11.64) is a linear cost function with the slope of $\frac{dC_{adj,kc-m}(Q_{C,kc-m})}{dQ_{C,kc-m}} = \frac{\$13.24}{100\ M\ \mathrm{var}\ hour}$, which can be approximately represented as:

$$C_{adj}(\Delta Q_{C,kc-m}) = \Delta Q_{C,kc-m}{*}(13.24/100)\$/hr \qquad (11.65)$$

Method 1 - Objective of minimization of line overflows only

Step 1. Real power generation redispatch $\Delta P_{g,k-m}$, reactive power generation redispatch $\Delta Q_{g,k-m}$ and change in capacitor reactive support $\Delta Q_{C,kl-m}$ are computed for each valid generation schedule in population, where $k =$ generating unit no., $kc =$ capacitor unit no. and $m =$ no. of generation schedule in population.

Step 2. Correspondingly, redispatch costs of real power generation $C_{adj}(\Delta P_{g,k-m})$, reactive power generation $C_{adj}(\Delta Q_{g,k-m})$ and change in capacitor reactive support $C_{adj}(\Delta Q_{C,kcm})$ are computed as per expressions (11.62), (11.63), and (11.65), respectively.

Step 3. Fitness of each generation schedule in a population is calculated as:

$$Fitness = \frac{1}{A{*}line\_flow\_sum} \qquad (11.66)$$

where, $A =$ numerical constant.

Step 4. Finally values of real and reactive power generation schedule, transformers tapping values, bus voltages, capacitor reactive support values and line flows calculated in last generation of GA-Fuzzy based optimization approach.

Method 2 - Objective of minimization of line overflows along with (real power generation + reactive generation) re-dispatch cost and change in capacitor support cost

1. Step1 and Step 2 of method-1 are followed.
2. Fitness of each generation schedule in a population is calculated as:

$$Fitness = \frac{e^{-B\times\left(\sum\limits_{g}^{NG}C_{adj}(\Delta P_{g,k-m})+\sum\limits_{g}^{NG}C_{adj}(\Delta Q_{g,k-m})+\sum\limits_{c}^{NC}C_{adj}(\Delta Q_{C,kl-m})\right)}}{A\times line\_flow\_sum}$$

(11.67)

   where A and B are numerical constants.
3. Step 4 of method-1 is followed.

Method 3 – Objective of minimization of line overflows along with (real power generation + reactive generation) redispatch cost, change in capacitor support cost and load curtailment

1. Step1 of method-1 is followed.
2. If real loads connected on load buses under congestion are termed as base load values, then load cutailment is done by reducing base load values in three different groups (G-1, G-2 and G-3). G-1, G-2 and G-3 refer to groups of loads (consumers) which are paying fee (*willingness to pay*) for load curtailment upto 80, 60 and 40 of their base case load values respectively, in a congestion state. Load values after curtailment $(P_{d,kl-m,gr-i})$ in three different groups (G-1, G-2 and G-3) are computed.
3. Step2 of method-1 is followed.
4. Fitness of each generation schedule in a population is calculated as:

$$Fitness = \frac{\begin{aligned}e^{-}B\times\Bigg(&\sum\limits_{g}^{NG}C_{adj}(\Delta P_{g,k-m})+\sum\limits_{g}^{NG}C_{adj}(\Delta Q_{g,k-m})+\sum\limits_{c}^{NC}C_{adj}(\Delta Q_{C,kc-m})\\&+\sum\limits_{i=1}^{3}K_{i}(\sum\limits_{kl}^{NL}(P_{d,kl-m,gr-i})-\sum\limits_{kl}^{NL}(P_{d,kl-m,base-i}))^{2}\Bigg)\end{aligned}}{A\times line\_flow\_sum}$$

(11.68)

   where A, B and K_i are numerical constants.
5. Step4 of method-1 is followed.

11.5.4 Test Results

The proposed methods are implemented on modified IEEE 30 bus system. The busdata and linedata are given in Appendix F. Line (8,28) get congested (exceeding flow limit of 12 MVA) if outage of line (6,28) is considered.

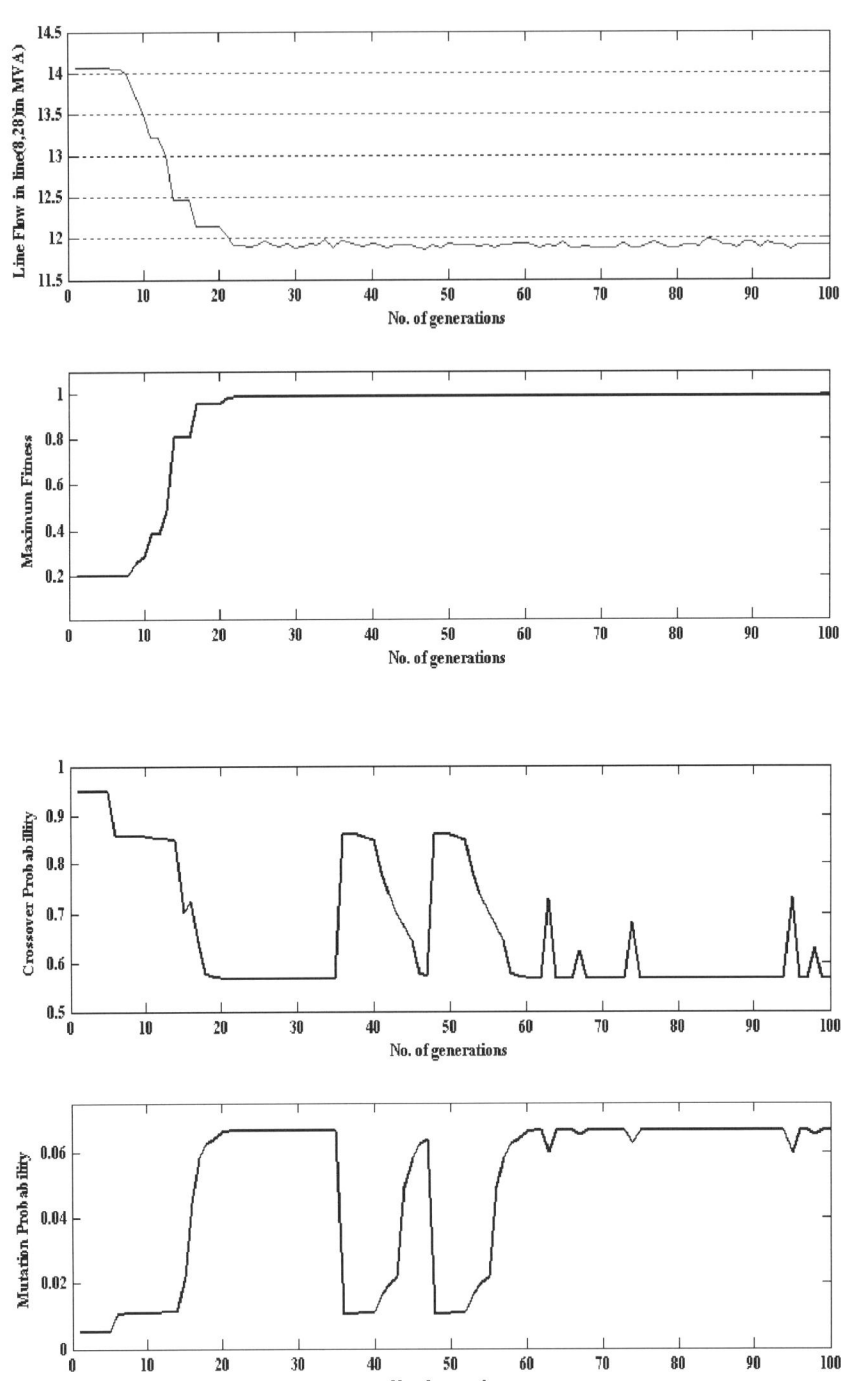

Fig. 11.20. Convergence of different parameters, crossover probability and mutation probability variations using GA-Fuzzy approach for Method-1

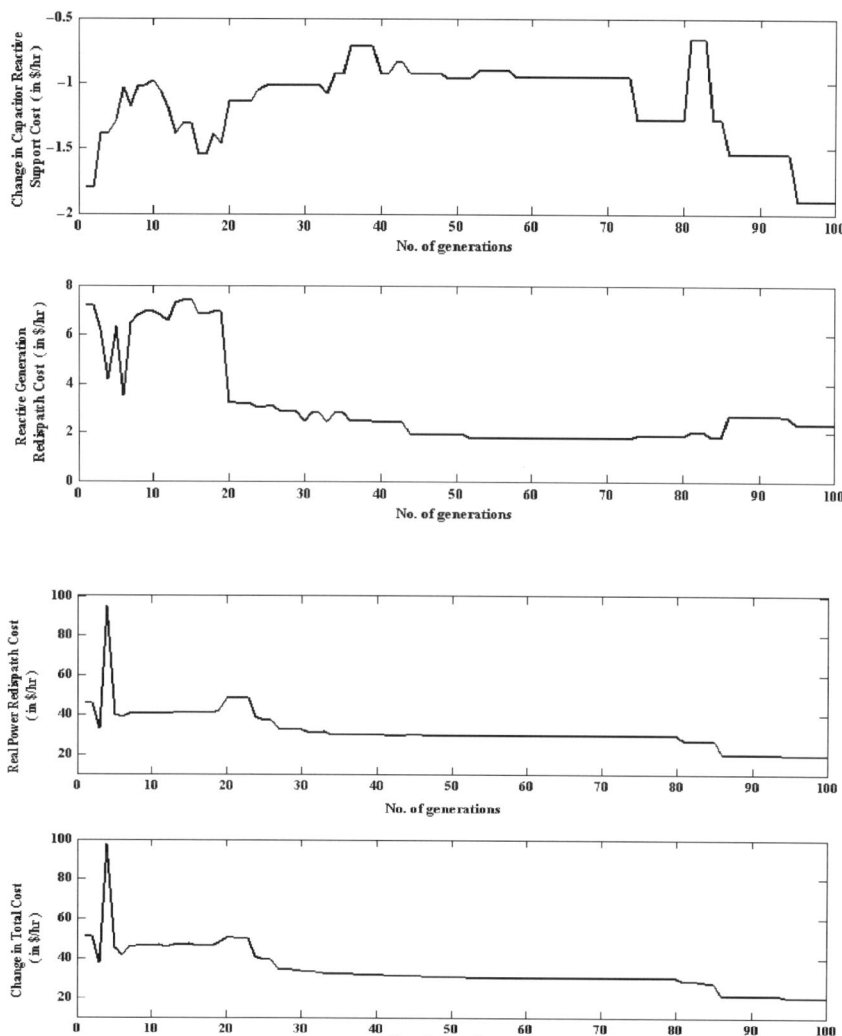

Fig. 11.21. Convergence of different parameters, crossover probability and mutation probability variations using GA-Fuzzy approach for Method-2

Figures 11.20, 11.21 and 11.22 show the convergence of different parameters along with crossover probability and mutation probability variations.

Figures 11.23–11.26 and Table 11.27 represent bus voltage profile for different methods.

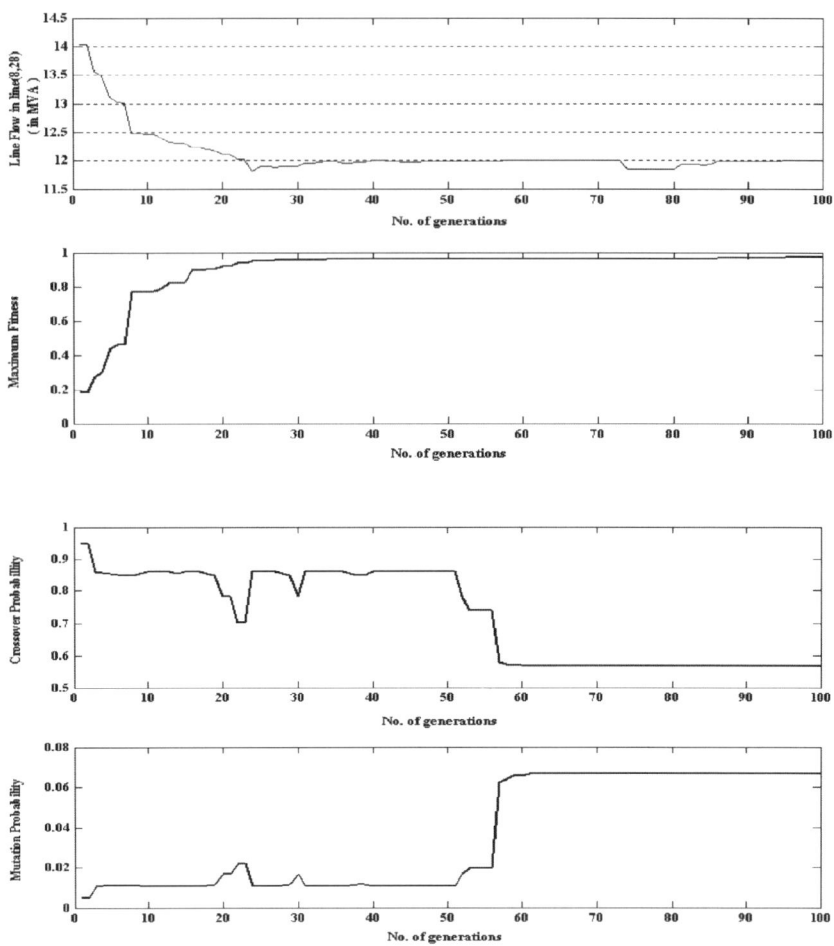

Fig. 11.21. (*Continued*)

11.5.5 Conclusions

The results tabulated in Table 11.28a shows optimal values of active power generation, reactive power generation and capacitor reactive support to avoid congestion for method-1 and method-2. Method-1 is found to be superior than method-2 so far controlling of power overflow is concerned. In table 28b method-2 seems to be more economical than method-1. The differences in performance of both the methods are due to modeling of their respective fitness function. In method-1, emphasis is only on control of power overflow on the Lines, whereas control of power overflow along with redispatch costs of (real power + reactive power) generation and change in capacitor

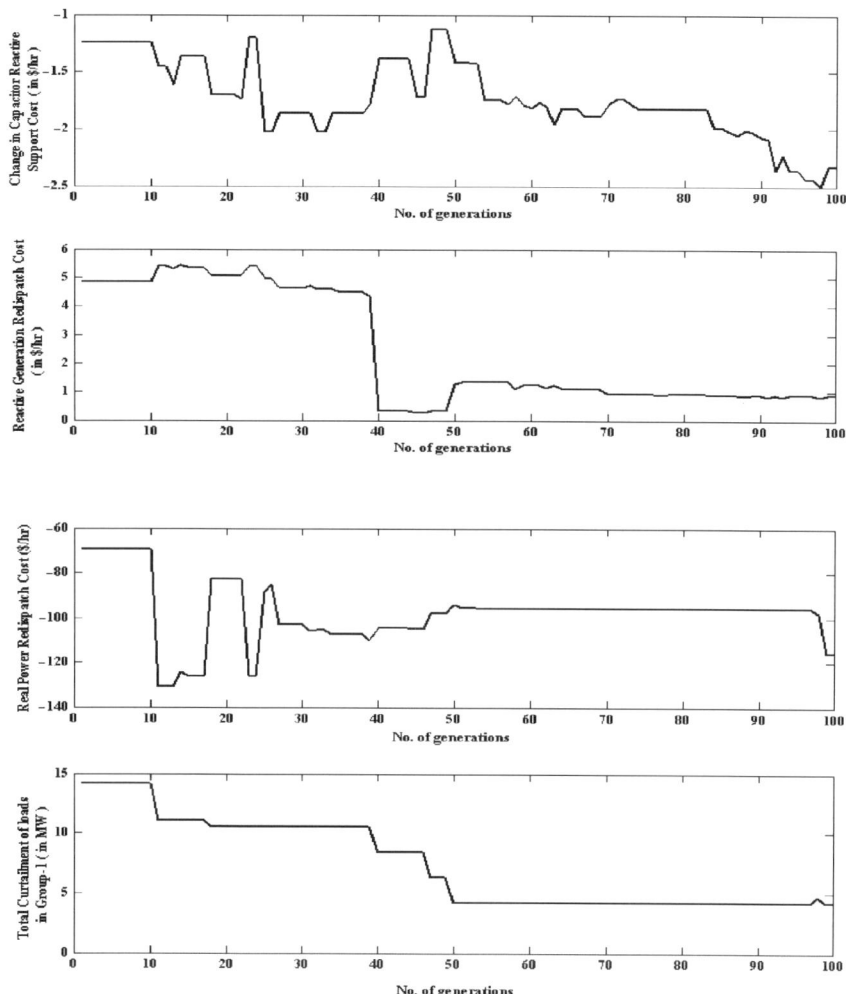

Fig. 11.22. Convergence of different parameters, crossover probability and mutation probability variations using GA-Fuzzy approach for Method-3

reactive support cost are intermingled in method-2. It is also clear from Fig. 11.21 for method-2 that a controlling action to check power overflow is dominant over economic redispatchment cost feature throughout the GA-Fuzzy based optimization procedure. From the results it is seen that slightly lesser load bus voltage variation (i.e. between maximum and minimum load bus voltages) with very small increment in average system voltage value (i.e. average of all bus voltages of the system). It means that from voltage point of view, method-2 is not inferior than method-1, although this particular aspect requires verification for other power systems also. Therefore,

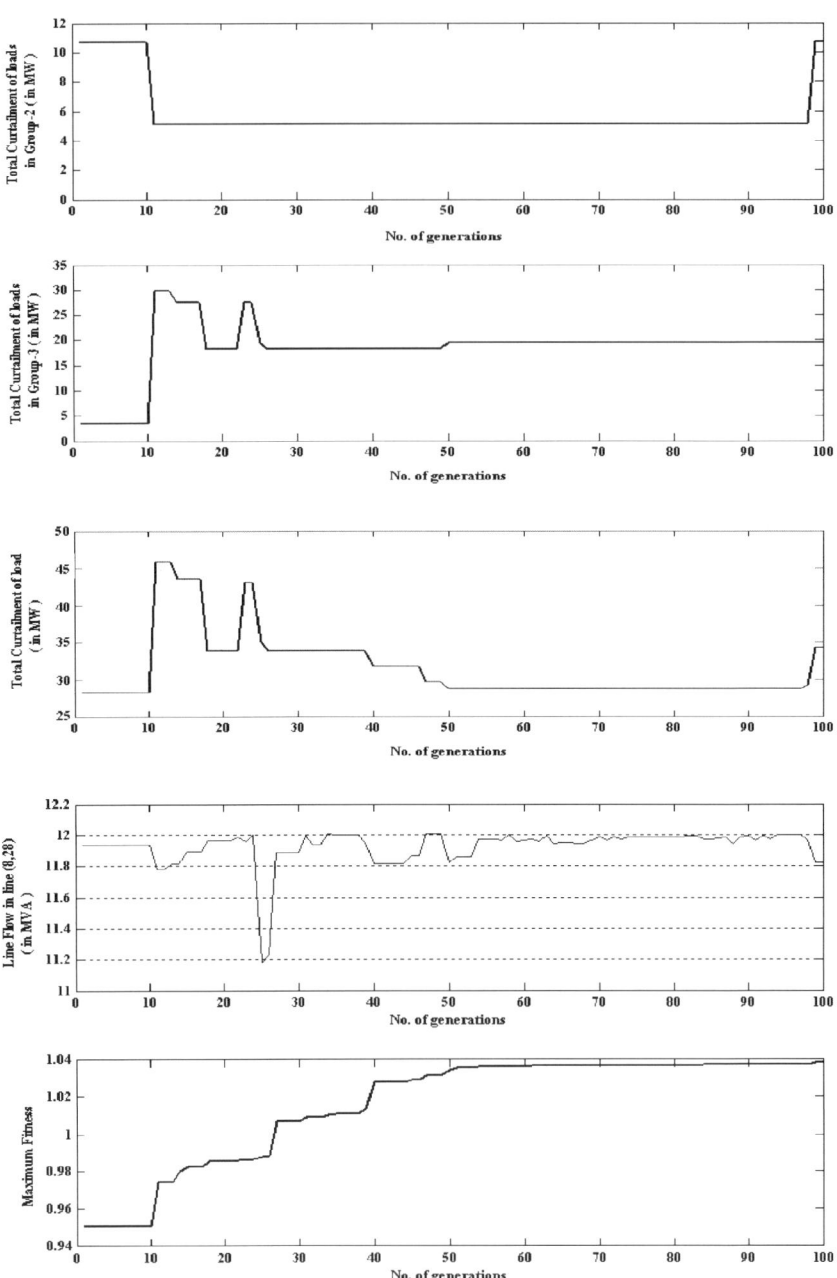

Fig. 11.22. (*Continued*)

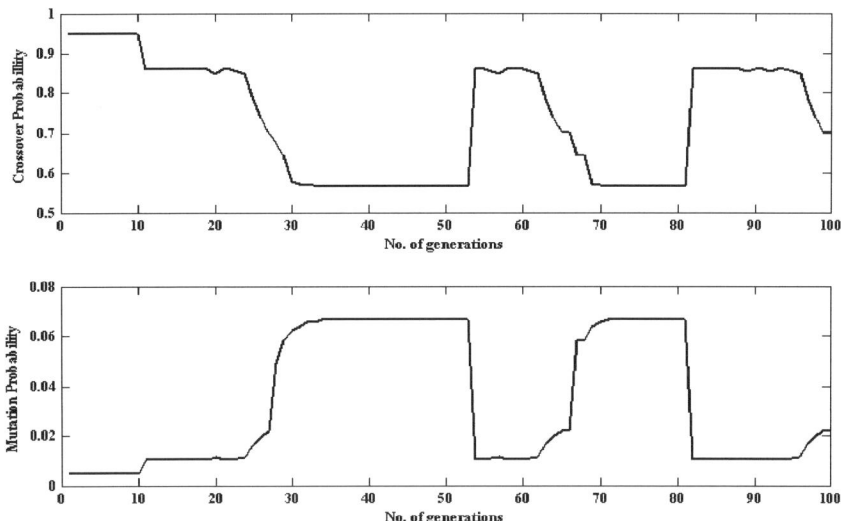

Fig. 11.22. (*Continued*)

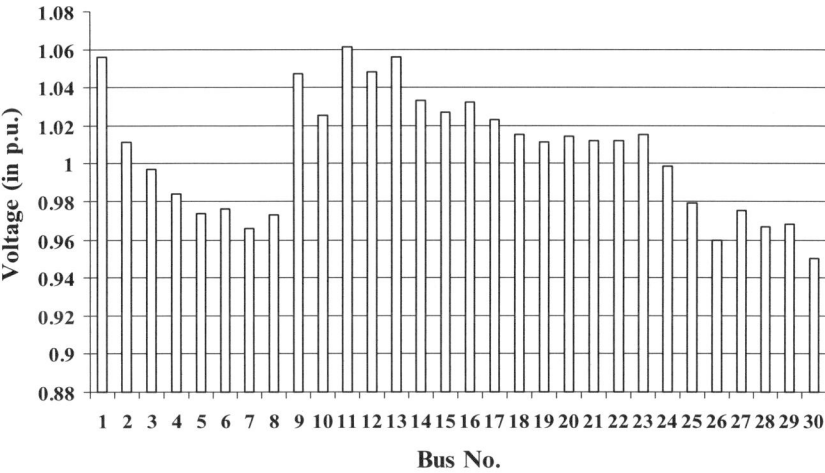

Fig. 11.23. Bus voltage profile using congestion management method-1

method-1 and method-2 both have applicability from congestion management view point.

Method-3 is developed for a scenario different from one in which method-1 and method-2 work. In this method, a load curtailment feature is also added in fitness function by mathematical modeling. This feature enables pool customers to pay extra charges in order to avoid congestion as shown in Ta-

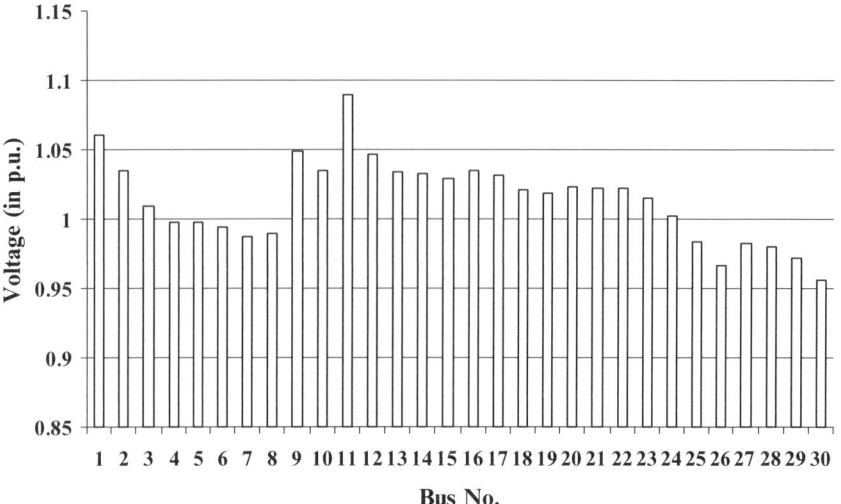

Fig. 11.24. Bus voltage profile using congestion management method-2, when kprofit = 5%

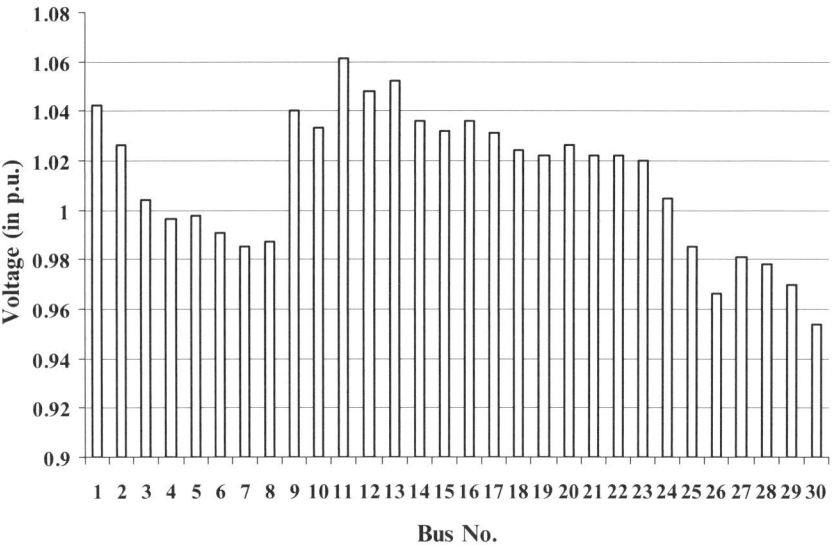

Fig. 11.25. Bus voltage profile using congestion management method-2, when kprofit = 10%

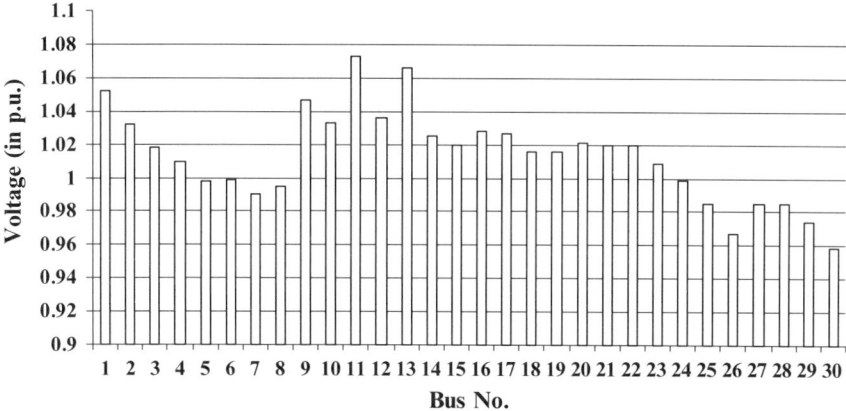

Fig. 11.26. Bus voltage profile using congestion management method-3

Table 11.27. Comparison of maximum and minimum voltage levels at Load buses and average system voltages for proposed methods of congestion management

		Method-1	Method-2		Method-3
			Kprofit = 5%	Kprofit = 10%	
	Maximum	Bus 12: 1.048 p.u.	Bus 9: 1.049 p.u.	Bus 12: 1.048 p.u.	Bus 9: 1.047 p.u.
load bus	Minimum	Bus 30: 0.95 p.u.	Bus 30: 0.956 p.u.	Bus 30: 0.954 p.u.	Bus 30: 0.959 p.u.
	Difference	0.098 p.u.	0.093 p.u.	0.094 p.u.	0.088 p.u.
Average value of system voltage		1.005533 p.u.	1.0139 p.u.	1.012433 p.u.	1.0135 p.u.

ble 11.28c. This method can be applicable in deregulated environment as it seems to be fair, transparent and consumer satisfaction to great extent.

A hybrid strategy having two stages is also formed on the basis of three methods developed and tested on modified IEEE 30 bus system. In first stage, method-1 or method-2 can be used. If congestion is still not avoidable then under second stage method-3 with load-curtailment and willingness to pay feature can be used.

11.5.6 Bibliography and Historical Notes

The application of genetic algorithms for altering membership functions of fuzzy controllers to make it adaptive Karr and Gentry 1993; Park et al. 1994).

The idea of fuzzifying genetic algorithms emerged in 1990s. Various ways of integrating fuzzy systems and genetic algorithms were proposed by Sanchez (1993), Xu and Vukovich (1993) and Buckley and Hayashi (1994a).

El-Hawary (1998) has shown various fuzzy system applications to Electric Power Applications in deregulated Environment. Iyer (2003) mentioned an integrated fuzzy-neural approach to electricity spot-price forecasting in a deregulated electricity market. Ming et al. (2004) used an ARIMA approach to forecasting electricity price. Saini et al. (2006) explained the GA-Fuzzy integrated System Approach to solve OPF problem and help in congestion management. Ravikumar et al. (2007) paper deals with the intelligent approach for fault diagnosis using support vector machines.

Table 11.28a. Comparison of redispatchment of P, Q, change in capacitor reactive power support and line flow at line (8,28) for method-1 and method-2

Real power generation redispatch cost $(in\ \$h) = \sum_g^{NG} C_{adj}(\Delta P_{g,k-m})$ $Method-1\ Fitness = \dfrac{1}{A \times line\_flow\_sum}$

Reactive power generation redispatch cost $(in\ \$h) = \sum_g^{NG} C_{adj}(\Delta Q_{g,k-m})$ $Method-2\ fitness$

$$= \frac{e^{-B \times \left(\sum_g^{NG} C_{adj}(\Delta P_{g,k-m}) + \sum_g^{NG} C_{adj}(\Delta Q_{g,k-m}) + \sum_c^{NC} C_{adj}(\Delta Q_{C,kl-m}) \right)}}{A \times line - flow - sum}$$

Change in capacitor support cost $(in\ \$h) = \sum_c^{NC} C_{adj}(\Delta Q_{c,kc-m})$ $Method-3\ Fitness$

$$= \frac{e^{-B \left(\sum_g^{NG} C_{adj}(\Delta P_{g,k-m}) + \sum_g^{NG} C_{adj}(\Delta Q_{g,k-m}) + \sum_c^{NC} C_{adj}(\Delta Q_{C,kc-m}) + \sum_{i=1}^{3} K_i \left(\sum_{kl}^{NL}(P_{d,kl-m,gr-i}) + \sum_{kl}^{NL}(P_{d,kl-m,base-i}) \right)^2 \right)}}{A \times line\_flow\_sum}$$

S. no.	Generation at	In congestion state		Method-1		Congestion management method			
						Method-2			
						When kprofit = 5%		When kprofit = .10%	
		P_g(in MW)	Q_g(in MVAr)	P_g(in MW)	Q_g(in MVAr)	P_g(in MW)	Q_g(in MVAr)	P_g(in MW)	Q_g(in MVAr)
1.	Bus 1	175.165	4.624	156.568	72.914	163.962	32.23	156.348	7.958
2.	Bus 2	48.941	29.345	33.647	-4.262	41.882	39.692	48.471	41.596
3.	Bus 5	21.176	28.21	33.941	25.8	20.765	33.403	21.039	39.725
4.	Bus 8	22.647	40.559	12.549	29.386	11.667	27.073	12.451	29.301
5.	Bus 11	12.588	17.124	26.706	8.078	26.471	22.08	21.059	11.095
6.	Bus 13	12.00	10.263	28.8	6.634	28.031	-8.37	33.082	3.108
	Total	292.517	130.125	292.211	138.551	292.778	146.107	292.45	132.782

Table 11.28a. (*Continued*)

S. no.	Generation at	In congestion state			Congestion management method								
					Method-1			Method-2					
								When kprofit = 5%			When kprofit = .10%		
		P_g(in MW)	Q_g(in MVAr)	Q_C(in MVAr)	P_g(in MW)	Q_g(in MVAr)	Q_C(in MVAr)	P_g(in MW)	Q_g(in MVAr)	Q_C(in MVAr)	P_g(in MW)	Q_g(in MVAr)	Q_C(in MVAr)
S.No.	Capacitor at												
1.	Bus 10			3.982			1.585			2.299			4.012
2.	Bus 12			0.02			2.916			1.037			0.02
3.	Bus 15			4.149			0.998			4.354			3.836
4.	Bus 17			4.99			3.043			1.937			3.562
5.	Bus 20			4.432			1.526			2.114			4.658
6.	Bus 21			4.354			0.802			0.773			3.748
7.	Bus 23			4.54			3.503			1.155			3.464
8.	Bus 24			4.687			4.237			1.791			2.485
9.	Bus 29			2.097			4.355			3.387			2.903
Total				33.251			22.965			18.847			28.686
Line flow at line (8,28)(*in MVA*)				16.64			11.921			11.993			11.9715

Table 11.28b. Comparison of congestion relief charges for method-1 and method-2

S. No.	Generation at	Congestion management method				
		Method-1		Method-2		
				When kprofit = 5%		When kprofit = .10%
		$C(\Delta P_g)$(in $ h)	$C(\Delta Q_g)$(in $ h)	$C(\Delta P_g)$(in $ h)	$C(\Delta Q_g)$(in $ h)	$C(\Delta P_g)$(in $ h)	$C(\Delta Q_g)$)(in $ h)
1.	Bus 1	−59.823	2.0764	−36.0217	0.3343	−60.5312	0.0097
2.	Bus 2	−48.4633	1.6359	−23.3393	0.1525	−1.5495	0.4278
3.	Bus 5	58.2958	0.0211	−1.3936	0.0978	−0.4239	0.9631
4.	Bus 8	−35.7547	0.3496	−38.7967	0.515	−36.0927	0.7102
5.	Bus 11	55.1735	0.3117	54.1465	0.0925	31.6587	0.2745
6.	Bus 13	67.8	0.0412	64.5317	1.1247	87.5395	0.3215
Total(in $ h)		37.2283	4.4359	19.1269	2.3169	20.6009	2.706975

S. No.	Capacitor at	$C(\Delta Q_c)$(in $/hr)	$C(\Delta Q_c)$(in $/hr)	$C(\Delta Q_c)$(in $/hr)
1.	Bus 10	−0.3174	−0.2228	0.004
2.	Bus 12	0.3834	0.1347	0
3.	Bus 15	−0.4172	0.0271	−0.0414
4.	Bus 17	−0.2578	−0.4042	−0.1891
5.	Bus 20	−0.3848	−0.3069	0.0299
6.	Bus 21	−0.4703	−0.4741	−0.0802
7.	Bus 23	−0.1373	−0.4482	−0.1425
8.	Bus 24	−0.0596	−0.3834	−0.2915
9.	Bus 29	0.299	0.1708	0.1067
Total(in $ h)		−1.3619	−1.9071	−0.6041
Grand total(in $ h)		40.3023	19.5367	22.703775

Table 11.28c. Redispatchment of (Real power + reactive power generation), change in capacitor reactive power support and line flow at line (8,28) for method-3

Method-3

S. no.	Generation at	P_g(in MW)	Q_g(in MVAr)
1.	Bus 1	150.747	13.316
2.	Bus 2	42.353	32.196
3.	Bus 5	19.529	30.004
4.	Bus 8	10.098	31.131
5.	Bus 11	15.02	13.246
6.	Bus 13	19.027	23.381
Total		256.774	143.273

S.No.	Capacitor at	Q_C(in MVAr)
1.	Bus 10	2.329
2.	Bus 12	2.808
3.	Bus 15	1.693
4.	Bus 17	0.401
5.	Bus 20	1.986
6.	Bus 21	1.027
7.	Bus 23	1.115
8.	Bus 24	1.115
9.	Bus 29	3.16
Total		*15.636*

Table 11.28c. (*Continued*)

Method-3

S. no.	Generation at — Load Group-1	P_l (in MW) In Congestion	Method-3	Load Group-2	P_l (in MW) In Congestion	Q_g (in MVAr) Method-3	Load Group-3	P_l (in MW) In Congestion	Method-3
1.	Bus 7	22.8	21.6232	Bus 4	7.6	7.0116	Bus 2	21.7	14.56
2.	Bus 8	30.0	28.4516	Bus 5	94.2	86.9071	Bus 3	2.4	1.6103
3.	Bus 17	9.0	8.5355	Bus 15	8.2	7.5652	Bus 10	5.8	3.8916
4.	Bus 18	3.2	3.0348	Bus 16	3.5	3.229	Bus 12	11.2	7.5148
5.	Bus 26	3.5	3.3194	Bus 21	17.5	16.1452	Bus 14	6.2	4.16
6.	Bus 29	2.4	2.2761	Bus 23	3.2	2.9523	Bus 19	9.5	6.3742
7.	Bus 30	10.6	10.0529	Bus 24	8.7	8.0265	Bus 20	2.2	1.4761
	Total	*81.5*	*77.2935*	*Total*	*142.9*	*131.8369*	*Total*	*59.0*	*39.587*

Line flow at line (8,28) (in MVA)   *11.823*

Group G1- load group 1
Group G2- load group 2
Group G3- load group 3

Integration of Neural Networks and Fuzzy Systems

Neuro-Fuzzy and Soft Computing, as a mature and extensive coverage of neuro-fuzzy soft computing, demonstrates a paradigm shift in managing complexity, uncertainty and subjectivity.

Irena Nagisetty
*Advanced Manufacturing Technology Development
Ford Motor Company

12.1 Introduction

The literature reveals that the ANN and Fuzzy set theoretic approaches have been often used for non-linear and complex problems such as load forecasting in power system. The integration of these approaches gives improved results as compared to conventional techniques. Both the modeling techniques have their own merits and demerits as follows:

1. Fuzzy models possess large power in representing linguistic and structured knowledge by fuzzy sets and performing fuzzy reasoning by fuzzy logic in qualitative manner and usually rely on the domain experts to provide the required knowledge for a specific problem. Further, the compensatory operators in the fuzzy models as connectives are found quite suitable and produce results, which are very close to the actual results (Mizumoto 1989).
2. On the other hand, neural network models are particularly good for non-linear mappings and for providing parallel processing facility to simulate complex system. The neural network models are developed via training.
3. Furthermore, while the behavior of fuzzy models can be understood easily due to their logical structure and step-by-step inference procedures. Neural network models act normally as a black box, without providing explanation facility.

D.K. Chaturvedi: *Soft Computing Techniques and its Applications in Electrical Engineering,*
Studies in Computational Intelligence (SCI) **103**, 479–499 (2008)
www.springerlink.com   © Springer-Verlag Berlin Heidelberg 2008

From these investigations, it is quite natural to consider the possibilities of integrating the two paradigms, in order to utilize the desired strength of both types of models to produce improved results. There are three possibilities for combining fuzzy systems and neural networks to work together competitively and/or co-operatively:

1. Fuzzy systems work at higher level in hierarchy and neural networks perform the lower level computations as shown in Fig. 12.1.
2. The task in hand is divided broadly into two categories, i.e. problems need qualitative modeling and other part needs quantitative modeling. The qualitative work is done by fuzzy systems and quantitative work is done by ANN. Both of these techniques work in co-operation as shown in Fig. 12.2.

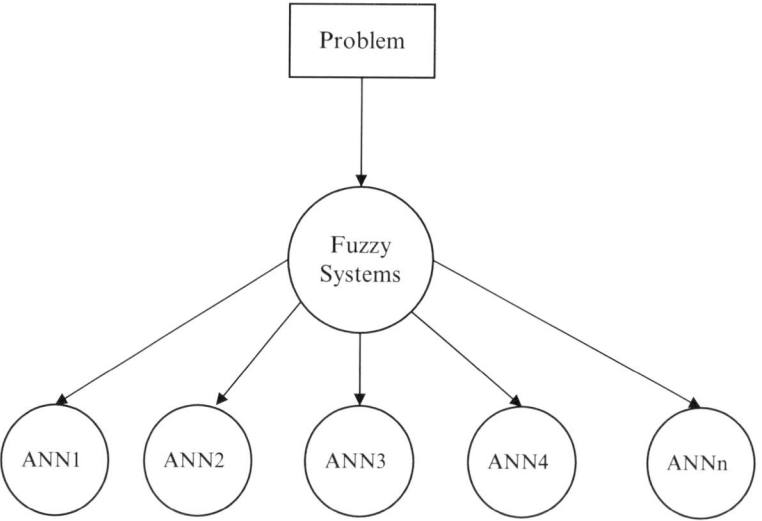

Fig. 12.1. Hierarchal Combination of Fuzzy Systems and Neural Network

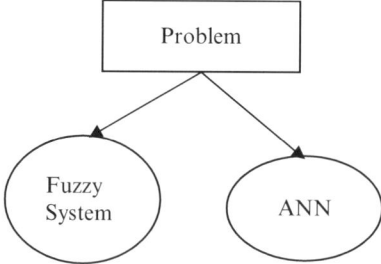

Fig. 12.2. ANN and Fuzzy System working in co-operation.

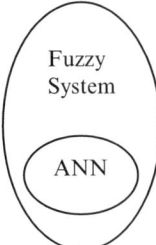

Fig. 12.3a. Neuralizing the Fuzzy System

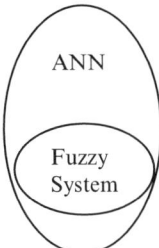

Fig. 12.3b. Fuzzifying ANN

3. Emerging Fuzzy System and ANN techniques:

(a) By neuralizing fuzzy systems, i.e. the introduction of neural network concepts in fuzzy systems. Technically, it may be realized by mapping out fuzzy systems into neural network, either functionally or structurally as shown in Fig. 12.3a.

(b) Or fuzzifying neural networks by introducing fuzzy concepts in neural networks as shown in Fig. 12.3b. The fuzzy neural networks retain the basic properties and architecture of neural network and simply fuzzify some of their elements. It is well recognized that fuzzy systems are logical based with fuzzy set representation and flexible fuzzy logic operations. Thus the resulting fuzzy neural system may include minimum, maximum or compensatory operators, apart from usual sum and product operators found in neural computing.

12.2 Adaptive Neuro-Fuzzy Inference Systems

In Sects. A and B, ANN and fuzzy systems have been explained in the detail. However, the design of fuzzy systems relies on two important factors:

1. *Knowledge acquisition* – To acquire knowledge, an appropriate knowledge acquisition technique is required.
2. *Human expert* – The expert in a particular field is also equally important to share his knowledge.

Table 12.1. Comparison of Mamdani and Sugeno models

S. No.	Characteristics	Mamdani model	Sugeno model
1.	Output membership function	Could be of any continuous shape	Only spikes can be used
2.	Aggregation, implication, or compositional methods	Any method can be used	Almost fixed structure for Sugeno Models
3.	Defuzzification method	Any method could be used	Only weighted average method can be used
4.	Complexity	Less complex	Complex
5.	Adaptibility	Not adaptive system	adaptive system
6.	Training data	No need of training data	Requires

These factors restrict the application of fuzzy systems. The adaptive neuro-fuzzy inference systems (ANFIS) could be used to overcome the effect of these factors to a certain limit (Jang 1992, 1993, Jang and Sun 1995).

The ANFIS architecture can identify the near optimal membership functions of Fuzzy systems for achieving desired output. It is generated by the fuzzy toolbox available in MATLAB, which allows to optimize standard Sugeno Fuzzy model, which was introduced in 1985 (Sugeno 1985). The difference between Sugeno model and Mamdani model is that in Sugeno model the output membership functions are singleton spikes rather than a distributed fuzzy sets. The comparision of Mamdanoi and Sugeno models are given in Table 12.1.

In case of Sugeno models, singleton membership functions of the output of fuzzy system simplify the defuzzification step. The typical rule in a Sugeno model has the form

If $X_1 = x$ and $X_2 = y$ then output is $z = ax + by + c$

The output level z_i of each rule is weighted by the firing strength W_i of the rule. In the above case the firing strength is $W_i = \text{AndMethod}(F1(x), F2(y))$ as shown in Fig. 12.4 along with its output surface in Fig. 12.4b.

The final output of the system is the weighted average of all the rule outputs computed as

$$\text{Final output} = \frac{\sum\limits_{i=1}^{n} W_i Z_i}{\sum\limits_{i=1}^{n} W_i}$$

Luckily, it is frequently the case that singleton output functions are completely sufficient for the needs of a given problem. Sugeno models is very well suited to the task of smoothly interpolating the linear gains that would be applied across the input space; it is natural and efficient systems.

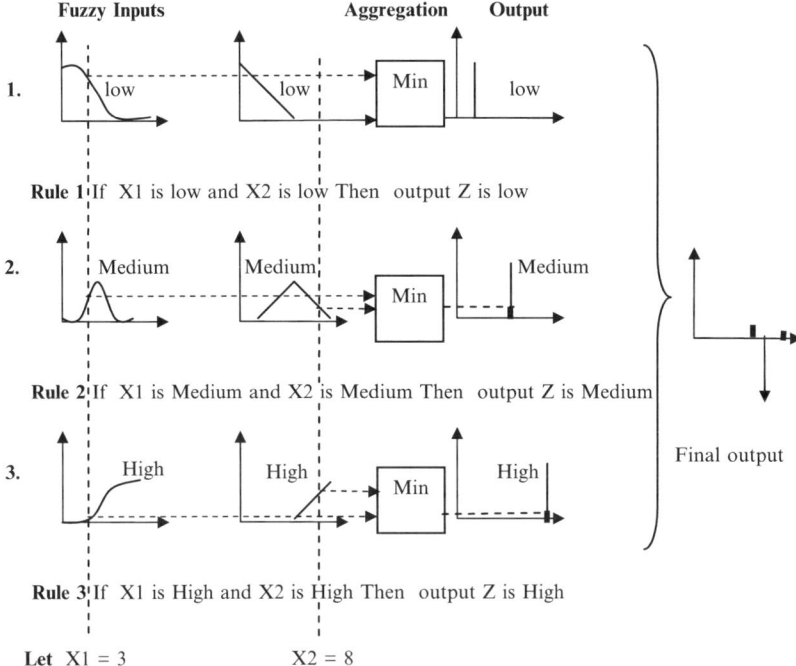

Fig. 12.4. Graphical representation of Inference with Sugeno Model

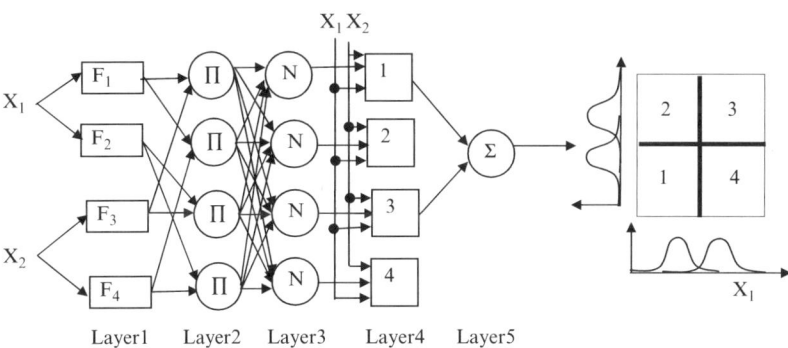

Fig. 12.5. ANFIS architecture based on Two input Sugeno Fuzzy model.

In the ANFIS, ANN uses a back propagation gradient descent method for training fuzzy system membership function parameters to emulate a given training data set. Figure 12.5 shows a simple two input ANFIS architecture for a Sugeno Fuzzy model. It is composed of five layers. The output of the first layer can be mathematically written as

$$O_i^1 = \mu_{Ai}(X_1) \tag{12.1}$$

where μ is the membership function for that particular input.

The output of second layer is the product of two membership functions

$$O_i{}^2 = \prod O_i{}^1 = \mu_{Fi}(X_1) \quad \mu_{Fi}(X_2) \tag{12.2}$$

Output of Layer 3

$$O_i{}^3 = \frac{O_1^2}{O_1^2 + O_2^2} \tag{12.3}$$

Output of Layer 4

$$O_i{}^4 = O_i{}^3(P_i{}^1 * X_1 + P_i{}^2 * X_2) \tag{12.4}$$

Output of Layer 5

$$O_i{}^5 = \sum_i O_i{}^4 \tag{12.5}$$

This ANFIS structure can update its parameters according to the gradient descent approach.

The layers from 1 to 5 are called: input linguistic layer, condition layer, rule layer, consequent layer and output linguistic layer. The fuzzification of the inputs and the defuzzification of the outputs are respectively performed by the input linguistic and output linguistic layers respectively while the fuzzy inference is collectively performed by the rule, condition and consequence layers.

12.3 Constraints of ANFIS

1. It is much more complex than simple fuzzy system. Hence, it is suitable for first or zeroth order Sugeno type systems.
2. It only supports Sugeno models. Therefore, it can only be developed for single output systems.
3. In ANFIS only weighted average method of defuzzification can be used.
4. The strength of every rule has unity weight.
5. It is less flexible.

12.4 HIV/AIDS Population Model Using Neuro-Fuzzy Approach

The Human Immunodeficiency Virus/Acquired Immunodeficiency syndrome (HIV/AIDS) is spreading rapidly in all regions of the world. But in India it is only 20 years old. Within this short period it has emerged as one of the most serious public health problems in the country, which greatly affect the socio-economical growth. The HIV problem is very complex and ill defined from the modeling point of view. Keeping in the view the complexities of the HIV infection and its transmission, it is difficult to make exact estimates of

HIV prevalence. It is more so in the Indian context, with its typical and varied cultural characteristics, and its traditions and values with special reference to sex related risk behaviors. Therefore, it is necessary to develop a good model which will help in making exact estimates of HIV prevalence that may be used for planning HIV/AIDS prevention and control programs. In this paper Neuro-Fuzzy approach has been used to develop dynamic model of HIV population of Agra region.

12.4.1 Introduction

Mathematical models of transmission dynamics of HIV play an important role in better understanding of epidemiological patterns and methods for disease control as they provide short and long term prediction of HIV and AIDS incidence and its dependence on various factors. The modeling study is also helpful in determining the demographic and economic impact of the epidemic, which in turn helps us to develop reasonable, scientifically, and socially sound intervention plans in order to reduce the spread of the infection. Mathematical and statistical models can serve as tools for understanding the epidemiology of HIV and AIDS if they are constructed carefully. Here an attempt is made to model the spread of HIV in a comprehensive manner with limited data of Agra Region.

12.4.2 Roots of HIV/AIDS

Investigators have assessed high-risk patients by studying long-distance truck drivers (Bwayo et al. 1991), female sex workers Kreiss et al. (1993), STD clinics (Bwayo et al. 1991; Bollinger et al. 1997; Rodrigues et al. 1995), and tuberculosis patients.

It has been estimated that there are at least several million female sex workers (FSW) in India. The number of clients is of course much higher. In India, it has been examined that the most men with sexually transmitted infections (STI) probably acquired their infections from sex workers. In India, from current knowledge, approximately 80% of sexually transmitted infections are first generation infections derived from sex work. Also, HIV infection in monogamous women (Gangakhedkar et al. 1997) is probably linked to their husbands having visited sex workers. This problem is more severe in rural areas where the people do not open their mouth and allow this disease to spread silently. It has long been recognized that sexual behavior is very heterogeneous, most people have few partners, while a minority (the core group) have many and therefore account for a much of the transmission of HIV and STI in a population. Prevention of mother-to-child transmission is possible through peripartum antiretroviral treatment of mother and child and followed by non-breastfeeding (Gibb and Tess 1999).

Three different male-female partnerships were considered.

1. *Commercially sex worker (CSW) - client relationships.* The risk of transmission during a single unprotected sex worker-client contact is determined by the risk factors of the risk of transmission from female-to-male and from male-to-female, respectively. The number of sex contacts between sex workers and clients is determined by the demand for it and the number of available sex workers.

$$\text{Commercially} \underset{\alpha_2}{\overset{\alpha_1}{\rightleftarrows}} \text{Client} \xrightarrow{\alpha_3} \begin{array}{l} \text{Life} \\ \text{Partner} \end{array}$$

2. *Spousal relationships.* "Spousal" partnerships between low risk (non-client) men and low risk (non CSW) women were considered to be more risky than above case, as these partnerships usually involve several or many sexual encounters. However, per sexual act these partnerships are implicitly assumed to be safer than individual sex worker contacts. The rate by which women form such partnerships is determined by the rate by which men form such partnerships. This simply ensures that the number of partnerships formed by men equals that formed by women, and it does not aim to be reflective of any realistic pattern of partnership formation. Transmission can occur when one of the two partners is HIV positive and enters into such a relationship.

$$^{+}\text{Non-CSW} \xrightarrow{\alpha_4} \text{Non-client Person}$$

3. In addition to these new "spousal" partnerships formed at (presumably) a low rate between low risk men and low risk women, all other sexual relationships between men and women. We modeled the effect of transmission occurring as a result of such relationships and subsequent HIV transmission to existing "spousal" partnerships and other contacts by allowing HIV positive individuals to "leak" infection to low-risk individuals of the opposite sex. Most individuals are married or have other "spouse-like" sex partners among the low risk population. If these people become HIV infected from other partners (via sex worker or other routes) they have a high risk of "leaking" the infection to their spouses or other partners.

Vertical transmission

HIV infected women were considered as fertile as uninfected women. The fraction of births of infected women also infected with HIV.

The latest estimate for the HIV/AIDS infected adult population in the country is 3.8 million in 2000. HIV/AIDS is not a disease which spreads randomly and is transmitted as a consequence of a specific behavioral pattern and has strong socio-economic implications. It not only costs huge sum of money in terms of controlling the opportunistic infections such as TB, Pneumonia and Cryptococcus meningitis, but seriously affects individuals in their prime productive years causing serious economic loss to them and their Families. To study the socio-economic effect of these diseases on the country, it is very important to predict the correct values of HIV/AIDS population. The information gathered from the infected people is not 100% accurate and correct (but it has ambiguity and vagueness). Most of the time, the infected individuals hide the information due to number of reasons like society's fear, etc. To overcome the problem of dealing with imprecise and vague informations, Fuzzy Logic Based Approach works well. Fuzzy Set Theoretic Approach is the best technique to deal with such uncertainties due to vagueness, impreciseness, or incomplete information. On the other hand, a number of NGOs and GOs are collecting the data for HIV/AIDS, these data may be numeric or non-numeric in nature. Unfortunately, Fuzzy systems could not deal with numerical data. Hence, artificial neural network is used along with Fuzzy System to use numeric as well as non-numeric data in modeling.

12.4.3 Neuro-Fuzzy Approach of Modeling

In recent years, various modeling studies have been conducted to describe the transmission dynamics of HIV (Ram Naresh 2000; Srinivasa Rao 2003; Nagelkerke and Jha 2001; Vanhowe 1999; Chaturvedi et al. 2001; Bhave et al. 1995; Aggarwal et al. 1997; Brookmeyer et al. 1996; Ghani et al. 2002; Garnett et al. 2002; Stover et al. 2002). In particular, Anderson et al. (1986) described some preliminary attempts to use mathematical models for transmission of HIV in a homosexual community. In 1987 May and Anderson showed that if the probability of developing AIDS increases linearly with time since infection then the distribution of the AIDS incubation period is a Weibull distribution. Nowak et al. (1991) analyzed a model where the mean rate of acquisition of new partners depends on the size of the sexually active population. Srinivasa Rao et al. developed a Mathematical model of AIDS epidemic. Baily et al.presented a model for HIV infection and AIDS in which infected people proceed through a sequence of stages to AIDS and then to death. Most of the above mentioned models consider only one population but HIV transmission takes place in populations that are heterogeneous in a variety of ways. The models incorporating demographic factors has also been studied by May et al., Anderson et al. (1986). The incubation period of AIDS in India estimated through deconvoluting HIV epidemic density and reporting AIDS cases, is between 8 and 12 years. Quantitative information on female commercial sex (FSW) activity in India is available through various sources, but it is just impossible to get accurate information. Due to lack of information, it is not possible to develop an exact model using conventional method. Hence, there is need to develop non-conventional (Neuro-Fuzzy) modeling technique to model this problem.

In the earlier models the population was taken constant and the growth rate is not taken into account, but this aspect is very important in the spread of disease due to long incubation period during which the population might have doubled particularly in developing countries.

Further, there is strong argument that variable infectivity, the nature and type of social/sexual mixing structures and the long and variable periods of infectiousness are key factors in modeling of HIV+ population dynamics.

12.4.3.1 Survey Results

The investigator surveyed the rural and urban areas of Agra region in which HIV is spreading rapidly and also surveyed the Blood-Banks of Agra region.

1. *Blood-banks data*
   The survey results of blood banks are tabulated in Table 12.2 and graphically represented in Fig. 12.6.

High risk group:

- Commercial Sex workers (C.S.W.)
- Clients:

  ○ Pre marital relation
  ○ Extra marital relation

Table 12.2. Blood bank information

Year	Total donation/ consumption of blood 10^6 unit/year	No. of HIV+in Agra region
2000	3,298	3
2001	3,221	10
2002	3,282	15
2003	11,098	30

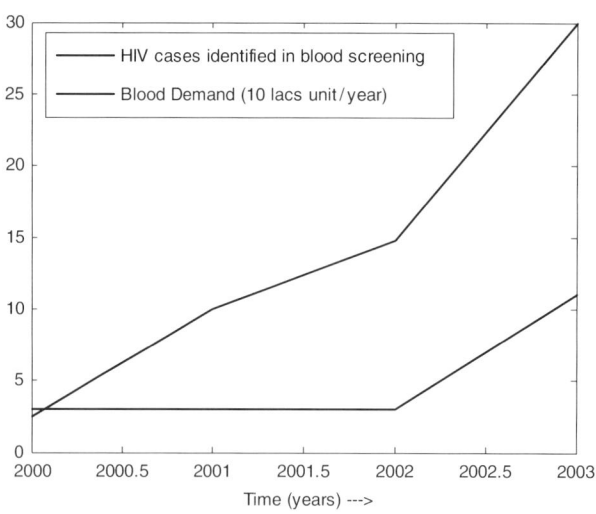

Fig. 12.6. Trend of blood demand and HIV+ cases

- Blood recipient
- Children of HIV+ (mother to child)
- Policeman and military personal

Low risk group:

- Students
- General mass

Main castes responsible for spreading HIV+ in Agra Region are Bediya, Bhantu, Kabutare. Main Red light areas in Agra city are Seb ka bazaar, Mal ka bazaar, Kashmiri bazaar, Panni gali, Sikandra and in Rural areas of Fatehabad such as Chaurahe mauhalla, Kanoon goyan, Ambedkar nagar, Village ai and lodhai; Shamsabad: Kans tila, Gopalpura; Bah: Basaur, Sabora, Rai khilla, Nagla Swaroop Lal; Fatehpur sikri: Churiyari, Shringarpur, Korai; Zarar: Manikpura, Pinahat; Jagner; Jalesar.

No of HIV+ detected in rural areas = 60 cases.

These results show that 67% HIV+ population is in the age from 18 to 45 years as shown in Fig. 12.7, which can contribute in national economy. Almost all HIV+ individuals are of low or medium income group (i.e. labour, or farmers) as shown in Table 12.3.

Main reason of HIV+ infection in Agra region

- Sexual contact (unsafe) = 80%
- Mother to child = 10%
- Blood transmission = 6%
- Using infected injections = 4%

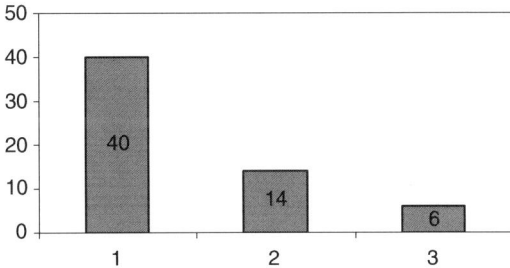

Fig. 12.7. HIV Infected male, female and children population

Table 12.3. Statistics of HIV spread in rural area

S. No.	Person	No. of HIV	Age group (years)	Income level	Married/unmarried
1.	Male	40	18–45	Labour & farmer	70% married 30% unmarried
2.	Female	14	18–45	Labour	100% married
3.	Infants	6	6–9	–	–

Table 12.4. Literacy level of HIV+ people in Agra

Sex	% of literacy	Primary level (%)	High sec. level (%)	Inter-mediate level (%)
Male	95	60–75	25	1
Female	10	100	–	–

Table 12.5. Population and literacy level of Agra region (according to 2001 survey data)

	Rural		Urban	
	Population	Literacy	Population	Literacy
Male	1,114,971	675,043	834,804	614,428
Female	938,985	292,586	722,541	373,444

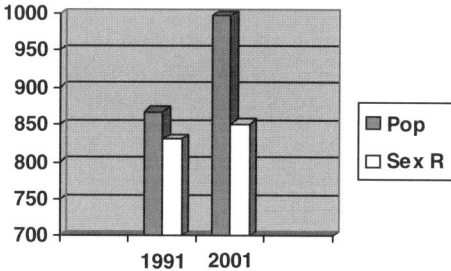

Fig. 12.8. Population Density and Sex Ratio of Agra region

This shows that most of the HIV+ people are educated only upto primary level and a very few educated upto higher secondary level and almost none upto intermediate and higher education level as shown in Table 12.4. Another important observation from these data is that the percentage of female literacy level is very low specially in rural areas as mentioned in Table 12.5. This is also an important reason of high rate of HIV infection. Hence, attention is required to improve the literacy level of female. Also the sex ratio is 85.2% and population density is 89.7% of Agra region as per the survey results of Govt. of India 2001 (ref. Fig. 12.8).

Rehabilitation

There is no community care center in Agra region as far as author's knowledge is concern. Doctors prefer the HIV+ patients for home care and they provide medicines only for opportunistic infections (e.g. fever, cough, etc).

12.4.3.2 Model Development Phase

For modeling and simulation of HIV+ Population using Neuro-Fuzzy Approach, the following steps have to be followed.

Table 12.6. Variables and their ranges

S. no.	Variable name	Range (%)
1.	FSW	1–5
2.	Literacy level	20–70
3.	Low income group	10–50
4.	Migration	0–8
5.	Awareness level	0–100
6.	Blood demand	0–20,000 unit/year
7.	Population density	800–950
8.	Sex ratio	80–110%
9.	Risk	0–100%

Step-1 The first step in model development is the identification of key variables. The key variables identified for the HIV+ infected population are: Risk, Female Sex Worker (FSW) of different caste as Bediya, Kabutare and Bhantu in Agra region, Blood demand (BD), Literacy Level, Low Income Group (LIG), Migration (M), Infected population (IP), Awareness level (AL), Rate of Awareness (RA), Government Support (GS), Man Power available for IEC (MP), Population Density (PD), Sex Ratio (SR), Susceptible Population (SP), Rural Population, Urban Population, Rate of Infection (RIP), Social Economy (SE), Injecting drug users (IDU) but there is no case identified of IDU in Agra as per the records. The range of these variables are given in Table 12.6.

Step-2 Causal links have been developed between a pair of variables under cetris paribus conditions (i.e. keeping other variables constant) as in the System Dynamic methodology. From these causal links a causal loop diagram is drawn as shown in Fig. 12.9.

Step-3 Development of fuzzy knowledge base from causal link developed in Step-2.

Fuzzy Knowledge Base

Rule–1 If FSW is high then Risk is high Else if FSW is medium then Risk is medium Else if FSW is low then Risk is low.

Rule–2 If Literacy is low then Risk is high Else If Literacy is medium then Risk is medium Else if Literacy is high then Risk is low.

Rule–3 If LIG is high then Risk is high Else if LIG is medium then Risk is medium Else if LIG is low then Risk is low.

Rule–4 If M is high then Risk is high Else if M is medium then Risk is medium Else if M is low then Risk is low.

Rule–5 If AL is high then Risk is low Else if AL is medium then Risk is medium Else if AL is low then Risk is high.

Rule–6 If PD is high then Risk is low Else if PD is Medium then Risk is Medium Else if PD is low then Risk is low.

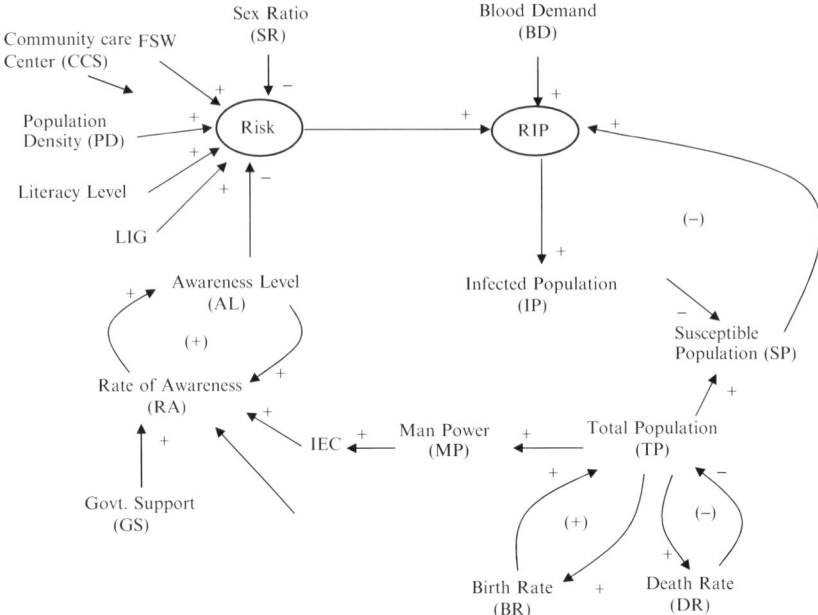

Fig. 12.9. Causal Loop Diagram for HIV/AIDS model

Rule–7 If SR is high then Risk is high Else if SR is Medium then Risk is Medium Else if SR is low then Risk is high.

Rule–8 If Risk is high then IP is high Else if risk is medium then IP is medium Else if risk is low then IP is low.

Rule–9 If BD is high then RIP is high Else if BD is medium then RIP is medium Else if BD is low then RIP is low.

Rule–10 If SP is high then RIP is high Else if SP is medium then RIP is medium Else if SP is low then RIP is low.

Rule–11 If RIP is low then IP is low Else if RIP is medium then IP is medium Else if RIP is high then IP is high.

Rule–12 If IP is high then SP is high Else if IP is medium then SP is medium Else if IP is low then SP is low.

Rule–13 If IP is high then SE is low Else if IP is medium then SE is medium Else if IP is low then SE is high.

Rule–14 If TP is low then SP is low Else if TP is medium then IP is medium Else if TP is high then IP is high.

Rule–15 If BR is high then TP is high Else if BR is medium then TP is medium Else if BR is low then TP is low.

Rule–16 If DR is high then TP is low Else if DR is medium then TP is medium Else if DR is low then TP is high.

Rule–17 If RA is high then AL is high Else if RA is medium then AL is medium Else if RA is low then AL is low.

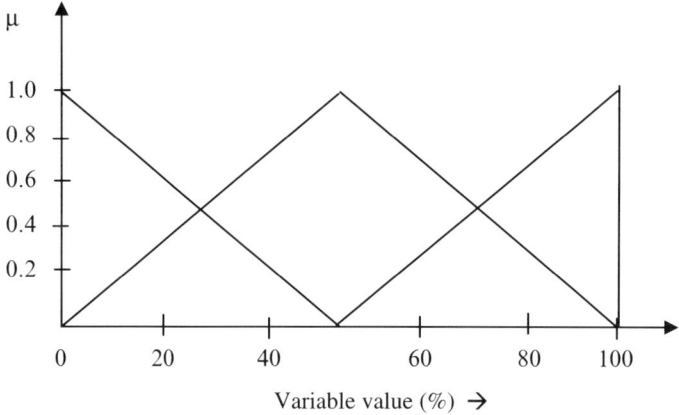

Fig. 12.10. Fuzzy Membership functions

Rule–18 If IEC is high then RA is high Else if IEC is medium then RA is medium Else if IEC is low then RA is low.

Rule–19 If MP is high then RA is high Else if MP is medium then RA is medium Else if MP is low then RA is low.

Rule–20 If GS is high then RA is high Else if GS is medium then RA is medium Else if GS is low then RA is low.

Rule–21 IF GS is high then CCS is high else if GS is low CCS is low.

Rule–22 If CCS is high then RA is high else if CCS is low RA is low.

The suitable triangular fuzzy sets have been defined for the identified variables as shown Fig. 12.10. The rule matrices have been developed for different rules using discrete fuzzy sets and max min fuzzy composition, e.g.

$$Rule\ 1 = H_{FSW} \times H_{Risk} + M_{FSW} \times M_{Risk} + L_{FSW} \times L_{Risk}$$

$= [0/2.9, 0.25/3.375, 0.5/3.85, 0.75/4.225, 1.0/4.8] \times [0/50, 0.25/62.5,$

$0.5/75, 0.75/87.5, 1.0/100] + [0/1, 0.25/1.475, 0.5/1.95, 0.75/2.425, 1/2.9,$

$0.75/3.375, 0.5/3.85, 0.25/4.225, 0/4.8] \times [0/0, 0.25/12.5, .5/25, 0.75/37.5,$

$1/50, 0.75/62.5, 0.5/75, 0.25/87.5, 0/100] + [1/1, 0.75/1.475, 0.5/1.95,$

$0.25/2.425, 0/2.9] \times [1/0, 0.75/12.5, .5/25, 0.25/37.5, 0/50]$

The rule 1 may also be represented in the matrix form as given in Table 12.7.

Let us find the effect of FSW on Risk.

Given Crisp % FSW = 1.475

Fuzzy % FSW = [1/1, 0.75/1.475, 0.5/1.95, 0.75/2.425, 1/2.9, 0.75/3.375, 0.5/3.85, 0.25/4.225, 0/4.8]

After getting fuzzified valued of % FSW, determine the risk using rule #1 and max-min composition.

Table 12.7. Representation of rule 1 in matrix form

	0	12.5	25	37.5	50	62.5	75	87.5	100
1	1	0.75	0.5	0.25	0	0	0	0	0
1.475	0.75	0.75	0.5	0.25	0.25	0.25	0.25	0.25	0
1.95	0.5	0.25	0.5	0.5	0.5	0.5	0.5	0.25	0
2.425	0.25	0.25	0.25	0.75	0.75	0.75	0.5	0.25	0
2.9	0	0.25	0.5	0.75	1.0	0.75	0.5	0.25	0
3.375	0	0.25	0.5	0.75	0.75	0.75	0.5	0.25	0.25
3.85	0	0.25	0.5	0.5	0.5	0.5	0.5	0.5	0.5
4.225	0	0.25	0.25	0.25	0.25	0.25	0.5	0.75	0.75
4.8	0	0	0	0	0	0.25	0.5	0.75	1

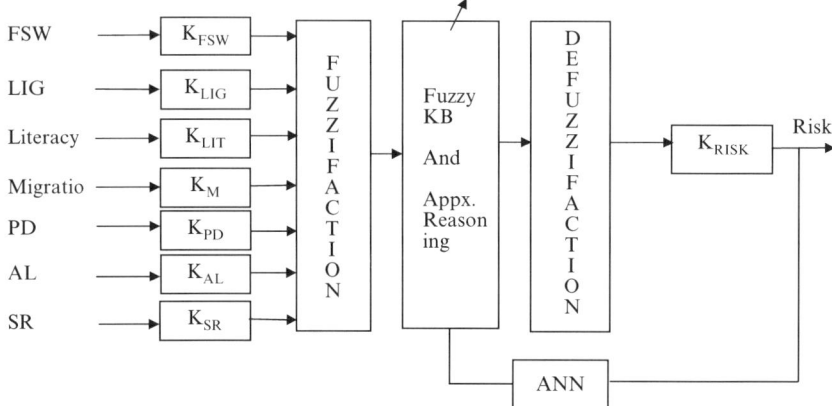

Fig. 12.11. Block diagram of ANFIS model of risk

$$Risk_{FSW} = Fuzzy \% \ FSW \ o \ Rule1$$
$$= [1/0, \ 0.75/12.5, \ 0.5/25, \ 0.75/37.5, \ 1/50, \ 0.75/62.5, \ 0.5/75,$$
$$0.5/87.5, \ 0.5/100]$$

Similarly risk from other factors also calculated and then finally combined risk is calculated.

Step-4 The fuzzy knowledge base developed in step-3 is fixed and does not have adaptability. Hence, Neuro-fuzzy approach has been used to make it more flexible and adaptive. The ANFIS model of risk is shown in Fig. 12.11, which is developed, in Fuzzy toolbox of Matlab Ver. 7.0. Similarly the Infected population model is developed.

12.4.3.3 Simulation Phase

Step-5 The above-developed ANFIS Model is simulated and the results have been compared with the actual results. The model is tuned for error (i.e. deviation in the model response from actual response).

Step-6 Tuned Model is used for prediction purposes and the results are shown in Figs. 12.12–12.18.

12.4.4 Conclusions

It has been accepted that HIV/AIDS is not a health problem alone, but a problem of such magnitude that it affects every facet of human life. The country will have serious socio-economic consequences if sufficient and necessary steps are not taken. In the paper, the dynamic model of HIV/AIDS population has been developed by establishing the cause – effect (causal) relationship between a pair of variables and then from these relationships a Fuzzy knowledge

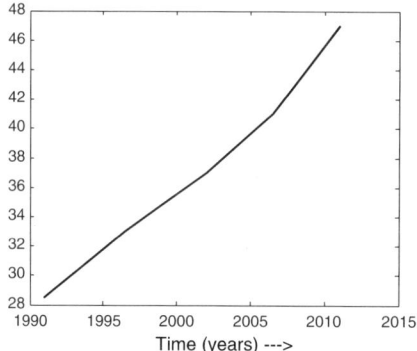

Fig. 12.12. Estimated population

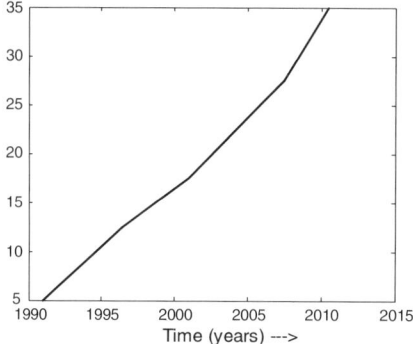

Fig. 12.13. Risk of HIV infection

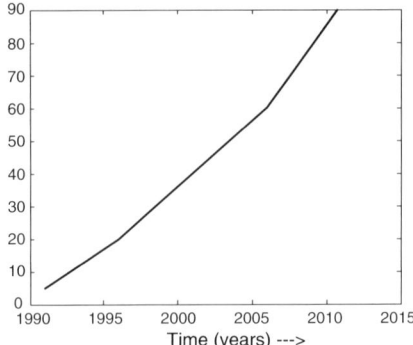

Fig. 12.14. Awareness Level among the common man

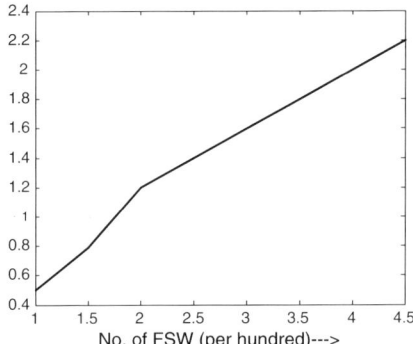

Fig. 12.15. Effect of FSW on risk

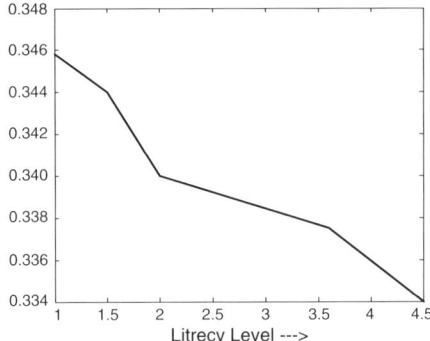

Fig. 12.16. Effect of literacy on risk

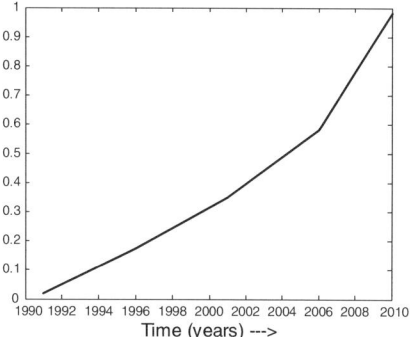

Fig. 12.17. Trend of IEC for future

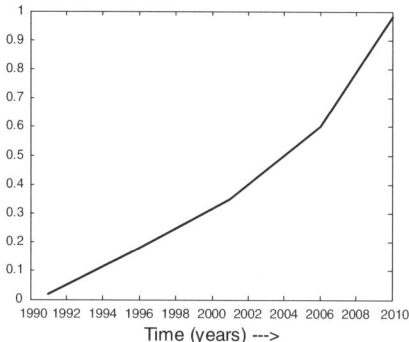

Fig. 12.18. Trend of infected population

base is prepared. For the variables used in causal relations, the information is gathered through the survey. The problems faced during survey are mentioned below:

1. There is no information obtained for HIV+ unmarried female due to typical Indian culture.
2. People do not want to disclose that he/she is HIV+ due to stigma.
3. Some private blood banks do not have full proof HIV testing facility.

Due to the above mentioned difficulties, the information gathered is not quite accurate and correct, which in turn affect the model. Hence, fuzzy system is used for modeling of HIV population dynamics to include the uncertainty in the data. To include adaptability in the developed model, ANN is used to change the rules and membership functions of fuzzy sets. The neuro-fuzzy model developed for HIV/AIDS population dynamics is quite suited for this application. The model includes the variable growth rate of population, diffusion of population due to migration, transmission in a heterosexual community, role of exposed classes, demographic factors, effect of education and intervention programmes.

12.5 Summary

Some of the main topics covered in this chapter are mentioned below:

1. The merits and demerits of ANN and fuzzy systems are mentioned.
2. The different approaches to combine these two techniques are also described.
3. The ANFIS model of neuro-fuzzy system is described in detail.
4. ANFIS always works with Sugeno model of Fuzzy inferencing. Hence Sugeno model is also explained in this chapter.
5. Sugeno model is compared with Mamdani model.
6. Backpropagation gradient descent learning algorithm is used to identify the parameters of fuzzy system.

12.6 Bibliographical and Historical Notes

Takagi and Hayashi did pioneer work in developing neuro-fuzzy technology in the late 1980s. They have also worked closely with leading Japanese industries and developed many consumer products using neuro-fuzzy systems. P. Werbos published a paper in which he mentioned several means of integrating ANN and fuzzy systems. Jang did his Ph.D. on neuro-fuzzy systems in 1990s and developed ANFIS for Matlab of Mathworks. Jang, Sun and Mizutani later wrote a book on neuro-fuzzy systems and soft computing. Abraham and Nath (2001) used neuro-fuzzy approach for modeling electricity demand. Denai et al. (2000) and Nounou and Rehman (2000) used this approach for modeling and control of ac machines. Jain and Kumar (2007) demonstrated the hybrid approach for forecasting problems. Patidar et al. (2007) used hyrbid appraoch for voltage contingency analysis.

12.7 Exercise

1. Discuss the merits and demerits of ANN, Fuzzy systems and neuro-fuzzy systems?
2. Discuss the similarities and difference between Mamdani model and Sugeno model of inference.
3. Explain the different possible integration of ANN and fuzzy systems.
4. Consider the following non-linear model $Y = 1 - \exp(-k^*t) + 0.5^* \sin(w^*t) + e(k)$
   Where e(k) is a zero mean Gaussian noise with variance 0.1. Generate 500 simulated data points from the model. Construct a neuro-fuzzy model using the first 400 data points and then test the model using the remaining 100 data points. Compare the accuracy of neuro-fuzzy model with mathematical model.

5. What is the difference between neuralizing the fuzzy systems and fuzzifying the ANN.

6. Let two universe of discourse defined as U = [1, 2, 3, 4, 5] and V = [0, 10, 20 30].

The fuzzy set small = {(2, 0.3), (3, 0.8), (4, 0.1)} and Large = {(3, 0.2), (4, 0.5), (5, 1.0)}.

Consider the following Sugeno model Rule 1 If x is small then y = 2x + 2, Rule 2 If x is Large then y = 10 − x.

What is the model's output for x = 2, 3, 4, 5.

13

ANN – GA-Fuzzy Synergism and Its Applications

The feed-forward backpropagation artificial neural networks (ANN) are widely used to control the various industrial processes, modelling and simulation of systems and forecasting purposes. The backpropagation learning has various drawbacks such as slowness in learning, stuck in local minima, requires functional derivative of aggregation function and thresholding function to minimize error function etc. Various researchers suggested a number of improvements in simple back-propagation learning algorithm developed.

In this paper, a program is developed for feed-forward artificial neural network with genetic algorithm (GA) as the learning mechanism to overcome some of the disadvantages of backpropagation learning mechanism to minimize the error function of ANN.

Genetic algorithm (GA) simulates the strategy of evolution and survival of fittest. It is a powerful domain free approach integrated with ANN as a learning tool. The ANN – GA integrated approach is applied to different problems to test this approach. It is well known that the GA optimization is slow and depending on the number of variables. To improve the convergence of GA, a modified GA is developed, in which, the GA parameters are modified using five fuzzy rules and concentration of genes is suggested.

13.1 Introduction

The optimization algorithms may be classified as follows:

1. The algorithms that are deterministic, with specific rules for moving from one solution to the other like gradient descent optimization. These algorithms have been successfully applied to many engineering design problems.
2. There has been a new and very popular method for the optimization emerged now and that is the optimization with Genetic Algorithm. The difference between GA and traditional methods of optimization is, GA works with a binary strings representing variables instead of the actual

D.K. Chaturvedi: *Soft Computing Techniques and its Applications in Electrical Engineering*, Studies in Computational Intelligence (SCI) **103**, 501–508 (2008)
www.springerlink.com   © Springer-Verlag Berlin Heidelberg 2008

values of those variables. The advantage of the binary strings is that the discontinuous function can be handled with no extra cost. This allows GA to be applied to a wide variety of problems. Another advantage is that the GA operators exploit the similarities in string structures to make an effective search. The most striking difference between GA and many traditional optimization methods is that GA works with a population of points instead of a single point. Because there are more than one string being processed simultaneously, it is very likely that the expected GA solution may be global solution. In GA previously found good information is emphasized using reproduction operator and propagated adaptively through crossover and mutation operator.

The genetic algorithms takes the same number of strings as the population size and among them it selects the better strings of the population for the training ANN. The error function of ANN is taken as the objective function and the adjustment of the weights is done in such a way that while training, the error should be minimized.

The comparison of GA and traditional back-propagation training may be given as:

1. The generalized delta rule which gets the optimal value, reduces the search space for the next iterations, thereby some good solutions are likely to be left behind in that process. Whereas in the GA search space is never reduced and it takes all possible good population for every iteration. Hence the possibility of leaving any good solution is negligibly small.
2. The most important advantage of using GA as the training algorithm is that the discontinuous function can be handled very easily, whereas the general back-propagation algorithm could not train the ANN; if the aggregation function and/or thresholding function are not a continuous one, because it requires a derivative of the function.

13.2 Training of ANN

The major development in learning of neuron occurred, after the development of neuron in 1943 by Mc-Culloh Pits, when D.O. Hebb (Baldi and Hornik 1989) proposed a learning mechanism for the brain that became the starting point for artificial neural network learning (training) algorithms. He postulated that as brain learns, it changes its connectivity patterns. More specifically, his learning hypothesis is as follows:

"When the axon of cell A is near enough to excite cell B and repeatedly or persistently takes part in firing it, some growth process or metabollic change takes place in one or both cells such that the A's efficacy, as one of the cell firing cell B, is increased".

Hebb further proposed that if one cell B repeatedly assists in firing another, the knobs of the synapse, or the junction, between the cells would grow so as to increase the area of contact. The Hebb's learning hypothesis is schematically given in Chap. 3.

Although this technique has been used in neural nets after certain modifications suggested by Bernard Widrow in 1962, which was called "delta rule". Later on a chain rule was developed called generalized delta rule or standard back-propagation learning algorithm for artificial neural networks.

Drawbacks of back-propagation learning algorithm

The back-propagation learning algorithm has various drawbacks as follows:

1. The slowness of the learning process, especially when large training sets or large networks have to be used.
2. Network may stuck in local minima (Chaturvedi et al. 1996, Chaturvedi and Das 2000).
3. The learning rate and momentum have great effect of the training time. With constant step size of learning rate and momentum the convergence of back-propagation tends to be very slow and often yields sub-optimal solutions.
4. Initial weights also affect the training time.
5. The neural network may not Converge at all if the initial weights are not selected properly (Hinton 1996).
6. The threshold function should be differentiable and non decreasing (Fu, 1994).
7. The training time is a function of the error function (Humpert 1994).
8. The normalization range of training data and input output mapping also affect the training time and accuracy (Chaturvedi et al. 1996).
9. Network complexity, i.e. number of hidden layers and number of hidden units and their interconnections with neuron of other layers greatly affects the learning speed of ANN.

13.3 Advantages of GA

The central theme of research on genetic algorithms has been robustness, the balance between efficiency and efficacy necessary for survival in many different environments. The following are the advantages of GA:

1. The genetic algorithms are a set of sophisticated search and improve procedures based on the mechanics of natural genetics; the search is absolutely blind, but guided by pre-designated precise operators.
2. The potentials of genetic algorithms as a problem solving especially in finding near optimal solution (Humpert 1994).
3. Genetic algorithms search from a population of points, not a single point.
4. Genetic algorithms use pay off information's (objective function), not derivatives or auxiliary knowledge.
5. GA uses probabilistic transition rules, not deterministic rules.

6. Genetic algorithms work with a coding of the parameter themselves.
7. GA uses probabilistic procedure to select input to produce outputs for the next generations that include only fittest among the input and output.

13.4 ANN Learning Using GA

In the ANN learning, GA model replaces the back propagation mechanism of learning. The error, which is caused due to the difference between the desired output and the actual output as obtained from the ANN, is fed-back to GA model and with the help of GA the weights are optimized to obtain minimum error as shown in Fig. 13.1. A computer program for ANN-GA is developed in MATLAB Ver. 7.1.

The simple genetic algorithm, for which the program has been formulated, contains non-overlapping string populations. The population as an array of individuals where each individual contains the coded parameter, the bit strings and the fitness (objective function) value along with the auxiliary information. To obtain new population three operators namely select, crossover, and mutation is performed. In conventional GA, the selection is done according to the roulette wheel, which is developed on the basis of fitness value of strings. In this selection criterion there is no guaranty that the best strings will be selected. Therefore, the convergence time will be affected. To overcome this problem, the stochastic remainder selection is performed as suggested by Deb (1995).

The objective function is taken as the error function and is given by

$$F = 1/(1 + e^2); \text{ where e is the sum squared error.}$$

The program GA-ANN uses binary strings to representing the parameter values, there it makes to clear that decoded value of string by which a parameter is represented is not equal to the parameter value. A coefficient named, normalizing coefficient is used to calculate the actual value of a parameter

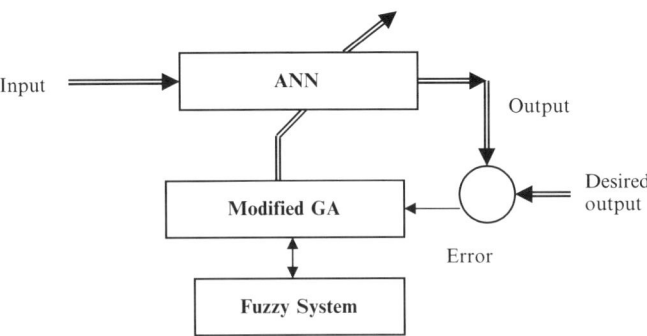

Fig. 13.1. GA as learning tools for ANN

represented by a particular string. The value of coefficient can be calculated by following formula:

$$\text{Coeff} = 2^{\text{lchrom}} - 1/(\text{ub} - \text{lb})$$

Where lchrom is the length of chromosome of a string (number of bits in string), ub and lb are the upper and lower limits of parameter value (range of parameter).

The important GA parameters which are reqiured by GA program are listed below:

Population size (popsize)   Maximum no of generations (maxgen)
= 20 to 50   = 10–50
Crossover probability (pcross)   Mutation probability (pmutation)
= .25 to .6   = .001 to .003

13.5 Validation and Verification of ANN-GA Model

The earlier developed improved GA (GA-Fuzzy technique) has been used to train ANN for the four different types of problems. The ANN used for training these problems are shown in Fig. 13.2.

The weights of neural network

$$W(p, 1) = \{w(p, 11), w(p, 12), w(p, 13), \text{bais1}\},$$
$$W(p, 2) = \{w(p, 21), w(p, 22), w(p, 23), \text{bais2}\}$$

and

$$W(p, 3) = \{w(p, 31), w(p, 32), w(p, 33), \text{bais3}\}.$$

Initially, Improved Genetic Algorithm has been used to optimize the neural network set of weights like W(p,1), W(p,2) and W(p,3). Then keeping any two sets of weights constant (e.g. W(p,2) and W(p,3)) and optimize the remaining weights (eg. w(p,11), w(p,12) and w(p,13)). This process is repeated till error is not become zero or near zero.

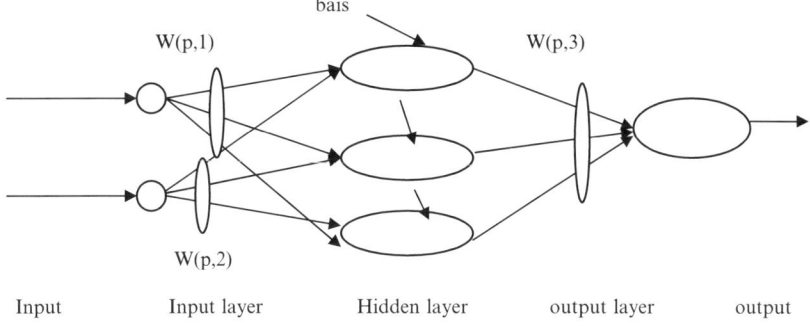

Fig. 13.2. Three Layered ANN

The following four problems have been considered as shown in Table 13.1 for training and testing of ANN-GA model.

The results of ANN training using improved GA have been shown in Table 13.2.

The test results for all four problems are shown in Table 13.3.

Table 13.1. Different problems considered for validation of ANN-GA approach

Problem no.	Problem	Expression
1.	EX-OR problem	$X + Y = \text{output}$
2.	Square function	$X^2 + Y^2 = \text{output}$
3.	Product function	$X \cdot Y = \text{output}$
4.	Power function	$X^Y = \text{output}$

Table 13.2. ANN weights obtained for different problems after training of ANN using improved genetic algorithms

	Problem – 1	Problem – 2	Problem – 3	Problem – 4
x(p,11)	−15.0000	−1.4032	−8.2258	−7.2581
x(p,12)	−4.3548	7.2580	0.4838	3.3871
x(p,13)	−15.000	−1.1129	−3.3870	−3.3871
x(p,21)	−13.0645	−13.0645	3.3871	−3.3871
x(p,22)	−13.0645	−13.0645	3.3871	−3.3871
x(p,23)	−13.0645	−13.0645	3.3871	−2.4194
x(p,31)	13.0645	13.0645	2.4194	3.3871
x(p,32)	13.0645	13.0645	2.4194	1.4516
x(p,33)	13.0645	13.0645	2.4194	2.4194
Fitness value	0.9999	0.9983	0.9416	0.9519

Table 13.3. Simulation results of ANN-GA

Problem 1 – Ex-OR function

X	Y	Actual value	ANN-GA
0	0	0	0.0001
0	1	1	0.996
1	0	1	0.994
1	1	0	0.001
0	0	0	0.0001

Problem 2 – Square function

0.1	0.1	0.02	0.0019
0.2	0.2	0.08	0.0078
0.3	0.3	0.18	0.1790
0.4	0.4	0.32	0.3210
0.5	0.5	0.5	0.5010
0.6	0.6	0.72	0.7200
0.7	0.7	0.98	0.9800
0.8	0.8	1.28	1.2801
0.9	0.9	1.62	1.6201
1	1	2	2.0010

Table 13.3. (*Continued*)

Problem 3 – Product function			
0.1	0.1	0.01	0.0101
0.2	0.2	0.04	0.0401
0.3	0.3	0.09	0.0899
0.4	0.4	0.16	0.1598
0.5	0.5	0.25	0.2479
0.6	0.6	0.36	0.3610
0.7	0.7	0.49	0.4890
0.8	0.8	0.64	0.6401
0.9	0.9	0.81	0.8088
1	1	1	0.9960
Problem 4 – Power function			
0.1	0.1	0.7943	0.7939
0.2	0.2	0.7247	0.7246
0.3	0.3	0.6068	0.60679
0.4	0.4	0.6931	0.6930
0.5	0.5	0.7071	0.7070
0.6	0.6	0.7360	0.7361
0.7	0.7	0.7790	0.7788
0.8	0.8	0.8365	0.8364
0.9	0.9	0.9095	0.9095
1	1	1	0.9960

13.6 Summary

The ANN can be trained through exposure to a set of samples of input and output. There are many different procedures that might be used for training of an ANN. Improved Genetic Algorithm has been used to train artificial neural networks. Learning is achieved by adjusting the weights (by optimizing the variables (weights of ANN) of GA optimization function i.e. sum squared error function) of the network until its action computing performance is acceptable. The results show that the neural network with the help of IGA performs well with non-derivative learning mechanism. It helps us to minimize the error very near to zero (i.e. the fitness value of objective function reaches near to one).

The approach may be further extended in the following directions:

1. More complex problem may be tried.
2. Fuzzy aggregation operators like t-norm, t-co-norm, averaging or compensatory operators may be used as the aggregation function(s) of ANN-GA model.
3. The IGA training algorithm could be made more adaptive by selecting length of chromosome and maximum number of generations using fuzzy system for better accuracy.

13.7 Bibliography and Historical Notes

There are numerous books on the subject of integration of ANN, fuzzy systems and evolutionary (genetic) algorithms. An excellent compendium on the subject of neuro-fuzzy integration is in the books by Lin and Lee (1996), Jang et al. (1997) and Abraham et al. (2002). An excellent collection of papers on the subject of integration of these techniques is in the books Bezdek and Pal (1992), Carpenter and Grossberg (1996), and Frank Hoffman (2005).

The relation of genetic algorithms and soft computing is explored in Herrera and Verdegay (1996), Jose Manuel Benítez (2003), Wang (2003), Rutkowski (2004) and Melin (2005).

In addition to all these there are a number of journals that publish papers on all aspects of softcomputing. IEEE publishes three transactions: Transactions on Neural Networks, Transactions on Fuzzy Systems, and Transactions on Evolutionary Algorithms. Springer is also publishing International Journal of Soft Computing: A Fusion of Foundations, Methodologies and Applications. Other includes: Evolutionary comutation, Fuzzy Sets and Systems, Softcomputing.

Ang et al. (2003, 2005) proposed a good combination of Fuzzy and ANN. Antsaklis (1990), Anderson (1988) mentioned ANN applications for control. Buckley (1994b, 1994c), Sun (1994), Kosko (1992), Brill (1992), Chang (1998), Chaturvedi et al. (Dec. 1999, 2001, 2005) discussed about synergism of ANN, fuzzy based systems and GA.

References

Abdelazim, Tamer and Malik, O.P., "An adaptive power system stabilizer using on-line self-learning fuzzy system", Proceedings, IEEE Power Engineering Society 2003 General Meeting, July 13–17, 2003, Toronto, Canada.

Abido, M.A., Optimal power flow using tabu search algorithm. *Journal of Electric Power Components and Systems*, 30: 469–483, 2002.

Abido, M.A. and Abdel-Magid, Y.L., Tuning of power systems stabilizers using fuzzy basis function networks. *Electric Machines and Power Systems*, 27: 865–877, 1999.

Abraham, A., Ruiz-del-Solar, J., and Kčoppen, M., *Soft Computing Systems: Design, Management and Applications*. IOS Press, 2002.

Abraham, A. and Nath, B., A neuro-fuzzy approach for modelling electricity demand in Victoria, *Applied Soft Computing*, 1(2): 127–138, 2001.

Ackley, D.H., Hinton, G.E., and Sejnowski, T.J., A learning algorithm for Boltzmann Machine. *Cognitive Science*, 9: 147–169, 1985.

Adam, G.K., Hybrid neural controller of a stepper motor for a manipulator arm, *Proceedings of the Fourth International Workshop on Robot Motion and Control*, pp. 321–326, 2004.

Aggarwal, O.P., Sharma, A.K., and Indrayan, A. *HIV/AIDS Research in India*. National AIDS Control Organisation, New Delhi, 1997.

Aguado José, A., Víctor, H., Quintana, M.M., and Rosehart, W.D. Coordinated spot market for congestion management of inter-regional electricity markets, *IEEE Transactions on Power Systems*, 19(1): 180–187, 2004.

Ahluwalia Sanjeev, S. and Bhatiani, Tariff setting in the electric power sector: base paper on Indian case study, *TERI Conference on Regulation in Infrastructure Services*, pp. 2–34, Nov., 2000.

Ai-Ping, L., Jia, Y., and Quan-Yuan, W.U., Study on fuzzy aggregation operator in multi-criteria decision systems. *Proceedings of International Conference on Hybrid Information Technology*, 1: 380–386, 2006.

Akbarzadeh, T.M.R., et al., Evolutionary fuzzy speed regulation for a D.C. motor, *29th Southeastern Symposium on System Theory, Cookeville, TN*, March 1997.

Akbarzadeh, T.M.R., et al., Genetic algorithms and genetic programming: combing strengths in one evolutionary strategy, *Proceedings of the Joint Conference on the Environment, Albuquerque, NM*, 1998.

Akins, K., A bat without qualities? In Davies, M. and Humphreys, G. (eds.) *Consciousness: Psychological and Philosophical Essays*. Blackwell, Oxford, 1993.

Akins, K., Lost the plot? Reconstructing Dennett's multiple drafts theory of consciousness. *Mind and Language*, 11: 1–43, 1996.

Albert, V., Aruldoss, T., and Jeyakumar, E.A., A simulated annealing based hybrid optimization approach for a fuzzy modelled unit commitment problem. *International Journal of Emerging Electric Power Systems*, 3(1), Article 1054, 2005.

Alomoush, M.I. and Shahidehpour, S.M., Contingency-constrained congestion management with a minimum number of adjustments in preferred schedules, *Electrical Power and Energy Systems*, 22: 277–290, 2000.

Alomoush M.I., Performance indices to measure and compare system utilization and congestion severity of different dispatch scenarios. *Electric Power Systems Research*, 74: 223–230, 2005.

Alsac, O. and Stott, B., Optimal load flow with steady-state security. *IEEE Transactions on Power Apparatus Systems*, PAS-93, 745–751, 1974.

Al-Shakararchi, M.R. and Ghulaim, M.M., Short term load forecasting for Bangladesh electricity region. *Electric Machines and Power Systems*, 28: 355–371, 2000.

Amari, S., *A Theory of Adaptive Pattern Classifiers*, IEEE Trans. On Electronic Computers, Vol. EC-16, 299–307, June 1967.

Amizadeh, F. and Jamshidi, M., *Soft Computing, Fuzzy Logic, Neural Networks, and Distributed Artificial Intelligence*, vol. 4. Prentice Hall, Englewood Cliffs, NJ, 1994.

Anderson, J.A., A simple neural network generating an interactive memory. *Mathematical Biosciences*, 4: 197–220, 1972.

Anderson, K.P. and McCarthy, A., Transmission pricing and expansion methodology: lessons from Argentina, *Utility Policy*, 8: 199–211, 1999.

Anderson, P.M. and Fouad, A.A., *Power System Control and Stability*. Iowa State University Press, 1977.

Anderson, C.W., Learning to control an inverted pendulum with connectionist networks. *Proceedings of the American Controls Conference*, 2294, 1988.

Anderson, J., *The Architecture of Cognition*. Harvard University Press, Cambridge, MA, 1983.

Anderson, R.M; Medley, G.F., May, R.M., Johnson, A.M., A preliminary study of the transmission dynamics of the human immunodeficiency virus (HIV), the causative agent of AIDS. IMA J Math Appl Med Biol, 3(4): 229–63, 1986.

Ang, K.K. and Quek, C. RSPOP: rough set-based pseudo outer-product fuzzy rule identification algorithm. *Neural Computation*, 17(1): 205–243, 2005.

Ang, K.K., Quek, C., and Pasquier, M. POPFNN-CRI(S): pseudo outer product based fuzzy neural network using the compositional rule of inference and singleton fuzzifier. *IEEE Transactions on Systems, Man and Cybernetics*, Part B, 33(6): 838–849, 2003.

Antsaklis, P., Special issue on neural networks for control system, *IEEE Control Systems Magazine*, 10(3): 8, 1990.

Antsaklis, P.J. and Passino, K.M. (eds.), *An Introduction to Intelligent and Autonomous Control*. Kluwer Academic Publishers, Norwell, MA, 1993.

Apolloni, B., Avanzini, G., Cesa-Bianchi, N., and Ronchini, G., Diagnosis of epilepsy via backpropagation, *Proceedings of the International Joint Conference on Neural Networks*, Washington, DC, vol. 2, p. 571, 1990.

Arbib, M.A., *Brain Machines and Mathematics.* Springer Verlag, New York, 1987.

Arbib M.A. (ed.), *The Handbook of Brain Theory and Neural Network*, MIT Press, Cambridge, MA, 1995.

Arbib, M.A. (ed.), *Handbook of Brain Theory and Neural Networks*, 2nd edn. MIT Press, Cambridge, MA, 2003.

Armstrong, D., What is consciousness? In *The Nature of Mind.* Cornell University Press, Ithaca, NY, 1981.

Armstrong, D., *A Materialist Theory of Mind.* Routledge and Kegan Paul, London, 1968.

Arriaga, I.J.P., Rudnick, H., and Stadlin, W.O. International power system transmission open access experience. *IEEE Transactions on Power Systems*, 10(1): 554–561, 1995.

Arriaga, I.J.P., Rubio, J., Puetra, F.J., Arceluz, J., Marin, J., Marginal pricing of transmission services: an analysis of cost recovery. *IEEE Transactions on Power Systems*, 15(1): 448–454, 1995b.

Hiroshi, A. and Tsukamato, Y., Transmission pricing in Japan, *Utilities Policy*, 6(3): 203–210, 1997.

Ashraf, S.A., Khan, M.R., and Basu, K.P., Load forecasting using artificial neural network. *National Systems Conference at IIT, Kanpur*, pp. 127–130, 1993.

Auger, A. and Hansen, N., A restart CMA evolution strategy with increasing population size, *IEEE Congress on Evolutionary Computation*, 2: 1769–1776, 2005b

Auger, A. and Hansen, N., Performance evaluation of an advanced local search evolutionary algorithm, *IEEE Congress on Evolutionary Computation*, 2: 777–1784, 2005a.

Avadhanlu, T.V. and Chaturvedi, D.K., Evaluation of supply conditions and performance in an inverter fed induction motor drive system. *International Journal of Modelling, Measurements and Control, France*, A, 43(4): 23–39, 1992.

Azzam-ul-Asar, J.R., McDonald, J.R., and Khan, M.I., A specification of neural network applications in the load forecasting problem, *First IEEE Conference on Control Applications, New York*, l: 577–582, 1992.

Baars, B. *A Cognitive Theory of Consciousness.* Cambridge University Press, Cambridge, 1988.

Bäck, T. Optimal mutation rates in genetic search, *Proceedings of the Fifth International Conference in Genetic Algorithms.* Morgan Kaufmann, Los Altos, CA, 1993.

Back, T., Hammel, U. and Schwefel, H.P., Evolutionary computation: comments on the hstory and current state. *IEEE Transactions on Evolutionary Computation*, 1(1): 3–17, 1997.

Baily, G.G., Moore, C.B., Essayag, S.M., de Wit, S., Burnie, J.P., Denning, D.W. Candida inconspicua, a fluconazole-resistant pathogen in patients infected with human immunodeficiency virus. *Journal of Clinical Infectious Disease*, 25(1): 161–163, 1997.

Baker, J.E., Reducing bias and inefficiency in the selection and algorithm, *Proceedings of the Second International Conference on Genetic Algorithms and Their Applications*, Erlbaum, Cambridge, MA, 1987.

Bakirtzis, A.G., Biskas, P.N., Christoforos, E.Z., and Petridis, V., Optimal power flow by enhanced genetic algorithm. *IEEE Transactions on Power Systems*, 17(17): 229–236, 2002.

Baldi, P. and Hornik, K., Neural networks and principal component analysis learning from examples and local minima. Neural Networks, 2: 53–58, 1989.

Balog, K., Conceivability, possibility, and the mind–body problem. *Philosophical Review*, 108: 497–528, 1999.

Bandemer, H. and Gottwald, S., Fuzzy sets, *Fuzzy Logic and Fuzzy Methods with Applications*. Wiley, NewYork, 1995.

Barres, W. and Cormic, M.C., *Aerodynamics, Aeronautics and Flight Mechanics*. Wiley, New York, pp. 386, 1995.

Kosko, B., *Neural Networks and Fuzzy Systems – A Dynamical Systems Approach to Machine Intelligence*. Prentice Hall, NJ, 1992.

Basu, M., An interactive fuzzy satisfying – based simulated annealing technique for economic emission load dispatch with nonsmooth fuel cost and emission level functions. Electric Power Components and Systems, 32(2): 163–173, 2004.

Basu, T.K., Load forecasting – state of the art. In Sen Gupta, D.P. (ed.) *Recent Advances in Control and Management Energy System*. Interline Publishing, Bangalore, India, pp. 1–18, 1993.

Bath, S.K., Dhillon, J.S., and Kothari, D.P., Security constrained stochastic multi-objective optimal power dispatch. *International Journal of Emerging Electric Power Systems*, 8(1), Article 7, 2007.

Baughman, M.L. and Siddiqi, S.N., Real time pricing of reactive power: theory and case study results. *IEEE Transactions on Power Systems*, 6(1): 23–29, 1991.

Bawazeer, K.H., Prediction of crude oil product quality parameters using neural networks. *M.S. Thesis, Florida Atlantic University, Boca Raton, FL*, 1996.

Bear, M.F., Connors, B.W., and Paradiso, M.A., *Neuroscience: Exploring the Brain*. Williams and Wilkins, Baltimore, MD, 1996.

Beaufays, F., Abdel-Magid, Y., and Widrow, B., Application of neural networks to load frequency control in power systems, *Neural Networks*, 7(1): 405–415, 1999.

Benaouda, D. and Murtagh, F. Neuro-wavelet approach to time-series signals prediction: an example of electricity load and pool-price data. *International Journal of Emerging Electric Power Systems*, 8(2), Article 5, 2007.

Benaouda, D., Murtagh, F., Starck, J.L., and Renaud, O., Wavelet- based nonlinear multiscale decomposition model for electricity load forecasting. *International Journal of Neurocomputing* (Elsevier), 70: 139–154, 2006.

Bernard, J.A., Use of rule based system for process control. *IEEE Control Systems Magazine*, 8(5): 3–13, 1988.

Beyer, H.-G., *The Theory of Evolution Strategies*. Springer, Berlin Heildelberg New York, 2001.

Beyer, H.-G. and Schwefel, H.-P. Evolution strategies: a comprehensive introduction. *Journal Natural Computing* 1(1): 3–52, 2002.

Bezdek, J. and Pal, S.K., *Fuzzy Models for Pattern Recognition*. IEEE Press, New York, 1992.

Bezdek, J., *Pattern Recognition with Fuzzy Objective Function Algorithms*. Plenum Press, New York, 1981.

Bhat, N., Minderman, P., McAvoy, T., and Wang, N., Modeling chemical process systems via neural network computation. *IEEE Control Systems Magazine*, 10(3): 24, 1990.

Bhattacharya, B., Bollen, M.H.J., and Daalder, J.E. *Operation of Restructured Power Systems*. Kluwer, Dordrecht, 2001.

Bhave, G., Lindan, C.P., Hudes, E.S., Desai, S., Wagle, U., Tripathi, S.P., Mandel, J.S., Impact of an intervention on HIV, sexually transmitted diseases, and condom use among sex workers in Bombay, India. *AIDS*, 9, (Suppl 1): S21–S30, 1995.

Bialek, J., Allocation of transmission supplementary charge to real and reactive loads. *IEEE Transactions on Power Systems*, 13(3): 749–654, 1998.

Bialek, J., Topological generation and load distribution factors for supplement charge allocation in transmission open access, *IEEE Transactions on Power Systems*, 12(3): 1185–1193, 1997.

Bingul, Z., Adaptive genetic algorithms applied to dynamic multiobjective problems, *Applied Soft Computing*, 7(3): 791–799, 2007.

Blackemore, C., *Mechanics of Mind*. BBC, UK, 1977.

Bloch, I. Information combination operators for data fusion: a comparative review with classification. IEEE Transactions on System, Man and Cybernetics, 26(1): 52–67, 1996.

Block, N. Mental paint and mental latex. In Villanueva, E. (ed.) *Perception*. Ridgeview, Atascadero, CA, 1996.

Block, N. Troubles with functionalism. In Block, N. (ed.) *Readings in the Philosophy of Psychology*, vol. 1. Harvard University Press, Cambridge, MA, pp. 268–305, 1980a.

Block, N. and Stalnaker, R. Conceptual analysis, dualism, and the explanatory gap. *Philosophical Review*, 108/1: 1–46, 1999.

Block, N. Inverted earth. In Tomberlin, J. (ed.) *Philosophical Perspectives*, vol. 4. Ridgeview Publishing Company, Atascadero, CA, 1990.

Block, N. On a confusion about the function of consciousness, *Behavioral and Brain Sciences*, 18: 227–247, 1995.

Block, N. What is Dennett's theory a theory of? *Philosophical Topics*, 22/1–2: 23–40, 1994.

Block, N. Are absent qualia impossible? *Philosophical Review*, 89/2: 257–274, 1980b.

Bollinger, K.E., Laha, A., Hamiliton, R. and Haras, T., PSS design using root locus methods, *IEEE Transactions on Power Apparatus and System*, PAS-94: 1484–1488, 1975.

Bollinger, R.C., Brookmeyer, R.S., Mehendale, S.M., et al., Risk factor and clinical presentation of acute primary HIV infection in India. *JAMA*, 278: 2085–2059, 1997.

Bollinger K.E., Laha A., Hamilton R and Harras T, "PSS Design Using Root Locus Methods", *IEEE Trans. on Power Apparatus and Systems* Vol. PAS-94, PP. 1484–1488, 1978.

Bompard, E., Correia, P., Gross, G., and Amelin, M., Congestion-management schemes: a comparative analysis under a unified framework. *IEEE Transactions on Power Systems*, 18(1): 346–352, 2003.

Bompard E., Lu, W., and Napoli, R., Network constraint impacts on the competitive electricity markets under supply-side strategic bidding. *IEEE Transactions on Power Systems*, 21(1): 160–170, 2006.

Booker, L., Improving search in genetic algorithms. *Genetic Algorithms and Simulated Annealing*. Pitman, London, 1987.

Borman, S., Neural Network Applications in Chemistry Begin to Appear. *Chemical and Engineering News*, 67(17): 24, 1989.

Boyd, R., Materialism without reductionism: what physicalism does not entail. In Block, N. (ed.) *Readings in the Philosophy of Psychology*, vol. 1. Harvard University Press, Cambridge, MA, 1980.

Braae, M. and Ruthford, D.A., Fuzzy relations in a control setting. *Kybernetics*, 7(3): 185–188, 1978.

Braga António Silvestre, D. and João Tomé, S., A multiyear dynamic approach for transmission expansion planning and long-term marginal costs computation. *IEEE Transactions on Power Systems*, 20(3): 1631–1639, 2005.

Bråten, J., Transmission pricing in Norway. *Utilities Policy*, 6(3): 219–226, 1997.

Braun, H., On solving travelling salesman problems by genetic algorithms. *Lecture Notes in Computer Sciences*, 496: 129–133, 1990.

Brest, J., Greiner, S., Boškovic, B., Mernik, M., and Zumer, V. Self-adapting control parameters in differential evolution: a comparative study on numerical benchmark problems. *IEEE Transactions on Evolutionary Computation*, 2006.

Bridgeman, B., *The Biology of Behaviour and Mind*. Wiley, New York, 1988.

Brill, F.Z., Brown, D.E., and Martin, W.M., Fast genetic selection of features for neural network classifier. *IEEE Transactions on Neural Networks*, 324–333, 1992.

Brookmeyer, R., Mehendale, S.M., Pelz, R.K., Shepherd, M.E., Quinn, T., Rodrigues, J.J., and Bollinger, R.C. Estimating the rate of occurrence of new HIV infections using serial prevalence surveys: the epidemic in India. *AIDS*, 10: 924–925, 1996.

Brown, P.J., Forecasts and public policy, *The Statician*, 33: 51–63, 1983.

Buckles, B.P. and Petry, F.E. (eds.), Genetic algorithms. *IEEE Computer Society*, Los Alamitos, CA, 1992.

Buckley, J.J. and Hayashi, Y., Can fuzzy neural net approximate continuous fuzzy functions? *Fuzzy Sets and Systems*, 61(1): 43–52, 1994c.

Buckley, J.J. and Hayashi, Y., Fuzzy genetic algorithms and applications, *Fuzzy Sets and Systems*, 61(2): 129–136, 1994a.

Buckley, J.J. and Hayashi, Y., Fuzzy neural network: a survey. *Fuzzy Sets and Systems*, 66(1): 1–13, 1994b.

Bunn, D.W., Forecasting loads and prices in competitive power markets. *IEEE Proceedings*, 88(2): 163–169, 2000.

Bwayo, J., Plummer, F., Omari, A.M., Mutere, A.N., et al., Long distance truck-drivers: 1. Prevalence of sexually transmitted diseases (STDs). *East African Medical Journal*, 68: 425–429, 1991.

Byrne, A, Intentionalism defended. *Philosophical Review*, 110: 199–240, 2001.

Byrne, A, Some like it HOT: consciousness and higher-order thoughts. *Philosophical Studies*, 2: 103–129, 1997.

Sun, C.T., Rulebase structure identification in an adaptive network based fuzzy inference system, *IEEE Transactions on Fuzzy Systems*, 2(1): 64–73, 1994.

Campbell, J., *Past, Space, and Self*. MIT Press, Cambridge, MA, 1994.

Campbell, K., *Body and Mind*. Doubleday, New York, 1970.

Cannella, M., Disher, E., and Gagliardi, R., Beyond the contract path: a realistic approach to transmission pricing. *The Electricity Journal*, 1996.

Caprihan, R., Kumar, S., and Wadhwa, S., Fuzzy systems for control of flexible machines operating under information delay. *International Journal of Production Research*, 35(5): 1331–1348, 1997.

Caramanis, M.C., Roukos, N., and Schweppe, F.C., WRATES: A tool for evaluating the marginal cost of wheeling. IEEE Transactions on Power Systems, 4(2): 594–605, 1989.

Caramanis, M.C., Brown, R.E., and Schweppe, F.C., The cost of wheeling and optimal wheeling rates, *IEEE Transactions on Power Systems*, PWRS-1(1): 63–73, 1986.

Carpantier, J.L., Optimal power flows: uses, methods and developments. *IFAC, Electric Energy System, Brazil*: 11–21, 1985.

Carpenter, G.A. and Grossberg, S., ART2: self organisation of stable category recognition codes for analog input patterns. Applied Optics, 26(23): 4919–4930, 1987.

Carpenter, G.A. and Grossberg, S., ART3: hierarchical search using chemical transmitters in self organising pattern recognition architectures. Neural Networks, 3(2): 129–152, 1990.

Carpenter, G.A. and Grossberg, S., Learning, categorization, rule formation, and prediction by fuzzy neiural networks. In Chen, C.H. (ed.) Fuzzy Logic and Neural Network Handbook. McGraw Hill Inc., New York, pp. 1.3–1.45, 1996.

Carpenter, G.A. and Grossberg, S., The ART of adaptive pattern recognition by self organization neural network. IEEE Computer, 21(3): 77–88, 1988.

Carruthers, P., *Phenomenal Consciousness*. Cambridge University Press, Cambridge, 2000.

Central Electricity Regulatory Commission, Concept paper on "Open access in interstate transmission", New Delhi, 2003.

Chalmers, D., Does conceivability entail possibility? In Gendler, T. and Hawthorne, J. (eds.) *Conceivability and Possibility*. Oxford University Press, Oxford, 2002.

Chalmers, D., The content and epistemology of phenomenal belief. In: Jokic, A., and Smith, Q. (eds.) *Consciousness: New Philosophical Perspectives*. Oxford University Press, Oxford, 2003.

Chalmers, D. and Jackson, F., Conceptual analysis and reductive explanation. *Philosophical Review*, 110/3: 315–360, 2001.

Chalmers, D., *The Conscious Mind*. Oxford University Press, Oxford, 1996.

Chalmers, D. Facing up to the problem of consciousness. *Journal of Consciousness Studies*, 2: 200–219, 1995.

Chandy, K.M. and Mishra, J., *Parallel Programming Design – A Foundation*. Addison Wesley, Reading, MA, 1988.

Chang, C.S., Fu, W., and Wen. F., Load frequency controller using genetic algorithm based fuzzy gain scheduling of PI controller. *Institute of Electric Machines and Power Systems*, 26: 39–52, 1998.

Changaroon, B., Srivastava, S.C., Thukaram, D., A neural network based power system stabilizer suitable for on-line training – a practical case study for EGAT system, *IEEE Transaction on Energy Conversion*, 15(1): 103–109, 2000.

Chao, H. and Peck, S., A market mechanism for electric power transmission, *Journal of Regulatory Economics*, 10: 25–29, 1996.

Charache, S., Dover, G.J., Moore, R.D., et al., Hydroxyurea: effects on hemoglobin F production in-patients with sickle cell anemia. *Blood*, 79: 10, 1992.

Charache, S., Terrin, L.M., Moore, R.D., et al., Effect of hydroxyurea on the frequency of painful crises in sickle cell anemia. *New England Jorunal of Medicine*, 332: 1995.

Charnail, E. and McDermott, D., *Introduction to Artificial Intelligence*. Addison-Wesley, Reading, MA, 1985.

Chaturvedi, D.K., *Modelling and Simulation of Power Systems: An Alternative Approach*, Doctoral Thesis at Dayalbagh Educational Institute (Deemed University), Dayalbagh, Agra, 1992.

Chaturvedi, D.K., Das, V.G., Satsangi, P.S. and Kalra P.K., "Effect of Different Mappings And Normalizations of Neural Network Models", *Nineth National Power Systems Conference at Indian Technological Institute, Kanpur*, 2: 377–386, 1996.

Chaturvedi, D.K., Satsangi, P.S., and Kalra, P.K., Load frequency control: a generalized neural network approach. *Electric Power and Energy Systems*, Elsevier Science, 21: 405–415, 1999.

Chaturvedi, D.K., Satsangi, P.S., and Kalra, P.K., Short term load forecasting using generalized neural network (GNN) approach. *Journal of The Institute of Engineers (India)*, 78: 83–87, 1997.

Chaturvedi D.K. and Das V.S., "Optimization of Genetic Algorithm Parameters", *National Conference on Applied Systems Engineering and Soft Computing (SAS-ESC – 2000), Organized by Dayalbagh Educational Institute, Dayalbagh, Agra, India*, pp.194–198, Jan 2000.

Chaturvedi, D.K., Mishra, R.K., and Agarwal, A., Load forecasting using genetic algorithms. *Journal of The Institution of Engineers (India)*, 76: 161–165, 1995.

Chaturvedi DK, Malik OP and Kalra PK, Experimental Studies of Generalized Neuron Based Power System Stabilizer, *IEEE Trans. On Power systems*, Vol. 19 No. 3, August 2004, pp. 1445–1453.

Chaturvedi, D.K., Better Worldliness. *D.E.I. Magazine*, June 2006.

Chaturvedi, D.K., Mohan, M., and Saxena, A.K., Development of HIV infected population model using system dynamics technique. *International Journal of Modelling, Measurements, and Control, France*, 22(2): 1–14, 2001.

Chaturvedi, D.K., Modelling and simulation of electrical power systems: an alternative approach. Doctoral Thesis at Dayalbagh Educational Institute, Dayalbagh, Agra, India, 1997.

Chaturvedi, D.K., Singh, P., Mohan, M., Gaur, S.K., and Mishra, D.S., Development of HIV model and its simulation. *Journal of Health Management*, 3(1): 65–84, 2001.

Chaturvedi, D.K. and Malik, O.P., A generalized neuron based adaptive power system stabilizer for multimachine environment. *IEEE Transactions on Power Systems*, 20(1): 358–366, 2005.

Chaturvedi, D.K. and Malik, O.P., Experimental Studies of a generalized neuron based adaptive power system stabilizer. *International Journal of Soft Computing – A Fusion of Foundations, Methodologies and Applications*. Springer, Berlin Heidelberg New York, 11(2): 149–155, 2007.

Chaturvedi, D.K. and Sharma, R.K., Modelling and simulation of a parachute deceleration devices, *AMSE Journal on the Advancement of Modelling and Simulation*, 2007.

Chaturvedi, D.K. and Malik, O.P., Generalized neuron based PSS and adaptive PSS, *International Journal of Control Engineering Practice (Special issue on Power Plants and Power System Control), IFAC*, 13(12): 1507–1514, 2005.

Chaturvedi, D.K., Dynamic model of HIV/AIDS population of Agra region, *International Journal of Environmental Research and Public Health, USA*, 2(3): 420–429, 2005.

Chaturvedi, D.K., Malik, O.P., and Kalra, P.K., Experimental studies of generalized neuron based power system stabilizer. *IEEE Transactions on Power Systems*, 19(3): 1445–1453, 2004.

Chaturvedi, D.K., Malik, O.P., and Kalra, P.K., A generalized neuron based adaptive power system stabilizer, *IEE Proceedings of Generation, Transmission & Distribution*, 15(2): 213–219, 2004.

Chaturvedi, D.K., Malik, O.P., and Kalra, P.K., A generalized neuron based PSS in a multi-machine power system, *IEEE Transactions of Energy Conversions, USA*, 19(3): 625–632, 2004.

Chaturvedi, D.K., Malik, O.P., and Kalra, P.K., Application of generalized neuron based power system stabilizer in a multi-machine power system (Research Note), *International Journal of Engineering Transactions B, Iran*, 17(2): 131–140, 2004.

Chaturvedi, D.K., Malik, O.P., and Kalra, P.K., Generalized neuron based power system stabilizer, *International Journal of Electric Power Components and Systems*, 32(5): 467–490, 2004.

Chaturvedi, D.K., Malik, O.P., and Kalra, P.K., Neural network controller for power system stabilization, *Journal of the Institution of Engineers (India)*, EL, 85: 138–145, Dec. 2004.

Chaturvedi, D.K., Malik, O.P., and Kalra, P.K., Power system stabilizer using generalized neural network, *Proceedings of 34th North American Power Symposium (NAPS - 2002), Arizona, USA*, pp. 320–327, Oct. 14–16, 2002.

Chaturvedi, D.K. and Malik, O.P., Comparison of generalized neuron based PSS and adaptive PSS, *International Federation of Automatic Control (IFAC) Symposium on Power Plants and Systems Control, Seoul*, pp. 988–994, June 9–12, 2003.

Chaturvedi, D.K., Mohan, M., Singh, R.K., and Kalra, P.K., Improved generalized neuron model for short term load forecasting, *International Journal on Soft Computing – A Fusion of Foundations, Methodologies and Applications*, 8(1): 10–18, 2004.

Chaturvedi, D.K., Satsangi, P.S., and Kalra, P.K., Load frequency control: a generalized neural network approach. *International Journal of Electric Power and Energy Systems*, 21: 405–415, 1999.

Chaturvedi, D.K., Singh, P., Mohan, M., Gaur, S.K., and Mishra, D.S., Development of HIV model and its simulation. Jorunal of Health Management, 3(1): 65–84 (2001).

Chaturvedi, D.K, Gaur, S.K., Mishra, D.S., and Kalra, P.K., Generalised neural networks with genetic algorithm based learning, *International Conference on Modelling, Simulation and Communication (CMSC'99), Jaipur*, India, pp. t–108, Dec. 1–3, 1999.

Chaturvedi, D.K. and Lajwanti, Education that really works. *International Conference on Education Culture in the 21st Century: Knowledge, Teacher and Technology, Silchar Silchar (Assam), India*, Jan. 29–31, 1999.

Chaturvedi, D.K. and Satsangi, P.S., System dynamics modelling and simulation of basic commutating electric machines: an alternative approach. *Journal of the Institution of Engineers (India)*, EL, 73: 6–10, 1992.

Chaturvedi, D.K. and Gupta, B.R., Simulation of temperature variation in parachute inflation. *Journal of the Institution of Engineers (India)*, AS, 76: 29–31, 1995.

Chaturvedi, D.K. and Satsangi, P.S., Innovative approach for predicting the performance characteristics of synchronous generators. *Journal of the Institution of Engineers (India)*, El, 74: 109–113, 1993.

Chaturvedi, D.K. and Sharma, R.K., Modeling and simulation of force generated in Stanchion system of aircraft arrester barrier system, *International Journal of Modeling, Measurements, and Control, France*, B, 64(2): 33–51, 1996.

Chaturvedi, D.K., Mohan, M., and Kalra, P.K., Development of flexible neural network, *Journal of the Institution of Engineers (India)*, CP, 83: 1–5, 2002.

Chaturvedi, D.K., Misra, R.S., and Agarwal, A.K., Load Forecasting using genetic algorithms, *Journal of the Institution of Engineers (India)*, EL, 76(3): 161–165, Nov. 1995.

Chaturvedi, D.K., Neural networks: a simulation tool for basic commutating machine. *Journal of the Institution of Engineers (India)*, El, 75: 83–85, 1994.

Chaturvedi, D.K., Kumar, R., Kalra, P.K., Artificial neural network learning using improved genetic algorithms, *Journal of The Institution of Engineers (India)*, CP, 82: 1–8, 2001.

Chaturvedi, D.K., Chauhan, R., and Kalra, P.K, Applications of generalised neural network for aircraft landing control system. *International Journal on Soft Computing – A Fusion of Foundations, Methodologies and Applications*. Springer, Berlin Heidelberg New York, 6(6): 441–448, 2002.

Chaturvedi, D.K., Gaur, S.K., Mishra, D.S., and Kalra, P.K., Load frequency control: an integrated approach. *International Conference on Artificial Intelligence (ICAI'99), Durban, South Africa*, 24–26, Sept. 1999.

Chaturvedi, D.K., Satsangi, P.S., and Kalra, P.K., A fuzzy simulation model of basic commutating electrical machines. International Journal of Engineering Intelligent Systems, 6(4): 225–236, 1998.

Chaturvedi, D.K., Satsangi, P.S., and Kalra, P.K., Applications of generalized neural network to load frequency control problem. *Journal of The Institution of Engineers (India)*, EL, 80: 41–47, 1999.

Chaturvedi, D.K., Satsangi, P.S., and Kalra, P.K., Development of fuzzy simulator for DC machine modelling. *Journal of the Institution of Engineers (India)*, EL, 80: 53–58, 1999.

Chaturvedi, D.K., Satsangi, P.S., and Kalra, P.K., Flexible neural network models for electrical machine, *Journal of the Institution of Engineers (India)*, EL, 80: 13–16, May 1999.

Chaturvedi, D.K., Satsangi, P.S., and Kalra, P.K., Fuzzified neural network approach for load forecasting problems, *International Journal on Engineering Intelligent Systems, CRL Publishing, UK*, 9(1): 3–9, 2001.

Chaturvedi, D.K., Satsangi, P.S., and Kalra, P.K., New neuron model for simulating rotating electrical machines and load forecasting problems. *International Journal on Electric Power System Research*, Elsevier Science, Ireland, 52: 123–131, 1999.

Chaturvedi, D.K., Satsangi, P.S., and Kalra, P.K., Short term load forecasting using generalized neural network approach. *Journal of the Institution of Engineers (India)*, EL, 78: 83–87, 1997.

Chaturvedi, D.K., Satsangi, P.S., and Kalra, P.K., Load forecasting using flexible new neuron models, *International Conference on Power Generation, System Planning and Operation, in I.I.T. Delhi, India*, pp. 148–155, Dec. 12–13, 1997.

Chaturvedi, D.K., Satsangi, P.S., and Kalra, P.K., Load frequency control of power system using generalized neural network. *International Conference on Energy Management and Power Drives, Singapore*, 1998.

Chaturvedi, D.K., Kumar, V., Satsangi, P.S., and Kalra, P.K., Application of fuzzy logic in modelling and simulation of rotating electrical machines. *International Conference on Computer Application in Electrical Engineering, Roorkee, India*, pp. 151–156, Sept. 8–11, 1997.

Chaturvedi, D.K., Kumar, V., Satsangi, P.S., and Kalra, P.K., Development of fuzzy simulator for modelling and simulation of power systems components. *International Power Engineering Conference, Singapore*, May 23–27, 1997.

Chaturvedi, D.K., Kumar, V., Satsangi, P.S., and Kalra, P.K., Development of back-propagation with new models of neurons and its applications to power systems modelling, *International Power Engineering Conference, Singapore*, May 23–27, 1997.

Chaturvedi, D.K., Kumar, V., Satsangi, P.S., and Kalra, P.K., Electrical machines modelling new neuron models of backprop, *Second International Conference on Power Electronics and Drive Systems, PEDS '97, Singapore*, May 26–29, 1997.

Chaudhary, U.K., Chaturvedi, D.K., and Ibraheem, A fuzzy logic based variable gain power system stabilizer. *IEEE Conference on Recent Advanced Applications of Computers in Electrical Engineering(RACE)*, 24–25 March, 2007.

Chen, G.P., Malik, O.P., Hope, G.S., Qin, Y.H., and Xu, G.Y., An adaptive power system stabilizer based on self optimizing pole shifting control strategy. *IEEE Transactions on Energy Conversion*, 8(4): 639–645, 1993.

Chen, L., Suzuki, H., and Wachi, T., and Shimura, Y., Components of nodal prices for electric power systems. *IEEE Transactions on Power Systems*, 17(1): 41–49, Feb 2002.

Chen, P.-H., and Chang, H.-C., Large-scale economic dispatch by genetic algorithm. *IEEE Transactions on Power Systems*, 10(4): 1919–1926, 1995.

Chin-Teng, L. and George Lee, C.S., *Neural Fuzzy Systems: A Neuro-Fuzzy Synergism to Intelligent Systems*. Prentice Hall, New Jersey, 1996.

Chira, A., *Genetic Algorithms Applications to Economic Load Dispatch*. Master Thesis, AIT Bangkok, Thailand, August 1997.

Choi, J.Y., Rim, S., and Park, J. Optimal real time pricing of real and reactive powers. *IEEE Transactions on Power Systems*, 13(4): 1226–1231, 1998.

Choudhary, S.D., Kalra, P.K., Srivastava, S.C., and Kumar, D.M.V. Short-term load forecasting using artificial neural network. *ESAP*, 1993.

Chow, M., Zhu, J., and Tram, H., Application of fuzzy multi-objective decision making in spatial load forecasting. *IEEE Transactions on Power System*, 13(3): 1185–1190, 1998.

Chowdhury, B.H. and Rahman, S., A review of recent advances in economic dispatch. *IEEE Transactions on Power Apparatus and Systems*, 5(4): 1248–1257, 1990.

Christiaanse, W.R., Short term load forecasting using general exponential smoothing. *IEEE Transactions on PAS*, 90(2), March–April 1971.

Christie, R.D. and Bose, A., Load frequency issues in power systems operation after deregulation. *IEEE Transactions on Power Systems*, 11(3): 1191–1200, 1996.

Chung, F.L. and Lee, T., Fuzzy competitive learning, *Neural Networks*, 7(3): 539–551, 1994.

Churchland, P.M., Reduction, qualia, and direct introspection of brain states. *Journal of Philosophy*, 82: 8–28, 1985.

Churchland, P.M., *The Engine of Reason and Seat of the Soul*. MIT Press, Cambridge, MA, 1995.

Churchland, P.S., Consciousness: the transmutation of a concept". *Pacific Philosophical Quarterly*, 64: 80–95, 1983.

Churchland, P.S., On the alleged backwards referral of experiences and its relevance to the mind body problem. *Philosophy of Science*, 48: 165–81, 1981.

Churchland, P.S., The hornswoggle problem. *Journal of Consciousness Studies*, 3: 402–408, 1996.

Ciamp, A. and Zhang, F., A new approach to training back-propagation arti_cial neural networks: empirical evaluation on ten data sets from clinical studies. *Statistics in Medicine*, 21: 1309–1330, 2002.

Clark, A. *Sensory Qualities*. Oxford University Press, Oxford, 1993.

Clark, G. and Riel-Salvatore, J., Grave markers, middle and early upper paleolithic burials. *Current Anthropology*, 42/4: 481–90, 2001.

Clarke, D.W., Introduction to self tunning controllers. In Harris, C.J. and Billings, S.A. (eds.), Self-tuning adaptive control: theory and applications. Peter Peregrinus Ltd., pp. 36–69, 1981.

Cleeremans, A. (ed.) *The Unity of Consciousness: Binding, Integration and Dissociation*. Oxford University Press, Oxford, 2003.

Codvin Fred, H., *Aircraft Hand Book*. McGraw-Hill, New York, 1942.

Combettes, P.L., The foundations of set theoretic estimation. *Proceedings of the IEEE*, 81(2): 182–208, 1993.

Conejo Antonio, J., Milano, F., and García-Bertrand, R., Congestion management ensuring voltage stability. *IEEE Transactions on Power Systems*, 21(1): 357–364, 2006.

Constantin Von, A., *Fuzzy Logic & NeuroFuzzy Applications Explained – The Practical Hands-on Guide to Building Fuzzy Logic Applications*. Prentice Hall, New Jersey, 1995.

Cordon, O., Herrera, F., Hoffmann, F., and Magdalena, L., *Genetic fuzzy systems: evolutionary tuning and learning of fuzzy knowledge bases*. World Scientific Publishing Co. Pte. Ltd., Singapore, 2001.

Cotter, N.E., The Stone–Weierstrass theorem and its application to neural-networks. *IEEE Transaction of Neural Networks*, 1(4), December 1990.

Courtland, D., Perkins, and Hage, R.E., *Aero Plane, Performance, Stability & Control*. Wiley, New York, 1949.

Coxe, R. and Ilic, M., System planning under competition. In Ilic, M., Galiana, F., and Fink, L. (eds.), *Power Systems Restructuring – Economics and Engineering*. Kluwer, Dordrecht, 1998.

Crick, F. and Koch, C. Toward a neurobiological theory of consciousness. *Seminars in Neuroscience*, 2: 263–275, 1990.

Crick, F.H. *The Astonishing Hypothesis: The Scientific Search for the Soul*. Scribners, New York, 1994.

Croall, I.F. and Mason, J.P. (eds.), *Industrial Applications of Neural Networks*. Springer-Verlag, New York, 1991.

Cubillo, S., Trillas, E. and Castro, J.L., Conjunction and disjunction on (0,1). *Fuzzy Sets and Systems*, 72: 155–165, 1995.

Czogala, E. and Drewniak, J., Associative monotonic operations in fuzzy set theory. *Fuzzy Sets and Systems*, 12: 249–269, 1984.

Dai, Y., Ni, Y.X., Shen, C.M., Wen, F.S., Han, Z.X., Wu, F.F., A study of reactive power marginal price in electricity market. *Electric Power Systems Research*, 57: 41–48, 2001.

Damasio, A. *The Feeling of What Happens: Body and Emotion in the Making of Consciousness*. Harcourt, New York, 1999.

Daneshdoost, M., et al., Neural network with fuzzy set based classification for short term load Forecasting. *IEEE Transactions on Power Systems*, 13(4), 1998.

Dash, P.K., et al., A fuzzy neural network for electric load forecasting, *Ninth National Convention of Electrical Engineers at The Institution of Engineers (India) Karnataka State Center, Banglore*, Nov. 1993.

Dash, P.K., et al., A real time short term load forecasting system using functional link network. *IEEE Transactions on Power Systems*, 12(2): 675–681, May 1997.

David, A.K., Dispatch methodologies for open access transmission systems. *IEEE Transactions on Power Systems*, 13(1): 46–53, Feb. 1998.

David, G. and Gubina, F., Congestion management approach after deregulation of the Slovenian power system. *IEEE Transactions on Power Systems*, 0-7803-7519-X/02, pp. 1661–1665, 2002.

Davies, M. and Humphreys, G., *Consciousness: Psychological and Philosophical Essays*. Blackwell, Oxford, 1993.

Davis L. (ed.), *Handbook of Genetic Algorithms*, Von Nostrand Reinhold, New York, 1991.

Davison, E.J. and Tripathi, N.K., The optimal decentralized control of a large power system: load frequency control. *IEEE Transaction on Automatic Control*, AC-23(2): 312–315, 1978.

De Baets, B. and Kerre, E.E., Fuzzy relational compositions. *Fuzzy Sets and Systems*, 60: 109–120, 1993.

De Baets, B. and Kerre, E.E., The generalized modus ponens and the triangular fuzzy data model. *Fuzzy Sets and Systems*, 59: 305–317, 1993.

De Jong, K.A. and Spears, W.M., An analysis of interacting role of popsize and crossover in genetic algorithms. *Lecture Notes in Computer Sciences*, vol. 496. Springer Verlag, New York, 1990.

DeMello, F.P., Mills, R.J., and Rells, W.F.B., Automatic generation control: part 1 – process modelling. *IEEE Transactions on Power Apparatus and Systems*, PAS-92: 710–715, 1973.

Deb, K., *Optimisation for Engineering Design*. Prentice Hall of India, New Delhi, 1995.

DeMello, F.P., Hannett, L.N., and Undrill, J.M., Practical approaches to supplementary stabilizing from accelerating power. *IEEE Transactions on Power Apparatus and Systems*, PAS-97: 1515–1522, 1978.

DeMello, F.P. and Laskowski, T.F., Concepts of power system dynamic stability. *IEEE Transactions on Power Apparatus and Systems*, PAS-94: 827–833, 1979.

Demirli, K. and Turksen, I.B., A review of implications and the generalized modus ponens. *Proceedings of IEEE Conference on Fuzzy Systems*, 2: 1440–1445, 1994.

Demiroren, A., Zeynelgil, H.L., and Sengor, N.S., The application of ANN technique to load – frequency control for three-area power system, *IEEE Porto Power Tech Proceedings*, vol. 2, 2001.

Denai, M.A., Palis, F., Zeghbib, A., Modeling and control of non-linear systems using soft computing techniques. *Applied Soft Computing*, 7(3): 728–738, 2000.

Dennett, D.C., Quining qualia. In Lycan, W. (ed.) *Mind and Cognition*. Blackwell, Oxford, pp. 519–548, 1990.

Dennett, D.C., The self as the center of narrative gravity. In Kessel, F., Cole, P., and Johnson, D.L. (eds.) *Self and Consciousness: Multiple Perspectives*. Lawrence Erlbaum, Hillsdale, NJ, 1992.

Dennett, D.C. and Kinsbourne, M., Time and the observer: the where and when of consciousness in the brain. *Behavioral and Brain Sciences*, 15: 187–247, 1992.

Dennett, D.C., *Brainstorms*. MIT Press, Cambridge, 1978.

Dennett, D.C., *Consciousness Explained*. Little, Brown and Company, Boston, 1991.

Dennett, D.C., *Elbow Room: The Varieties of Free Will Worth Having*. MIT Press, Cambridge, 1984.

Dennett, D.C., *Freedom Evolves*. Viking, New York, 2003.

Descartes, R., *The Principles of Philosophy*. Translated by E. Haldane and G. Ross. Cambridge University Press, 1644/1911.

Devouge, C., Convex error functions for multilayered perceptrons, *International Joint Conference on Neural Networks, Seattle*, vol. 2, pp. 898, 1991.

Diamond, P. and Kloeden, P., *Metric spaces of fuzzy sets: theory and applications*. World Scientific, Singapore, 1994.

Diaz-Robainas, R., Zilouchian, A., and Huang, M., Fuzzy identification on finite training-set with known features. *International Journal of Automation Soft Computing*.

Dillon, T.S., Sestito, S., and Leung, S., An adaptive neural network approach in load forecasting in a power system, *IEEE Proceedings of the First International Forum on Applications of Neural Networks to Power Systems*, New York, NY, USA, pp. 17–21, 1991.

Ding, F. and David Fuller, J., Nodal, uniform, or zonal pricing: distribution of economic surplus. *IEEE Transactions on Power Systems*, 20(2): 875–882, 2005.

Djukanovic, M., Novicevic, M., Sobajie, D.J., and Pao, Y.P., Conceptual development of optimal load frequency control using artificial neural networks and fuzzy set theory. *International Journal of Engineering Intelligent Systems for Electrical Engineering and Communications*, 3(2): 95–108, 1995.

Dommel, H.W. and Tinney, W.F., Optimal power solutions. *IEEE Transactions on Power Systems*, 87: 1866–1876, Oct. 1968.

Douglas, A.P., et al., The impacts of temperature forecast uncertainty on Bayesian load forecasting. *IEEE Transactions on Power System*, 13(4): 1507–1513, 1998a.

Douglas, A.P., et al., Risk due to load forecasting uncertainty in short term power system planning. *IEEE Transactions on Power System*, 13(4): 1514–1520, 1998b.

Draeger, A., Engell, S., and Ranke, H., Model predictive control using neural networks. *IEEE Control Systems*, 15(5), 61, 1995.

Dretske, F., Conscious experience. *Mind*, 102: 263–283, 1993.

Dretske, F., Differences that make no difference. *Philosophical Topics*, 22/1–2: 41–58, 1994.

Dretske, F., *Naturalizing the Mind*. The MIT Press, Bradford Books, Cambridge, MA, 1995.

Drezga, I. and Rahaman, S., Input variable selection for ANN based short term load forecasting. *IEEE Transactions on Power Systems*, 13(4): 1238–1244, 1997.

Drezga, I. and Rahaman S., Short load forecasting with local ANN predictors forecasting. *IEEE Transactions on Power System*, 14(3): 844–857, 1999.

Dubois, D. and Prade, H., A review of fuzzy set aggregation connectives. *Information Sciences*, 36: 85–121, 1985.

Dubois, D. and Prade. H. Combination of fuzzy information in the framework of possibility theory. In Abidi, M.A. and Gonzalez, R.C. (eds.) *Data Fusion in Robotics and Machine Intelligence*. Academic Press, New York, 1992.

Dubois, D. and Prade, H., *Fuzzy sets and systems: theory and applications*. Academic Press, New York, 1980.

Dubois, D. and Prade, H., Fuzzy sets in approximate reasoning, Part I and II: Inference with possibility distributions. *Fuzzy Sets and Systems*, 40(1): 143–202, 1991.

Duda, R., Hart, P., and Stork, D., *Pattern Classification*, 2nd Edn. Wiley, New York, 2001.

Dudai, Y., *The Neurobiology of Memory*. Oxford University Press, UK, 1989.

Earl, C. *The Fuzzy Systems Handbook – A Practitioner's Guide to Building, Using, and Maintaining*, Fuzzy Systems, 2nd ed. AP Professional,MA, 1998.

Eccles, J. and Popper, K., *The Self and Its Brain: An Argument for Interactionism*. Springer, Berlin, 1977.

Echauz, J. and Vachtsevanos, G., Neural network detection of antiepileptic drugs from a single EEG trace, *Proceedings of the IEEE Electro/94 International Conference*, Boston, MA, p. 346, 1994.

Eckmiller, R., Neural nets for sensory and motor trajectories, *IEEE Control Systems*, 9(3): 53, 1989.

Edelman, G., *The Remembered Present: A Biological Theory of Consciousness*. Basic Books, New York, 1989.

Eiben, A.E. and Smith, J.E., *Introduction to Evolutionary Computing*. Springer-Verlag, Berlin Heidelberg New York, 2003.

Elgred, O.I. and Fosha, C.E., Optimum megawatt frequency control of multi-area energy systems. *IEEE Transactions on Power Apparatus and Systems*, PAS–89: 556–563, 1970.

El-Hawary, M.E., *Electric Power Applications of Fuzzy Systems*, Wiley-IEEE Press Series on Power Engineering, June 1998.

El-Metwally K.A., Hancock G.C. and Malik O. P., 'Implementation of a fuzzy logic PSS using a micro-controller and experimental test results', *IEEE Trans. on Energy Conversion*, vol. 11, no. 1, pp. 91–96, March 1996.

Elkateb, M.M. and Desouky, A.A., Hybrid adaptive techniques for electric-load forecast using ANN and ARIMA. *IEE Proceedings*, 147(4): 213–217, 2000.

Enrique H.-V. and Pasi, G., Soft approaches to information retrieval and information access on the Web: An introduction to the special topic section: special topic section on soft approaches to information retrieval and information access on the web. *Journal of the American Society for Information Science and Technology*, 57(4): 511–514, 2006.

Epstein, C.M., *Introduction to EEG and Evoked Potentials*. J. B. Lippincot Co., ISBN 0-397-50598-1, 1983.

Fahlman, S.E., Fast learning variations on backpropagation: an empirical study, *Proceedings of the Connectionist Models Summer School*, SanMateo, CA, pp. 38–51, 1988.

Fang, R.S. and David, A.K., Transmission congestion management in an electricity market. *IEEE Transactions on Power Systems*, 14(3): 877–883, 1999.

Farah, M., *Visual Agnosia*. MIT Press, Cambridge, 1990.

Farmer, E.D., Perera, B.L.P.P., and Cory, B.J., Optimal pricing of transmission services: application to large scale power systems, *IEE Proceedings on Generation, Transmission and Distribution*, 142(3): pp. 263–267, 1995.

Fausett L., *Fundamentals of Neural Networks, Architecture, Algorithms, and Applications*. Prentice Hall, Englewood Cliffs, NJ, 1994.

Federal Energy Regulatory Commission, Inquiry Concerning the Commission's Pricing Policy for Transmission Services Provided by Public Utilities Under Federal Power Act, June 30, 1993.

Federal Energy Regulatory Commission (FERC). *Regional Transmission Organizations*, Washington, DC, Docket RM99-2-000, Dec., 1999.

Fentress, S.W., *Exaptation as a Means of Evolving Complex Solutions*. MA Thesis, University of Edinburgh, 2005.

Ferrero, R.W., Rivera, J.F., and Shahidehpour, S.M., Application of games with incomplete information for pricing electricity in deregulated power pools, *IEEE Transactions on Power Systems*, 13(1): 184–189, 1998.

Finny John, D., Othman, H.A., and Riutz, W.L., Evaluating transmission congestion constraints in system planning. *IEEE Transactions on Power Systems*, 12(3): 1143–1149, 1997.

Flanagan, O., *Consciousness Reconsidered*. MIT Press, Cambridge, MA, 1992.

Fleten, S.-E. and Pettersen, E., Constructing bidding curves for a price-taking retailer in the Norweigian electricity market, *IEEE Transactions on Power Systems*, 20(2): 701–708, 2005.

Flohr, H., An information processing theory of anesthesia. *Neuropsychologia*, 33/9: 1169–1180, 1995.

Flohr, H., Glade, U., and Motzko, D., The role of the NMDA synapse in general anesthesia. *Toxicology Letters*, 100–101: 23–29, 1998.

Fodor, J., Special sciences. *Synthese*, 28: 77–115, 1974.

Fodor, J., *The Modularity of Mind*. MIT Press, Cambridge, MA, 1983.

Fogarty, T.C., Varying the probability of mutation in the genetic algorithm, *Proceedings of the Third International Conference in Genetic Algorithms and Applications*, Arlington, VA, pp. 104–109, 1981.

Fogarty, T.C., Varying the probabilities of mutation in the Genetic Algorithm, *Proceedings of the third Int. Conf. on Genetic Algorithms*, Edited by Schaffer, J., Morgan Kaufmann Publishers, Los Altos, CA, 1989.

Fogel, D.B., *Evolutionary Computation: Towards a New Philosophy of Machine Intelligence*. IEEE Press, New York, 2000.

Forrester, J.W., *Industrial Dynamics*. MIT Press, Cambridge, MA, 1961.

Fosa, C.E. and Elgerd, O.I., The Megawatt-frequency control problem, *IEEE Transactions PAS-89*: 563–577, 1970.

Foster, J., *The Immaterial Self: A Defence of the Cartesian Dualist Conception of Mind*. Routledge, London, 1996.

Foster, J., A defense of dualism. In Smythies, J. and Beloff, J. (eds.) *The Case for Dualism*. University of Virginia Press, Charlottesville, VA, 1989.

Frank, H. (EDT)., *Soft Computing: Methodologies and Applications*. Springer, Berlin Heidelberg New York, 2005.

Freeman, J.A. and Skapura, D.M., *Neural Networks Algorithms, Applications, and Programming Techniques*. Addison-Wesley, MA, pp. 89–105, 1992.

Fu, J. and Lamont, J.W., A combined framework for service identification and congestion management. *IEEE Transactions on Power Systems*, 16(1): 56–61, 2001.

Fu, L., *Neural Networks in Computer Intelligence*. McGraw Hill, New York, 1993.

Fu, Limin, *Neural Networks in Computer Intelligence*. McGraw Hill Inc., New York, 1994.

Fujitec, F., *FLEX-8800 series elevator group control system*. Fujitec Co. Ltd. Osaka, Japan, 1988.

Furuhashi, T., Fusion of fuzzy/neuo/evolutionary computing for knowledge acquisition. *IEEE Proceedings*, 89(9): 2001.

Gaines, B.R. and Shaw, M.L.G., From fuzzy logic to expert systems. *Information Sciences* 36: 5–16, 1985.

Gaines, B.R., Foundation of fuzzy reasoning. *International Journal of Man–Machine Studies*, 8: 623–668, 1976.

Gaines, B.R., Precise pat – fuzzy future. *International Journal of Man–Machine Studies*, 19: 117–134, 1983.

Galetovic, A. and Montecinos, C., The new chilean transmission charge scheme as compared with current allocation methods. *IEEE Transactions on Power Systems*, 21(1): 99–107, 2006.

Galiana, F.D. and Illic, M., A mathematical framework for tha analysis and management of power transactions under open access. *IEEE Transactions on Power Systems*, 13(2): 681–687, 1998.

Galiana Francisco, D. and Phelan, M., Allocation of transactions losses to bilateral contracts in a competitive environment. *IEEE Transactions on Power Systems*, 15(1): 143–150, 2000.

Galiana, G.G., Short term load forecasting. *Proceedings IEEE*, 75(12): 155–1573, 1987.

Galiana, F.D., et al., Identification of stochastic electric load models from physical data. *IEEE Transactions on Automatic Control*, AC-19(6), 1974.

Gallistel, C., *The Organization of Learning*. MIT Press, Cambridge, MA, 1990.

Gan, D. and Bourcier, D.V., Locational market power screening and congestion management: experience and suggestions. *IEEE Transactions on Power Systems*, 17(1): 180–185, 2002.

Gang, D. and Dong, Z.Y., Bai, W., and Wang, X.F., Power flow based monetary flow method for electricity transmission and wheeling pricing. *Electric Power Systems Research*, 74: 293–305, 2005.

Gangakhedkar, R.R., Bentley, M.E., Divekar, A.D., Gadkari, D., Mehendale, S.M., Shepherd, M.E., Bollinger, R.C., Quinn, T.C. Spread of HIV infection in married monogamous women in India. *Journal of the American Medical Association*, 278: 2090–2092, 1997.

Gardiner, H., *The Mind's New Science*. Basic Books, New York, 1985.

Garnett, G.P., Bartley, L., Grassly, N.C., and Anderson, R.M. (01/07/2002) Antiretroviral therapy to treat and prevent HIV/AIDS in resource – poor settings. *Nature Medicine* 8(7): 651–654 (issn: 1078–8956).

Garrison, G. and Tyrrell, A.M., *Introduction to Evolvable Hardware: A Practical Guide for Designing Self-Adaptive Systems*. Wiley, New York, 2006.

Gawish, S.A., Khalifa, F.A., Sabry, W., Badr, M.A.L. "Large power systems stability enhancement using a delayed operation fuzzy logic PSS" *J. Electrical Machines and Power systems*, Vol. 27. No. 2 PP. 157–168, 1999.

Gazzaniga, M., *Mind Matters: How Mind and Brain Interact to Create our Conscious Lives*. Houghton Mifflin, Boston, 1988.

Gennaro, R., *Consciousness and Self-Consciousness: A Defense of the Higher-Order Thought Theory of Consciousness.* John Benjamins, Amsterdam and Philadelphia, 1995.

Gennaro, R. (ed.) *Higher-Order Theories of Consciousness.* John Benjamins, Amsterdam and Philadelphia, 2004.

Georg Gross Galiana, Short Term Load Forecasting, Proceedings of IEEE, Col. 75, No. 12, Dec. 1987, 1558–1573.

Ghalia, M.B. and Alouani, A.T., Artificial neural networks and fuzzy logic for system modeling and control: a comparative study. *27th Southeastern Symposium on System Theory (SSST'95)*, p. 258, 1995.

Ghani, A.C., Ferguson, N.M., Fraser, C., Donnelly, C.A., Danner, S., Reiss, P., Lange, J., Goudsmit, J., Anderson, R.M., De Wolf, F., Viral replication under combination antiretroviral therapy: a comparison of four different regimens. *Journal of Acquired Immuno-Deficiency Syndrome* 30(2): 167–76 (issn: 1525–4135), 2002.

Gibb, D.M. and Tess, B,H. Interventions to reduce mother-to-child transmission of HIV infection: new developments and current controversies. *AIDS*, 13(Suppl A): S93–102, 1999.

Gibbard M.J. "Coordinated design of Multi M/C PSS based on damping Torque Concepts" *IEE Proc-C*, Vol. 135, No. 4, PP. 276–284, 1988.

Gil, H., da Silva, E., and Galiana, F., Modeling competition in transmission expansion. *IEEE Transactions on Power Systems*, 17(4): 1043–1049, 2002.

Gil Hugo, A., Galiana, F.D., and da Silva, E.L., Nodal price control: a mechanism for transmission network cost allocation. *IEEE Transactions on Power Systems*, 21(1): 3–10, 2006.

Glanchant, J.M. and Pignon, V., *Nordic Electricity Congention's Arrangements as a Model for Europe: Physical Constraints or Operator's Opportunity.* University of Cambridge, Department of Applied Economics, Tech. Report, 2002.

Glavitsch, H. and Alvarado, F., Management of multiple congested conditions in unbundled operation of power system. *IEEE Transactions on Power Systems*, 13: 1013–1019, 1998.

Goldberg, D.E. and Holland, J.H., *Genetic Algorithms and Machine Learning*, vol. 3. Kluwer, Dordrecht, 1988.

Goldberg, D.E. and Holland, J.H., *Genetic Algorithms, in Search, Optimization and Machine Learning.* Addison-Wesley, Reading, MA, 1989.

Gray, J., The contents of consciousness: a neuropsychological conjecture. *Behavior and Brain Sciences*, 18(4): 659–722, 1995.

Green, R., Ilic, M., Galiana, F., and Fink, L.H., The political economy of the pool. In Power System Restructuring – Engineering and Economics. Kluwer, Dordrecht, 2000.

Grefenstette, J.J., Optimization of control parameters for genetic algorithms. *IEEE Transactions on Systems, Man and Cybernetics*, SMC-16(1): 122–128, 1981.

Grefenstette, J.J., Optimization of control parameters for Genetic Algorithms. *IEEE Trans. On System Man, and Cybernetics*, 16(1) : 122–128, 1986.

Grossberg, S., Neural expectations: cerebellar and ratinal analogs of cells fired by learnable or unlearned pattern classes. *Kybernetics*, 10: 49–57, 1972.

Grossberg, S., *Neural Networks and Natural Intelligence.* MIT Press, Cambridge, MA, 1988.

Grossberg, S., Nonlinear difference – differential equations in prediction and learning theory, *Proceedings of the National Academy of Science*, USA, 58: 1329–1334, 1967.

Guan. L., Cheng. S and Zhow. R "Artificial Neural Network Power System Stabilizer with an Improved BP Algorithm", *IEE Proc. Part-C*, Vol. 143 PP. 135–141, 1996.

Guo, J., Wu, Q.H., Turner, D.R., Wu, Z.X., and X.X. Zhou, Revenue reconciliation for spot pricing: implementation and implication. *Electric Power Components and Systems*, 32: 53–73, 2004.

Gupta, M.M. and Yamakawa, T. (eds.) *Fuzzy computing: theory, hardware, and applications*, North Holand, New York, 1988.

Gupta, N.L, Computer aided power system planning. *Electrical. India*, 24(7): 19–22, 1994.

Gupta, M. and Sinha, N. (eds.), *Intelligent Control Systems: Theory and Applications*. IEEE Press, Piscataway, NJ, 1996.

Gupta, M., Saridis, G., and Gaines, B., *Fuzzy Automatica and Decision Processes*. North-Holland, New York, 1977.

Guyon, I., Application of neural networks to character recognition. *International Journal of Pattern Recognition and Artificial Intelligence*, 5(1 and 2): 353, 1991.

Hadipriono, Fabian C. and Sun, Kerning 'Angular fuzzy set models for linguistic values', Civil Engineering and Environmental Systems, 7:3, 148–156, (1990).

Hagan, M.T., The time series approach to short term load forecasting. *IEEE Transactions on Power System*, 2(3): 785–791, 1987.

Hameroff, S., Quantum computation in brain microtubules? The Penrose–Hameroff "Orch OR" model of consciousness. *Philosophical Transactions Royal Society London*, A 356: 1869–1896, 1998.

Handelman, D., Lane, S., and Gelfand, J., Integrating Neural Networks and Knowledge-Based Systems for Intelligent Robotic Control, *IEEE Control Systems Magazine*, 10(3): 77, 1990.

Handschim E., Heine, M., Konig, D., Nikoden, T., Seibt, T., and Palma, R., Object-oriented software engineering for transmission planning in open access schemes. *IEEE Transactions on Power Systems*, 13: 94–100, 1998.

Happ, H.H., Optimal power dispatch – a comprehensive survey. *IEEE Transactions on Power Apparatus and Systems*, 96: 841–854, 1977.

Happ, H.H., Cost of wheeling methodologies. *IEEE Transactions on Power Systems*, 9: 272–278, 1994.

Hardin, C., Physiology, phenomenology, and Spinoza's true colors. In Beckermann, A., Flohr, J., and Kim, J. (eds.) *Emergence or Reduction?: Prospects for Nonreductive Physicalism*. De Gruyter, Berlin and New York, 1992.

Hardin, C. *Color for Philosophers*. Hackett, Indianapolis, 1986.

Harman, G., The intrinsic quality of experience. In Tomberlin, J. (ed.) *Philosophical Perspectives*, vol. 4. Ridgeview Publishing, Atascadero, CA, 1990.

Hartshorne, C., Panpsychism: mind as sole reality. *Ultimate Reality and Meaning*, 1: 115–129, 1978.

Hashimoto, H., Kubota, T., Kudou, M., and Harashima, F., Self-organization visual servo system based on neural networks, *IEEE Control Systtems Magazine*, 12(2): 31, 1992.

Hasker, W. *The Emergent Self*. Cornell University Press, Ithaca, NY, 1999.

Hassoun, M.H., *Fundamentals of Artificial Neural Networks*. Prentice Hall of India Private Limited, New Delhi, 1997.

Hassan M.A.M., Malik O.P. and Hope G.S ., 'A Fuzzy Logic Based Stabilizer for a Synchronous Machine', *IEEE Trans. on Energy Conversion*, vol. 6, no. 3, pp. 407–413, 1991.

Haykin, S., *Neural Networks: A Comprehensive Foundation*. Macmillan College Publishing Company Inc., New York, 1994.

Hebb, D.O., *The Organization of Behaviour: A Neuropsychological Theory*. Wiley, New York, 1949.

Hecht-Nielsen R., Applications of counterpropagation networks. *Neural Networks*, 1: 131–139, 1988.

Hecht-Nielsen R., Counterpropagation networks, *Applied Optics*, 26: 4979–4984, 1987.

Hecht-Nielsen R., *Neuro Computing*. Addison Wesley, Reading, MA, 1990.

Heidegger, M., *Being and Time (Sein und Zeit)*. Translated by J. Macquarrie and E. Robinson. Harper and Row, New York, 1927/1962.

Heistermann, The application of a genetic approach as an algorithm. *Lecture Notes in Computer Sciences*, 496: 297–307, 1990.

Heller William, J., Jansen, P.J., Silverman, L.P., The new electric industry: what's at stake?" *The McKinsey Quarterly*, 3: 86, 1996.

Hellman, G. and Thompson, F., Physicalism: ontology, determination and reduction. *Journal of Philosophy*, 72: 551–564, 1975.

Henderson, D., An advanced electronic load governor for control of microwave hydroelectric generation. *IEEE Transactions on Energy Conversion*, 13(3): 300–304, 1998.

Herrera, F. and Verdegay, J.L., *Genetic Algorithms and Soft Computing*. Physica-Verlag, Wurzburg, 1996.

Hertz, J., Krogh, A., and Palmer, R.G., *Introduction to the Theory of Neural Computation*. Addison-Wesley, Reading, MA, 1991.

Hill, C., Imaginability, conceivability, possibility, and the mind–body problem. *Philosophical Studies*, 87: 61–85, 1997.

Hill, C. and McLaughlin, B., There are fewer things in reality than are dreamt of in Chalmers' philosophy. *Philosophy and Phenomenological Research*, 59/2: 445–54, 1998.

Hill, C., *Sensations: A Defense of Type Materialism*. Cambridge University Press, Cambridge, 1991.

Hinton, G.E., Learning distributed representations of concepts, *In Proceedings of the Cognitive Science Society*, Ambherst, MA, 1996.

Hippert, H.S., Pedreira, C.E., and Souza, R.C., Neural networks for short-term load forecasting: a review and evaluation. *IEEE Transaction on Power System*, 16(1): 44–55, February 2001.

Hirota, K. and Ozawa, K., The concept of Fuzzy Flip Flop. *IEEE Transactions on Systems, Man, and Cybernetics*, 19(5): 980–997, 1989.

Hirst, E. and Kirby, B., *Ancillary Service Details: Operating Reserves*. Oak Ridge National Laboratory Report, 1996, ORNL/CON-452.

Hiyama, T. and Lim, C.M., Application of fuzzy logic vontrol dcheme for stability enhancement of a power system, *Proceedings of IFAC Symposium on Power Systems and Power Plant Control*, Singapore, Aug. 1989.

Hiyama, T., Oniki, S., Nagashima, H., "Experimental studies on micro computer based fuzzy logic power system stabilizer" *Proc. of Second International forum on Applications of Neural networks to power systems*, ANNPS'93, PP. 212–217, 1993.

Ho, K.L., Short term load forecasting of Taiwan power system using a knowledge based expert system. *IEEE Transactions on Power Systems*, 5(4): 1214–1220, Nov. 1990.

Hofer, D.S., Neumerkel, D., and Hunt, K., Neural control of a steel rolling mill. *IEEE Control Systems*, 13(3): 69, 1993.

Hogan, W.W., Contract networks for electric power transmission. *Journal of Regultory Economics*, 4: 211–242, 1992.

Holland, J.H., *Adaptation in Natural and Artificial Systems*. University of Michigan Press, Ann Arbor, 1975.

Holland, J,H., Genetic algorithms and eth optimization allocation of trails. *SIAM Journal of Computing*, 2: 88–180, 1973.

Holland, J.H., Outline for a logical theory of adaptive systems. *Journal of ACM*, 3: 297–314, 1962.

Holmes, A.S., A review and evaluation of selected wheeling arrangements and a proposed general wheeling tariff. *FERC Paper*, September 1983.

Hong-Tzer, Y. and Huang, C.M., A new short term load forecasting approach using self organizing fuzzy ARMAX models. *IEEE Transactions on Power Systems*, 13(1): 217–225, 1998.

Hopfield, J.J. and Tank, D.W., Neural computation of decisions in optimization problems. *Biology and Cybernetics*, 52: 141–152, 1985.

Hopfield, J.J., Neural networks and physical systems with emergent collective computational capabilities, *Proceedings of National Academic of Sciences (USA)*, 79: 2554–2558, 1982.

Hosseinzadeh N. and Kalam, A., "A direct adaptive fuzzy power system stabilizer," IEEE Trans. Energy Conversion, vol. 14, no. 4, pp. 1564–1571, Dec. 1999.

Horgan, T., Jackson on physical information and qualia. *Philosophical Quarterly*, 34: 147–183, 1984.

Horgan, T. and Tienson, J., The intentionality of phenomenology and the phenomenology of intentionality. In Chalmers, D.J. (ed.) *Philosophy of Mind: Classical and Contemporary Readings*. Oxford University Press, New York, 2002.

Hornik, K., Stinchombe, M., and White, H., Multilayer feedforward networks are universal approximators. Neural Networks 2: 359–366, 1989.

Hosseinzadeh, N. and Kalam, A., A direct adaptive fuzzy power system stabilizer. *IEEE Transaction on Energy Conversion*, 14(4): 1564–1571, 1999.

Houghton and Corpenter, *Aerodynamics for Engineering students*. CBS Publishers Delhi, 1996.

Hsu, Y.-T. and Chao-Rong, C., Tuning of power system stabilizer using an artificial neural network. *IEEE/PES 1991, Winter Meeting, New York*, Feb 3–7, 1991.

Hsu, Y.Y. and Cheng, C.H., "Design of fuzzy PSS for multi M/C Power systems" *IEE Proc. Part-C*, Vol. 137, No. 3, PP.233–238, 1990.

Huang, S.J., Hydroelectric generation scheduling – an application of genetic – embedded fussy system approach. *Electric Power Systems Research* 48: 65–72, 1998.

Huang T.L., Chen S.C., Hwang T.Y. and Yang W.T. "Power system output feedback stabilizer design via optimal sub eigen structure assignment", Int. journal of energy research, Vol. 15. No. 6, PP. 497–507, Aug. 1991.

Huber, D.W., Runtz, K.J., Hope, G.S., and Malik, O.P., Digital AVR for use in computer control of a synchronous machine. *IEEE Transactions on PES*, 579–601, 1972.

Hume, D. In Selby-Bigge, L. (ed.) *A Treatise of Human Nature*. Oxford University Press, Oxford, 1739/1888.

Humpert, B., Improving backpropagation with a new Error function. *Neural Network*, 7(8): 1191–1192, 1994.

Humphreys, G. (eds.) *Consciousness: Psychological and Philosophical Essays*. Blackwell, Oxford.

Humphreys, N., *A History of the Mind*. Chatto and Windus, London, 1992.

Humphreys, N., *Consciousness Regained*. Oxford University Press, Oxford, 1982.

Hunt, S. and Shuttleworth, G., *Competition and Choice in Electricity*. Chichester, Wiley, 1996.

Hurley, S., *Consciousness in Action*. Harvard University Press, Cambridge, MA, 1998.

Husserl, E., *Cartesian Meditations: an Introduction to Phenomenology*. Translated by Dorian Cairns. The Hague: M. Nijhoff, 1929/1960.

Husserl, E. *Ideas: General Introduction to Pure Phenomenology (Ideen au einer reinen Phänomenologie und phänomenologischen Philosophie)*. Translated by W. Boyce Gibson. MacMillan, New York, 1913/1931.

Hwang, J.-N. and Seokyong, M., Temporal difference method for multi-step prediction: application to power load forecasting, Proceedings of the First International Forum on Applications of Neural Networks to Power Systems, IEEE, NY, USA, pp. 41–45, 1991.

Iba, K., Reactive power optimization by genetic algorithm. *IEEE Transactions on PWRS*, 9: 685–692, 1994.

Ichalewiccz, Z., *Genetic Algorithms+ Data Structure= Evolution Programs*. Springer Verlag, Berlin Heidelberg New York, 1992.

IEEE Committee Report, Load forecasting: bibliography, phase-1. *IEEE Transactions on Power Apparatus & Systems*, PAS-99(1): 53–55, 1980.

IEEE Committee Report, Load forecasting: bibliography, phase-2. *IEEE Transactions on Power Apparatus & Systems*, PAS-99(1): 3217–3220, 1981.

Illic, M. and Ramanan, P., Transmission pricing of distributed multilateral energy transactions to ensure system security and guide economic dispatch. *IEEE Transactions on Power Systems*, 18(2): 428–434, 2003.

Iracleous, D.P. and Alexandridis, A.T., Fuzzy tuned PI controllers of series connected D.C. motor drives. *Proceedings of the IEEE International Symposium on Industrial Electronics*, Athens, Greece, 1995.

Irie, B. and Miyake, S., Capabilities of three-layered perceptrons. *Proceedings of the IEEE International Conference on Neural Networks*, San Diego, CA, 1: 641–647, 1988.

Ishibashi, K., Komura, T., Ueki, Y., Nakanishi, Y., and Matsui, T., Short-term load forecasting using an artificial neural network, *Second International Conference on Automation, Robotics and Computer Vision*, vol. 3, pp. INV 11.1/1–5, Nanyang Technol. Univ. Singapore, 1992.

Itoh, O., Gotoh, K., Nakayama, T., and Takamizawa, S., Applications of fuzzy control to activated slidge process. In *Proceedings of the Second IFSA Congress*, Tokyo, Japan, July, pp. 282–285, 1987.

Iyer, V., Fung, C.C., and Gedeon, T., A fuzzy-neural approach to electricity load and spot-price forecasting in a deregulated electricity market, *TENCON'3, Conference on Convergent Technologies for Asia-Pacific Region*, vol. 4, pp. 1479–1482, 2003.

Jackson, F., Mind and illusion. In Ludlow, P., Nagasawa, Y., and Stoljar, D. (eds.) *There's Something About Mary: Essays on the Knowledge Argument*. MIT Press, Cambridge, MA, 2004.

Jackson, F., Armchair metaphysics. In Leary-Hawthorne, J.O. and Michael, M. (eds.) *Philosophy of Mind*. Kluwer, Dordrecht, 1993.

Jackson, F., Epiphenomenal qualia. *Philosophical Quarterly*, 32: 127–136, 1982.

Jackson, F., Postscript on qualia. In Jackson, F. (ed.) *Mind, Method and Conditionals*. Routledge, London, 1998.

Jackson, F., What Mary didn't know. *Journal of Philosophy*, 83: 291–295, 1986.

Jacobs, R.A., Increased rates of convergence through learning rate adaptation. Neural Networks, 1(4): 295–307, 1988.

Jain, A. and Kumar, A.M., Hybrid neural network models for hydrologic time series forecasting. *Applied Soft Computing, 7(2): 585–592, 2007.*

Jaleey, N., et. al, Understanding automatic generation control. *IEEE Transactions Power Systems, PAS*, 89(4), 1990.

James, W., *The Principles of Psychology*. Henry Holt and Company, New York, 1890.

Jamshidi, M., Vadiee N., and Ross, T.J. (eds.), *Fuzzy Logic and Control: Software and Hardware Applications*. Englewood Cliffs, NJ, Prentice Hall, 1993.

Jang, J.R., ANFIS: Adaptive-network-based fuzzy inference systems. *IEEE Transactions on Systems, Man and Cybernetics*, 23(3): 665–685, 1993.

Jang, J.S. and Sun, C., Neuro-fuzzy modeling and control. *Proceedings of IEEE* 83(3): 378–406, 1995.

Jang, J.S., Self-learning fuzzy controllers based on temporal back propagation. *IEEE Transactions on Neural Networks*, 3(5), 1992.

Jhy-Shing Roger, Sun Chuen-Tsai, J., and Eiji, M., *Neuro-Fuzzy and Soft Computing: A Computational Approach to Learning and Machine Intelligence*. Prentice Hall, New Jersery, 1997.

Jaynes, J. *The Origins of Consciousness in the Breakdown of the Bicameral Mind*. Houghton Mifflin, Boston, 1974.

Jin, Y., Fuzzy modeling of high-dimensional systems: complexity reduction and interpretability improvement. *IEEE Transactions on Fuzzy Systems*, 8(2): 212–221, 2000.

Jones, W.P. and Hoskins, J., Backpropagation: a generalized delta rule. *BYTE*, 155–163, 1987.

Jose, M.B. (ed.), et al., *Advances in Soft Computing: Engineering Design and Manufacturing*. Springer Verlag, Berlin, 2003.

Jose Manuel, B., *Advances in Soft Computing: Engineering Design and Manufacturing*. Springer, Berlin Heidelberg New York, 2003.

Judieth, D. *Neural Network Architectures: An Introduction*. Van Nostrand, Reinhold, New York, 1991.

Jung, S. and Wen John, T., Nonlinear model predictive control for the swing up of a rotary inverted pendulum. Rensselaer Polytechnic Institute, Troy, NY, 12180, USA, April 2003.

Kalra, P.K., Srivastava, S.C., and Chaturvedi, D.K., Possible applications of neural nets to power system operation and control. *International Journal of Power System Research*, 25: 83–90, 1992.

Kalra, P.K, et al., Neural Network – A Simulation tool', National Conference on Paradigm of ANN for optimization process Modeling and Control at Indian Oil, Faridabad, Sept. 7–9, 1995.

Kandel, A. and Langholz, G. (eds.). *Hybrid Architectures for Intelligent Systems*. CRC Press, FL, 1992.

Kandel, B.R. and Schwartz, J.H., Molecular biology of learning: modulation of transmitter release. *Science*, 218: 433–443, 1982.

Kandel, B.R., Schwartz, J.H., and Jessel, T.M., *Principles of Neural Science*, 4th Edn. Appleton and Lange, New York, 2000.

Kandel Eric R., *Principles of Neural Science*, McGraw Hill, 2000.

Kant, I. *Critique of Pure Reason*. Translated by N. Kemp Smith. MacMillan, New York, 1787/1929.

Karr, C.L. and Gentry, E.J., Fuzzy control of ph using genetic algorithms. *IEEE Transactions on Fuzzy Systems*, 1(1): 46–53, 1993.

Kartam, N., Ian, F., Garrett James, H., Girish, A., *Artificial Neural Networks for Civil Engineers: Fundamentals and Applications*. ASCE Publications, 1997.

Kasai, Y. and Morimoto, Y., Electronically controlled continuous variable transmission system. In *Proceedings of International Congress on Transportation Clectronics*, Dearborn, MI, 1988.

Kaufmann, A. and Gupta, M. (eds.), *Introduction to Fuzzy Arithmetic Theory and Applications*. Van Nostrand Reinhold, New York, 1985.

Kaymak, U. and Sousa, J.M., Weighted constraint aggregation in fuzzy optimization. *Journal on Constraints*. Springer Netherlands, 8(1): 61–78, , 2003.

Kazarlis, S.A., Bakirtzis, A.G. and Petridis, V., A genetic algorithm solution to the unit commitment problem. *IEEE Transactions on Power Systems*, 11(1): 83–92, 1996.

Keib, A.A. El- and Ma, X., Calculating short-run marginal costs of active and reactive power production. *IEEE Transactions on Power Systems*, 12(2): 559–565, 1997.

Kercel, S.W., Softer than Soft Computing. *IEEE International Workshop on Soft Computing in Industrial Applications*. Binghamton University, Binghamton, New York, June 23–25, 2003.

Kermode, A.C., *Mechanics of Flight*. Himalayan Books, New Delhi, 1984.

Kermode, A.C., *Flight Without Formulae*. Pitman Publishing Limited, London, 1970.

Khalid, M. and Omatu, S., A neural network controller for a temperature control system. *IEEE Control System Magazine*, 58–64, June 1992.

Khincha, H.P. and Krishnan, N., Short term load forecasting using neural network for a distribution project, *National Conference on Power Systems (NPSC'96)*, at Indian Institute of Technology, Kanpur, pp. 417–421, Dec. 1996.

Kim, Y.T. and Baek, S.H., The speed regulation of a D.C. motor drive system with a PI, PID and command matching controllers, Dongguk, Journal, (in Korean), 29: 525–541, 1995.

Kim, J., The myth of non-reductive physicalism. *Proceedings and Addresses of the American Philosophical Association*, 1987.

Kim, J. *Mind in Physical World*. MIT Press, Cambridge, 1998.

Kinsbourne, M., Integrated field theory of consciousness. In Marcel, A. and Bisiach, E. (eds.) *Consciousness in Contemporary Science.* Oxford University Press, Oxford, 1988.

Kinzel, J., et al., Modification of genetic algorithms for design and optimizing fuzzy controllers. *IEEE International Conference on Fuzzy Systems,* 28–33, 1994.

Kirk, R., Why shouldn't we be able to solve the mind–body problem? *Analysis,* 51: 17–23, 1991.

Kirk, R., Zombies vs materialists. *Proceedings of the Aristotelian Society,* Supplementary Volume 48, pp. 135–152, 1974.

Klir, G.J. and Yuan, B. *Fuzzy Sets and Fuzzy Logic, Theory and Applications.* Prentice Hall, Englewood Cliffs, NJ, 1995.

Klir, G.J. and Folger, T.A., *Fuzzy Sets, Uncertainty and Information.* Prentice Hall of India, New Delhi, 2000.

Klir, G.J. and Folger, T.A., *Fuzzy sets, Uncertainty and Information,* Prentice Hall, Englewood Cliffs, NJ. 1988.

Klir, G.J. and Yaun, B., *Fuzzy Sets and Fuzzy Logic Theory and Applications,* Prentice Hall Inc., NJ, 1995.

Koch, C., *Biophysics of Computation: Information Processing in Single Neuron.* Oxford University Press, New York, 1999.

Köffka, K., *Principles of Gestalt Psychology.* Harcourt Brace, New York, 1935.

Köhler, W., *Gestalt Psychology.* Liveright, New York, 1929.

Kohonen, T., Correlation matrix memories. *IEEE Transactions on Computers,* c-21:353–359, 1972.

Kohonen, T., *Self Organizing Maps,* 2nd Edn. Springer-Verlag, Berlin, 1997.

Kohonen, T., *Self-Organization and Associative Memory,* 3rd Edn. Springer-Verlag, New York, 1988.

Kosko, B., Bidirectional associative memory, *Applied Optics,* 26(23): 4947–4960, 1987.

Kosko, B., Bidirectional associative memory, *IEEE Transactions on Systems, Man and Cybernetics,* 18(1): 49–60, 1988.

Kosko, B., *Fuzzy Engineering,.* Prentice Hall, Upper Saddle River, NJ, 1997.

Kosko, B., *Neural Network for Signal Processing.* Prentice-Hall, Englewood Cliffs, NJ, 1992.

Kothari M.L., Nanda. J and Bhattcharya K., "Design of variable structure power system stabilizers with desired eigen values in the Sliding Mode", *IEE Proc., part-C,* Vol. 140–4, PP. 263–268, July 1993.

Kovacs, R.R. and Leverett, A.L., A load flow based method for calculating embedded, incremental & marginal cost of transmission capacity. *IEEE Transactions on Power Systems,* 9, 272–278, 1994.

Kramer Alan, H. and Sangiovanni-Vincentelli, A., *Advances in neural information processing systems.* Morgan Kaufmann, San Francisco, CA, USA, pp. 40–48, 1989.

Kreiss, J.K., Hopkins, S.G., The association between circumcision status and human immunodeficiency virus infection in homosexual men. *J Infect Dis* 168, 1404–8 1993.

Kristiansen, T., Congestion management, transmission pricing and area price hedging in the Nordic region. *Electrical Power and Energy Systems* 26 (2004), pp. 685–695.

Krunic, M.S.S.C. and Rajakovik, N., An improved neural network application for short term load forecasting in power system. *Electric Machines and Power Systems*, 28: 703–721, 2000.

Kumar, A., Srivastava, S.C., and Singh, S.N., A zonal congestion management approach using real and reactive power rescheduling. *IEEE Transactions on Power Systems*, 19(1): 554–561, 2004.

Kumar, A., Srivastava, S.C. and Singh, S.N., Congestion management in competitive power market: a bibliographical survey. *Electric Power Systems Research*, 76: 153–164, 2005.

Kupperstien, M. and Rubinstein, J., Implementation of an adaptive neural controller for sensory-motor coordination. *IEEE Control Syst*, 9(3): 25, 1989.

Lai, L.L., Ma, J.T., Yokoyama, R., and Zhao, M., Improved genetic algorithms for optimal power flow under both normal and contingent operation states. *Electrical Power and Energy Systems*, 19(5): 287–292, 1997.

Lajwanti and Chaturvedi, D.K., Education and computerisation, *40th WEF International Conference on Educating for a Better World: Vision to Action*, Launceston, Australia, pp. 5–10, Jan. 1999.

Lajwanti, C.D.K., Nandita, S., Satsangi, S.P., Role of computers in higher education, *World Conference on Education India: The Next Millennium*, New Delhi, India, pp. 54, Nov. 12–14, 1997.

Lajwanti, C.D.K. and Soami, D.V., Teaching and learning through computer games, *International Conference on Educational Management, Technology and Values, Udaipur (Raj.), India*, Souvenir, p. 56, Dec. 10–12, 1999.

Lakhmi C.J., and Jain, R.K. (eds.). *Hybrid Intelligent Engineering Systems*. World Scientific Publishing, Co. Pte. Ltd., USA, 1997.

Lamont, J.W. and Fu, R., Cost analysis of reactive power support. *IEEE Transactions on Power Systems*, 14(3): 890–898, 1999.

Langdon, W.B. and Poli, R., *Foundations of Genetic Programming*. Springer-Verlag, Berlin, 2002.

Larsen, P.M., Industrial applications of Fuzzy Logic Control, *International Journal of Man–Machine Studies*, 12(1), 3–10, 1980.

Larsen, E.V. and Swann, D.A., Applying Power System Stabilizer, *IEEE Transactions on Power Apparatus and Systems*, Vol. PAS-100, 3017–3046, 1981.

Lawrence, D., *Handbook of Genetic Algorithms*. Von Nostrand Reinhold, New York, 1991.

LeCun, Y., Jackel, L.D., et al., Handwritten Digital Recognition: Application of Neural Network Chips and Automatic Learning. *IEEE Communications Magazine*, 41, 1988.

Lee, K.Y., Park, Y.M., and Ortiz, J.L., A united approach to optimal real and reactive power dispatch. *IEEE Transactions on Power Apparatus and Systems*, PAS-104(5): 1147–1153, 1985.

Lee, M. and Esbensen, H., Evolutionary algorithms based multiobjective techniques for intelligent systems design. *FUZZ-IEEE*, 360–364, 1996.

Lee, M.A. and Takagi, H., Integrating design stages of fuzzy systems using genetic algorithms, *Proceedings of the Second IEEE International Conference on Fuzzy Systems, Sanfransisco*, pp. 612–617, 1993.

Lee, S.K., Son, K.M. and Park, J.K., Voltage profile optimization using a pyramid genetic algorithm. *Proceedings ISAP 97*, Seoul, Korea, pp. 407–414, July 6–10, 1997.

Lee, Y., Oh, S.H., and Kim, M.W., The effect of initial weights on premature saturation in backpropagation learning. IEEE International Joint Conference on Neural Networks, (Seattle), New York, 1: 765–770, 1991.

Lee, D.G., Lee B.W., and Chang S.H., Genetic programming for long term forecasting of electric power demand. *Electric Power System Research*, 40: 17–22, 1997.

Lee, K.Y., Choi, T.I., Ku, C.C., and Park, J.H., Short-term load forecasting using diagonal recurrent neural network, *IEEE Proceedings of the Second International Forum on Applications of Neural Networks to Power Systems, pp. 227–32*, New York, NY, USA, 1993.

Leibniz, G.W., *Discourse on Metaphysics*. Translated by D. Garter and R. Aries. Hackett, Indianapolis, 1686, 1991.

Levine, J., Materialism and qualia: the explanatory gap. *Pacific Philosophical Quarterly*, 64: 354–361, 1983.

Levine, J., On leaving out what it's like". In Davies, M. and Humphreys, G. (eds.) *Consciousness: Psychological and Philosophical Essays*. Blackwell, Oxford, 1993.

Levine, J. Out of the closet: a qualophile confronts qualophobia. *Philosophical Topics*, 22/1–2: 107–26, 1994.

Levine, J. *Purple Haze: The Puzzle of Conscious Experience*. The MIT Press, Cambridge, MA, 2001.

Lewis, D., Psychophysical and theoretical identifications. *Australasian Journal of Philosophy*, 50: 249–258, 1972.

Lewis, D., What experience teaches. In Lycan, W. (ed.) *Mind and Cognition: A Reader*. Blackwell, Oxford, 1990.

Li, G. and Manli, X., Changing topological artificial neural network for load forecasting. *Automation of Electric Power Systems*, 17(11): 28–33, 1993.

Li, T. and Shahidehpour, M., Strategic bidding of transmission-constrained GENCOs with incomplete information. *IEEE Transactions on Power Systems*, 20(1): 437–447, 2005.

Li, Y.Z. and David, A.K., Optimal multi-area wheeling. *IEEE Transactions on Power Systems*, 9(1): 288–294, 1993.

Li, Y.Z. and David, A.K., Wheeling rates of reactive power flow under marginal cost pricing. *IEEE Transactions of Power Systems*, 4(2): 594–605, 1989.

Li, Y. and Flynn, P.C., Deregulated power prices: changes over time. *IEEE Transactions on Power Systems*, 20(2): 565–572, 2005.

Liang and Cheng, C.C., Combined regression Fuzzy approach for short-term load forecasting. *IEE Proceedings*, 147(4): 261–265, 2000.

Libet, B., Subjective antedating of a sensory experience and mind-brain theories. *Journal of Theoretical Biology*, 114: 563–570, 1982.

Libet, B., Unconscious cerebral initiative and the role of conscious will in voluntary action. *Behavioral and Brain Sciences*, 8: 529–66, 1985.

Lima, J.W.M. and De Oliveira, E.J., Long-term impact of transmission pricing. *IEEE Transactions on Power Systems*, 30(4): 1514–1520, 1998.

Lima, J.W.M., Allocation of transmission fixed charges: an overview. *IEEE Transactions on Power Systems*, 11(3): 1409–1417, 1996.

Lima, J.W.M. and Pereira, M.V.F., An integrated framework for cost-allocation in a multi-owned transmission system. *IEEE Transactions on Power Systems*, 10(2): 971–977, 1995.

Lin, C.T. and Lee, C.S.G., Neural Fuzzy Systems: A Neuro-Fuzzy Synergism to Intelligent Systems. Prentice Hall, Upper Saddle River, NJ, 1996.

Linkens, D.A. (ed.), *Intelligent Control in Biomedicine*. Taylor & Francis, London, 1994.

Lipmann, R.P., Pattern Classification Using Neural Networks. *IEEE Communications Magazine*, Nov., 1989.

Lippmann, R.P., An Introduction to Computing with Neural Network. *IEEE Acoustic, Speech, and Signal Processing Magazine*, 4, 1987.

Lister, R., Bakker, P., and Wiles, J., Error signals, exceptions, and back propagation, *Proceedings of International Joint Conference on Neural Networks, Nagoya*, vol. 1, pp. 573–557 (1993).

Lister, R., and Stone, J.V., An empirical study of the time complexity of various error functions with conjugate gradient backpropagation, *Proceedings IEEE International Conference on Neural Networks*, 1: 237–241, 1995.

Liu, H., Iberall, T., and Bekey, G., Neural network architecture for robot hand control. *IEEE Control Systems*, 9(3): 38, 1989.

Liu W., Moorthy G.K.V., Wunsch D.C. II. "Adaptive Neural Network Based Power System Stabilizer Design" *IEEE Trans. Power Systems*, PP. 2970–2975, 2003.

Llinas, R. *I of the vortex: from neurons to self*. MIT Press, Cambridge, MA, 2001.

Lo, K.L. and Zhu, S.P., Wheeling and marginal wheeling rates: theory and case study results. Electric Power Systems Research, 27: 11–26, 1993.

Lo, K.L. and Zhu, S.P., A theory of pricing wheeled power. *Electrical Power Systems Research*, 28: 191–200, 1994.

Lo, C.H., Chan, P.T., Wong, Y.K., Rad, A.B., Cheung, K.L., Fuzzy-genetic algorithm for automatic fault detection in HVAC systems. *Applied Soft Computing, 7(2): 554–560, 2007*.

Loar, B., Phenomenal states. In Block, N., Flanagan, O., and Guzeldere, G. (eds.) *The Nature of Consciousness*. MIT Press, Cambridge, MA, 1997.

Lockwood, M., *Mind, Brain, and the Quantum*. Oxford University Press, Oxford, 1989.

Lorenz, K., *Behind the Mirror (Rückseite dyes Speigels)*. Translated by R. Taylor. Harcourt Brace Jovanovich, New York, 1977.

Lozano, J.A., Larrañaga, P., Inza, I., and Bengoetxea, E. (eds.), *Towards a New Evolutionary Computation: Advances on Estimation of Distribution Algorithms* Series: Studies in Fuzzy & Soft Computing, Springer, 2006.

Lycan, W., The superiority of HOP to HOT. In Gennaro, R. (ed.) *Higher-Order Theories of Consciousness*. John Benjamins, Amsterdam and Philadelphia, 2004.

Lycan, W., *Consciousness and Experience*. MIT Press, Cambridge, MA, 1996.

M{u}hlenbein H., Parallel genetic algorithms, population genetics and combinatorial optimization. *Proceedings of Third International Conference on Genetic Algorithms*. Morgan Kaufmann, Los Altos, CA, 1989.

Ma, X., Sun, D. and Cheung, K., Evolution toward standardized market design. *IEEE Transactions on Power Systems*, 18(2): 460–469, 2003.

Ma X., Sun, D., and Ott, A., Implementation of the PJM financial transmission rights auction market system. In *Proceedings of the Power Engineering Society Summer Meeting*, 2002, IEEE, 3: 1360–1365, 2002.

Maha Somnath, S.P., Roy, K., and Banerjee, S., Improved genetic algorithm for channel allocation with channel borrowing in mobile computing. *IEEE Transactions on Mobile Computing*, 5(7): 884–892, July 2006.

Mahalanbis, A.K., Kothari, D.P., and Jahson, S.I., *Computer Aided Power System Analysis and Control*, I; Tata McGraw-Hill Publishing Company Ltd., New Delhi, pp. 93–124, 1991.

Majors, M., Stori, J., and Cho, D., Neural network control of automotive fuel-injection systems. *IEEE Control Systems*, 14(3): 31, 1994.

Mamdani, E.H. and Assilian, S., An Experiment in Linguistic Synthesis with a Fuzzy logic controller. *International Journal of Man–Machine Studies*, 7(1): 1–13, 1975.

Mamdani, E.H., Advances in the linguistic synthesis of fuzzy controller. *International Journal of Man–Machine Studies*, 8(6): 669–678, 1976.

Mamdani, E.H., Approximations of fuzzy algorithms for simple dynamic plant. *Proceedings of IEE* 21(12): 1585–1588, 1974a.

Mamdani, E.H., Application of fuzzy algorithms for control of simple dynamic plant. *Proceedings of IEE*, 121(12), 1974b.

Man Mohan Chaturvedi, D.K., Satsangi, P.S., and Kalra, P.K., Neuro-fuzzy approach for development of new neuron model. *International Journal of Soft Computing – A Fusion of Foundations, Methodologies and Applications*. Springer, Berlin Heidelberg New York, 8(1), 19–27, Oct. 2003.

Man, M., Chaturvedi, D.K., and Kalra, P.K., Development of new neuron structure for short term load forecasting. *International Journal of Modeling and Simulation, AMSE Periodicals*, 46(5): 31–52, 2003.

Mandal S.K. and Agarwal, A., Fuzzy Neural Network based short term Peak and average Load forecasting system with Network Security, *Proceedings of International Conference on Computer Applications in Electrical Engineering at University of Roorkee*, 257–261, Sept. 8–11, 1997.

Mandal, P., Tomonobu, S., Naomitsu, U., and Toshihisa, F., Electricity price and load short-term forecasting using artificial neural networks. *International Journal of Emerging Electric Power Systems*, 7(4), Article 4, 2006.

Mandler, G. *Mind and Emotion*. Wiley, New York, 1975.

Maria, G.A. and Findlay, J.A., A Newton optimal power flow program for Ontario hydro EMS. *IEEE Transactions on Power Apparatus and Systems*, 2(3): 576–584, 1987.

Marks, II R. (ed.), *Fuzzy Logic Technology and Applications*, IEEE Press, Piscataway, NJ, 1994.

Marshall, I. and Zohar, D. *The Quantum Self: Human Nature and Consciousness Defined by the New Physics*. Morrow, New York, 1990.

Mathewman, P.D. and Nicholson, H., Techniques for Load Prediction in Electric Supply Industry. Proc. IEEE, 115(10), Oct. 1968.

Mathewman, P.D. and Nicholson, H., Techniques for load prediction in electric supply industry. Proceedings IEEE, 115(10), Oct. 1986.

Mathewman, P.D. and Nicholson, H., Techniques for load prediction in electric supply industry. *Proceedings IEEE*, 115(10), Oct. 6., 1992.

Matsumoto, T., Kitamura, S., Ueki, Y., Matsui, T., Short-term load forecasting by artificial neural networks using individual and collective data of preceding years, *Proceedings of the Second International Forum on Applications of Neural Networks to Power Systems*, pp. 245–250, IEEE, New York, NY, USA, 1993.

Matsuoka, K. and Yi, J., Backpropagation based on logarithmic error functions and elimination of local minima, *Proceedings of the IJCNN-91*, Singapore, 2: 1117–1122, 1991.

Mbamalu, M.E. and Hawary, E.L., Load forecasting aia sub optimal seasonal auto regressive models and iterative reweighted least squares estimation. *IEEE Transactions on PWRS*, 1, Feb. 1993.

McClelland, J.L. and Rumelhart, D.E., The PDP Research Group, Parallel Distributed Processing – Explorations in the Microstructure of Cognition, vol. 2, *Psychological and Biological Models*. MIT Press, MA, 1986.

McCulloch, W.W. and Pitts, W., A logical calculus of ideas imminentin nervous activity. *Bulletin of Mathmatical Biophysics*, 5, 115–133, 1943.

McGinn, C., Can we solve the mind–body problem? *Mind*, 98: 349–366, 1989.

McGinn, C., Consciousness and space. In Metzinger, T. (ed.) *Conscious Experience*. Ferdinand Schöningh, Paderborn, 1995.

McGinn, C. *The Problem of Consciousness*. Blackwell, Oxford, 1991.

Medsker, L. and Jay, L., *Design and Development of Expert Systems and Neural Networks*. Macmillan College Publishing Company, Inc., USA, 1994.

Medsker, L., *Hybrid Intelligent Systems*. Kluwer, Dordrecht, 1995.

Medsker, L., *Hybrid Neural Network and Expert Systems*. Kluwer, Dordrecht, 1994.

Melin, P., and Castillo, O., *Hybrid Intelligent Systems For Pattern Recognition Using Soft Computing*. Springer, Berlin Heidelberg New York, 2005.

Méndez, R., and Rudnick, H., Congestion management and transmission rights in centralized electric markets. *IEEE Transactions on Power Systems*, 19(2): 889–896, 2004.

Mendis, B.S.U., Gedeon, T.D., Botzheim, J., Kóczy, L.T., Generalised weighted relevance aggregation operators for hierarchical fuzzy signatures, *International Conference on Computational Intelligence for Modelling Control and Automation (CIMCA'06)*, p. 198, 2006.

Merrill Hyde, M. and Erickson, B.W., Wheeling rates based on marginal-cost theory. *IEEE Transactions on Power Systems*, 4(4): 1445–1451, 1989.

Metzinger, T. (ed.) *Neural Correlates of Consciousness: Empirical and Conceptual Questions*. MIT Press, Cambridge, MA, 2000.

Metzinger, T. (ed.) *Conscious Experience*. Ferdinand Schöningh, Paderborn, 1995.

Michalewiccz, Z., *Genetic Algorithms + Data Structure = Evolution Programs*. Springer, Berlin Heidelberg New York, 1992.

Mill, J.S., *An Analysis of Sir William Hamilton's Philosophy*. London, 1865.

Miller, W.T., Real-time neural network control of a biped walker robot. *IEEE Control Systems*, 14(1), 1994.

Miller, W.T., Sutton, R., and Werbos, P., *Neural Networks for Control*. MIT Press, MA, 1990.

Ming, Z., Yan, Z., Ni, Y., Li, G., An ARIMA approach to forecasting electricity price with accuracy improvement by predicted errors. *Power Engineering Society General Meeting*, 1: 233–238, 2004.

Mingye, H., et al. Neural network modelling of flight control system, *IEEE AES Systems Magazine*, 27–29, 1998.

Minsky, M.L. and Papert, S.A., *'Perceptron'*. MIT Press, Cambridge, MA, 1969.

Minsky, M.L., Steps toward artificial intelligence. *Proceedings of Institute of Radio Engineers*, 49: 8–30, 1961.

Minsky, M.L., Theory of *Neural – Analog Reinforcement Systems and Its Application to the Brain-Model Problem*. Ph.D. Thesis, Princeton University, Princeton, NJ, 1954.

Minsky, M.L. and Papert, S.A., *Perceptron: An Introduction to Computational Geometry.* MIT Press, MA, 1969.

Miranda, V., Srinivasan, D., and Proenca, L.M., Evolutionary computation in power systems, A Survey Paper, *Proceedings of 12th PSCC, Dresden, Germany,* 19–23, August 1996, pp. 25–40.

Miranda, V., Ranito, J.V., and Proenca, L.M., Genetic algorithms in optimal multistage distribution network planning. *IEEE Transactions on PWRS,* 9: 1927–1933, 1994.

Mizumoto, M., Pictorial representations of fuzzy connectives, Part II: cases of compensatory operators and self-dual operators. *Fuzzy Sets and Systems,* 32: 45–79. North-Holland, Amsterdam, 1989.

Mohamad, H.H., *Fundamentals of artificial neural networks.* MIT Press, Cambridge, MA, 1995.

Moya, O., Marginal cost of transmission system adequacy for spot pricing. *Electric Power Systems Research,* 61: 89–92, 2002.

Muchayi, M. and El-Hawary, M.E., A method for optimal pricing of electric supply including transmission system considerations. *Electric Power Systems Research* 50: 43–46, 1999.

Murostsu, Y., Tsujio, S., Sendo, K., and Hayashi, M., Trajectory control of flexible manipulators on a free-flying space robot. *IEEE Control Systems Magazine,* 12(3): 51, 1992.

Murtagh, F., Starck, J.L., and Renaud, O., On neuro-wavelet modeling, *Decision Support Systems,* 37: 475–484, 2004.

Muzhlenbein, H., Parallel genetic algorithms, population genetics and combinatorial optimization, *Proceedings of Third International Conference on Genetic Algorithms.* Morgan Kaufmann, Los Altos, CA, 1989.

Nagaraja, N.S., *Elements of Electronic Navigation.* Tata McGraw-Hill, New Delhi, 1975.

Nagel, T., In Nagel, T. (ed.) Panpsychism. *Mortal Questions.* Cambridge University Press, Cambridge, 1979.

Nagel, T., What is it like to be a bat? *Philosophical Review,* 83: 435–456, 1974.

Nagelkerke, N. and Jha, P., *Modelling the HIV/AIDS epidemic in India and Botswana: the effect of interventions*", CMH, Working Paper Series, 2001.

Naida, S., Zafiriou, E., and McAvoy, T., Use of neural networks for sensor failure detection in a control system. *IEEE Control Systems Magazine,* 10(3): 49, 1990.

Nakanishi, H., Turksen, I.B., and Sugeno, M., A review and comparison of six reasoning methods. *Fuzzy Sets and Systems,* 57(3): 257–294, 1993.

Nanda, J., Kothari, D.P., and Srivastava, S.C., A new optimal power dispatch algorithm using quadratic programming method. *IEE Proceeding,* Part 'C', 136(3): 153–161, May 1989.

Narenda, K. and Parthasarathy, K., Identification and control of dynamic systems using neural networks. *IEEE Transactions on Neural Networks,* 1(4), 1990.

Narendra, K.S., Balakrishnan, J., and Cliliz, K., Adaptation and learning using multiple models, switching and tuning. *IEEE Control Systems,* 15(3): 37, 1995.

Natsoulas, T., Concepts of consciousness. *Journal of Mind and Behavior,* 4: 195–232, 1983.

Nazarko, J. and Zalewski, W., The fuzzy regression approach to peak load estimation in power distribution system, Forecasting. *IEEE Transactions on Power System,* 14(3): 809–814, 1999.

Nedugadi, A., A fuzzy logic based robot control. *Jorunal of Intelligent and Fuzzy Systems*, 1(3): 243–251, 1993.

Negoita, C.V., *Expert Systems and Fuzzy Systems*. Benjamin/Cummings, Menlo Park, CA, 1985.

Neil D.V.S., *'Modern Airmanship'*. Van Nostrand Company, Inc., New Jersey, 1957.

Nekovie, R. and Sun, Y., Back-propagation network and its configuration for blood vessel detection in angiograms. *IEEE Transactions on Neural Networks*, 6(1), 64, 1995.

Nelkin, N., Unconscious sensations. *Philosophical Psychology*, 2: 129–141, 1989.

Nelson, N., *Flight Stability & Automatic Control*. McGraw-Hill, New York, 1990.

Nemirow, L., Physicalism and the cognitive role of acquaintance. In Lycan, W. (ed.) *Mind and Cognition: A Reader*. Blackwell, Oxford, 1990.

Newton, R.T. and Xu, Y., Neural network control of a space manipulator. *IEEE Control Systems Magazine*, 14, 1993.

Nguyen, D., and Widrow, B., Neural networks for self-learning control systems. *IEEE Control Systems Magazine*, 18–23, 1990.

Nguyen, H., Sugeno, M., Tong, R., and Yager, R., *Theoretical Aspects of Fuzzy Control*. Wiley, New York, 1995.

Nguyen, L., Patel, R., and Khorasani, K., Neural network architectures for the forward kinematics problem in robotics, *Proceedings of the IEEE International Conference on Neural Networks*, p. 393, 1990.

Nida-Rümelin, M., What Mary couldn't know: belief about phenomenal states. In Metzinger, t. (ed.) *Conscious Experience*. Ferdinand Schöningh, Paderborn, 1995.

Nie, J. and Linkens, D., *Fuzzy – Neural Control: Principles, Algorithms and Applications*. Prentice Hall of India, New Delhi, 1995.

Nielsen, R.H., *Neurocomputing*. Addison-Wesley, Reading, MA, 1990.

NikRavesh, M., BISC decision support system: fuzzy logic-GA based decision and risk analysis, *Proceedings Annual Meeting of the North American Fuzzy Information Processing Society*, pp. 338–343, 2002.

Nounou, H.N., Rehman, H.U., Application of adaptive fuzzy control to ac machines. *Applied Soft Computing, 7(3): 899–907, 2000.*

Nowack, M. and Schuster, P., Error thresholds of replication in finite populations mutation frequencies and the Onset of Muller's Rutchet. *Journal of Theoritical Biology*, 137: 375–395, 1989.

Nowak, M.A. and May, R.M., Mathematical biology of HIV infections: antigenic variation and diversity threshold. *Mathematical Bioscience*, 106(1): 1–21, 1991.

Oderiz, J., Rubio, I., and Prez-Arriaga, Marginal pricing of transmission services: a comparative analysis of network cost allocation methods. *IEEE Transactions on Power Sytems*, 10(2): 1125–1142, 2000.

Ohtsuka, K.S., Yokokama, H.T., and Doi, H., A multivariable optimal control system for a generator. *IEEE Transactions on Energy Conversion*, EC-1(2): 88–98, 1986.

Olmos, L. and Neuhoff, K., Identifying a balancing point for electricity transmission contracts. *IEEE Transactions on Power Systems*, 21(1): 91–98, 2006.

Ongasakul, W. and Chayakulkheeree, K., Coordinated fuzzy constrained optimal power dispatch for bilateral contract, balancing electricity and ancillary services markets. *IEEE Transactions on Power Systems*, 21(2): 593–604, 2006.

Orero, S.O. and Irving, M.R., A genetic algorithm for generator scheduling in power systems. *International Journal on Electric Power and Energy Systems*, 18(1): 19–26, 1996.

Orero, S.O. and Irving, M.R., Genetic algorithms modelling framework and solution technique for short term for optimal hydrothermal scheduling. *IEEE Transactions on Power Systems*, 13(2): 501–518, 1998.

Ortmeyer, T.H., Hiyama T., "Frequency Response Characteristics of the Fuzzy Polar Power System Stabilizer", *IEEE Trans. On Energy Conversion*, Vol. 10, No. 2, pp. 333–338, June 1995.

Osman, M.S., Abo-Sinna, M.A., and Mousa, A.A., A solution to the optimal power flow using genetic algorithm. *Applied Mathematics and Computation*, 155(2): 391–405, 2004.

Osowski, S. and Siwek, K., Regularization of neural networks for improved load forecasting in power system, *IEE Proceedings – Generation Transmission and Distribution*, 149(3), 2002.

Ovaska, S.J., VanLandingham, H.F., and Kamiya, A., Fusion of soft computing and hard computing in industrial applications: An overview. *IEEE Transactions on Systems, Man, and Cybernetics – Part C: Applications and Reviews*, 32(2): 72–79, 2002.

Ozcalik, H.R. and Kucuktufekci, A., An efficient neural controller for a DC servo motor by using ANN and PLR identifiers. *IEEE International Conference on Artificial Intelligence Systems*, 224–227, 2002.

Pal, S.K., Soft data mining, Computational theory of perceptrons, and rough-fuzzy approach, Inf. Sci. 163(1–3): 5–12, 2004.

Pallet, E.H.J., *Automatic Flight Control*, 2nd Edn. Granada Publishing Limited, 1983.

Palm, R., Control of redundant manipulatorusing fuzzy rules, *Fuzzy Sets and Systems*, 45(3): 279–298, 1992.

Panksepp, J. *Affective Neuroscience*. Oxford University Press, Oxford, 1998.

Papadakis, S.E., et al., A novel approach to short term load forecasting using fuzzy neural network forecasting. *IEEE Transactions on Power System*, 13(2): 480–491, 1998.

Papalexopoulos, A.D., Hao, S., and Peng, T.-M., Short-term system load forecasting using an artificial neural network, *Proceedings of the Second International Forum on Applications of Neural Networks to Power Systems*, pp. 239–44, IEEE, New York, NY, USA, 1993.

Papalexopoulos, A.D., Shangyou Hao, and Tie-Mao Peng, An implementation of a neural network based load forecasting model for the EMS. *IEEE Transactions on Power Systems*, 9(4): 1956–1962, 1994.

Papineau, D., The antipathetic fallacy and the boundaries of consciousness. In Metzinger, T. (ed.) *Conscious Experience*. Ferdinand Schöningh, Paderborn, 1995.

Papineau, D., *Philosophical Naturalism*. Blackwell, Oxford, 1994.

Papineau, D., *Thinking About Consciousness*. Oxford University Press, 2002.

Paranjothi, S.R. and Anburaja, K., Optimal power flow using refined genetic algorithm. *Electric Power Components and Systems*, 30: 1055–1063, 2002.

Parikh, J., Bhattacharya, K., Reddy, S., and Parikh, K., 'Chapter 5 – energy system: need for new momentum. *India Development Report*. Oxford University Press, 1997.

Parikh, J. and Chattopadhyay, D., A multi-area linear approach for analysis of Economic Operation of the Indian Power System, 11(1): 52–58, 1995.

Park, D., Electric load forecasting using an artificial neural network. *IEEE Transactions on Power Systems*, 6: 442–449, 1991.

Park D., Kandel, A., and Langholz, G., Genetic based new fuzzy reasoning model with application to fuzzy controls, *IEEE Transactions on Systems, Man and Cybernetics*, 24(1): 39–47, 1994.

Parker, T.S. and Chau, L.O., Chaos: a tutorial for engineers. *Proceedings IEEE*, 75: 982–1008, 1987.

Parlos, A.G., Chong, K.T., and Atiya, A.F., Application of recurrent neural multilayer perceptron in modeling complex process dynamic. *IEEE Transactions on Neural Networks*, 5(2): 255, 1994.

Parniani M. and Lesani H., 'Application of Power System Stabilizer at Bandar-Ablias Power Station', *IEEE Trans. on Power Systems*, vol. 9, no. 3, pp. 1366–1370, Aug. 1994.

Passino, K. and Yurkovich, S., *Fuzzy Control*. Addison-Wesley, Menlo Park, CA, 1998.

Patidar, N.P. and Sharma, J., A hybrid decision tree model for fast voltage contingency screening and ranking. *International Journal of Emerging Electric Power Systems*, 8(4), Article 7, 2007.

Patrick, H.W., *Artificial Intelligence*, 3rd Edn. Addison-Wesley, Reading, MA, 1992.

Patterson, D.W., *Artificial Neural Networks – Theory and Practice*. Prentice-Hall, New York, 1995.

Peacocke, C., *Sense and Content*. Oxford: Oxford University Press, 1983.

Pearson, M.P., *The Archeology of Death and Burial*. College Station, Texas A&M Press, Texas, 1999.

Penfield, W., *The Mystery of the Mind: a Critical Study of Consciousness and the Human Brain*. Princeton University Press, Princeton, NJ, 1975.

Peng, T.M., et al., Advancement in the application of neural network for short term load forecasting. *IEEE Transactions on Power Systems*, 7(1): 250–257, 1992.

Peng, T.M., Hubele, N.F., and Karady, G.G., An adaptive neural network approach to one-week ahead load forecasting. *IEEE Transactions on Power Systems*, 8(3): 1195–1203, 1993.

Penrose, R., *Shadows of the Mind*. Oxford University Press, Oxford, 1994.

Penrose, R., *The Emperor's New Mind: Computers, Minds and the Laws of Physics*. Oxford University Press, 1989.

Pereira, B.L.P.P., Farmer, E.D., and Cory, B.J., Revenue reconciled optimum pricing of transmission services. *IEEE Transactions on Power System*, 11(3): 1419–1426, 1996.

Perry, J., *Knowledge, Possibility, and Consciousness*. MIT Press, Cambridge, MA, 2001.

Philpott Andy, B. and Pettersen, E., Optimizing demand-side bids in day-ahead electricity markets. *IEEE Transactions on Power Systems*, 21(2): 488–498, 2006.

Pierre, D.A., A perspective on adaptive control of power systems. *IEEE Transactions on Power Systems*, PWRS-2(2): 387–396, 1987.

Pin, F.G., Step towards sensor based vehicle navigation in outdoor environments using a fuzzy behaviourist approach. Journal of Intelligent and Fuzzy Systems, 1(2): 95–107, 1993.

Place, U.T., Is consciousness a brain process? *British Journal of Psychology*, 44–50, 1956.

Plazas Miguel, A., Conejo, A.J., and Prieto, F.J., Multimarket optimal bidding for a power Producer. *IEEE Transactions on Power Systems*, 20(4): 2041–2050, 2005.

Portilla, M.I., Burillo, P., and Eraso, M.L., Properties of the fuzzy composition based on aggregation operators. *Fuzzy Sets and Systems*, 110(2): 217–226, 2000.

Potter, D., Matthews, M., and Ali, M., Industrial and engineering applications of artificial intelligenceand expert system. Proceedings of 10th International Conference, Atlanta, Georgia, USA, 1997.

Putnam, H., Philosophy and our mental life. In Putnam, H. (ed.) *Mind Language and Reality: Philosophical Papers*, vol. 2. Cambridge University Press, Cambridge, 1975.

Putnam, H. and Oppenheim, P., Unity of science as a working hypothesis. In Fiegl, H., Maxwell, G., and Scriven, M. (eds.) *Minnesota Studies in the Philosophy of Science II*. University of Minnesota Press, Minneapolis, 1958.

Quek, C., and Zhou, R.W., The POP learning algorithms: reducing work in identifying fuzzy rules. *Neural Networks*, 14(10), 1431–1445, 2001.

Rabelo, L.C. and Avula, X., Hierarchical Neuo-controller architecture for robotic manipulation. *IEEE Control Systems Magazine* 12(2): 37, 1992.

Radivojevic, I., Herath, J., and Gray, S., High-performance DSP architectures for intelligence and control applications. *IEEE Control Systems*, 11(4): 49, 1991.

Radko, M., Mayor, G., and Calvo, T., Aggregation operators: new trends and applications. *Journal of Electrical Engineering* 12: 29–32, 2000.

Rahaman, S. and Bhatanagar, R., An expert systems based algorithm for short term load forecast. *IEEE Transactions on Power Systems*, 3(2): 392–399, 1988.

Rahaman S. and Hazim O., A generalized knowledge-based short term load forecasting technique. *IEEE Transactions on Power system*, 8: 508–514, 1993.

Rahaman, S. and Shrestha, A priory vector based technique for load forecasting. *IEEE Transactions on Power System*, 6: 1459–1464, 1993.

Raisinghani, S.C., Ghosh A.K., and Kalra, P.K., Two new techniques for aircraft parameter estimation using neural networks. *Journal of Aeronautical*, pp. 25–30, 1998.

Rajurkar, K.P. and Nissen, Data-dependent system approach to short-term load forecasting. *IEEE Transactions on Systems, Man & Cybernetics*, vol. SMC-15(4), 1985.

Ralescu, A. (ed.), *Applied Research in Fuzzy Technology*. Kluwer, Boston, 1994.

Ram, N., *The Transmission Dynamics of HIV/AIDS: Some Basic Models*. Narosa Publishing House, New Delhi, 2000.

Rauch, H.E. and Winarske, T., Neural networks for routing communication traffic. *IEEE Control Systems Magazine*, 8(2): 26, 1988.

Ravikumar, B. Dhadbanjan, T., and Khincha, H.P., Intelligent approach for fault diagnosis in power transmission systems using support vector machines. *International Journal of Emerging Electric Power Systems*: 8(4), Article 3, 2007.

Ray, G.D., Dey, S., and Bhattacharyya, T.K., Design of variable structure controller using fuzzy PI type sliding surface: an application to load-frequency control problem. *International Journal of Emerging Electric Power Systems*: 7(4), Article 5, 2006.

Read, E.G., Transmission Pricing in New Zealand. *Utilities Policy*, 6: 227–235, 1997.

Renaud, O., Starck, J.L., and Murtagh, F., Prediction based on a multiscale decomposition, *International Journal of Wavelets, Multiresolution and Information Processing*, 1: 217–223, 2003.

Rescher, N., *Many-valued logic*. McGraw Hill, New York, 1969.

Rey, G., A question about consciousness. In Otto, H. and Tuedio, J. (eds.) *Perspectives on Mind*. Kluwer, Dordrecht, 1986.

Rios, J.J.A., The liberalization of the electricity industry: reasons and alternative structures, the organization of wholesale electricity markets. *Proceedings of Liberalization and Modernization of Power System: Operation and Control problems Conference*, pp. 6–18, 2000.

Roa-Sepulveda, C.A. and Pavez-lazo, B.J., A solution to the optimal power flow using simulated annealing. *Electrical Power and Energy Systems*, 25: 47–57, 2003.

Robert, N.B., *Weather Flying*, Revised Edition. MacMillan, New York, 1978.

Robinson, D., Epiphenomenalism, laws, and properties. *Philosophical Studies*, 69: 1–34, 1993.

Robinson, H., *Matter and Sense: A Critique of Contemporary Materialism*. Cambridge University Press, Cambridge, 1982.

Rochester, N., Holland, J.H., Haibt, L.H., and Duda, W.L., Tests on cell assembly theory of the action of the brain, using a large digital computer. *IRE Transactions on Information Theory*, vol. IT-2, 80–93, 1956.

Rodrigues, J.J., Mehendale, S.M., Shepherd, M.E., Divekar, A.D., Gangakhedkar, R.R., Quinn, T.C., Paranjape, R.S., Risbud, A.R., Brookmeyer, R.S., Gadkari, D.A., et al. Risk factors for HIV infection in people attending clinics for sexually transmitted diseases in India. *British Medical Journal*, 311: 283–286, 1995.

Rogers, G.J., The application of power system stabilizers to a multigenerator plant", *IEEE Transactions on Power Systems*, 15(1): 350–355, 2000.

Ronald, R.Y. and Zadeh, L.A. (eds.). *Fuzzy Sets, Neural Networks, and Soft Computing*. Van Nostrand Reinhold, New York, 1994.

Rosenberg, G., *A Place for Consciousness: Probing the Deep Structure of the Natural World*. Oxford University Press, New York, 2004.

Rosenblatt, F., *Principles of Neurodynamics*. Spartan Press, Washington, DC, 1961.

Rosenblatt, F., The Perceptron: a probabilistic model for information storage and organization in the brain. *Psychological Review*, 65: 386–408, 1958.

Rosenthal, D., First person operationalism and mental taxonomy. *Philosophical Topics*, 22/1–2: 319–350, 1994.

Rosenthal, D., The independence of consciousness and sensory quality. In Villanueva, E. (ed.) *Consciousness*. Ridgeview Publishing, Atascadero, CA, 1991.

Rosenthal, D., Two concepts of consciousness. *Philosophical Studies*, 49: 329–359, 1986.

Rosenthal, D.M., A theory of consciousness. In Block, M., Flanagan,O., and Guzeldere, F. (eds.) *The Nature of Consciousness*. MIT Press, Cambridge, MA, 1997.

Rosenthal, D.M., Thinking that one thinks. In Davies, M. and Humphreys, G. (eds.) *Consciousness: Psychological and Philosophical Essays*. Blackwell, Oxford, 1993.

Ross T.J., *Fuzzy Logic with Engineering Applications*. McGraw Hill Inc., New York, 1995.

Rubio, F.O. and Prez-Arriage, I., Marginal pricing of transmission cost allocation methods. *IEEE Transactions on Power Sytems*, 15(0885–8950), 448–454, 2000.

Rudnick, H. and Raineri, R., Transmission pricing practices in South America. *Utilities Policy*, 6(3): 211–218, 1997.

Rudnick, H., Palma, R., Cura, E., and Silva, C., Economically adapted transmission systems in open access schemes-applications of genetic algorithms. *IEEE Transactions on Power Systems*, 11(3): 1427–1440, 1996.

Rumelhart, D.E., Hilton, G.E. and Willams, R.J., Learning internal representation by error propagation. *Parallel Distributed Processing*, vol. 1. MIT Press, Cambridge, pp. 318–362, 1986.

Rumelhart, D.E., Hinton, G.E., and Williams R.J., Learning internal representation by error propagation. *Parallel Distributed Processing*, vol. 1. MIT Press, Cambridge, MA, pp. 318–362, 1986.

Russel, S. and Norvig, P., *Artificial Intelligence: A Modern Approach*. Prentice Hall, Upper Saddle River, NJ, 1995.

Russell, B. *The Analysis of Matter*. Kegan Paul, London, 1927.

Rutkowski, L., *Artificial Intelligence and Soft Computing – ICAISC 2004: 7th International*. Springer, Berlin Heidelberg New York, 2004.

Ryle, G., *The Concept of Mind*. Hutchinson and Company, London, 1949.

Fukami, S., Mizumoto, M., and Tanaka, K., Some considerations on Fuzzy conditional inference. *Fuzzy Sets and Systems*, 4: 243–273, 1980.

Saadat, H., *Power System Analysis*. Tata McGraw-Hill Publishing Co. Ltd., Ed., 2002.

Saade, J.J. and Schwarzlander, H., Application of Fuzzy Hypothesis testing to signal detection under uncertainity. *Fuzzy Sets and Systems*, 62(1): 9–19, 1994.

Saini, A., Chaturvedi, D.K., Saxena, A.K., and Kalra, P.K., Transmission pricing model under deregulated environment. *IEEE Symposium on Technological Advances and IT Applications for Indian Power Sector*, Bangalore, 2006, pp. 1–10.

Saini, A., Chaturvedi, D.K., Saxena, A.K., Optimal load flow: a GA-fuzzy system approach. *International Journal of Emerging Electrical Power System Research*, 5(2): 1–21, 2006.

Sanchez, E., Fuzzy genetic algorithms in soft computing environment. *Proceedings V IFSA, World Congress, Seoul*, 1: XLIV-L, 1993.

Sankar, K.P., Mitra, S., Mitra, P., Rough-fuzzy MLP: modular evolution, rule generation, and evaluation. *IEEE Transactions on Knowledge and Data Engineering*, 15(1): 14–25, 2003.

Sankar, K.P., Soft data mining, computational theory of perceptions, and rough-fuzzy approach. *International Journal on Information Sciences*, 163(1–3): 5–12, 2004.

Sanner, R. and Akin, D., Neuromorphic pitch attitude regulation of an underwater Telerobot. *IEEE Control Systems Magazine*, 10(3): 62, 1990.

Sartori, M. and Antsaklis, P., Implementation of learning control systems using neural networks. *IEEE Control Systems Magazine*, 12(2): 49–57, 1992.

Sasson, A.M. and Merril, H.M., Some applications of optimization technique to power system problem. *Proceedings IEEE*, 62(7): 954–972, 1974.

Schacter, D., On the relation between memory and consciousness: dissociable interactions and consciousness." In H. Roediger and F. Craik eds. *Varieties of Memory and Consciousness*. Erlbaum, Hillsdale, NJ, 1989.

Schaffer, J.D., Caruna, R.A., Eshelman, L.J., and Das, R., A study of control parameters affecting online performance of genetic algorithms for function optimization. In Schaffer, J.D. (ed.), Proceedings of the Thrid International Conference Genetic Algorithms & Applications, Arlington, VA, pp. 51–60, 1989.

Schaffer, J.D., Caruna, R.A., Eshelman, L.J., and Das, R., A study of control parameters affecting on line performance of schalkoff R.J.. *Artificial Neural Networks.* McGraw Hill, New York, 1997.

Schneider, W. and Shiffrin, R., Controlled and automatic processing: detection, search and attention. *Psychological Review,* 84: 1–64, 1977.

Schuster, P., Effect of Finite Population Size and other Stochastic Phenomenon in Molecular Evolution., *Complex System Operational Approaches, Neurobiology, Physics and Computers.* Springer, Berlin, New York, Heidelberg, 1985.

Schweppe, F.C., Caramanis, M.C., Tabors, R.D., and Bohn, R.E., *Spot Pricing of Electricity.* Kluwer, Norwell, MA, 1988.

Seager, W., Consciousness, information, and panpsychism." *Journal of Consciousness Studies,* 2: 272–88, 1995.

Searle, J.R., Consciousness, explanatory inversion and cognitive science. *Behavioral and Brain Sciences,* 13: 585–642, 1990.

Searle, J. *The Rediscovery of the Mind.* MIT Press, Cambridge, MA, 1992.

Sedaghati, A., Cost of transmission system usage based on an economic measure. IEEE Transactions on Power Systems, 21(2): 466–473, 2006.

Segal, R., Kothari, M.L., and Madnani, S., Radial basis function (RBF) network adaptive power system stabilizer. *IEEE Transactions on Power Systems,* 15(2): 722–727, 2000.

Sevilla, S., Importance of input data normalization for the application ofneural networks to complex industrial problems. *IEEE Transactions on Nuclear Science,* 44(3): 1464–1468, 1997.

Shallice, T., *From Neuropsychology to Mental Structure.* Cambridge University Press, Cambridge, 1988.

Shantiswarup, K. and Satish, B., Integrated approach to forecast load. *IEEE Transactions on Computer Applications in Power Systems,* 46–49, 2002.

Sharma, K.L.S. and Mahalanabis, A.K., Recursive short term load forecasting algorithm. *Proceedings IEE* 121(1): 59–62, 1974.

Shear, J., *Explaining Consciousness: The Hard Problem.* MIT Press, Cambridge, MA, 1997.

Sheble, G.B. and Britigg, K., Refined genetic algorithm-economic dispatch example. *IEEE Transactions on Power Systems,* 10(1): 117–124, 1995.

Shi, Y., et al., Implementation of evolutionary fuzzy systems. *IEEE Transactions on Fuzzy Systems,* 7(2): 109–119, 1999.

Shigeo, A., *Neural Networks and Fuzzy Systems – Theory and Applications.* Kluwer, Norwell, MA, 1997.

Shirmohammadi, D. and Thomas, C.L., Valuation of the transmission impact in a resource bidding process. *IEEE Transactions on PWRS,* 6(2): 316–323, 1991.

Shirmohammadi, D., et al., Some fundamental, technical concepts about cost based transmission pricing. *IEEE Transactions on Power Systems,* 11(2): 1002–1008, 1996.

Shirmohammadi, D., An engineering perspective of transmission access and wheeling, *Proceedings of the Third International Symposium of Specialists in Electric Operational and Expansion Planning (SEPOPE III),* Belo-Horizonte, Brazil, 1992.

Shirmohammadi, D., Rajagopalam, C., Alward, E.R. and Thomas, C.L., Cost of transmission transactions: an introduction. *IEEE Transactions on Power Systems*, 6(4): 1546–1560, 1991.

Shirmohammadi, D., Wollenberg, B., Vojdani, A., Sandrin, P., Pereira, M., Rahimi F., Schneider, T., and Stott, B., Transmission dispatch and congestion management in emerging market structures. *IEEE Transactions on Power Systems*, 13(4): 1466–1474, 1998.

Shoemaker, S., Absent qualia are impossible. *Philosophical Review*, 90: 581–599, 1981.

Shoemaker, S., Functionalism and qualia. *Philosophical Studies*, 27: 291–315, 1975.

Shoemaker, S., Qualities and qualia: what's in the mind. *Philosophy and Phenomenological Research*, Supplement 50: 109–131, 1990.

Shoemaker, S., The inverted spectrum. *Journal of Philosophy*, 79: 357–381, 1982.

Shoemaker, S., Two cheers for representationalism. *Philosophy and Phenomenological Research*, 1998.

Short term Course on Artificial Neural Network Applications to Power Systems, IIT, Kanpur, June 14–19, E1–E34, 1993.

Siewert, C., *The Significance of Consciousness*. Princeton University Press, Princeton, NJ, 1998.

Silberstein, M., Converging on emergence: consciousness, causation and explanation. *Journal of Consciousness Studies*, 8: 61–98, 2001.

Silberstein, M., Emergence and the mind-body problem. *Journal of Consciousness Studies*, 5: 464–482, 1998.

Silva, E.L., Mesa, S.E.C., and Morozwski, M., Transmission access pricing to wheeling transactions: a reliability based approach. *IEEE Transactions on Power Systems*, 13(4): 1481–1486, 1998.

Simpson, P.K., *Artificial Neural Systems: Foundations, Paradigm, Applications and Implementation*. Pergamon Press, Elmsford, NY, 1990.

Simpson, P.K., Fuzzy min-max classification with neural networks. *Heuristics Journal of Knowledge Engineering*, 4: 1–9, 1991.

Singer, P., *Animal Liberation*. Avon Books, New York, 1975.

Singer, W., Neuronal synchrony: a versatile code for the definition of relations. *Neuron*, 24: 49–65, 1999.

Singh, H., Hao, S., and Papalexopoulos, A., Transmission Congestion Management in Competitive Electricity Markets. *IEEE Transactions on Power Systems*, 13(2): 672–680, 1998.

Singh, S.N. and David, K., Congestion management in dynamic security constrained open power markets. *Computers and Electrical Engineering*, 29: 575–588, 2003.

Singh, S.N., Chandramouli, A., Kalra, P.K., Srivastava, S.C., and Mishra, D.K., Optimal reactive power dispatch using genetic algorithms. *Proceedings of the International Symposium on Electricity Distribution and Energy Management*, Singapore, pp. 464–469, 1993.

Sinha, A.K., Short-term load forecasting using artificial neural networks, *Proceedings of IEEE* 0-7803-5812, pp. 548–552, 2000.

Sinha, M., Kumar, K., and Kalra, P.K., Some new neural network architectures with improved learning schemes. *Journal of Soft Computing*, 4(4), 2000.

Sinha, S., Power sector privatization: lessons of Orissa, *Workshop on Restructuring and Financing of Power Sector*, IIT Kanpur, 26–30, 2001.

Skinner, B.F., *Science and Human Behavior*. Macmillan, New York, 1953.

Smart, J., Sensations and brain processes. *Philosophical Review*, 68: 141–156, 1959.

Smith, J.E. and Fogarty, T.C., Operator and parameter adaptation in genetic algorithms. *Soft Computing*: 1, 81–87, 1997.

Smith, J.S., Evolving a better solution. *Developers Network Journal*, March 2002.

Song, Y.H. and Xuan, Q.Y., Combined heat and power economic dispatch using genetic algorithm based penalty function method. *Electric Machines and Power Systems*, 26: 363–372, 1998.

Sood Yog, R., Padhy, N.P., and Gupta, H.O., Wheeling of power under deregulated environment of power system – a bibliographical survey. *IEEE Transactions on Power Systems*, 17(3), 2002.

Srinivas A and Ramar K. "New Time Domain Method for Stable PSS Design with Output Feedback", *J. Electrical Machines and Power Systems* Vol. 13, No.2, PP. 123–133, 1981.

Srinivas, M. and Patnaik, L.M., Adaptive probabilities of crossover and mutation in genetic algorithms. *IEEE Transactions on Systems, Man and Cybernetics*, 24(4): 656–667, 1994.

Srinivas, M. and Patnaik, L.M., Genetic algorithms: a survey. *IEEE Computer*, 17–26, 1994a.

Srinivasa Rao, S.R., Mathematical modeling of AIDS epidemic in India. *Current Science*, 84(10), 2003.

Stanton, K.N. and Gupta, P.C., Forecasting annual or seasonal peak demand in electric utility system. *IEEE Transactions*, PAS-90: 591–599, 1971.

Stanton, K.N. and Gupta, P.C., Forecasting annual or seasonal peak demand in electric utility system. *IEEE Transactions*, PAS-90: 591–599, 1971.

Stapp, H., *Mind, Matter and Quantum Mechanics*. Springer, Berlin Heidelberg New York, 1993.

Steck, J.E., Rokhsaz, M., and Shue, S., Linear and neural network feedback for flight control decoupling. *IEEE Control Systems*, 16(4): 22, 1996.

Stefan, S., Steele, R., Dillon, T.S., and Chang, E., Fuzzy trust evaluation and credibility development in multi-agent systems. *Applied Soft Computing*, 7(2): 492–505, 2007.

Stoft S., *Power System Economics*. IEEE Press, Piscataway, pp. 6–16, 2002.

Stoljar, D., Two conceptions of the physical. *Philosophy and Phenomenological Research*, 62: 253–281, 2001.

Stott, B. and Hibson, E., Power system security control calculations using linear programming, Parts I and II. *IEEE Transactions on Power Apparatus and Systems*, PAS-97: 1713–1731, 1978.

Stott, B. and Marinho, J.L., Linear programming for power system network security applications. *IEEE Transactions on Power Apparatus and Systems*, 98(3): 837–848, 1979.

Stover, J., Walker, N., Garnett, G.P., Salomon, J.A., Stanecki, K.A., Ghys, P.D., Grassly, N.C., Anderson, R.M., and Schwartländer, B., (06/07/2002) '*Can we reverse the HIV/AIDS pandemic with an expanded response?*. Lancet 360(9326): 73–77 (issn: 0140–6736).

Strawson, G., *Mental Reality*. MIT Press, Cambridge, MA, 1994.

Styblinski M.A., B.D. Meyer, Signal flow graphs vs fuzzy cognitive maps in application to qualitative circuit analysis, International Journal of Man-Machine Studies, 15(2), 175–186, September 1991.

Sugeno, M. (Ed.), "Industrial Applications of Fuzzy Control", North Holland, New York, 1985.

Sugeno, M. and Kang, G.T., Structure identification of Fuzzy model. *Fuzzy Sets and Systems*, 28: 15–33, 1988.

Suh, Y.H. and Van, G., Incorporating heuristic information into genetic search. *Proceeding of Second Interenational Conference on Genetic Algorithms*, pp. 100–107. Lawrence Emlbaum Associates, 1987.

Sun, D.I., Ashley, B., Brewer, B., Hughes, A., and Tinney, W.F., Optimal power flow by Newton approach. *IEEE Transactions on Power Apparatus and Systems*, 103(10): 2864–2880, 1984.

Sun, J., Grosky, W.I., Hassoun, M.H., A fast algorithm for finding global minima of error functions in layered neural networks. International Joint Conference on Neural Networks, 1: 715–720, 1990.

Sundararajan, N. and Saratchandran, P., *Parallel Architectures for Artificial Neural Networks: Paradigms and Implementations*. Wiley-IEEE Computer Society Press, December 1998.

Surgeno, M. (ed.), *Industrial Applications of Fuzzy Control*. North-Holland, Amsterdam, 1985.

Suzuki, Y., (ed.), et. al., *Soft Computing in Industrial Applications*. Springer, Berlin, Heidelberg New York 2000.

Svenska, K., The Swedish electricity market reforms and its implications for Svenska Kraftnät", 2nd Edition, 1997.

Swartz, B.E., Timeline of the history of EEG and associated fields. *Electroencephalography and Clinical Neurophysiology*, 106: 173–176, 1998.

Swarup, K.S., Yoshimi, M., Shimano, S., and Izni, Y., Optimization methods using genetic algorithms for reactive power planning in power systems, *Proceedings of 12th PSCC*, Dresden, Germeny, vol. 1, pp. 483–491, 1996.

Swidenbank, E., McLoone, S., Flym, D., Irwin, G.W., Brown, M.D., and Hogg, B.W., Neural network based control for synchronous generators. *IEEE Transactions on Energy Conversion*, 14(4): 1673–1679, 1999.

Swinburne, R., *The Evolution of the Soul*. Oxford University Press, Oxford, 1986.

Tabors, R.D., Transmission system management and pricing. *IEEE Transactions on Power Systems*, 9(1): 206–214, 1994.

Tagaki, T. and Sugeno, M., Fuzzy identification of systems and its applications to modeling and control. *IEEE Transations on Systems, Man, and Cybernetics*, 15: 116–132, 1985.

Takagi, and Hayashi, I., NN driven fuzzy reasoning, *Interantional Journal of Approximate Reasoning*, 5(3): 91–212, 1991.

Tao, S. and Gross, G., A congestion management allocation mechanism for multiple transactions networks. *IEEE Transactrions on Power Systems*, 17(3): 826–833, Feb. 2002.

Taraore, C.O., Torres, G.L., Lagoce, P.J., and Mukhedkar, D., Applications of neural networks for load forecasting. *Sixth National Power System Conference, Bombay, TMH*, pp. 417–422, 1990.

Tasos, F., The impact of the Error Functions in neural network based classifiers. *Proceedings of IEEE International Joint Conference on Neural Networks*, pp. 1799–1804, 1999.

Taylor, E.T. and Parmar, H.A., *Ground Studies for pilots*. Crosby Lockwood & Son Ltd., London, 1983.

Taylor, W.K., Electrical simulation of some nervous system functional activities. *Information Ttheory*, vol. 3. Butterworths, London, pp. 314–328, 1956.

Terano, T., Sugeno, M., and Tsukamoto, Y., Ning in management by fuzzy dynamic programming. In Sanchez (ed.) pp. 381–386, 1984.

Thukaram, D. and Yesuratnam, G., Fuzzy expert approach with curtailed controllers for improved reactive power dispatch. *International Journal of Emerging Electric Power Systems*: 8(3), Article 7, 2007.

Timothy, M., *Practical neural network recipe in C++*, Academic Press, Inc. 1993.

Tolat, V., An adaptive broom balancer with visual inputs. *Proceedings of IEEE International Conference on Neural Networks*, p. 641, 1998.

Travis, C., The silence of the senses. *Mind*, 113: 57–94, 2004.

Triesman, A. and Gelade, G., A feature integration theory of attention. *Cognitive Psychology*, 12: 97–136, 1980.

Trillas, E.C., Alsina, and Valverda, L. On some logical connectives for fuzzy set theory. *Journal of Methematical Analysis and its Applications*, 93: 15–26, 1983.

Tsukamoto, Y. and Iyoda, I. Allocation of fixed transmission cost to wheeling transactions by co-operative game theory. *IEEE Transactions on Power Systems*, 11(2): 1427–1440, 1996.

Tye, M., Blurry images, double vision and other oddities: new troubles for representationalism? In Jokic, A., and Smith, A. (eds.). *Consciousness: New Philosophical Perspectives*. Oxford University Press, Oxford, 2003.

Tye, M., *Consciousness, Color, and Content*. MIT Press, Cambridge, MA, 2000.

Tye, M., *Ten Problems of Consciousness*. MIT Press, Cambridge, MA, 1995.

Van Gulick, R., Dennett, drafts and phenomenal realism. *Philosophical Topics*, 22/1–2: 443–456, 1994.

Van Gulick, R., Higher-order global states HOGS: an alternative higher-order model of consciousness. In Gennaro, R. (ed.) *Higher-Order Theories of Consciousness*. John Benjamins, Amsterdam and Philadelphia, 2004.

Van Gulick, R., Inward and upward: reflection, introspection and self-awareness. *Philosophical Topics*, 28: 275–305, 2000.

Van Gulick, R., Maps, gaps and traps. In Jokic, A. and Smith, A. (eds.) *Consciousness: New Philosophical Perspectives*. Oxford University Press, Oxford, 2003.

Van Gulick, R., Nonreductive materialism and intertheoretical constraint. In Beckermann, A., Flohr, H., and Kim, H. (eds.) *Emergence and Reduction*. De Gruyter, Berlin and New York, 157–179, 1992.

Van Gulick, R., Physicalism and the subjectivity of the mental. *Philosophical Topics*, 13: 51–70, 1985.

Van Gulick, R., What would count as explaining consciousness? In Metzinger, T. (ed.) *Conscious Experience*. Ferdinand Schöningh, Paderborn, 1995.

Vanhowe, R.S., Circumcision and HIV infection: review of the literature and meta analysis. *Journal of STD& AIDS*, 1999.

Varela, F., Neurophenomenology: a methodological remedy for the hard problem. *Journal of Consciousness Studies*, 3: 330–349, 1995.

Varela, F. and Maturana, H., *Cognition and Autopoiesis*. D. Reidel, Dordrecht, 1980.

Varela, F. and Thomson, E., Neural synchronicity and the unity of mind: a neurophenomenological perspective." In Cleermans, A. ed. *The Unity of Consciousness: Binding, Integration, and Dissociation*. Oxford University Press, Oxford, 2003.

Varian, H.R., *Intermediate Microeconomics*, 3rd Edn. W.W. Norton and Company, New York, 1990.

Velmans, M., How could conscious experiences affect brains? *Journal of Consciousness Studies*, 9: 3–29, 2003.

Velmans, M., Is human information processing conscious? *Behavioral and Brain Sciences*, 14/4: 651–668, 1991.

Verma, K.S. and Gupta, H.O., Impact on real and reactive power pricing in open power market using unified power flow controller. *IEEE Transactions on Power Systems*, 21(1): 365–371, 2006.

Verma, K.S., Singh, S.N., and Gupta, H.O., Location of unified power flow controller for Congestion Management. *Electric Power Systems Research*, 58: 89–96, 2001.

Vipnik, V.N., *Statistical Learning Theory*. Wiley, New York, 1998.

Vitela, J.E. and Reifman, J., Premature saturation in backpropagation networks mechanism and necessary conditions. *Neural Network*, 10(4): 721, 1997.

Vojdani, A.F., Imparato, C.F., Saini, N.K., Wollenberg, B.F., and Happ H.H., Transmission access issues. *IEEE Transactions on Power Systems*, 11(1): 41–51, 1996.

Vries L. de, Capacity allocation in a restructured electricity market: technical and economic evaluation of congestion management methods on interconnectors, In *Proceedings of the Power Tech. Proceedings, IEEE Porto*, 1: 6–11, 2001.

Wakefield, R.A., Graves, J.S., and Vojdani, A.F., A transmission services costing framework. *IEEE Transactions on Power Systems*, 12(2): 549–558, 1997.

Walters, D.C. and Sheble, G.B., Genetic algorithm solution of economic dispatch with valve point loading. *IEEE Transactions on Power Systems*, 8(3): 1325–1332, 1993.

Wang, L.X. and Mandel, J.M., Fuzzy adaptive filters, with non-linear channel equalization. *IEEE Transactions on Fuzzy Systems*, 1(3): 161–170, 1993.

Wang, L.X. and Mandel, J.M., Generating fuzzy by learning through examples. *IEEE Transactions on Systems Man and Cybernetics*, 22(6): 1414–1427, 1992.

Wang, L.X., *Adaptive fuzzy systems and control: design and stability analysis*. Prentice Hall, Englewood Cliffs, NJ, 1994.

Wang, L.X., Stable adaptive fuzzy control to non-linear systems. *IEEE Transactions on Fuzzy Systems*, 1(2): 146–155, 1993.

Wang, L., *Soft Computing in Communications*. Springer, Berlin Heidelberg, New York, 2003.

Wang, S., Generating fuzzy membership functions: A monotonic neural network model. *Fuzzy Sets and Systems*, 61(1): 71–82, 1994.

Wang H.F., Swift. F.J., Hogg. B.J., Chen and Hang. G "Rule based variable gains PSS" *IEE Proc. Part-C*, Vol. 142–1, PP. 29–32, 1995.

Wang, Z., Generating pseudo-t-norms and implication operators. *Fuzzy Sets and Systems*, 157(3): 398–410, 2006.

Wasserman Philip, D., *Neural Computing: Theory and Practice*. Van Nostrand Reinhold, New York, 1989.

Watson, J., *Behaviorism*. W. W. Norton, New York, 1924.

Weaver, W., Science and Complexity. *American Scientist*, 36: 536–544, 1948.

Webb, J., *Fly The Wing*. The Iowa State Univ. Press, Ames, Iowa, USA, 1971.

Webb, A.R., Functional approximation by feed forward network: a least square approach to generalization. *IEEE Transactions on Neural Networks*, 5: 363–371, 1994.

Weber, S., Measures of fuzzy sets and measure of fuzziness. *Fuzzy sets and Systems*, 13: 247–271, 1984.

Weedy, B.M. and Cory, B.J., Electric Power Systems: Chapter 12, Basic Power-System Economics and Management, 4th Ed.. Wiley, New York, 2002.

Wegner, D., *The Illusion of Conscious Will*. MIT Press, Cambridge, MA, 2002.

Wei, Q., Dayawansa, W.P., and Levine, W.S., Nonlinear controller of an inverted pendulum having restricted travel. *Automatica* 31: 841–850, 1995.

Welstead, S.T., *Neural Networks and Fuzzy Applications in C/C++*. Wiley, New York, 1994.

Wen F., Chang, C.S., and Fu, W., New approach to alarm processing in power systems based on the set covering theory and a refined genetic algorithm. *Electric Machine and Power System*, 26, 53–67, 1998.

Wen, F., Chang, C.S., and Fu, W., New approach to alarm processing in power systems based on the set covering theory and a refined genetic algorithm. *Electric Machine and Power System*, 26: 53–67, 1998.

Werbos, P.J., Backpropagation through time: What it does and How to do it [Special issue on Neural Networks]. *Proceedings of IEEE*, 2: 1550–1560, 1990.

Werbos, P.J., *The roots of backpropagation: From Ordered Derivatives to Neural Networks and Political Forecasting*. Wiley, New York, 1994.

Werbos, P.J., *Beyond regression: New tools for prediction and analysis in the behavioural sciences*. PhD Thesis, Harvard University, Cambridge, MA, 1974.

Wernerus, J., A model for studies related to deregulation of an electricity market. Licentiate Thesis, Kungl Tekniska Högskolan, 1995.

Widrow, B. and Hoff, M.E., Jr., Adaptive Switching Circuits', IRE Western Electric Show and Convention Record, Part 4, 96–104 (Aug. 23), 1960.

Widrow, B., Generalization and information storage in networks of Adaline 'Neurons', in SelfOrganizing Systems (M.C. Yovitz, G.T. Jacobi and G. Goldstein, eds.) Washington, DC: Spartan Books, 435–461, 1962.

Widrow, B. and Lehr, M.A., 30 years of adaptive neural networks: perceptrons, madaline, and backpropagation. *Proceedings of IEEE*, 78(9): 1415–1442, 1990.

Wilkes, K.V., Is consciousness important? *British Journal for the Philosophy of Science*, 35: 223–243, 1984.

Wilkes, K.V., Losing consciousness. In Metzinger (ed.) *Conscious Experience*. Ferdinand Schöningh, Paderborn, 1995.

Wolak, F.A. and Patrick, R.H., Industry structure and regulation in the England and Wales Electricity market. *Pricing and Regulatory Innovations Under Increasing Competition*, 3rd Edn. Kluwer, Boston, pp. 65–90, 1999.

Woolenberg, B.F. and Wood, A.J., *Power Generation, Operation and Control*. Wiley, New York, 1996.

Wright, R., Can Machines Think? *Time*, 147(13), March 1996.

Xin Yao, A review of evolutionary artificial neural networks. *International Journal of Intelligence Systems*, 8: 539–567, 1993.

Xu, H.Y. and Vukovich, G., A fuzzy genetic algorithms with effective search and optimization. *Proceedings of the Joint International Conference on Neural Network*, 2967–2970, 1993.

Yablo, S., Concepts and consciousness. *Philosophy and Phenomenological Research*, 59: 455–463, 1998.

Yagar, R., Ovchinnikov, S., Tong, R.M., and Nguyen, H.T., *Fuzzy Sets and Applications*. Wiley, New York, 1987.

Yager, R.R. and Kelman, A., Fusion of fuzzy information with considerations for compatibility, partial aggregation, and reinforcement. *International Journal of Approximate Reasoning*, 15(2): 93–122, 1996.

Yager, R.R., Concepts, theory and techniques, A new methodology for ordinal multi-objective decision based on fuzzy sets. *Decision Science*, 12: 589–600, 1981.

Yager, R.R., Connectives and quantifiers in fuzzy sets. *Fuzzy Sets and Systems*, 40: 39–75, 1991.

Yager, R.R., Multiple objective decision making using fuzzy sets. *International Journal of Man–Machine Studies*, 9: 375–382, 1977.

Yager, R.R., On ordered weighted averaging aggregation operators in multicriteria decision making. *IEEE transactions on Systems, Man and Cybernetics*, 18(1): 183–190, 1988.

Yager, R.R. and Filev, D.P., *Essentials of Fuzzy Modelling and Control*. Wiley, New York, 2002.

Yager, R.R. and Filev, D.P., SLIDE: a simple adaptive defuzzification method. *IEEE Transactions on Fuzzy Systems*, 1(1): 69–78, 1993.

Yager, R. and Zadeh, L.A. (eds.), *An introduction to fuzzy logic applications in intelligent systems*. Kluwer, Boston, 1992.

Yagishita, O., Itoh, O., and Sugeno, M., Application of Fuzzy reasoning to the water purification process. In Industrial Applications of Fuzzy Control, M. Sugeno Ed. North-Holland, Amsterfdam, pp. 19–40, 1985.

Yamakawa, T., A simple fuzzy computer hardware system employing min and max operations – A challenge to 6th generation computer", in *Proceedings of the 2nd IFSA Congress, Tokyo, Japan*, July 1987.

Yamakawa, T., Fuzzy controller hardware system. In Proceedings Second IFSA Congress, Tokyo, Japan, July 1987.

Yamakawa, T., High speed fuzzy controller hardware system. *in Proceedings. 2nd Fuzzy System Sympo*, Japan, pp. 122–130, 1986.

Yamakita, M., Nonaka, K., Sugahara, Y., and Ad Furita, K., Robust state transfer control of double pendulum. *IFAC Advances in Control Education, Tokyo, Japan*, pp. 205–208, 1994.

Yang, S., *Evolutionary Computation in Dynamic and Uncertain Environments*. Springer, Berlin Heidelberg New York, 2007.

Yao, X., Evolving artificial neural networks. *Proceedings of IEEE*, 87(9): 423–1447, 1999.

Yaochu, J. and Branke, J., Evolutionary optimization in uncertain environments. *IEEE Transactions on Evolutionary Computation*, 9(3): 303–317, 2005.

Yaochu, J. and Rasheed, K., Fitness approximation in evolutionary computation, *A Tutorial on Genetic and Evolutionary Computation Conference*, June 26, Washington DC, 2005.

Yaochu J., (ed.) *Multi-Objective Machine Learning*. Springer, Berlin, Heidelberg, 2006.

Yaochu, J., *Advanced Fuzzy Systems Design and Applications*. Physica/Springer, Heidelberg, 2003.

Yaochu, J., Evolutionary computation in dynamic and uncertain environments, *CEC'04, Portland*, July 2004.

Yaochu, J., Fitness approximation in evolutionary computation. *Soft Computing*, 9(1): 3–12, 2005.

Yasunobu S. and Miyamoto T.S., Automatic train operation by predictive fuzzy control. In Industrial Applications of Fuzzy Control. Sugeno, M. (ed). North-Holland, Amsterdam, pp. 1–18, 1985.

Yasunobu, S., Miyamoto, S., and Ihara, H., Fuzzy control for automatic train operation system. In *Proceedings of the 4th IFAC/IFIP/IFORS International Congress on Control in Transportation Systems*, Baden-Baden, April 1983.

Yasunobu, S., Sekino, S., and Hasegawa T., Automatic train operation and automatic crane operation system based on predictive fuzzy control", in *Proceedings of 2nd IFSA Congress Tokyo, Japan*, pp. 835–838, July 1987.

Ying, M., Implication operators in fuzzy logic, *IEEE Transactions on Fuzzy Systems*, 10(1): 88–91, Feb 2002.

Yongcai, X., Masami, I., and Katsuhisa, F., Time Optimal Swing – Up Control of Single Pendulum. *Journal of Dynamic Systems, Measurement and Control on Transactions of the ASME*, 123, 2001.

Yongjun, R., and Francisco D.G., Pay-as-bid versus marginal pricing – part I: strategic generator offers. *IEEE Transactions on Power Systems*, 19(4): 1771–1776, 2004.

Yousef, H. and Khalil, H.M., Fuzzy logic based control of series D.C. motor drives, *Proceedings of the IEEE International Symposium on Industrial Electronics, Athens, Greece*, 1995.

Yu, C.W. and David, A.K., Pricing transmission services in the context of industry deregulation. *IEEE Transactions on Power Systems*, 12(1): 503–510, 1997.

Yu, C.W., Long-run marginal cost based pricing of interconnected system wheeling. *Electric Power Systems Research*, 50: 205–212, 1999.

Yu Yao-Nan Li Quing & Hua "Pole Placement PSS Design of an Unstable Nine Machine System" *IEEE Trans on Power System* Vol-5, No. 2, PP. 353–358, 1990.

Yuryevich, J. and Wong, K.P., Evolutionary programming based optimal power flow algorithm. *IEEE Transactions on Power Systems*, 14(4): 1245–1250, 1999.

Zadeh, L.A., A rationale for fuzzy control. *Transactions of ASME, Journal Dynam. Syst. Measure. Control*, 92: 3–4, 1972.

Zadeh, L.A., Fuzzy algorithm, *Informat. Control*, 12: 94–102, 1968.

Zadeh, L.A., Toward a theory of Fuzzy Systems. In Kalman, R.E. and Declarks, N., (ed.) *Aspects of Network and System Theory*. Holt Rinehart and Winston, New York, pp. 469–490, 1971.

Zadeh, L.A., Outline of a new approach to the analysis of complex systems and decision processes. *IEEE Transactions on Systems, Man and Cybernetics*, 3: 28–44, 1973.

Zadeh, L.A., The concept of a linguistic variable and its application to approximate reasoning, Part 1,2 and 3, *Information Science*, 8: 199–249, 8: 301–357, 9: 43–80, 1975.

Zadeh Lotfi, A., From computing with numbers to computing with words – from manipulation of measurements to manipulation of perceptions (abstract). *IEEE Transactions on Circuits and Systems – I: Fundamental Theory and Applications*, 45(1): 105–119, 1999.

Zadeh, L.A., A critical view of our research in automatic control. *IRE Transactions on Automatic Controls*, AC-7, 74, 1962.

Zadeh, L.A., A rationale for fuzzy control. *Journal of Dynamic System, Meas. and Control*, 94, Series G, 3, 1972.

Zadeh, L.A., Fuzzy Sets, *Information and Control*, 8, 338, 1965.

Zadeh, L.A., Making the computers think like people. *IEEE Spectrum*, 1994.

Zadeh, L.A., The evolution of systems analysis and control: a personal perspective, *IEEE Control Magazine*, 16(3): 95, 1996.

Zeng, Y. and Trauth, K.M., Internet-based fuzzy multicriteria decision support system for planning integrated solid waste management. *Journal Of Environmental Informatics*, 6(1): 1–15, 2005.

Zeynelgil, H.L., Demiroren, A., Sengor, N.S., The Application of ANN Technique to Automatic generation Control for Multi Area Power System. *Electric Power and Energy Systems*, 24: 345–354, 2002.

Zhang, Y., Chen, G.P., Malik, O.P., and Hope, G.S., A Multi-input power system stabilizer based on artificial neural network. *Proceedings, IEEE WESCANEX 93*, May 17–18, Saskatoon, Canada, 240–246, 1993.

Zhang, Y., Chen, G.P., Malik, O.P., and Hope, G.S., An artificial neural network based adaptive power system stabilizer. *IEEE Transactions on Energy Conversion*, EC-8(1): 71–77, 1993.

Zhang, Y., Malik, O.P. and Chen, G.P., Artificial neural power system stabilizers in multi-machine environment. *IEEE Transactions on Energy Conversion*, 10(1): 147–155, 1995.

Zhang, Y., Malik, O.P., Hope, G.S., and Chen, G.P., Application of an inverse input/output mapped ANN as a power system stabilizer. *IEEE Transactions on Energy Conversion*, 9(3): 433–441, 1994.

Zhang, Y., Sen, P., and Hearn, G., An on-line trained adaptive neural controller. *IEEE Control Systems*. 15(5): 67–75, 1995.

Zheng, T., Wang, X., Hiroki, T., Masahiro, I., An algorithm of supervised learning for multilayer neural networks. *Neural Computation*, 15(5): 1125–1142, 2003.

Zhou, J. and Civco, D.L. Using genetic learning neural networks for spatial decision making in GIS. *Photogrammetric Engineering and Remote Sensing*, 62(11): 1287–1295, 1996.

Zhu, J.Z., Chang, C.S., and Xu, G.Y., A new model and algorithm of secure and economic automatic generation control. Electric Power System Research System, 45: 119–127, 1998.

Zimmermann, H., *Fuzzy Set Theory and its Applications*. Kluwer, Boston, 1991.

Zobian, A. and Illic, M.D., Unbundling of transmission and ancillary services Part-II: cost-based pricing framework. IEEE Transactions on Power Systems, 12(2): 549–558, 1997.

Zurada, J., *Introduction to Artificial Neural Systems*. West Publishing Co., St. Paul, MN, 1992.

Glossary

I. Artificial Neural Network

Activation/threshold function	It is a function which controls the neuron output.
Artificial neuron	Mimic the behaviour of biological neuron with the help of electronic circuit.
Axon	output channel.
Back propagation	ANN models where error at output layer propagated back to modify the weights.
Bias	It is connection strength for a fixed input.
Biological neuron	The tiny processing cell in the human brain.
Cell Body	Accumulator (with threshold function)
Cerebrum	It is the most complex part of human brain, comprising of sheet of neurons (arranged into layers) that folds at the gyri and sulci. It localized the important functions of human body.
Dendrites	Input receptors in neuron.
Epoch/Iteration	A cycle of processing in a neural network, which contains, forward calculation for determining Neural output as well as backward calculation to update the weights.
Global Minima	There is no other value of x in the domain of the function f, where the value of the function is smallest.
Hidden Layer	An array of neurons positioned between the input and output layers.
Input layer	An array of neurons to which an external input or signal is presented.
Local minima	During learning of ANN the network could not reach to its absolute minima is called local minima.
Noise	A distortion of an input.

Output Layer	An array of neurons to which output from the network is taken.
Percetron	A neural network for linear pattern matching. It is a single neuron of feedforward binary threshold type.
Supervised learning	A learning process where the out for the given input is known and used for modifying the weights. In this learning examples are used.
Synapses	Communication between neurons takes place through specialized contact points between neurons called synapses.
Training	The process of changing weights or rather refining weights is called learning/training.
Unsupervised Learning	Learning in the absence of external information on outputs.
Weight	It is connection strength between different neuron situated at different layers.
Working Memory	A component or a place of computer system where the intermediate results of intelligent system are temporarily stored.

II. Fuzzy Systems

Artificial Intelligence	The discipline devoted to producing systems that perform tasks which would require 'intelligence' if performed by a human being.
Automatic knowledge Acquisition	A branch of machine learning devoted to explicating the principles of the induction of rule bases.
Backtracking	The process of backing up through a series of inferences in the face of unacceptable results.
Backward chaining	An inference mechanism which works from a goal and attempts to satisfy a set of initial conditions. Also referred to as goal – directed chaining.
Cognition	A intelligent process by which knowledge is gained about perceptions or ideas.
Convex Fuzzy Set	Fuzzy sets whose α cuts are crisp sets for all $\alpha \in [0, 1]$.
Data base system	A system which marries the properties of database systems with the properties of expert systems.
Domain	A bounded area of knowledge. A pool of values used to define columns of a relation.
Domain Expert	The person who provides the expertise on which a knowledge base is modified.
Equivalence relation	Relation that is reflexive, symmetric, and transitive.

Expert System	A computer system that achieves high levels of performance in areas that for human beings require large amounts of expertise.
Expertise	A set of capabilities underlying skilled performance in some task area.
Fact	A relationship between objects.
Forward Chaining	An inference mechanism which works from a set of initial conditions to a goal. Also referred to as a data – directed chaining. Making inferences by matching the condition sides of the IF-THEN rules to the facts at hand. It works from facts to conclusions, it also called antecedent mode, event-driven mode, or data driven mode of inference.
Fuzzy aggregation operator	It is an operator which aggregated two or more fuzzy sets. $h[0, 0, 0, \ldots 0] = 0$ and $h[1, 1, 1 \ldots, 1] = 1$. $H[0, 1]n \rightarrow [0, 1]$, where $n \geq 2$ and h is a continuous monotonically increasing function.
Fuzzy Systems	A system whose variables (Fuzzy) values are linguistic terms.
Heuristic	A rule of thumb. A mechanism with no guarantee of success.
Inference Engine	That part of a knowledge base system which makes inferences from the knowledge base.
Inference	The process of generating conclusions from conditions or new facts from known facts.
Information	It is the interpreted data. Information is data with attributed meaning in context.
Knowledge	It is derived from information by integrating information with existing knowledge. The same data may be interpreted differently by different people depending on their existing knowledge.
Knowledge base System	A system containing knowledge which can perform tasks that require intelligence if done by human beings.
Knowledge base	An artificial intelligence data base that is made up not merely of files of uniform content, but of facts, inferences, and procedures corresponding to the type of information needed for problem solution.
Knowledge Engineering	A person, analogous to the system analyst in traditional computing, who builds a knowledge base system.
Knowledge representation	The process of mapping the knowledge of some domain into a computational medium.
Knowledge source	Any source for knowledge – documentats, manuals, tape recording etc.

Natural Language processing	Processing of natural language (English, for example) by a computer to facilitate human communication with the computer – or the other purposes, such as language translation.
Parallel Processing	Simultaneous processing, as opposed to the sequential processing in a conventional (Von-Neumann) type of computer architecture.
Production rules	An IF-THEN rule having a set of conditions and a set of consequent conclusions.
Rule	A mechanism for generating new facts.
Syllogism	A deductive argument in logic whose conclusion is supported by two premises.

III. Genetic Algorithms

Allele	One of a pair or series of alternative genes that occur at a given locus in chromosome: one constraining form of genes (bit value or feature value).
Carriers	A heterozygous individual with recessive and dominant both alleles in allelic pair.
Chromosome	Microscopically observable thread like structures which are the main carriers of hereditary information (coded string).
Crossover	A genetic process which results gene exchange by combining the different chromosome selected from previous generation (parents).
Deoxyribonucleic acid (DNA)	A chemical known as genetic material from which the genes are composed.
Diploid	An organism or cell having a set of two genome.
Dominance	Applied one member of an allelic pair that has the ability to manifest itself at the exclusion of the expression of the other alleles.
Fitness	Survivability of a living being in a particular environment. It is the objective function value.
Gametes	They are reproductive cells.
Gene	A hereditary determiner specifying a biological function; a unit of inheritance located in a fixed place on chromosome. It has feature, character or detector.
Genome	A complete set of chromosomes inherited as a unit from one parent. It is the complete string of all variables.
Genotype	Actual gene constitution for a trait (String structure).
Heterozygous	An organism carrying unlike alleles in allelic pair.

Homozygous	An organism carrying same alleles in allelic pair.
Lethals	An allele which has an influence on viability of an organism in such a way that the organism is unable to live, known as lethal gene. (String which disappears under specified conditions).
Mutation	Sudden change in genetic material or gene in chromosome. It is just flipping a bit within a string.
Phenotype	Visible expression of traits. In GA it is the set of parameters or a decoded string.

A

Power System Model and its Parameters

A.1 Single Machine Infinite Bus System

1. The generating unit is modeled by seven first order non-linear differential equations for power system stabilizer application as given below:

$$\frac{d\delta}{dt} = \omega_0(\omega - 1), \quad \frac{d\omega}{dt} = \frac{1}{2H}(T_m - T_e + k_d{}^*\frac{d\delta}{dt} + g)$$

$$\frac{d\lambda_d}{dt} = e_d + r_a{}^*i_d + \omega_0\omega\lambda_q, \quad \frac{d\lambda_q}{dt} = e_q + r_a{}^*i_q + \omega_0\omega\lambda_d,$$

$$\frac{d\lambda_f}{dt} = e_f - r_f{}^*i_f, \quad \frac{d\lambda_{kd}}{dt} = -r_{kd}{}^*i_{kd}, \quad \frac{d\lambda_{kq}}{dt} = -r_{kq}{}^*i_{kq}$$

2. The AVR and exciter used in the system have the transfer functions, respectively:

$$v_a = \frac{k_a}{(1 + sT_a)}(v_{ref} - v_t + u_{pss}), \quad v_f = \frac{k_e}{(1 + s * TT_e)}v_a$$

3. The governor used in the system has the transfer function

$$g = \frac{T_{ref} + k_g{}^*\omega}{(1 + s^*T_g)}, \quad T_m = \frac{g}{(1 + s * T_s)}$$

5. Parameters used in the simulation studies are given below:

5.1 Machine parameters

$$r_a = 0.00528, \ r_f = 0.00116, \ r_{kd} = 0.0179, \ r_{kq} = 0.0179,$$
$$x_{md} = 1.74, \ x_{mq} = 1.65, x_f = 0.16, \ x_{kd} = 0.09, \ x_{kq} = 0.146,$$
$$H = 5.83, \ k_d = 0.027$$

5.2 Governor Parameters $\quad T_g = 0.1, \ T_s = 0.3, \ k_g = 0.0796$

5.3 AVR and Exciter Parameters $k_a = 0.001$; $T_a = 0.01$;
$k_e = 5.56$; $TT_e = 0.01$;

5.4 CPSS parameters   T1=1.14, T2=0.02, Tw=2.5, $K_\delta = 0.03$;

5.5 GNN based PSS Parameters
Learning Rate $\eta = 0.05$,   Gain Scale Factor $\lambda = 1.0$ Momentum factor $\alpha = 0.9$

5.6 Transmission Line Parameter - $R_t = 0.06$, $X_t = 0.25$

5.7 Transformer Parameter - $R_{tr} = 0.008$, $X_{tr} = 0.10$
All resistance and reactance are in per unit and time-constants in seconds.

A.2 Multimachine Power System

1. For a Multi-machine Power System Generator Parameters on a 100 MVA base:
for small generators #3 and #5:
xd = 1.026,   xq = 0.6580,   xd' = 0.3390,   xd" = 0.2690,   xq" = 0.3350,
H = 10;
for big generators #1, #2 and #4:
xd = 0.1026,   xq = 0.0658,   xd' = 0.0339,   xd" = 0.0269,   xq" = 0.0335,
H = 80;
Time constant for all generators: Tdo'=5.67, Tdo"=0.614,
Tqo"=0.723.

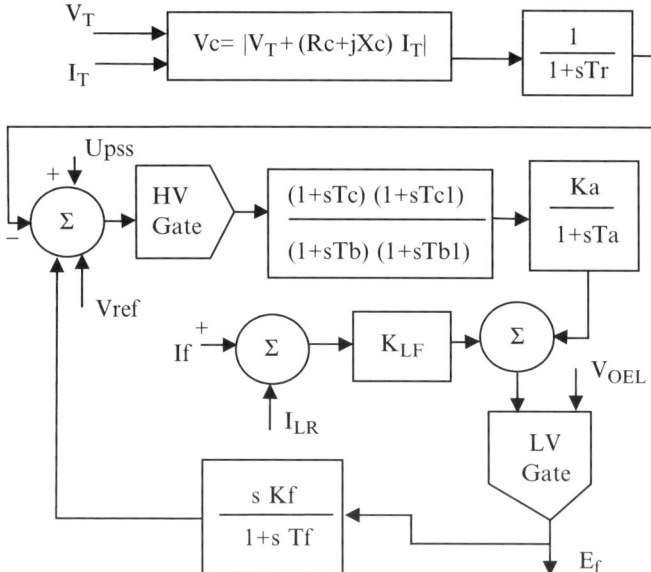

Fig. A.1. Schematic diagram of AVR and exciter model

2. Simplified IEEE standard type ST1A AVR and exciter model is shown in Fig. A1 and its parameters are:
Rc = 0.0, Xc = 0.0, Tr = 0.04, Ka = 190.0, Tf = 1.0, Kf = 0.05, Kc = 0.08, Ta = 0.0,
Tc = 1.0, Tc1 = 0.0, Tb = 10.0, Tb1 = 0.0, Voel = 999.

3. Governor Model: $g = \left[a + \frac{b}{1+sTg} \right] \frac{d}{dt} \delta$

   where Tg=0.25, a = -0.00133, b = -0.170 for generator #3 and #5.
   Tg=0.25, a = -0.00015, b = -0.0150 for generator #1, #2 and #4.

4. Operating conditions:

	G1	G2	G3	G4	G5
P, pu	5.4200	8.5835	0.8055	8.5670	0.8501
Q, pu	6.8177	4.3452	0.7280	4.4493	0.1608
V, pu	1.0750	1.1300	1.0250	1.0750	1.0250
δ, rad	0.0	0.4600	0.3060	0.1032	0.3101

5. Load admittance (pu):

$$L1 = 7.4 - j\ 4.0,\ L2 = 8.5 - j\ 5.0, L3 = 7.0 - j\ 4.0.$$

B

C-Code For Fuzzy System

B.1 Introduction

This appendix consisting of a C-program for the fuzzy simulation of power system. The program has the codes for the following tasks:

    I. Generate the Fuzzy sets for different variables from the given input. Fuzzy set may be linear or non-linear.

    II. Rule matrix is to be generated form these fuzzy sets, using any one implications, out of seventeen. There is also a choice that, with any implication, any one union and any one intersection operator can be taken.

    III. Fuzzy output is to be calculated from the given fact and the rule matrix which is generated. Any compositional rule may be used out of six. Also any Union and Intersection operator can be choosen for the selected compositional rule.

    IV. Defuzzification module is used to calculate defuzzified output from the fuzzified output of fuzzy simulator. Here also the flexibility to choose any one defuzzification method from COG method, Weighted area method, First max method, middle max or mean max method, and last max method.

    V. Inspite of these, there is a possibility to take compensatory operators instead of union and intersection operators. because in natural inferencing we never use exact max or min, but use combinations of both (i.e. compensatory operators say 20% max and 80% min).

B.2 Program for Fuzzy Simulation

```
/* Module I for fuzzification */
# define matlimit 20
# define limit 50
void main()
```

```
{
float useless,loadmat[matlimit][limit],endmat[limit],max();
float crisp,minima,step,loadval[limit],slope,maxmem;
int i,j,nmat,order,betweenfirst,present[limit],numpresent;
numpresent=betweenfirst=0;
for (i=0;i<=limit-1;++i)
{ present[i]=0;
endmat[i]=0; }
for (i=0;i<=8;++i)
scanf ("%f",&useless);
scanf ("%d",&nmat);
scanf ("%d",&order);
for (j=0;j<=nmat-1;++j)
{
for (i=0;i<=order-1;++i)
scanf ("%f",&loadmat[j][i]);
for (i=0;i<=order-1;++i)
scanf ("%f",&useless); }
scanf ("%f",&minima);
scanf ("%f",&step);
for (i=0;i<=order-1;++i)
scanf ("%f",&useless);
scanf ("%f",&crisp);
for (i=0;i<=order-1;++i)
loadval[i]=minima+i*step;
i=0;
while (loadval[i] < crisp)
++i;
if (crisp==minima) betweenfirst=0; else betweenfirst=--i;
for (i=0;i<=nmat-1;++i)
{
if (loadmat[i][betweenfirst] != 0 || loadmat[i]
   [betweenfirst+1] != 0)
{present[numpresent]=i;
++numpresent;
}
}
for (i=0;i<=numpresent-1;++i)
{
slope=(loadmat[present[i]][betweenfirst+1]
   -loadmat[present[i]][betweenfirst])/
   (loadval[betweenfirst+1] - loadval[betweenfirst]);
maxmem = slope*(crisp-loadval[betweenfirst])
   + loadmat[present[i]][betweenfirst];
```

```
{
for (j=0;j<=order-1;++j)
if (loadmat[present[i]][j] > maxmem)
   loadmat[present[i]][j] = maxmem;
}
}
for (i=0;i<=numpresent-1;++i)
{
for (j=0;j<=order-1;++j)
endmat[j]=max(endmat[j],loadmat[present[i]][j]);
}
for (i=0;i<=order-1;++i)
printf ("%f",endmat[i]);
printf ("\n");}
float max(a,b) float a,b; { if (a>b) return(a);
   else return(b);}
```

/* Module I - for the Generation of fuzzy set */

```
# define limit 100
# define matlim 10
void main() {
float step,slope,loadmat[2][limit][matlim],endmat[2][limit]
   [matlim];
float pts[2][limit][matlim],maxima,minima,min_param,
   max_param;
float imp_param,crisp,speed;
int currentorder,oldfinalorder,finalorder,totalorder[matlim],
   npoints,i,j;
int initialzero,nmat,initialj,k,j1,optmin, optmax, optimp,
   optcompos;
int optcomp,optdef;
scanf ("%d",&optdef);
scanf ("%d",&optcomp);
scanf ("%d",&optimp);
scanf ("%f",&imp_param );
scanf ("%d",&optmin );
scanf ("%d",&optmax );
scanf ("%f",&min_param );
scanf ("%f",&max_param );
scanf ("%d",&optcompos );
scanf ("%d",&nmat);
for (k=0;k<=1;++k) { if (k==0) initialj = 0;
else {
initialj=nmat;
nmat=nmat*2; }
```

```
scanf("%f",&step);
scanf("%f",&minima);
scanf("%f",&maxima);
for(j=initialj;j<=nmat-1;++j) {
scanf("%f%f",&pts[0][0][j],&pts[1][0][j]);
loadmat[0][0][j] = pts[0][0][j];
loadmat[1][0][j] = pts[1][0][j];
totalorder[j] = (maxima-minima)/step + 1;
oldfinalorder=1;
finalorder=npoints=1;
while (pts[0][npoints-1][j] != 100) {
scanf ("%f%f",&pts[0][npoints][j],&pts[1][npoints][j]);
if (pts[0][npoints][j] != 100) { currentorder
   = (pts[1][npoints][j]-pts[1][npoints-1][j])/step + 1;
finalorder = finalorder + currentorder - 1;
slope = (pts[0][npoints][j] - pts[0][npoints-1][j])/
   (pts[1][npoints][j]-pts[1][npoints-1][j]);
for (i=oldfinalorder;i<=finalorder-1;++i)
{ loadmat[1][i][j] = pts[1][npoints-1][j]
   + (i-oldfinalorder+1)*step;
loadmat[0][i][j] = (loadmat[1][i][j]-pts[1][npoints][j])
   *slope + pts[0][npoints][j];
}
oldfinalorder = finalorder; }
++npoints; }
if (loadmat[1][0][j]==0) initialzero=0;
else initialzero=(loadmat[1][0][j]-minima)/step;
for (i=0;i<=initialzero-1;++i)
 {
endmat[0][i][j]=0;
endmat[1][i][j]=minima + step*i;
 }
for (i=initialzero;i<=initialzero+finalorder-1;++i)
{ endmat[0][i][j]=loadmat[0][i-initialzero][j];
endmat[1][i][j]=loadmat[1][i-initialzero][j];
 }
for (i=initialzero+finalorder;i<=totalorder[j]-1;++i)
 { endmat[0][i][j]=0;
endmat[1][i][j]=minima+step*i;
 } } }
nmat = nmat/2;
printf ("%d \n",optdef);
printf ("%d \n",optcomp);
printf ("%d \n",optimp);
printf ("%f \n",imp_param );
```

```
printf ("%d \n",optmin );
printf ("%d \n",optmax );
printf ("%f \n",min_param );
printf ("%f \n",max_param );
printf ("%d \n",optcompos );
printf ("%d \n",nmat);
printf ("%d \n",totalorder[0]);
for (j=0;j<=nmat-1;++j) { { for (i=0;i<=totalorder[j]-1;++i)
printf ("%f",endmat[0][i][j]);
printf ("\n"); } { j1 = j+nmat;
for (i=0;i<=totalorder[j1]-1;++i)
printf ("%f",endmat[0][i][j1]);
printf ("\n"); } }
printf ("%f \n",minima);
printf ("%f \n",step);
for (i=0;i<=totalorder[0]-1;++i)
{ scanf ("%f",&speed);
printf ("%f",speed);
}
printf ("\n");
scanf ("%f",&crisp);
printf ("%f \n",crisp); }
```

/* Module - II for finding rule matrix */

```
# include <math.h>
# define limit 30
void main() { float mat[limit][limit],loadmat[limit],
   speedmat[limit];
float min(),max(),implication(),min_param, max_param,
   imp_param, speed,useless;
float compmax(),compmin(),comp(),compsum(),v;
int order,i,j,k,nmat,optmin, optmax, optimp,optcompos,
   optcomp,optdef;
scanf ("%d",&optdef);
scanf ("%d",&optcomp);
scanf ("%d",&optimp);
scanf ("%f",&imp_param );
scanf ("%d",&optmin );
scanf ("%d",&optmax );
scanf ("%f",&min_param );
scanf ("%f",&max_param );
scanf ("%d",&optcompos );
scanf ("%d",&nmat);
scanf ("%d",&order);
if (optcomp == 2) v=1.0; else v=0.0;
```

```
for (i=0;i<=limit-1;++i)
{          for (j=0;j<=limit-1;++j)
mat[i][j]=v;
}
for (j=0;j<=nmat-1;++j) {
for (i=0;i<=order-1;++i)
scanf ("%f",&loadmat[i]);
for (i=0;i<=order-1;++i)
scanf ("%f",&speedmat[i]);
for (i=0;i<=order-1;++i)
  { for (k=0;k<=order-1;++k)
  mat[i][k] = implication(mat[i][k],loadmat[i],speedmat[k],
  optmin,optmax,optimp,min_param,max_param,imp_param,
  optcomp);
  } }
printf ("%d \n",optdef);
printf ("%d \n",optcomp);
printf ("%d \n",optmin );
printf ("%d \n",optmax );
printf ("%f \n",min_param);
printf ("%f \n",max_param);
printf ("%d \n",optcompos );
printf ("%d \n",order);
for (i=0;i<=order-1;++i)
 {
{          for (j=0;j<=order-1;++j)
printf ("%f",mat[i][j]);
}
printf ("\n");
 }
scanf ("%f %f",&useless,&useless);
for (i=0;i<=order-1;++i)
{ scanf ("%f",&speed);
printf ("%f",speed);
}
printf ("\n"); }
float min(ua,ub,opt,w)
float ua,ub,w;
int opt; { float x,y,m;
switch(opt) {
case 1 : { if (ua<=ub) m=ua;
 else m=ub;
 break; }
case 2:
  { /* Yu min */
```

```
  x=(1.0+w)*(ua+ub-1.0)-w*ua*ub;
  m=max(0.0,x,1,0.0);
  break;}
case 3:
  { /* Hamcher min */
x= ua*ub;
y= w+(1.0-w)*(ua+ub-x);
m=x/y;
break;}
case 4:
  {/* schweizer & sklar 1*/
x=pow(ua,w)+pow(ub,w)-1.0;
m=pow(max(0.0,x,1,0.0),(1.0/w));
break;}
case 5:
  { /* Schweizer & sklar 2*/
x=pow((1.0-ua),w);
y=pow((1.0-ub),w);
m=1.0-pow((x+y-x*y),(1.0/w));
break;}
case 6:
  { /* Schweizer & Sklar 3 */
x=pow(abs(log(ua+0.12)),w);
y=pow(abs(log(ub+0.12)),w);
m=exp(-pow((x+y),(1.0/w)));
break;}
case 7:
  { /* Schweizer & Sklar 4 */
x=pow(ua,w);
y=pow(ub,w);
m=ua*ub/(pow((x+y-x*y),(1.0/w))+0.001);
break;}
case 8:
  { /* Dubois & Prade [1980] */
m=ua*ub/max(max(ua,ub,1,0.0),w,1,0.0);
break;}
case 9:
  { /* Yager min [1980] */
x=pow((1.0-ua),w);
y=pow((1.0-ub),w);
m=1.0-min(1.0,pow((x+y),(1.0/w)),1,0.0);
break;}
case 10:
  { /* Weber min [1983] */
x=ua+ub+w*ua*ub-1.0;
```

```
y=x/(1+w);
m=max(0.0,y,1,0.0);
break;}
case 11:
  { /* Dombi min [1982] */
x=pow(((1.0-ua)/(ua+0.01)),w);
y=pow(((1.0-ub)/(ub+0.01)),w);
m=pow((1.0+pow((x+y),(1.0/w))),(-1.0));
break;}
case 12:
  { /* Frank [1979] */
x = (pow(w,ua)-1.0)*(pow(w,ub)-1.0)/(w-1.0);
m = log(1.0+x)/log(w);
break;} }
return(m);
  }
  float max(ua,ub,opt,w) float ua,ub,w; int opt;
   { float x,y,m; switch(opt) {
case 1: { if (ua>=ub) m=ua;
       else m=ub;
       break;}
case 2:
  { /*Yu max [1985]*/
x=ua+ub+w*ua*ub;
m=min(1.0,x,1,0.0);
break;}
case 3:
  { /* Hamcher max [1978] */
y=(ua+ub+(w-2.0)*ua*ub)/(w-(1.0-w)*ua*ub);
m=y;
break;}
case 4:
  {/* schweizer & sklar 1 [1963]*/
  x=pow((1.0-ua),w)+pow((1.0-ub),w)-1.0;
  m=1.0-pow(max(0.0,x,1,0.0),(1.0/w));
  break;}
case 5:
  { /* Schweizer & sklar 2*/
  x=pow(ua,w);
  y=pow(ub,w);
  m=pow((x+y-x*y),(1.0/w));
  break;}
case 6:
  { /* Schweizer & Sklar 3 */
x=pow(abs(log(1.001-ua)),w);
```

```
y=pow(abs(log(1.001-ub)),w);
m=1.0-exp(-pow((x+y),(1.0/w)));
break;}
case 7:
  { /* Schweizer & Sklar 4 */
x=pow(((1.0-ua),w);
y=pow(((1.0-ub),w);
m=1.0-((1.0-ua)*(1.0-ub)/(pow((x+y-x*y),(1.0/w))+0.001));
break;}
case 8:
  { /* Dubois & Prade [1980] */
m=1.0-(1.0-ua)*(1.0-ub)/max(max((1.0-ua),(1.0-ub),1,0.0),
   w,1,0.0);
break;}
case 9:
  { /* Yager max [1980] */
x=pow(ua,w);
y=pow(ub,w);
m=min(1.0,pow((x+y),(1.0/w)),1,0.0);
break;}
case 10:
  { /* Weber max [1983] */
x=ua+ub-w*ua*ub/(1.0-w);
m=max(1.0,x,1,0.0);
break;}
case 11:
  { /* Dombi min [1982] */
x=pow(((1.0-ua)/(ua+0.01)),w);
y=pow(((1.0-ub)/(ub+0.01)),w);
m=pow((1.0+pow((x+y),(-1.0/w))),(-1.0));
break;}
case 12:
  { /* Frank [1979] */
x = (pow(w,(1.0-ua))-1.0)*(pow(w,(1.0-ub))-1.0)/(w-1.0);
m = 1.0 - log(1.0+x)/log(w);
break;}
}
return(m); }
   float  implication(umat,ua,ub,optmin,optmax,optimp,
   w1,w2,w3,optcomp)  float umat,ua,ub,w1,w2,w3; int optmin,
   optmax,optimp,optcomp; { float x,m; switch(optimp)
   { case 1: { /* Zadeh Implication [1973] */
x=compmax(compmin(ua,ub,optmin, w1,optcomp),1.0-ua,optmax,
   w2,optcomp);
m=compmax(x,umat,optmax ,w2,optcomp);
```

```
break;}
case 2:
  { /* Gaines - Rescher Implication [1969] */
if (ua<=ub) x=1.0;
else x=0.0;
m=compmax(x,umat,optmax, w2,optcomp);
break;}
case 3:
  { /* Godel Implication [1976] */
if (ua<=ub) x=1.0;
else x=ub;
m=compmin(umat,x,optmin ,w1,optcomp);
break;}
case 4:
  {/* Goguen Implication [1969] */
if (ua<=ub) x=1.0;
else x=ub/ua;
m=compmax(umat,x,optmax, w2,optcomp);
break;}
case 5:
  { /* Kleene - Dienes Implication [1949] */
x=compmax(1.0-ua,ub,optmax ,w2,optcomp);
m=compmax(umat,x,optmax ,w2,optcomp);
break;}
case 6:
  { /* Lukasiewicz Implication [1920] */
x=compmin(1.0,1.0-ua+ub,optmin ,w1,optcomp);
m=compmax(umat,x,optmax ,w2,optcomp);
break;}
case 7:
  { /* Pseudo -Lukasiewicz 1 Implication [1987] */
x=compmin(1.0,(1.0-ua+(1.0+w3)*ub)/(1.0+w3*ua),optmin,
   w1,optcomp);
m=compmax(umat,x,optmax ,w2,optcomp);
break;}
case 8:
  { /* Pseudo -Lukasiewicz 2 Implication [1987] */
x=compmin(1.0,pow(1.0-pow(ua,w3)+pow(ub,w3),(1.0/w3)),
   optmin,w1,optcomp);
m=compmax(umat,x,optmax ,w2,optcomp);
break;}
case 9:
  { /* Reichenbach Implication [1935] */
x=1.0-ua+ua*ub;
m=compmax(umat,x,optmax ,w2,optcomp);
```

```
break;}
case 10:
  { /* Willmott Implication [1980] */
x=compmin(compmax(1.0-ua,ub,optmax ,w2,optcomp),
   compmax(ua,1.0-ua,optmax ,w2,optcomp),
   compmax(ub,1.0-ub,optmax ,w2,optcomp),optmin ,w1,optcomp);
m=compmax(umat,x,optmax ,w2,optcomp);
break;}
case 11:
  { /* Wu Implication [1986] */
if (ua<= ub) x=1.0;
else x=compmin(1-ua,ub,optmin ,w1);
m=compmax(umat,x,optmax ,w2,optcomp);
break;}
case 12:
  { /* Yager [1980] */
if (ua==0 & ub==0) x=1;
else x=pow(ub,ua);
m=compmax(umat,x,optmax ,w2,optcomp);
break;}
case 13:
  { /* Klir&Yuan 1 [1994] */
x=1.0-ua+ub*pow(ua,2.0);
m=compmax(umat,x,optmax ,w2,optcomp);
break;}
case 14:
  { /* Klir & Yuan 2 [1994] */
if (ua==1.0) x=ub;
else if (ua!=1.0 & ub!=1.0) x=1.0-ua;
else x=1.0;
m=compmax(umat,x,optmax ,w2,optcomp);
break;}
case 15:
  { /* Mimdani Implication [1994] */
x=compmin(ua,ub,optmin ,w1,optcomp);
m=compmax(umat,x,optmax ,w2,optcomp);
break;}
case 16:
  { /* Stochastic min Implication [1994] */
x=compmin(1.0,1.0-ua+ua*ub,optmin ,w1);
m=compmax(umat,x,optmax ,w2,optcomp);
break;}
case 17:
  { x = ua*ub;
m = compmax(umat,x,optmax ,w2,optcomp);
```

```
break;}
}
return(m);
  } float compmax(ua,ub,optmax ,w,optcomp) float ua,ub,w;
   int optmax ,optcomp; { float x;
if (optcomp==2) x=comp(ua,ub,optmax ,w);
else if (optcomp==3) x=compsum(ua,ub,optmax ,w);
else x=max(ua,ub,optmax,w);
return(x); }
float compmin(ua,ub,optmin ,w,optcomp) float ua,ub,w;
   int optmin, optcomp; { float x;
if (optcomp==2) x=comp(ua,ub,optmin, w);
else if (optcomp==3) x=compsum(ua,ub,optmin, w);
else x=min(ua,ub,optmin,w);
return(x); }
float comp(ua,ub,optcompos ,w) float ua,ub,w; int optcompos;
   { float x,y,c; switch(optcompos) {
case 1 : x=pow(min(ua,ub,1,1.0),(1.0-w));
y=pow(max(ua,ub,1,1.0),w);
c=x*y;
break;
case 2 : x=pow((ua*ub),(1.0-w));
y=pow((ua+ub-ua*ub),w);
c=x*y;
break;
case 3 : x=pow(max(0,(ua+ub),1,1.0),(1.0-w));
y=pow(min(1.0,ua+ub,1,1.0),w);
c=x*y;
break;
case 4 : x=pow(min(ua,ub,1,1.0),(1.0-w));
y=pow((ua+ub-ua*ub),w);
c=x*y;
break;
case 5 : x=pow(max(ua,ub,1,1.0),(1.0-w));
y=pow((ua*ub),w);
c=x*y;
break;
case 6 : x=pow(min(ua,ub,1,1.0),(1.0-w));
y=pow(min(1.0,(ua+ub),1,1.0),w);
c=x*y;
break;
case 7: x=pow(max(ua,ub,1,1.0),(1.0-w));
y=pow(max(0,(ua+ub-1.0),1,1.0),w);
c=x*y;
break;
```

```
case 8: x=pow((ua*ub),(1.0-w));
y=pow(min(1.0,(ua+ub),1,1.0),w);
c=x*y;
break;
case 9: x=pow((ua+ub-ua*ub),(1.0-w));
y=pow(max(0,(ua+ub-1.0),1,1.0),w);
c=x*y;
break;
case 10: x=pow(max(0,(ua+ub-1.0),1,1.0),(1.0-w));
y=pow(min(ua,ub,1,1.0),w);
c=x*y;
break;
case 11: x=pow(max(ua,ub,1,1.0),w);
y=pow(min(1.0, (ua+ub),1,1.0),(1.0-w));
c=x*y;
break;
case 12 : x=pow((ua*ub),w);
y=pow(max(0,(ua+ub-1.0),1,1.0),(1.0-w));
c=x*y;
break;
case 13: x=pow(min(1.0,(ua+ub),1,1.0),(1.0-w));
y=pow((ua+ub-ua*ub),w);
c=x*y;
break;
case 14: x=pow((ua*ub),(1.0-w));
y=pow(min(ua,ub,1,1.0),w);
c=x*y;
break;
case 15: x=pow((ua+ub-ua*ub),(1.0-w));
y=pow(max(ua,ub,1,1.0),w);
c=x*y;
break;
case 16: x=pow(min(ua,ub,1,1.0),(1.0-w));
y=pow(((ua+ub)/2.0),w);
c=x*y;
break;
case 17: x=pow(max(ua,ub,1,1.0),(1.0-w));
y=pow(((ua+ub)/2.0),w);
c=x*y;
break;
case 18: x=pow((ua*ub),(1.0-w));
y=pow(((ua+ub)/2.0),w);
c=x*y;
break;
case 19: x=pow((ua+ub-ua*ub),(1.0-w));
```

```
y=pow(((ua+ub)/2.0),w);
c=x*y;
break;
case 20: x=pow(max(0,(ua+ub-1.0),1.0),(1.0-w));
y=pow(((ua+ub)/2.0),w);
c=x*y;
break;
case 21: x=pow(min(1.0,(ua+ub),1,1.0),(1.0-w));
y=pow(((ua+ub)/2.0),w);
c=x*y;
break;
case 22: x=pow(min(ua,ub,1,1.0),(1.0-w));
y=pow(sqrt(ua*ub),w);
c=x*y;
break;
case 23: x=pow(max(ua,ub,1,1.0),(1.0-w));
y=pow((1.0-sqrt((1.0-ua)*(1.0-ub))),w);
c=x*y;
break;
case 24: x=pow(min(ua,ub,1,1.0),(1.0-w));
y=pow((1.0-sqrt((1.0-ua)*(1.0-ub))),w);
c=x*y;
break;
case 25: x=pow(max(ua,ub,1,1.0),(1.0-w));
y=pow(sqrt(ua*ub),w);
c=x*y;
break;
case 26: x=pow(sqrt(ua*ub),(1.0-w));
y=pow((1.0-sqrt((1.0-ua)*(1.0-ub))),w);
c=x*y;
break;
case 27: x=pow((2.0*ua*ub/(ua+ub)),(1.0-w));
y=pow(((ua+ub-2.0*ua*ub)/(2.0-ua-ub)),w);
c=x*y;
break;
case 28: x=pow(sqrt(ua*ub),w);
y=pow((ua+ub)/2.0,(1.0-w));
c=x*y;
break;
case 29: x=pow(((ua+ub)/2.0),(1.0-w));
y=pow((1.0-sqrt((1.0-ua)*(1.0-ub))),w);
c=x*y;
break;
case 30: x=pow(((ua+ub)/2.0),(1.0-w));
y=pow((2.0*ua*ub/(ua+ub)),w);
```

```
c=x*y;
break;
case 31: x=pow(((ua+ub)/2.0),(1.0-w));
y=pow((((ua+ub-2.0*ua*ub)/(2.0-ua-ub)),w);
c=x*y;
break;
case 32: x=pow((ua*ub*(ua+ub-ua*ub)),(1.0-w));
y=pow((ua+ub-ua*ub*(ua+ub-ua*ub)),w);
c=x*y;
break;
case 33: x=pow((min(ua,ub,1,1.0)*ua*ub),(1.0-w));
y=pow((ua+ub-ua*ub+max(ua,ub,1,1.0)
   -(ua+ub-ua*ub)*max(ua,ub,1,1.0)),w);
c=x*y;
break;
case 34: x=pow((ua*ub+min(ua,ub,1,1.0)
   -ua*ub*min(ua,ub,1,1.0)),(1.0-w));
y=pow(((ua+ub-ua*ub)*max(ua,ub,1,1.0)),w);
c=x*y;
break;
  }
return(c); } float compsum(ua,ub,optcompos ,w) float ua,ub,w;
   int optcompos ; { float x,y,c; switch(optcompos ){
case 1 : x=min(ua,ub,1,1.0)*(1.0-w);
y=max(ua,ub,1,1.0)*w;
c=x+y;
break;
case 2 : x=(ua*ub)*(1.0-w);
y=(ua+ub-ua*ub)*w;
c=x+y;
break;
case 3 : x=max(0,(ua+ub-1),1,1.0)*(1.0-w);
y=min(1.0,ua+ub,1,1.0)*w;
c=x+y;
break;
case 4 : x=min(ua,ub,1,1.0)*(1.0-w);
y=(ua+ub-ua*ub)*w;
c=x+y;
break;
case 5 : x=max(ua,ub,1,1.0)*(1.0-w);
y=(ua*ub)*w;
c=x+y;
break;
case 6 : x=min(ua,ub,1,1.0)*(1.0-w);
y=min(1.0,(ua+ub),1,1.0)*w;
```

```
c=x+y;
break;
case 7: x=max(ua,ub,1,1.0)*(1.0-w);
y=max(0,(ua+ub-1.0),1,1.0)*w;
c=x+y;
break;
case 8: x=(ua*ub)*(1.0-w);
y=min(1.0,(ua+ub),1,1.0)*w;
c=x+y;
break;
case 9: x=(ua+ub-ua*ub)*(1.0-w);
y=max(0,(ua+ub-1.0),1,1.0)*w;
c=x+y;
break;
case 10: x=max(0,(ua+ub-1.0),1,1.0)*(1.0-w);
y=min(ua,ub,1,1.0)*w;
c=x+y;
break;
case 11: x=max(ua,ub,1,1.0)*w;
y=min(1.0, (ua+ub),1,1.0)*(1.0-w);
c=x+y;
break;
case 12 : x=(ua*ub)*w;
y=max(0,(ua+ub-1.0),1,1.0)*(1.0-w);
c=x+y;
break;
case 13: x=min(1.0,(ua+ub),1,1.0)*(1.0-w);
y=(ua+ub-ua*ub)*w;
c=x+y;
break;
case 14: x=(ua*ub)*(1.0-w);
y=min(ua,ub,1,1.0)*w;
c=x+y;
break;
case 15: x=(ua+ub-ua*ub)*(1.0-w);
y=max(ua,ub,1,1.0)*w;
c=x+y;
break;
case 16: x=min(ua,ub,1,1.0)*(1.0-w);
y=((ua+ub)/2.0)*w;
c=x+y;
break;
case 17: x=max(ua,ub,1,1.0)*(1.0-w);
y=((ua+ub)/2.0)*w;
c=x+y;
```

```
break;
case 18: x=(ua*ub)*(1.0-w);
y=((ua+ub)/2.0)*w;
c=x+y;
break;
case 19: x=(ua+ub-ua*ub)*(1.0-w);
y=((ua+ub)/2.0)*w;
c=x+y;
break;
case 20: x=max(0,(ua+ub-1.0),1.0)*(1.0-w);
y=((ua+ub)/2.0)*w;
c=x+y;
break;
case 21: x=min(1.0,(ua+ub),1,1.0)*(1.0-w);
y=((ua+ub)/2.0)*w;
c=x+y;
break;
case 22: x=min(ua,ub,1,1.0)*(1.0-w);
y=sqrt(ua*ub)*w;
c=x+y;
break;
case 23: x=max(ua,ub,1,1.0)*(1.0-w);
y=(1.0-sqrt((1.0-ua)*(1.0-ub)))*w;
c=x+y;
break;
case 24: x=min(ua,ub,1,1.0)*(1.0-w);
y=sqrt(1.0-((1.0-ua)*(1.0-ub)))*w;
c=x+y;
break;
case 25: x=max(ua,ub,1,1.0)*(1.0-w);
y=sqrt(ua*ub)*w;
c=x+y;
break;
case 26: x=sqrt(ua*ub)*(1.0-w);
y=(1.0-sqrt((1.0-ua)*(1.0-ub)))*w;
c=x+y;
break;
case 27: x=(2.0*ua*ub/(ua+ub))*(1.0-w);
y=((ua+ub-2.0*ua*ub)/(2.0-ua-ub))*w;
c=x+y;
break;
case 28: x=sqrt(ua*ub)*w;
y=((ua+ub)/2.0)*(1.0-w);
c=x+y;
break;
```

```
case 29: x=((ua+ub)/2.0)*(1.0-w);
y=(1.0-sqrt((1.0-ua)*(1.0-ub)))*w;
c=x+y;
break;
case 30: x=((ua+ub)/2.0)*(1.0-w);
y=(2.0*ua*ub/(ua+ub))*w;
c=x+y;
break;
case 31: x=((ua+ub)/2.0)*(1.0-w);
y=((ua+ub-2.0*ua*ub)/(2.0-ua-ub))*w;
c=x+y;
break;
case 32: x=(ua*ub*(ua+ub-2.0*ua*ub))*(1.0-w);
y=(ua+ub-ua*ub*(ua+ub-ua*ub))*w;
c=x+y;
break;
case 33: x=min(ua,ub,1,1.0)*ua*ub*(1.0-w);
y=(ua+ub-ua*ub+max(ua,ub,1,1.0)
   -(ua+ub-ua*ub)*max(ua,ub,1,1.0))*w;
c=x+y;
break;
case 34: x=(ua*ub+min(ua,ub,1,1.0)
   -ua*ub*min(ua,ub,1,1.0))*(1.0-w);
y=((ua+ub-ua*ub)*max(ua,ub,1,1.0))*w;
c=x+y;
break;
}
return(c); }
```

/* Module -III for

A3. Example

The contents of the input file should be as follows:

1. Option of defuzzification method
2. Option of max-min/compensatory
3. Option of implication method
4. Parameter value of intersection in implication option
5. Parameter value of union in implication option
6. Parameter value of Intersection for compositional rule
7. Parameter of union for compositional rule
8. Number of fuzzy sets
9. Step size for calculating intermediate values
10. Minimum value of variable

11. Maximum value of variable
12. Membership values and corresponding variable value

(There should be number of fuzzy sets times entries and each entry should be terminated by the entry 100 0). The entries of the first rule should be the first

The options for union operator are:

1. Ordinary min
2. Yu max[0,{(1+w1)*(ua+ub-1)-w1*ua*ub)}] w1 > -1
3. Hamcher - ua*ub/{w1+(1-w1)*(ua+ub-ua*ub)} w1 > 0
4. Schweizer & Sklar 1 {max(0,ua^w1+ub^w1-1)}^(1/w1)
   -inf to +inf w1 != 0
5. Schweizer & Sklar 2 {1-((1-ua)^w1+(1-ub)^w1-((1-ua)^w1)
   *((1-ub)^w1)}^(1/w1)   w1 > 0
6. Schweizer & Sklar 3
   exp[-(abs(log(ua)^w1+abs(log(ub)^w1)^(1/w1)] w1 > 0
7. Schweizer & Sklar 4 ua*ub/(ua^w1+ub^w1-(ua^w1*ub^w1))
   w1 > 0
8. Dubois & Prade ua*ub/max(ua,ub,w1) 0 <= w1 <= 1
9. Yager 1-min[1,{(1-ua)^w1+(1-ub)^w1}^(1/w1)] w1 > 0
10. Weber max(0,ua+ub+w1*ua*ub-1)/(1+w1) w1 > -1
11. Dombi [1+(((1/ua)-1)^w1+((1/ub)-1)^w1)^(!/w1)]^(-1)
    w1 > 0
12. Frank log(to base w1)[1+((w1^a-1)*(w1^b-1))/(w1-1)]
    w1 > 0 w1 != 1
The Intersection operators are as follows:
1. Ordinary max
2. Yu - min[1,(ua+ub+w2*ua*ub)] w2 > -1
3. Hamcher - (ua+ub+(2-w2)*ua*ub))/(w2+(1-w2)*ua*ub) w2 > 0
4. Schweizer & Sklar 1
   (1-max(0,(1-ua)^w2+(1-ub)^w2-1))^(1/w2)
   -inf to +inf w2 != 0
5. Schweizer & Sklar 2
   (ua^w2+ub^w2-(ua^w2)*(ub^w2))^(1/w2) w2 > 0
6. Schweizer & Sklar 3
   1-exp[-(abs(log(1-ua)^w2+abs(log(1-ub)^w2)^(1/w2)]
   w2 > 0 7. Schweizer & Sklar 4
   1-((1-ua*(1-ub)/((1-ua)^w2+(1-ub)^w2
   -((1-ua)^w2*(1-ub)^w2)^(1/w2)) w2 > 0
8. Dubois & Prade 1-(1-ua)*(1-ub)/max((1-ua),(1-ub),w2)
   0 <= w2 <= 1
9. Yager min[1,(ua^w2+ub^w2)^(1/w2)] w2 > 0
10. Weber min(1,ua+ub-w2*ua*ub/(1-w2)) w2 > -1
11. Dombi [1+(((1/ua)-1)^w2+((1/ub)-1)^w2}^(-1/w2)]^(-1)
    w2 > 0

12. Frank 1-log(to base w2)[1+((w2^(1-a)-1)*(w2^(1-b)-1))/
    (w2-1)] w2 > 0 w2 != -1

The options of operators are:

1 - max-min
2 - min-max
3 - min-min
4 - max-max
5 - Godel
6 - max-product
7 - max-average
8 - sum-product

The options of compensatory operators are from 1 to 34 (product)
The implications are:

1. Zadeh
2. Gaines-Rascher
3. Godel
4. Goguen
5. Kleenes-Diene
6. Lukasiewicz
7. Pseudo-Lukasiewicz 1
8. Pseudo-Lukasiewicz 2
9. Reichenbach
10. Willmott
11. Wu
12. Yager
13. Klir & Yuan 1
14. Klir & Yuan 2
15. Mimdani
16. Stochastic min
17. Correlation product

The options of defuzzification are:
1-centroid
2-weighted average
3-mean max
4-first max
5-last max

The options of max/min or compensatory product or compensatory sum are:
1-maxmin
2-product compensatory
3-sum compensatory

The input file for dc machine problem is given as follows -
3 2 2 1 1 1 1.0 0.7 1 5 1 0 12
1 0
0.2 4
100 0
0.2 2
1 4
0.2 6
100 0
0.2 4
1 6
0.2 8
100 0
0.2 6
1 8
0.2 10
100 0
0.2 8
1 12
100 0
1
0 12
0.2 8
1 12
100 0
0.2 6
1 8
0.2 10
100 0
0.2 4
1 6
0.2 8
100 0
0.2 2
1 4
0.2 6
100 0
1 0
0.2 4
100 0
    150.28 151.16 152.22 152.9651 153.71 154.195 154.94 155.5222 156.12
156.76 157.4 158.0175 158.65 11.5

C

Data For 26-Bus System

See Fig. C.1.

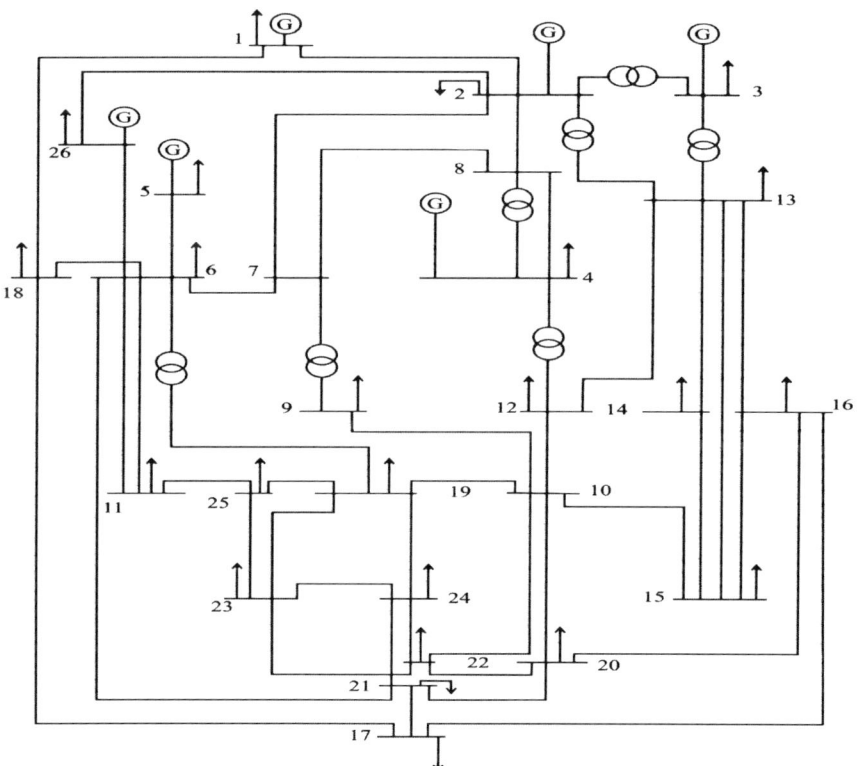

Fig. C.1. 26-bus system

	Generation data			
Bus no.	Voltage (p.u.)	Generation (MW)	Mvar limits	
			Min.	Max.
1	1.025	–	–	–
2	1.020	79.0	40.0	250.0
3	1.025	20.0	40.0	150.0
4	1.050	100.0	40.0	80.0
5	1.045	300.0	40.0	160.0
26	1.015	60.0	15.0	50.0

The generator's real power limits are:

Generator real power limits		
Gen.	Min. MW	Max. MW
1	100	500
2	50	200
3	80	300
4	50	150
5	50	200
26	50	120

$$C_1 = 240 + 7.0\,P_1 + 0.0070\,P_1^2 \qquad C_2 = 200 + 10.0\,P_2 + 0.0095\,P_2^2$$
$$C_3 = 220 + 8.5\,P_3 + 0.0090\,P_3^2 \qquad C_4 = 200 + 11.0\,P_4 + 0.0090\,P_4^2$$
$$C_5 = 220 + 10.5\,P_5 + 0.0080\,P_5^2 \quad C_{26} = 190 + 12.0\,P_{26} + 0.0075\,P_{26}^2$$

The line and transformer data containing the series resistance and reactance in per unit and one-half the total capacitance in per unit susceptance on a 100-MVA base are tabulated below.

					Line data					
Bus no.	Bus no.	R, pu	X, pu	$\frac{1}{2}$B, pu		Bus no.	Bus no.	R, pu	X, pu	$\frac{1}{2}$B, pu
1	2	0.0005	0.0048	0.0300		10	22	0.0069	0.0298	0.005
1	18	0.0013	0.0110	0.0600		11	25	0.0960	0.2700	0.010
2	3	0.0014	0.0513	0.0500		11	26	0.0165	0.0970	0.004
2	7	0.0103	0.0586	0.0180		12	14	0.0327	0.0802	0.000
2	8	0.0074	0.0321	0.0390		12	15	0.0180	0.0598	0.000
2	13	0.0035	0.0967	0.0250		13	14	0.0046	0.0271	0.001
2	26	0.0323	0.1967	0.0000		13	15	0.0116	0.0610	0.000
3	13	0.0007	0.0054	0.0005		13	16	0.0179	0.0888	0.001
4	8	0.0008	0.0240	0.0001		14	15	0.0069	0.0382	0.000
4	12	0.0016	0.0207	0.0150		15	16	0.0209	0.0512	0.000
5	6	0.0069	0.0300	0.0990		16	17	0.0990	0.0600	0.000
6	7	0.0053	0.0306	0.0010		16	20	0.0239	0.0585	0.000
6	11	0.0097	0.0570	0.0001		17	18	0.0032	0.0600	0.038
6	18	0.0037	0.0222	0.0012		17	21	0.2290	0.4450	0.000
6	19	0.0035	0.0660	0.0450		19	23	0.0300	0.1310	0.000

6	21	0.0050	0.0900	0.0226	19	24	0.0300	0.1250	0.002
7	8	0.0012	0.0069	0.0001	19	25	0.1190	0.2249	0.004
7	9	0.0009	0.0429	0.0250	20	21	0.0657	0.1570	0.000
8	12	0.0020	0.0180	0.0200	20	22	0.0150	0.0366	0.000
9	10	0.0010	0.0493	0.0010	21	24	0.0476	0.1510	0.000
10	12	0.0024	0.0132	0.0100	22	23	0.0290	0.0990	0.000
10	19	0.0547	0.2360	0.0000	22	24	0.0310	0.0880	0.000
10	20	0.0066	0.0160	0.0010	23	25	0.0987	0.1168	0.000

Shunt capacitors	
Bus no.	Mvar
1	4.0
4	2.0
5	5.0
6	2.0
11	1.5
12	2.0
15	0.5
19	5.0

Transformer tap	
Bus no.	Mvar
2–3	0.960
2–13	0.960
3–13	1.017
4–8	1.050
4–12	1.050
6–19	0.950
7–9	0.950

Load data					
Bus no.	Load		Bus no.	Load	
	MW	Mvar		MW	Mvar
1	51.0	41.0	14	24.0	12.0
2	22.0	15.0	15	70.0	31.0
3	64.0	50.0	16	55.0	27.0
4	25.0	10.0	17	78.0	38.0
5	50.0	30.0	18	153.0	67.0
6	76.0	29.0	19	75.0	15.0
7	0.0	0.0	20	48.0	27.0
8	0.0	0.0	21	46.0	23.0
9	89.0	50.0	22	45.0	22.0
10	0.0	0.0	23	25.0	12.0
11	25.0	15.0	24	54.0	27.0
12	89.0	48.0	25	28.0	13.0
13	31.0	15.0	26	40.0	20.0

D

Data For 6-Bus System

See Fig. D.1.

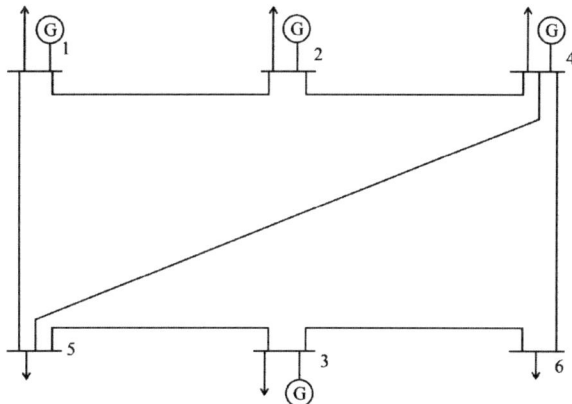

Fig. D.1. 6-bus system

Bus characteristics for six-bus system

Bus number	Load (MW)	Load (MVAR)	Min. generation (MW)	Max. generation (MW)
1	100	20	50	250
2	100	20	50	250
3	100	20	50	250
4	100	20	50	250
5	100	50	0	0
6	100	10	0	0

Economic information for six-bus system

Generator bus	$a\left[\frac{\$}{h}\right]$	$b\left[\frac{\$}{MWh}\right]$	$c\left[\frac{\$}{MW^2h}\right]$
1	105	12.0	0.012
2	96	9.6	0.0096
3	105	13.0	0.0130
4	94	9.4	0.0094

Line characteristics for six-bus system

From bus	To bus	Resistance (P.U.)	Reactance (P.U.)	Line charging (P.U.)	Line limit (MVA)
1	2	0.04	0.08	0.02	100
1	5	0.04	0.08	0.02	100
2	4	0.04	0.08	0.02	100
3	5	0.04	0.08	0.02	100
3	6	0.04	0.08	0.02	100
4	5	0.04	0.08	0.02	50
4	6	0.04	0.08	0.02	100

E

Data For IEEE 30-Bus System

See Fig. E.1.

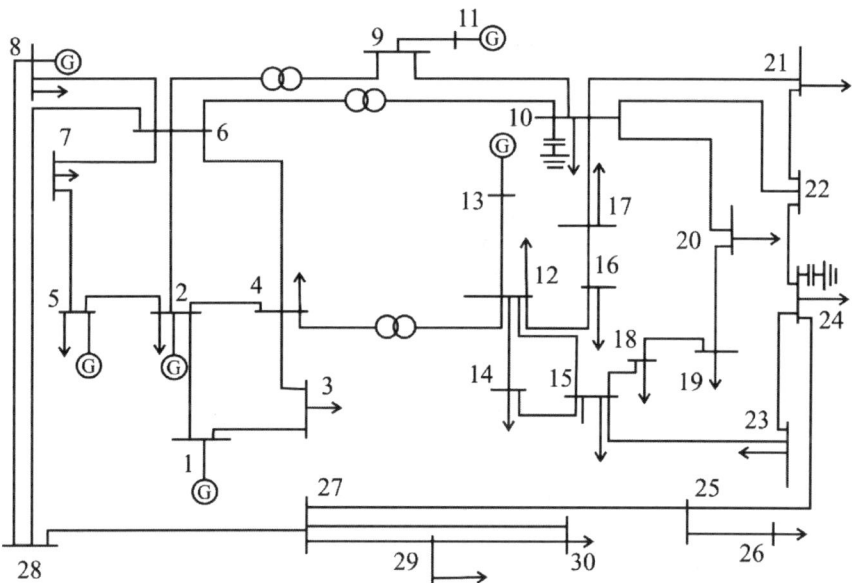

Fig. E.1. IEEE 30-bus system

Generator data

Bus no.	P_G^{min} (MW)	P_G^{max} (MW)	Q_G^{min} (MVAR)	S_G^{max} (MVA)	Coefficients		
					a	b	c
1	50	200	−20	250	0.0	2.0	0.00375
2	20	80	−20	100	0.0	1.75	0.0175
5	15	50	−15	80	0.0	1.0	0.0625
8	10	35	−15	60	0.0	3.25	0.00834
11	10	30	−10	50	0.0	3.0	0.025
13	12	40	−15	60	0.0	3.0	0.025

Generating cost $f_i = a_i + b_i P_{Gi} + c_i P_{Gi}^2 \, \pounds/\text{h}$.

Transformers with assumed tapping range of 10%. The assumed branch loading limits are for convenience taken to be the same in the base and contingency cases, and similarly for bus voltage–magnitude limits.

The lower voltage-magnitude limits at all buses are 0.95 p.u., and the upper limits are 1.1 p.u. for generator buses 2, 5, 8, 11, and 13, and 1.05 p.u. for the remaining buses including the reference bus 1.

Branch data

Branch no.	Bus no.	R p.u.	X p.u.	B (total) p.u.	Rating MVA
1	1–2	0.0192	0.0575	0.0264	130
2	1–3	0.0452	0.1852	0.0204	130
3	2–4	0.0570	0.1737	0.0184	65
4	3–4	0.0132	0.0379	0.0042	130
5	2–5	0.0472	0.1983	0.0209	130
6	2–6	0.0581	0.1763	0.0187	65
7	4–6	0.0119	0.0414	0.0045	90
8	5–7	0.0460	0.1160	0.0102	70
9	6–7	0.0267	0.0820	0.0085	130
10	6–8	0.0120	0.0420	0.0045	32
11	6–9	0.0	0.2080	0.0	65
12	6–10	0.0	0.5560	0.0	32
13	9–11	0.0	0.2080	0.0	65
14	9–10	0.0	0.1100	0.0	65
15	4–12	0.0	0.2560	0.0	65
16	12–13	0.0	0.1400	0.0	65
17	12–14	0.1231	0.2559	0.0	32
18	12–15	0.0662	0.1304	0.0	32
19	12–16	0.0945	0.1987	0.0	32
20	14–15	0.2210	0.1997	0.0	16
21	16–17	0.0824	0.1932	0.0	16
22	15–18	0.1070	0.2185	0.0	16

23	18–19	0.0639	0.1292	0.0	16
24	19–20	0.0340	0.0680	0.0	32
25	10–20	0.0936	0.2090	0.0	32
26	10–17	0.0324	0.0845	0.0	32
27	10–21	0.0348	0.0749	0.0	32
28	10–22	0.0727	0.1499	0.0	32
29	21–22	0.0116	0.0236	0.0	32
30	15–23	0.1000	0.2020	0.0	16
31	22–24	0.1150	0.1790	0.0	16
32	23–24	0.1320	0.2700	0.0	16
33	24–25	0.1885	0.3292	0.0	16
34	25–26	0.2544	0.3800	0.0	16
35	25–27	0.1093	0.2087	0.0	16
36	28–27	0.0	0.3960	0.0	65
37	27–29	0.2198	0.4153	0.0	16
38	27–30	0.3202	0.6027	0.0	16
39	29–30	0.2399	0.4533	0.0	16
40	8–28	0.0636	0.2000	0.0214	32
41	6–28	0.0169	0.0599	0.0065	32
42	10–10	0.0	−5.2600	–	–
43	24–24	0.0	25.0000	–	–

Load data

Bus no.	Load		Bus No.	Load	
	MW	MVAR		MW	MVAR
1	0.0	0.0	16	3.5	1.8
2	21.7	12.7	17	9.0	5.8
3	2.4	1.2	18	3.2	0.9
4	7.6	1.6	19	9.5	3.4
5	94.2	19.0	20	2.2	0.7
6	0.0	0.0	21	17.5	11.2
7	22.8	10.9	22	0.0	0.0
8	30.0	30.0	23	3.2	1.6
9	0.0	0.0	24	8.7	6.7
10	5.8	2.0	25	0.0	0.0
11	0.0	0.0	26	3.5	2.3
12	11.2	7.5	27	0.0	0.0
13	0.0	0.0	28	0.0	0.0
14	6.2	1.6	29	2.4	0.9
15	8.2	2.5	30	10.6	1.9

Base MVA = 100

F

Data For Modified IEEE 30-Bus System

See Fig. F.1.

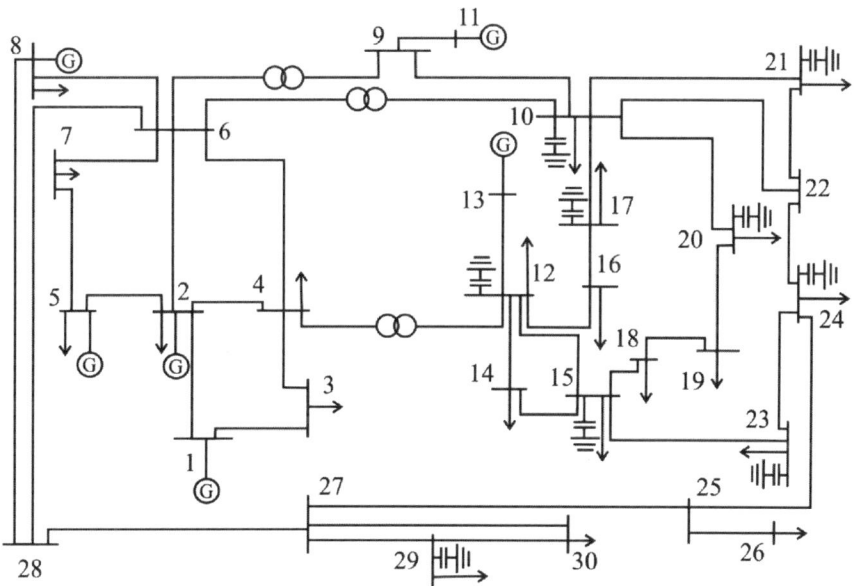

Fig. F.1. Modified IEEE 30-bus system

Variable limits and generator cost parameters

Power generation limits and fuel cost parameters ($S_B = 100\,MVA$)

Bus	1	2	5	8	11	13
P_g^{max}	2	0.8	0.5	0.35	0.3	0.4
P_g^{min}	0.5	0.2	0.15	0.1	0.1	0.12
Q_g^{max}	2	1	0.8	0.6	0.5	0.6
Q_g^{min}	−0.2	−0.2	−0.15	−0.15	−0.1	−0.15
a	0	0	0	0	0	0
b	200	175	100	325	300	300
c	37.5	175	625	83.4	250	250

Bus voltage limits (in p.u.)

V_g^{max}	V_g^{min}	V_{load}^{max}	V_{load}^{min}	Branch apparent power limit S_k^{max} (in MVA) Branch (8, 28)
1.1	0.95	1.05	0.95	12

Transformer tap setting limits

Branch	(6, 9)	(6, 10)	(4, 12)	(28, 27)
T_k^{max}	1.1	1.1	1.1	1.1
T_k^{min}	0.9	0.9	0.9	0.9

Capacitor/reactor installation limits (in MVAr)

Bus	10	12	15	17	20	21	23	24	29
Q_c^{max}	5	5	5	5	5	5	5	5	5
Q_c^{min}	0	0	0	0	0	0	0	0	0

Variable	Lower limit	Upper limit
P_1 (MW)	50	200
P_2 (MW)	20	80
P_5 (MW)	15	50
P_8 (MW)	10	35
P_{11} (MW)	10	30
P_{13} (MW)	12	40
Q_1 (MVAR)	−20	200
Q_2 (MVAR)	−20	100
Q_5 (MVAR)	−15	80
Q_8 (MVAR)	−15	60
Q_{11} (MVAR)	−10	50
Q_{13} (MVAR)	−15	60

Line data

Line number	From bus number	To bus number	Line impedance		tap setting
			R (p.u.)	X (p.u.)	
1	1	2	0.0192	0.0575	–
2	1	3	0.0192	0.1852	–
3	2	4	0.0452	0.1737	–
4	3	4	0.0570	0.0379	–
5	2	5	0.0132	0.1983	–
6	2	6	0.0472	0.1763	–
7	4	6	0.0581	0.0414	–
8	5	7	0.0119	0.1160	–
9	6	7	0.0460	0.0820	–
10	6	8	0.0267	0.0420	–
11	6	9	0.0120	0.2080	1.078
12	6	10	0.0000	0.5560	1.069
13	9	11	0.0000	0.2080	–
14	9	10	0.0000	0.1100	–
15	4	12	0.0000	0.2560	1.032
16	12	13	0.0000	0.1400	–
17	12	14	0.1231	0.2559	–
18	12	15	0.0662	0.1304	–
19	12	16	0.0945	0.1987	–
20	14	15	0.2210	0.1997	–
21	16	17	0.0824	0.1932	–
22	15	18	0.1070	0.2185	–
23	18	19	0.0639	0.1292	–
24	19	20	0.0340	0.0680	–
25	10	20	0.0936	0.2090	–
26	10	17	0.0324	0.0845	–
27	10	21	0.0348	0.0749	–
28	10	22	0.0727	0.1499	–
29	21	22	0.0116	0.0236	–
30	15	23	0.1000	0.2020	–
31	22	24	0.1150	0.1790	–
32	23	24	0.1320	0.2700	–
33	24	25	0.1885	0.3292	–
34	25	26	0.2544	0.3800	–
35	25	27	0.1093	0.2087	–
36	28	27	0.0000	0.3960	1.068
37	27	29	0.2198	0.4153	–
38	27	30	0.3202	0.6027	–
39	29	30	0.2399	0.4533	–
40	8	28	0.6360	0.2000	–
41	6	28	0.0169	0.0599	–

Load data

Bus no.	Load		Bus no.	Load	
	P (p.u.)	Q (p.u.)		P (p.u.)	Q (p.u.)
1	0.000	0.000	16	0.035	0.018
2	0.217	0.127	17	0.090	0.058
3	0.024	0.012	18	0.032	0.009
4	0.076	0.016	19	0.095	0.034
5	0.942	0.190	20	0.022	0.007
6	0.000	0.000	21	0.175	0.112
7	0.228	0.109	22	0.000	0.000
8	0.300	0.300	23	0.032	0.016
9	0.000	0.000	24	0.087	0.067
10	0.058	0.020	25	0.000	0.000
11	0.000	0.000	26	0.035	0.023
12	0.112	0.075	27	0.000	0.000
13	0.000	0.000	28	0.000	0.000
14	0.062	0.016	29	0.024	0.009
15	0.082	0.025	30	0.106	0.019

G

Data For Indian UPSEB 75-Bus System

The 75-bus UPSEB, Indian Power System is shown in Fig. G.1. The system data is taken from the project report of UPSEB. The generation (for pool and bilateral transactions), the generation (for multilateral transactions), pool demand, reactor/capacitor, transformer and transmission line data are provided in the tables. The data in per unit (p.u.) is on 100 MVA base. The voltage at all buses is maintained within the limits 0.9 and 1.1 p.u. The tap ratio of transformers is kept between 0.90 and 1.1. The voltage phase angle is maintained between −45 degree and +45 degree. For the analysis on this system, minimum limits for all generation, demand and non-firm transaction bids have been considered to be zero if not specified.

Generation data (pool and bilateral transactions)

Generator	Bus no.	Real power rating (p.u.)		Reactive power rating (p.u.)	
		Minimum	Maximum	Minimum	Maximum
G1	1	1.000	15.000	−1.000	4.000
G2	2	1.000	3.000	0.000	0.960
G3	3	0.400	2.000	0.000	0.830
G4	4	0.400	1.700	0.000	0.600
G5	5	0.000	2.400	0.000	0.310
G6	6	0.000	1.200	0.000	0.200
G7	7	0.000	1.000	0.000	0.190
G8	8	0.200	1.000	0.000	0.680
G9	9	0.600	5.700	0.000	2.500
G10	10	0.300	2.500	0.000	0.560
G11	11	0.400	2.000	0.000	1.050

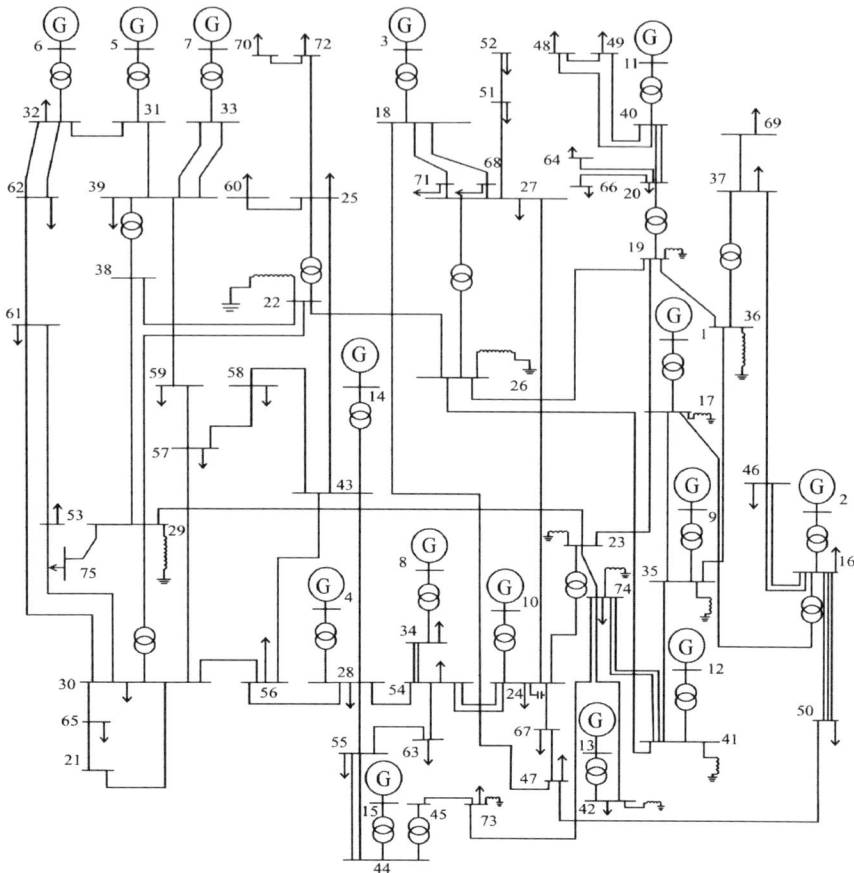

Fig. G.1. Indian UPSEB 75-bus system

Generation data (multilateral transactions)

Generator	Bus no.	Maximum real power rating (p.u.)	Reactive power rating (p.u.)	
			Minimum	Maximum
G12	12	13.000	0.000	3.440
G13	13	9.000	0.000	2.800
G14	14	1.500	−0.300	0.840
G15	15	4.540	−1.000	0.350

Cities of U.P. state at different buses

Bus no.	City	Bus no.	City	Bus no.	City
1	Obra	26	Lucknow	51	Sitapur
2	Obra	27	Lucknow	52	Shahajanpur
3	Unchchahar	28	Harduaganj	53	Barut
4	Harduaganj	29	Muradnagar	54	Mainpuri
5	Yamuna-I	30	Muradnagar	55	Agra
6	Yamuna-II	31	Yamuna-I	56	Khurja
7	Ramganga	32	Yamuna-II	57	Meerut
8	Paricha	33	Ramganga	58	Rishikesh
9	Anpara	34	Parichcha	59	Muzzafarnagar
10	Panki	35	Anpara	60	Nehtaur
11	Tanda	36	Azamgarh	61	Shamli
12	Singrauli	37	Azamgarh	62	Saharanpur
13	Rihand	38	Rishikesh	63	Ferozabad
14	NAPP	39	Rishikesh	64	Gonda
15	Auraiya	40	Tanda	65	Sahibabad
16	Obra	41	Singrauli	66	Phulpur
17	Obra	42	Rihand	67	Nebhasta
18	Unchchahar	43	Napp	68	Rai Bareily
19	Sultanpur	44	Auraiya	69	Jaunpur
20	Sultanpur	45	Auraiya	70	Badaun
21	Saharanpur	46	Mughal Sarain	71	CHNT
22	Moradabad	47	Fatehpur	72	Bareily
23	Panki	48	Basti	73	Agra
24	Panki	49	Gorakhpur	74	Kanpur
25	Moradabad	50	Allahabad	75	Dadri

The system generators G1–G11 are assumed to participate in power pool as well supplying power through bilateral transactions as shown in tables. The system generators at bus numbers 12, 13, 14 and 15 are considered not to participate in pool, but to supply power only through a multilateral transaction.

Transmission line data

Line no.	From bus	To bus	Series impedence (p.u.)		Half line charging susceptance (p.u.)	MVA rating (p.u.)	Length (km)	Annual Cost (ARR) 10^5 (Rs/year) or (Rs lakh/year)
			R	X				
33	16	46	0.01620	0.07760	0.07015	1.500	196.6	619.290
34	16	46	0.01620	0.07760	0.07015	1.500	196.6	619.290
35	16	50	0.02979	0.14238	0.12881	1.500	360.0	1134.000
36	16	50	0.02979	0.14238	0.12881	1.500	360.0	1134.000
37	16	50	0.29790	0.14238	0.12881	1.500	360.0	1134.000
38	17	19	0.00468	0.04770	0.62450	5.000	230.0	724.500
39	17	23	0.00785	0.07990	1.04738	5.000	386.6	1217.790
40	19	26	0.00294	0.02997	0.39206	5.000	145.0	456.750

41	47	50	0.01093	0.05221	0.18892	3.000	117.0	368.550
42	47	67	0.00662	0.03164	0.11451	3.000	38.6	121.590
43	24	27	0.00505	0.02416	0.08730	3.000	122.0	384.300
44	24	54	0.02582	0.12342	0.11164	1.500	311.0	979.650
45	24	54	0.02582	0.12342	0.11164	1.500	311.0	979.65
46	25	43	0.01270	0.06416	0.05220	1.500	155.0	488.250
47	54	28	0.01060	0.05060	0.18320	3.000	256.0	808.290
48	28	43	0.00580	0.02900	0.02370	1.800	20.0	63.000
49	28	56	0.00370	0.01780	0.06440	3.000	110.0	346.500
50	56	30	0.00490	0.02370	0.08590	3.000	95.0	299.250
51	30	57	0.00750	0.03840	0.03110	1.800	45.8	144.270
52	53	30	0.00679	0.03412	0.02782	1.800	41.0	129.140
53	53	61	0.00666	0.03390	0.02672	1.500	39.7	125.055
54	30	61	0.01440	0.07310	0.05850	1.500	86.6	272.790
55	57	58	0.00670	0.03390	0.02670	1.800	46.0	144.900
56	57	59	0.00583	0.02956	0.02346	1.800	50.0	157.500
57	59	39	0.01410	0.07180	0.05700	1.500	143.7	452.655
58	39	31	0.01440	0.07250	0.05900	1.500	96.6	304.290
59	54	63	0.00990	0.05090	0.04010	1.500	76.6	241.290
60	55	63	0.00780	0.03980	0.03140	1.800	30.0	94.500
61	61	62	0.01160	0.05830	0.04750	1.500	69.5	218.925
62	62	32	0.01380	0.07000	0.05630	1.500	85.0	267.750
63	62	32	0.01380	0.07000	0.05630	1.500	85.0	267.750
64	35	36	0.00479	0.04880	0.63614	5.000	241.9	761.985
65	46	37	0.01732	0.08784	0.06973	1.500	103.9	327.285
66	19	36	0.00254	0.02584	0.33798	5.000	125.5	395.325
67	17	35	0.00051	0.00517	0.06760	6.000	9.5	29.925
68	40	48	0.00830	0.04240	0.3340	1.500	52.7	166.005
69	74	41	0.00927	0.09429	1.23293	5.000	45.0	141.750
70	74	41	0.00833	0.08478	1.10855	5.000	45.0	141.750
71	26	41	0.00823	0.08375	1.09503	5.000	125.0	393.750
72	48	49	0.00930	0.04750	0.03740	1.500	56.1	176.715
73	49	40	0.01330	0.06680	0.05420	1.500	32.0	100.800
74	38	29	0.00370	0.03762	0.48870	5.500	182.7	575.505
75	38	22	0.00325	0.03307	0.43264	5.500	159.8	503.370
76	18	47	0.00437	0.02552	0.09399	3.600	113.0	355.950
77	30	65	0.00248	0.01186	0.04294	3.600	7.0	22.050
78	41	42	0.00031	0.00310	0.04056	6.000	10.0	31.500
79	42	74	0.00918	0.09306	1.21680	5.000	255.0	803.250
80	23	74	0.00015	0.00155	0.02704	12.000	25.0	157.500
81	24	67	0.00124	0.00593	0.02147	3.600	40.0	126.000
82	18	68	0.00336	0.01963	0.01808	1.800	108.0	340.200
83	18	71	0.01344	0.07852	0.07230	1.700	120.0	378.000
84	27	68	0.01344	0.07852	0.07230	1.700	122.0	384.300
85	27	71	0.00336	0.01963	0.01808	1.800	110.0	346.500
86	43	58	0.01315	0.06696	0.05278	1.500	83.0	261.450
87	43	56	0.00499	0.02397	0.08523	3.000	120.0	378.000
88	55	44	0.01996	0.09588	0.08523	1.500	90.0	283.500
89	55	44	0.01996	0.09588	0.08523	1.500	90.0	283.500
90	73	45	0.00121	0.01109	0.72815	10.000	90.0	567.000
91	29	22	0.00260	0.02646	0.34610	5.500	187.5	590.625
92	21	65	0.00830	0.00396	0.01431	3.500	868.0	2734.200
93	34	54	0.03540	0.17020	0.15130	3.000	38.31	120.677

94	34	54	0.03540	0.17020	0.15130	3.000	38.31	120.677
95	39	33	0.14100	0.07180	0.05700	1.500	250.0	787.500
96	39	33	0.14100	0.07180	0.05700	1.500	250.0	787.500
97	31	32	0.00050	0.00253	0.00805	3.600	10.0	31.500
98	20	40	0.01160	0.05880	0.04710	1.500	69.5	218.925
99	20	40	0.01160	0.05880	0.04710	2.400	69.5	218.925
100	21	30	0.00695	0.03500	0.02843	1.500	875.0	2756.250
101	28	55	0.01998	0.10127	0.08051	1.500	97.7	307.755
102	35	41	0.00031	0.00310	0.04056	3.600	9.5	29.925
103	37	69	0.01212	0.06100	0.04956	3.000	43.3	136.395
104	25	60	0.01660	0.8430	0.06720	1.500	67.7	213.255
105	51	52	0.01550	0.07940	0.06300	1.500	93.8	295.470
106	20	64	0.01830	0.09270	0.07390	1.500	110.0	346.500
107	70	72	0.00878	0.04430	0.03580	1.800	52.4	165.060
108	20	66	0.01325	0.06667	0.05416	1.500	77.3	243.495
109	29	75	0.00051	0.00517	0.06760	6.000	15.0	47.250
110	26	22	0.00650	0.06617	0.86521	5.000	294.0	926.100
111	23	29	0.00806	0.08169	1.06808	5.000	395.2	1244.880
112	74	73	0.00559	0.05686	0.74354	5.000	240.0	756.000
113	25	72	0.01598	0.08108	0.06436	1.500	75.5	237.825
114	27	51	0.01600	0.08100	0.06440	2.000	96.2	303.030

Cost coefficients for real and reactive generation costs for Indian UPSEB 75-bus system

Bus no.	Real generation cost $= a + bP_g + cP_g^2$ (in Rs/h)			Reactive generation cost $= a' + b'Q_g + c'Q_g^2$ (in Rs/h)		
	a	b	c	a'	b'	c'
1	0	1023	0.00375	0	0	0.00375
2	0	990	0.002	0	0	0.002
3	0	1010	0.00335	0	0	0.00335
4	0	1012	0.003375	0	0	0.003375
5	0	470	0.001	0	0	0.001
6	0	980	0.002375	0	0	0.002375
7	0	960	0.00175	0	0	0.00175
8	0	1035	0.00375	0	0	0.00375
9	0	1040	0.007	0	0	0.007
10	0	990	0.00675	0	0	0.00675
11	0	1025	0.00605	0	0	0.00605

Index